FUNDAMENTALS FOR CONTROL OF ROBOTIC MANIPULATORS

FUNDAMENTALS FOR CONTROL OF ROBOTIC MANIPULATORS

Antti J. Koivo
Purdue University

WILEY

JOHN WILEY & SONS, INC.
NEW YORK • CHICHESTER • BRISBANE • TORONTO • SINGAPORE

To
my parents: Elma and Niilo J. Koivuniemi
my family: Anne, Lilli and Allan Koivo
and
in memory of Jorma E. Salama

Library of Congress Cataloging-in-Publication Data:

Koivo, Antti J.
  Fundamentals for control of robotic manipulators / Antti J. Koivo.
    p. cm.
  Bibliography: p.
  ISBN 0–471–85714–9
  1. Robotics. 2. Manipulators (Mechanism) I. Title.
  TJ211.K65 1989
  629.8′92–dc19

Printed in the United States of America

10 9 8 7 6 5 4 3 2 1

PREFACE

Robotic manipulator systems are presently common in industrial applications and laboratories. On the basis of task specifications, the planner (planning algorithm) determines the desired trajectory for the manipulator motion. The controller must be so designed that the manipulator will track the desired trajectory as closely as possible. The actual trajectory values of the manipulator are fed back through feedback loops to determine possible tracking errors, and the controller will then react to compensate for the observed errors. Thus, the basic operations of a manipulator are controlled by computers that implement the designed algorithms. The construction of the control algorithms requires a thorough knowledge of manipulator dynamics as well as control theory. This textbook will give the reader the fundamentals on both robotics and controls. It will enable an engineer and scientist to apply the two disciplines to the design and operation of computer-controlled manipulators.

The material of the text represents a dual undergraduate–graduate level course for engineering students from various disciplines. It has been taught in the School of Electrical Engineering since 1984. In addition to the students of electrical engineering, the class population includes students from mechanical engineering, industrial engineering, aeronautical engineering, civil engineering, agricultural engineering, and computer science. The text has also been used in teaching practicing engineers and scientists from industry. In the School of Electrical Engineering, we cover the first half of the book (Chapters 1-5, and the main parts of Chapters 6-8) in a dual-level course in one semester. The second half of the book, which emphasizes control, is taught in a graduate-level course, which includes simulation studies with control algorithms. The text is equally well suited to self-study. It presents to practicing engineers guidelines, in the framework of manipulator control for designing control algorithms for industrial applications.

The text presents fundamentals of robotics to enable students to design and construct controllers for manipulator motion. Basic knowledge on vector-matrix manipulations and differential (difference) equations is assumed. Most engineering students by their senior level have acquired this knowledge. The students should also have had a basic course in control systems.

Each chapter is organized so that an introductory paragraph is followed by the presentation of the general approach or procedure. It is then illustrated by one or more examples. A summary and references are presented at the end of the chapter presentation. Analytical homework problems as well as computer problems related to the chapter material conclude each chapter.

The analysis and synthesis of robotic systems starts with the understanding of kinematic equations for the position and velocity of the end-effector expressed in the joint and Cartesian world coordinates. The dynamic model for the motion is obtained in Lagrange's or Newton–Euler's formulation. To facilitate the controller design, the equations of motion are linearized about a nominal trajectory. The desired trajectory for the manipulator motion is determined by off-line planning (trajectory generation). A control algorithm chosen by the designer is then constructed. We present the design of a controller for the motion of a robot manipulator in two stages: first a primary controller is constructed using *a priori* knowledge of the manipulator dynamics, and then a secondary controller is designed to compensate for tracking errors caused by modeling inaccuracies and disturbance effects. The same approach is applied to the controller design for the gross motion and the fine motion (force) control.

Chapter 1 deals with the basic terminology of robotics. It describes necessary components of a robotic system that are interconnected so that the resulting aggregate will function as a whole in a meaningful manner.

Chapters 2 and 3 describe the selection of the coordinate frames necessary to describe the position and orientation of the end-effector and the kinematic relations between these coordinate variables. The calculations of the solutions to the forward and inverse kinematic equations are discussed.

Chapter 4 establishes the kinematic velocity and acceleration relationships between the Cartesian base variables and the joint variables of the manipulator by means of the Jacobian matrix. The Jacobian matrix is also used to establish the relationship between the static external force and torque exerted by the manipulator on the environment and the generalized joint torques. The determination of the Jacobian matrix is presented.

Chapter 5 describes the equations of motion for a serial link manipulator. The dynamical model of the manipulator is presented both in Lagrange's and Newton–Euler's formulations.

Chapter 6 is devoted to the digital simulations of state-variable models and the linearization of the nonlinear dynamical models. Many general procedures for designing controllers are based on linear system models. Two basic procedures are described to obtain a linearized model for a set of nonlinear differential equations: One utilizes feedback and feedforward loops that generate appropriate nonlinear terms for the nonlinear model to cancel the inherent nonlinear effects. The other method is based on weak variations (perturbations) of the nonlinear equations about a nominal (desired) trajectory. Both methods lead to linear dynamical models for the manipulator motion. Some basic properties of the linearized manipulator models such as the stability and complete controllability are also discussed in this chapter.

Chapter 7 presents several kinematic approaches to plan the motion of a manipulator. The generation of the desired trajectory for the end-effector motion in the work space is performed by a planner off-line, which is on a high level in the control architecture. The planning of the motion for a manipulator depends on the specific task to be performed and the environment (the world model). Methods to generate the desired trajectory for smooth manipulator motion are described.

Chapter 8 presents the design of controllers on the low level of the control architecture. The construction of a primary and a secondary controller for the (unconstrained) motion of the end-effector is discussed when the manipulator motion is represented by a single-input single-output (SISO) model or by a multiple-input multiple-output (MIMO) model. The primary controller can be determined by calculating the solution to the inverse dynamics. The secondary controller may be chosen as a PID-controller (*p*roportional *i*ntegral *d*erivative). The tuning (determination) of the gains in this conventional controller is outlined in the framework of the eigenvalue assignment problem.

Chapter 9 presents the design of the adaptive self-tuning controller for robot manipulators. Such a controller is attractive when the dynamical model of the manipulator is completely unknown or only partially known, as when, for example, the numerical values of only some parameters are accurately known. The design is presented as the solutions to the LQG problem (*l*inear plant *q*uadratic criterion with *G*aussian noise) and to the pole-zero placement problem. In both cases, an autoregressive (AR) discrete time-series model describing the input–output relation of the manipulator system serves as the basis for the controller design.

Chapter 10 deals with the control of external forces and torques exerted by the end-effector on an environmental object. This problem occurs in many manufacturing tasks such as in assembly. In such a case, the motion of the manipulator is constrained. The dynamics of the constraint motion are discussed for the case that the model of the contact between the end-effector and the environment is soft (of mass-damper-spring type) and for the case that the contact is hard (described as a strict equality–inequality). The design of the system representing the constraint motion is presented in the framework of a primary and secondary controller design. Specifically, the design of a PID-controller such as the hybrid position-force controller and an adaptive self-tuning controller for the fine motion of constrained system in the case of soft contact is discussed. When the motion is subject to hard contraints, the DOF (degrees of freedom) of the entire system is less than the sum of the DOF of the subsystems. The determination of a reduced-order model for the entire system is presented.

ACKNOWLEDGMENTS

I would like to express my appreciation to the following people for their helpful comments and assistance in the preparation of this book: Charles P. Neuman, Fathi Salam, Heikki Koivo (my brother), Raimo Kankaanranta,

Jian-Hua Zhou, Senad Arnautovic, Michael Unseren, and Nasser Houshangi. In writing a textbook for a course, the students involved deserve particular thanks for the inspiration and feedback in many forms. They really have molded the text material through several stages to the final version.

I'd like to express my sincere thanks to Professor R. J. Schwartz, Head, School of Electrical Engineering and Professor H. W. Thompson, Assistant Head for Education for providing a peaceful atmosphere during the writing of this book.

My gratitude goes also to my family, Anne, Lilli and Allan, who gracefully stood by me in spite of my irritability while revising these chapters. I extend my special thanks to Jo Johnson who patiently has typed several versions of this text and often made helpful comments on these revisions.

I want to thank Christina Kamra, Richard Koreto and Sandra Russell of John Wiley & Sons for their collaboration, guidance and patience in finalizing this text.

I also thank National Science Foundation for the partial support when this book was being written.

ANTTI J. KOIVO

CONTENTS

CHAPTER ONE

ROBOTIC SYSTEMS

A basic unit in a robotic system includes a robot manipulator that performs a task often as the substitute of a human arm, and a digital computer that controls the manipulator motion acting as the brain of the system. Such a unit functions in some sense as an artificial man, of which mankind has dreamt a long time. It can perform mechanical tasks repeatedly with utmost accuracy, even in a hazardous and unhealthy environment caused, for example, by dust, fumes, heat, radiation, or noise. Many such applications can be found in the auto industry, for example, electric point-welding and spray painting of cars. Indeed, the sparks from welding and fumes of paint often render the environment hazardous and unhealthy to humans. Robot manipulators can often perform the work faster and with consistency; they do not become sick or even tired, they do not take coffee breaks, vacations, or dispute salary and pension; and they are not constrained by government and union regulations. The products made by robots are usually uniform in quality, satisfying the demanding requirements of tight tolerances. Robots often make better quality products than humans. Moreover, the production costs may be reduced considerably by means of robot manipulators, mainly because of recent wage inflation.

A robot system is mainly composed of mechanical components, a computer, and sensors. The understanding of the functions of these components, their connections, and interfaces in a robot system is a fundamental requirement for the construction of efficient and intelligent robot manipulator systems.

We will first describe some specific applications of robot manipulators. Then we will discuss the origin of the word robot, and the basis of classifying different types of manipulators. We will outline characteristics that must be specified when a robot manipulator is chosen for a specific application. We will present some interfacing and sensory components in robotic systems. In particular, we will describe sensors available for the designer to make robot manipulator systems adaptive and intelligent.

1.1 APPLICATIONS OF ROBOT MANIPULATORS

Robotic manipulators can presently be found to perform a variety of different tasks in factories, in which they replace one or often several workers. As an example, *Time* magazine [1] refers to a car-manufacturing plant in which 200 welders with their masks and welding guns have been replaced by only 50 welding robots that operate without rest in two shifts; as a consequence, the output of the assembly line has increased almost 20% since the installation of the robots. In such an example, the welder has to handle a heavy welding gun, which may weigh up to 40–45 kg. The gun is moved during the spot-welding process to certain positions so as to cause contact with the metal sheet for welding to occur. The metal sheet and the robotic welder must be positioned relative to each other with a sufficient accuracy so that the welding spot will be at the desired position. Usually there is no feedback to monitor and control the accuracy of the positioning of the gun relative to the metal sheet position. Hence, undesirable drifting in the spot welding may occur, which may require human adjustments.

If welding is to produce a continuous seam, the gun is required to move along a specific path at a desired distance from the pieces to be welded together so that the necessary welding arc is maintained. It has been estimated that a welding robot can sustain this arc almost 80% of the whole welding time, whereas a human welder averages approximately 25% [2]; moreover, welding seams produced by the robotic welders are uniform and look smooth. In most cases, the productivity of the process line is also increased considerably.

Another common, in fact, classical, application for robot manipulators is the loading and unloading of a workpiece. Different stages of this task may be described by means of the functions taking place when the workpiece is moved to the machine: loading the workpiece on the machine and affixing it accurately; selecting the cutting tool and inserting it into the machine; determining the setpoints such as operating speeds for the machine; controlling the motion of the cutting tool; sequencing the motions using different tools until all operations on the machine are finished; and finally, unloading the part from the machine. The sequence described illustrates a typical cycle in the process. At the very early stages of manufacturing, these steps were performed *manually*. As a step toward automating the process, the manual operation for determining the set speeds and for controlling the cutting was replaced by *numerically controlled* operations. Then, with the advent of digital computers and inexpensive microprocessors, the entire sequence of this batch production became a *computer-controlled* operation. It reduced the number of workers participating in the process. It also resulted in a higher output per unit time, that is, increase in productivity mainly because of a shorter processing time. The machines were now used very effectively; the time when the equipment was in operation ("up time") increased, while the time when the equipment was not operating ("down time") was reduced. The pieces were now processed faster, with consistent quality, and the amount of scrap decreased considerably. Moreover, the in-process inventory, that is, the inventory of unfinished pieces between the stages of the processing, was reduced.

In many plants, human intervention in the batch processing is minimal. Robot manipulators handle the raw or partly processed material to and from the machines, change appropriate tools for the processing machines, set proper speeds, and select the programs for the correct sequence of operations. Each phase of the material processing can be performed in a work cell. The entire manufacturing process can indeed be performed in a number of work cells, which are occupied by basic robotic units containing manipulators and computers.

A work cell can conveniently also be used for an assembly of processed pieces to make a product. For example, a work cell may contain a robot that can assemble a compressor valve consisting of 12 separate parts. It is relatively common to have several robots and additional fixtures to help in handling the parts for an assembly. An assembly task for a robot manipulator is a considerably more complicated task than that of materials handling.

1.2 WHAT IS A ROBOT?

We sometimes refer to a person who works mechanically and without thinking for himself (or herself) as a robot. A robot appears as an automaton. *Webster's* dictionary offers the following definition for a robot: "A robot is a machine in the form of a human being that performs mechanical functions of a human being." We should notice that this definition considers the work of humans as the basic function of a robot. Although the robot may not look or behave like a human, it needs an arm with a gripper for grasping and a computer (a brain) to perform its basic function.

An alternative definition for a robot is detailed by Robot Institute of America, which defines the word *robot* relative to its role in manufacturing. The Institute has adopted the following definition: "A robot is a reprogrammable multifunctional manipulator designed to move material, parts, tools, or specialized devices through variable programmed motions for the performance of a variety of tasks." This definition presumes that a robot must be *reprogrammable* and *multifunctional*, which implies that the operation of a robot is flexible. When a manipulator is used for a particular task such as spray painting, the manipulator may be reprogrammable depending on the computer, but it probably is not multifunctional; thus, the qualifications of being called a robot according to the foregoing definition may not be fulfilled. We observe that the definitions of a robot leave considerable freedom in these interpretations. A derivative of the word robot is *robotics*; it is commonly used in reference to the science and engineering dealing with robot manipulator systems.

The word *robot* seems to originate from the Czech language: *robotnik* means serf; *robota* is compulsory service. In a melodrama entitled *R. U. R.* (the abbreviation of *Rossum's Universal Robots*), which was published in 1921 and written by K. Capeck, robots refer to forced labor. The robots in this play looked like people. They worked very hard (workaholic), and performed unpleasant services, until they started rebelling against their makers. This setting appears to be common in many earlier fantasy stories. However, the movie *Star Wars* appears as an exception in the sense that R2D2 and C3PO become heroes in helping to rescue Princess Leia.

An industrial robot manipulator should be distinguished from a manipulator operated by a human in a closed-loop fashion [3]. Such manipulators have been used, for example, in handling radioactive materials at nuclear power plants and in deep underwater operations. A robotic system in which there is a man–machine interaction (human, computer, manipulator) is usually referred to as a *teleoperated manipulator*, or simply, a telerobot.

One of the basic units in a robotic system is a manipulator arm. It is made of several links connected usually in series by the joints to form an arm. The links can be numbered by starting from the base, which forms the zeroth link. A link is revolute or prismatic depending on the type of motion caused by the actuator attached to its joint. When the actuator of a joint causes rotational motion, the link is *revolute* (an articulated joint). When the actuator produces translational (linear) motion, the link is called *prismatic*. As an example, the cylindrical manipulator shown in Figure 1.1*a* has two prismatic links and one revolute link [4]. When the joints move, the links connected through joints also move. An end-effector, or a gripper, is attached to the arm by means of a wrist. A simple gripper usually has two opposing, moving plates for grasping an object.

The positions of the joints determine the *configuration of the arm*, which places the end-effector at a specific location in the environment. The function of the wrist is to orient the end-effector properly, for example, for the grasping of an object. The motion of the joints produced by actuators determines the position and orientation of the end-effector at any time. Transducers such as encoders can be used to provide information for determining the position and orientation of the end-effector and to control the manipulator motion.

WHAT IS A ROBOT?

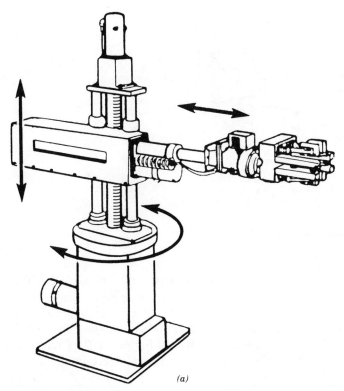

(a)

FIGURE 1.1*a*

Cylindrical manipulator with two prismatic links and one
revolute link.

The set of all points that can be reached by the end-effector of a manipulator arm forms the *workspace* of the manipulator. As an example, the
workspace of a cylindrical manipulator is shown in Figure 1.1*b*. The boundary surface of the workspace is referred to as the envelope; it contains the
points representing the maximum and minimum distances reachable by the
end-effector. The maximum distance of the end-effector from the base is
called the *reach* of the manipulator. The points that cannot be reached by
the end-effector form the *deadspace*. For example, points inside the base
belong to the deadspace; the boundary between the workspace and this
deadspace contains the point that is at the minimum distance from the base
and still reachable by the end-effector. The difference between the maximum
and the minimum distances reachable by the gripper is termed the *stroke* of
the manipulator movement. The maximum stroke on a horizontal plane is
called the *horizontal stroke;* in an analogous manner, the *vertical stroke* can be
defined.

The workspace of a manipulator represents all possible positions that the

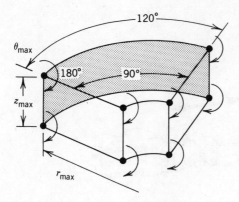

FIGURE 1.1*b*
Workspace of a cylindrical manipulator.

end-effector can assume. A particular position of the end-effector is specified by three independent coordinates, which represent three degrees of freedom (DOF). Similarly, a specific orientation of the end-effector is determined by three independent variables. Thus, six independent variables are needed to describe the position and orientation of the end-effector. They represent six DOF for a manipulator. The task of moving an object in the three-dimensional space through specified points at some particular orientation requires that three position and three orientation variables for the object are given. Thus, the DOF for the task is the same as that of the six-joint manipulator. If the arm contains more DOF than the number required for a class of tasks, the manipulator is termed *redundant*. For example, a seven-joint manipulator is a redundant manipulator for the foregoing task. The focus in this text is on nonredundant robot manipulators.

Industrial manipulators may be classified in several ways. On the basis of the joint actuator drives, a manipulator may be termed electric, hydraulic, or pneumatic. On the basis of the structure of a manipulator, the position of the end-effector can be described in an appropriate coordinate system, which may be rectangular, cylindrical, or spherical. The manipulators may be classified accordingly [4,5,6]:

Rectangular (Cartesian) Manipulator

When the arm of a manipulator moves in a rectilinear mode, that is, to the directions of the xyz coordinates of a Cartesian (rectangular right-hand) coordinate system, the arm is called a rectangular, or Cartesian manipulator. The moves may be referred to as travel, reach, and elevation. The workspace of the manipulator has the shape of a polyhedron, which is usually a right hexahedron, or a prism. This manipulator needs a large volume in which to operate. It has a rigid structure and provides an accurate position of the end-effector.

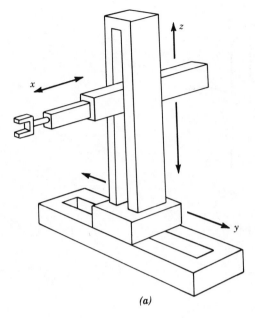

(a)

FIGURE 1.2*a*
Rectangular (Cartesian) robot manipulator.

A schematic diagram of the Cartesian manipulator is shown in Figure 1.2*a*. A special type of Cartesian robot manipulator is an overhead gantry robot shown schematically in Figure 1.2*b*. It is a rectangular coordinate robot, in which the orientation of the end-effector is determined by three rotating joints. This manipulator is large, versatile in its operation, and expensive.

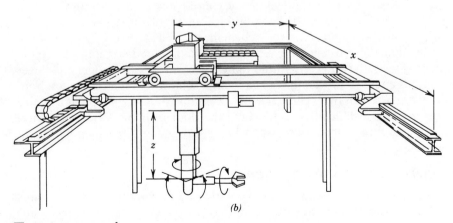

(b)

FIGURE 1.2*b*
Gantry robot manipulator.

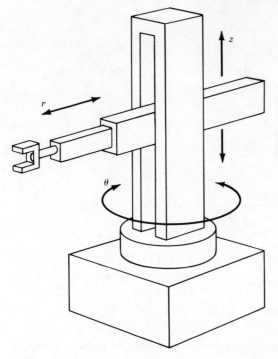

FIGURE 1.3
Cylindrical robot manipu-
lator.

Cylindrical Manipulator

When the arm of a manipulator possesses one revolute and two prismatic
links that correspond to three DOF, the points in its workspace can conveniently be specified by the cylindrical coordinates (radius r, angle θ, and
height z). This arm is termed a cylindrical manipulator. Such a manipulator is shown in Figure 1.1a and schematically in Figure 1.3. The moves of
the manipulator are called the reach, base rotation, and elevation. Since the
coordinates of the arm can assume values between specified upper and lower
limits, its end-effector can move in a limited volume that is a cut section from
the space between two cylinders with a common axis. A manipulator of this
type may have difficulties in touching the floor near the base. Cylindrical
manipulators are successfully used when a task requires reaching into small
openings.

Spherical (Polar) Manipulator

When the arm can change its configuration by moving two revolute joints
and one prismatic link, the arm position is conveniently described by means
of the spherical coordinates (r, θ, ψ). The arm is termed a spherical, or
polar, manipulator. The moves of the manipulator represent the reach, base

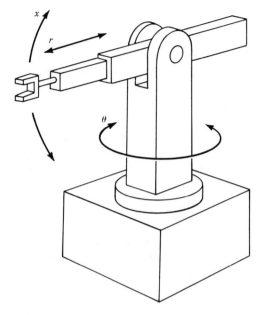

FIGURE 1.4
Spherical (polar) robot manipulator.

rotation, and elevation angles. It is shown schematically in Figure 1.4. The workspace of this manipulator occupies a large volume.

Revolute (Articulated) Manipulator

When the arm consists of links connected by revolute joints, it is called a revolute, or articulated, manipulator. It is described schematically in Figures 1.5a–1.5c. Figure 1.5d shows a picture of PUMA 500, a revolute manipulator. Its wrist position can be specified by three angles representing the base rotation, elevation, and reach. A revolute manipulator can successfully be used in a diversity of applications due to its flexibility.

In selecting a manipulator for a specific application, the following aspects should be considered:

The cost of the robotic system.

The tasks to be performed by the manipulator determine several factors: the size, the load carrying capacity, the number of joints (links), as well as the type of the manipulator and the power drives (actuators).

The workspace needed.

The speed and the movements of the end-effector necessary to perform the tasks.

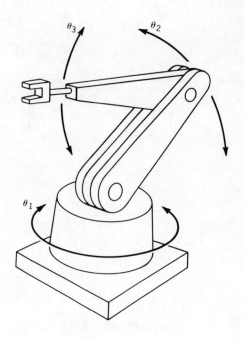

FIGURE 1.5*a*
Revolute (articulated) robot
manipulator.

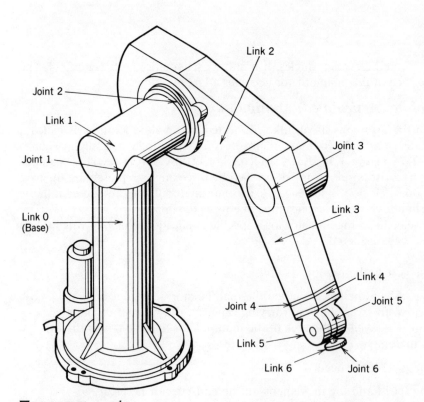

FIGURE 1.5*b*
PUMA (Programmable Universal Manipulator for Assembly) manipulator.

10

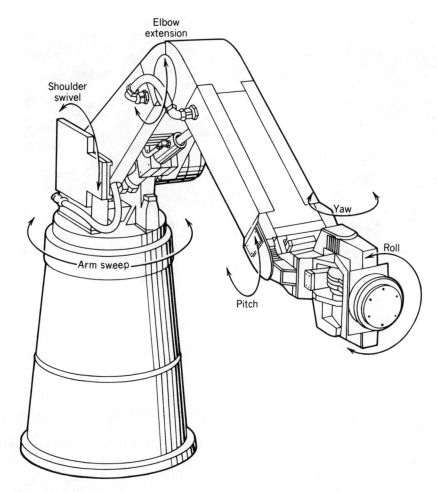

FIGURE 1.5*c*

A revolute manipulator (Cincinnati Milacron, T³).

The precision (resolution) in the positioning of the end-effector, and the repeatability of the motion. The tools that may be attached to the gripper may influence the positional accuracy.

The environmental conditions such as cleanliness of the air and ambient temperature. These factors also affect necessary maintenance.

The availability of sensory information for the controller.

The computer specifications: memory, programming (language), interfacing between high- and low-level computers.

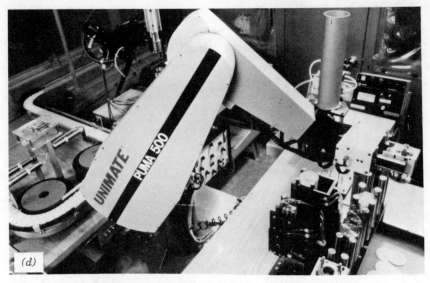

(d)

FIGURE 1.5d

PUMA 500 manipulator (Westinghouse/Unimation Inc.).

The repeatability mentioned refers to the measure of the neighborhood of a specified (desired) point to which the end-effector is placed in repetitive motions. The repeatability for the presently available good manipulators range usually from 0.02 to 1.2 mm.

A typical set of performance specifications provided by a manufacturer is illustrated in Table 1.1. It is for a four-axes DC motor driven cylindrical robot of high precision shown in Figures 1.6a and 1.6b.

The technical descriptions usually include (1) on the manipulator arm: configuration, the number of DOF, weight, actuators, power requirements, load carrying capacity (payload), joint velocities, repeatability, and operating ranges; (2) on the computer: programming language, memory, and external program storage. The maximum speed of a manipulator motion specified by the manufacturer depends strongly on the structure of the manipulator, such as the weights and the lengths of the links. The reach and the stroke are included in the specifications of the manipulator. It is also common to specify the ranges for the movements of the joints, as well as of the wrist, for example, by means of roll, pitch, and yaw angles. An end-effector is usually determined by the specific task that the manipulator is to perform. The maximum gripper opening and the gripping force may also be given by the manufacturer. It should be emphasized that the manufacturers usually provide additional information about their robotic systems on customer's request. The selection of a proper robot manipulator system for a specific

TABLE 1.1

PERFORMANCE SPECIFICATIONS		
Configuration		4 Axes
Payload (at maximum speed)		5 kg (11 lbs)
Arm range	*A:* Grip rotation	±999 deg
	r: Horizontal stroke	300 mm (11.81 in.)
	θ: Plane rotation	290 deg
	z: Vertical stroke	120 mm (4.72 in.)
Speed	*A:* Grip rotation	150 deg/sec
	r: Horizontal stroke	750 mm (30 in.)/sec
	θ: Plane rotation	90 deg/sec
	z: Vertical stroke	360 mm (14 in.)/sec
	Combined maximum	1400 mm (55 in.)/sec
Repeatability		±0.025 mm (0.0010 in.)
Resolution	*A:* Grip rotation	0.005 deg
	r: Horizontal stroke	0.025 mm (0.001 in.)
	θ: Plane rotation	0.003 deg
	z: Vertical stroke	0.012 mm (0.0005 in.)
Weight		108 kg (238 lbs)

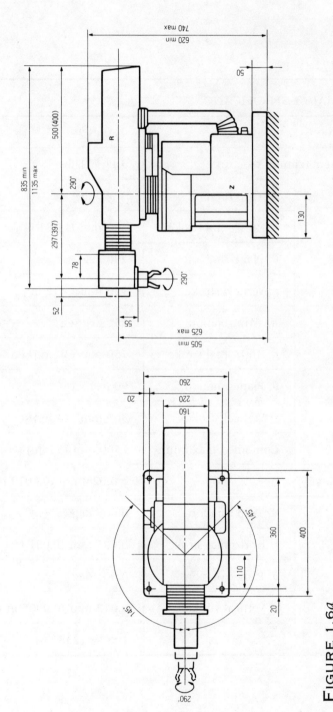

FIGURE 1.6a
Cylindrical robot manipulator (SEIKO Instruments Inc., RT 3000). Measurements in mm.

14

FIGURE 1.6*b*

Cylindrical robotic system (SEIKO RT 3000).

task is the result of cooperation and compromises between the customer and the manufacturer.

We will discuss next the other common components of a robotic system: the actuators of the joints, A/D and D/A converters, encoders, tachometers, and force/torque transducers.

1.3 COMPONENTS OF ROBOTIC SYSTEMS

The motion of a manipulator gripper is caused by the movements of the actuators driving the joints. The joint actuators can be electric or hydraulic motors, or pneumatically driven devices [2,4]. We will present the characteristics of the electric and hydraulic actuators which are relevant to the dynamics of a manipulator. Since there is a considerable time delay in the function of pneumatic actuators [3], they are not commonly used to drive manipulators at the present time; these actuators will not be discussed here. The function of A/D and D/A converters which play an important role in the interfacing of the digital computer and the analog components (e.g., motors) will be outlined. The position, and velocity of a joint shaft can be measured by means of an encoder, and a tachometer, respectively. These components provide necessary information about the actual values of the joint outputs. When the measurements are compared with the desired values using feedback loops, the errors can then be used to control the actuators so that the outputs will assume, or approach the desired values. We will discuss the basic function of the encoders and tachometers. For many tasks, the forces and torques exerted by a manipulator on the environment must be kept within acceptable ranges. These values are measured by means of force/torque sensors, which will be discussed. The transducers measuring the position, velocity, and force/torque in a manipulator system provide the control system with the basic ingredients to be able to function efficiently, even under the influence of disturbances.

1.3.1 AUCTUATORS FOR JOINT DRIVES AND THEIR CHARACTERISTICS

DC-Motor and Transmission Gear

The joints of a manipulator are commonly driven by electric motors such as DC motors or step motors. We will next discuss briefly the characteristics of these motors.

A circuit diagram of a DC motor with a separate excitation circuit is shown in Figure 1.7. The magnetizing circuit forms a magnetic field in the airgap between the stator and rotor of the motor; this magnetic field interacts with the electric current of the armature circuit producing a force and, thus, a torque, which causes the rotation of the rotor and the motor shaft.

The speed of a DC motor is usually controlled by means of the armature voltage or the armature current, when the magnetic flux is constant. In this manner, any value in the entire range 0–100% of the nominal speed can conveniently be chosen as the desired speed, that is, as a set–point value. In principle, the speed of the DC motor could also be controlled by holding the armature voltage or current constant and controlling the magnetizing current of the excitation winding. However, practical limitations restrict the range of the output speed to approximately 80–100% of the nominal speed, which represents only a small range for controlling the speed. This is due to the hyperbolic shape of the characteristic curve of the motor describing the angular velocity as a function of the excitation current. In constructing feedback control systems involving DC motors, the designer should also keep in mind that the inductance in the armature circuit of a DC machine is usually negligible, whereas that in the magnetizing circuit is quite large. Hence, the

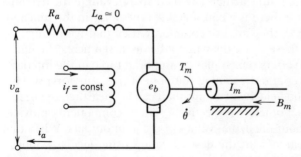

FIGURE 1.7
Schematic for separately excited DC-motor.

changes in the excitation circuit are associated with a large time constant, resulting in a sluggish time response, as compared with the responses of the armature circuit.

In a DC motor, the torque T_m generated is proportional to the product of the armature current i_a and the magnetizing current i_f:

$$T_m = (K_1 i_f) i_a \qquad (1.3.1)$$

If a DC motor has a constant magnetic flux, then the torque of the motor is

$$T_m = K_m i_a \qquad (1.3.2)$$

where K_m is a constant.

In many manipulators of the present, the motor shaft is mechanically coupled to a load through *gears*, as shown schematically in Figure 1.8a. The shaft of a gear train that is directly connected to the axis of a motor is referred to as the primary side (by the analogy of a transformer). The output shaft of the gear is on the secondary side. Let the angular speed of the shaft on the primary side be $\omega_1 = \dot{\theta}$ (the derivative of the angular position) and that on the secondary side ω_2. The power associated with the rotational motions of both shafts must be equal in an ideal gear train (no power losses). Then the effect of a load on the motor can be determined. The loading torque T_L on the secondary side is reflected to the shaft of the motor, that is, to the primary side as

$$T'_L = (\omega_2/\omega_1) T_L \qquad (1.3.3)$$

Thus, the loading torque T_L is reduced when it is reflected to the shaft of the motor if $\omega_2 < \omega_1$.

A coupling gear also affects the dynamical parameters of a motor. In addition to the explicit loading torque, the moment of the load inertia and the viscous friction on the secondary side of a gear are sensed on the primary side. In order to determine the reflected inertial moment I_{eq} corresponding to the moment of inertia I_L of the load on the secondary side, the energy

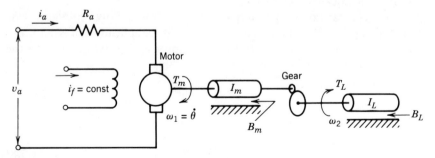

FIGURE 1.8a
A DC-motor, gear train, and load.

expressions associated with these second moments on both sides of the gear train are set equal:

$$\frac{1}{2}I_L\omega_2^2 = \frac{1}{2}I_{eq}\omega_1^2 \tag{1.3.4}$$

Equation 1.3.4 can be solved for the equivalent second mass moment I_{eq} on the motor shaft. Then, the total moment of the inertia I_t on the motor shaft is

$$I_t = I_m + \left(\frac{\omega_2}{\omega_1}\right)^2 I_L \tag{1.3.5}$$

where I_m represents the moment of the inertia of the rotating motor shaft. It should be emphasized that the inertial moment of load about the axis of joint motion in a multiple-link manipulator depends strongly on the configuration of the manipulator arm.

Using similar arguments as in Equation 1.3.4 the coefficient B_t of the total viscous friction can be determined:

$$B_t = B_m + \left(\frac{\omega_2}{\omega_1}\right)^2 B_L \tag{1.3.6}$$

where B_m and B_L specify the coefficients of viscous friction of the motor and of the load, respectively. The foregoing *viscous friction*, which is directly proportional to the velocity, describes fluid friction. Dry friction, which is often called *Coulomb friction*, is constant during the motion. At the start of the motion, however, the friction assumes a large value; it represents *static friction*. Any type of friction characteristically opposes the motion. Frictional effects on the motion of a manipulator can be very troublesome, particularly at low operating speeds.

Equations 1.3.3, 1.3.5 and 1.3.6 describe how a simple load on the shaft of the secondary side of a gear train can be referred to the shaft of the motor. Since these relations in a manipulator can be quite complex under usual operating conditions, their effects on the dynamics often necessitate the use of special sophisticated controllers.

The time constant of a separately excited DC motor shown in Figure 1.8b is mainly determined by the mechanical load. The application of Kirchhoff's voltage law to the armature circuit of the motor and Newton's law to the rotating system gives over a linear operating range:

$$V_a = R_a i_a + K_b \dot{\theta} \tag{1.3.7}$$

$$T = K_m i_a$$

$$= I_t \ddot{\theta} + B_t \dot{\theta} \tag{1.3.8}$$

where R_a is the armature resistance, and the armature inductance has been ignored, K_b is the constant of the back electromotive force (emf), K_m is the

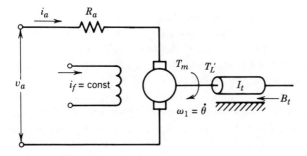

FIGURE 1.8b
DC-motor with an equivalent load.

torque constant, and θ is the angular position of the motor shaft. Equations 1.3.7 and 1.3.8 can be solved for the transfer function $G(s) = \theta(s)/V_a(s)$:

$$\frac{\theta(s)}{V_a(s)} = \frac{K}{s(1 + s\beta)} \tag{1.3.9}$$

where the static gain $K = K_m/(B_t R_a + K_m K_b)$, and the time constant $\beta = I_t/(B_t + K_m K_b/R_a)$. The latter expression shows how the time constant of the open loop system in Equation 1.3.9 is dependent upon the total moment of inertia I_t and the total frictional coefficient.

A gear train in robot manipulators is often realized by means of a harmonic drive. It provides a very high torque transmission, and it is precise (practically no backlash).

Although the links in many manipulators are driven through the gears, *direct drives* with DC motors are gaining popularity at the present. When the motors are directly connected to the joint shafts, they must generate very high torques. On the other hand, typical problems encountered in gear transmission such as friction and backlash can be avoided.

Step Motor

A step or stepping motor is an electromechanical device that accepts pulse and/or logic signals as the input and produces discrete repeatable steps in shaft angle position as its output. This position corresponds to one of the discrete levels resulting from the quantization of the range of the angular position.

The basic function of a step motor can be described by considering a simplified model in which the rotating part, the rotor, contains a permanent magnet or only one coil, and the stationary part, the stator, has many coils connected so that any two coils located 180° apart are in series but otherwise electrically isolated from other pairs. The moving part generates a magnetic field, whereas the current through a pair of serially connected coils in the stator is dependent on an electronic switch, and thus on the pulse input. When a particular pair of coils is energized, it generates also a magnetic field. The rotor winding will rotate then to an angular position such that the magnetic energy associated with the magnetic field generated by the moving coil and the field produced by the current in one pair of the stationary coils is minimized. Thus, the vector representing the magnetic field of the rotor will try to line up with the vector of the magnetic field produced by the current through the energized part of the stationary coils.

The input to the step motor usually consists of a sequence of pulses of constant amplitude over discrete time intervals. The output, the shaft position, depends on the characteristics of the step motor system. The output step response to a pulse over a sampling period should have settled down, that is, the transients have died out, before the next pulse is applied. If the transients have not settled down when the next pulse arrives, severe problems may arise. This mode of operation, the *slewing*, should be avoided, for example, by properly selecting the stepping rate.

A step motor functions as a positioning device, whose input is a pulse or digital signal, and the output a quantized angular position of the joint shaft. The accuracy of the angular position, that is, the resolution, depends on the number of the discrete levels resulting from quantizing the range of the angular position to equal steps and, thus, the number of the coil pairs.

A step motor from a manufacturer is usually delivered with a separate control unit. Its input controls the power into the step motor by means of electronic switches. The basic setup is shown in Figure 1.9. No additional equipment besides the control unit is usually needed to modify the input signal for the motor input. The interfacing between a digital computer, such as a microprocessor, and the motor is very convenient and straightforward.

The characteristics of a step motor are similar to those of a DC motor, if the quantization levels of the range of the angular position are sufficiently small. In fact, the effects of an external torque, the moment of the inertia, and the friction of a load on the shaft of a step motor can be described as discussed previously.

Hydraulic Motor

Hydraulically driven manipulators are common in applications requiring large power, for example, in lifting heavy objects. The function of a hydraulic motor is determined by the flow of oil from a pump. This flow can be decom-

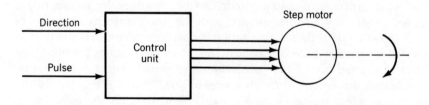

FIGURE 1.9
Step motor with its control unit.

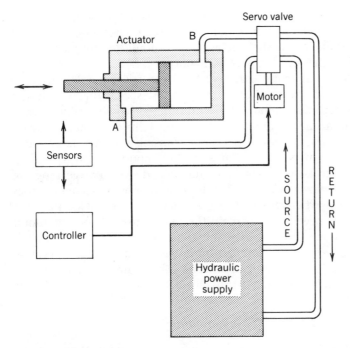

FIGURE 1.10a
Hydraulic transmission motor.

posed into three major components: flow into the motor, leakage flow, and the compressibility component. These components of the flow determine the dynamic behavior of the system, which depends on the hydraulic transmission and the mechanical (rotating) part. A schematic diagram for a hydraulic transmission is shown in Figures 1.10a and 1.10b, which display the paths for the oil flow in the hydraulic motor. The hydraulic flow circuit is closed and kept full by means of an oil-replenishing system. Figure 1.10b reveals the oil flow relation: oil flow from the pump equals the sum of the oil

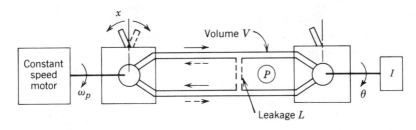

FIGURE 1.10b
Closed oil flow model.

flow to the motor, compressibility, and leakage flows. Under some restricted conditions, a linear model may be obtained for the input–output (stroke of pump–angular position of the shaft) relation. A hydraulic motor may exhibit a dynamic behavior that can be quite oscillatory.

A hydraulic drive has many attractive features. It is generally observed that the hydraulic drives run more smoothly at low speeds than do electric motors. Since the hydraulic motors have an inherent cooling system, they can be operated in a stall condition (short circuit) without damaging them. The speed/stroke characteristics of hydraulic transmission motors are such that the shaft of the motor rotates at a speed proportional to the pump stroke. The speed–stroke relationship is maintained over a large range of load variations. Specifically, the speed of the shaft in a hydraulic motor does not vary much even when the load torque changes over a wide range. A considerable force is needed to maintain the pump stroke when the motor is delivering a torque. Thus, a large amplification of the input signal is often required for the stroking mechanism in the pump. The amplification can be achieved electrically or hydraulically. The main drawback in the use of hydraulic motors is due to possible leaks in the closed oil flow system.

1.3.2 A/D AND D/A CONVERTERS

The input to a motor is in most cases an analog signal, and the input to a step motor unit is a pulse or digital signal. A signal that has a value at every time instant over the time interval under consideration is termed "analog." This signal is also referred to as a continuous-time signal. A signal that assumes well-defined values only at the sampling instances and is zero elsewhere on the sampling interval is called a "discrete-time signal." Such a signal is often modeled as a train of impulses to facilitate the mathematical analysis, although they have a small finite pulse-width in practical applications. If a signal is discretized (in amplitude) by means of quantization operations, the resulting signal can be expressed as a binary number. The number of the quantization levels determine the word length of the binary number. A discrete-time and quantized signal is often called a "digital signal."

A set of signals needs often to be processed in a digital computer. Before an analog signal can be fed into a digital computer, it must be sampled and converted into a digital form. The sampling operation is accomplished using a switch (with a clocking device), which is often assumed to be ideal, that is, it closes and opens "instantaneously" so that its output consists of impulses whose area is equal to the value of the signal at the sampling instant. The device that performs the quantization operation is called an *analog-to-digital converter* (ADC, or A/D converter). The output of a digital computer is in the form of pulses (or impulses). Before the signal of this type is applied to a plant such as a motor, the signal is modified so that it becomes an analog signal. The device that converts a digital signal to an analog form is termed a *digital-to-analog* converter (DAC, or D/A converter). We will outline here the basic function of the ADC and DAC [2,4].

A *D/A converter* is used as an interface between a digital computer and the plant (process) controlled by the computer. The function of a DAC is illustrated by the following example, which describes the basics of analog-to-digital conversion.

A simple DAC converter is shown in Figure 1.11. The inputs represent the binary number $b_5b_4b_3b_2b_1b_0$, where each $b_i, i = 1, \ldots, 5$, is 0 or 1. The most significant bit, b_5, represents the sign bit. The potentiometer settings can be adjusted so that when $b_0 = 1$ and all other bits zero, then the output of the operational amplifier is $0.1K$ Volts (K is a constant gain). Similarly, when b_1 is 1 and all other bits equal to zero, the output of the operational amplifier is $0.2K$ Volts, etc. If $b_i, i = 0, 1, 2, 3, 4$ equals zero, and the sign bit b_5 equals 1, the output of the amplifier is equal to $-3.1K$ (volts). The gain (K) of the

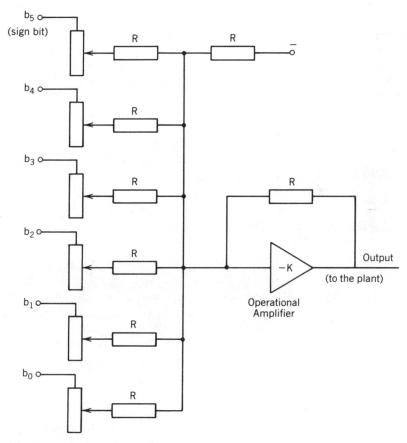

FIGURE 1.11
A simple D/A converter.

amplifier can be adjusted so as to obtain the desired output voltage of the operational amplifier.

The foregoing adjustments of the potentiometer settings and the output voltages are such that two's complement binary representation leads to correct analog output voltage. As a specific example, let the input to the DAC be 010110. Then the output is

$$[0 \cdot (-3.1) + 1 \cdot (1.6) + 0 \cdot (0.8) + 1 \cdot (0.4) + 1 \cdot (0.2) + 0 \cdot (0.1)]K \text{ Volts}$$

$$= 2.2K \text{ Volts}   (1.3.10)$$

If the input to the DAC is 110010, then the output equals

$$[-3.1 + 1.6 + 0 + 0 + 0.2 + 0]K \text{ Volts} = -1.3K \text{ Volts}   (1.3.11)$$

Thus, the binary input expressed in two's complement binary representation is converted to an analog voltage.

An *A/D converter* operates on an analog signal to generate a representative digital signal. It is accomplished by comparing an unknown analog input sample with the known voltage, relative to which the decision on the binary representation is made. After determining the closest quantized voltage level, the digital signal representation is obtained. This function can be implemented using comparators. An example of an A/D converter is a flash converter in which an unknown analog voltage sample is fed into comparators connected in parallel; its output is a binary number that represents the voltage level closest to the unknown analog input. A schematic diagram of the connections for a 3-bit A/D converter is shown in Figure 1.12. This encoder functions very fast, but it is also expensive because of the large number of the components used in the implementation.

1.3.3   ENCODERS AND TACHOMETERS MEASURE ANGULAR POSITIONS AND VELOCITIES

The angular position of a shaft can be measured by means of a simple *cylindrical potentiometer*, or an accurate encoder, that has a more complex structure. The resistance wire of a cylindrical potentiometer is wound in the shape of a cylinder. The center point of the potentiometer slides inside the cylindrical wall; its position depends on the angle of its shaft. If a constant voltage is connected to the end points of the potentiometer coil, the voltage between the sliding point and one of the end points will be proportional to the angle of the shaft attached to the sliding point. This type of potentiometer is small and simple, and it can be fitted to tight places such as the wrist and hand joints of a manipulator.

An *encoder* attached to the shaft of a joint is a transducer, whose output is a series of pulses (a binary number), representing the shaft position. The basic element in an encoder is a disc that converts the shaft position to

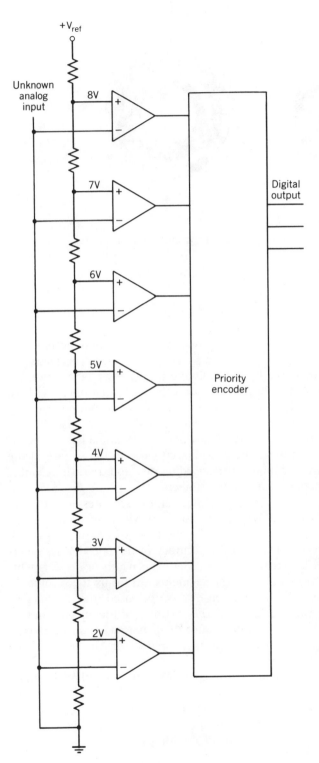

FIGURE 1.12
A three-bit A/D converter.

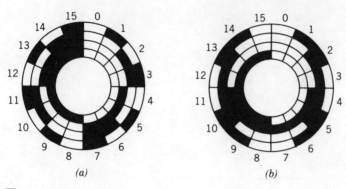

FIGURE 1.13*a* AND *b*
a An encoder disc with output as binary coded decimal
(BCD); e.g., $3_{10} = 0011$.
b An encoder disc with output in Gray code; e.g., $3_{10} = 0010$.

a binary number. Figures 1.13*a* and 1.13*b* demonstrate the basic principle
used in the encoding of the shaft angle to a binary number. Each segment of
the disc is separated by two adjacent radii and represents a binary number
with the light area corresponding to zero and the dark area one. The least
significant bit is the outside bit of a segment. The binary output representing
the position of the disc can be read using optical, magnetic, or brush devices.
An optical encoder has a segmented disc, a light source, and a light detector;
a magnetic encoder has a magnetic head placed close to the surface of the
disc; and a brush encoder contains the encoding disc and a brush assembly
that generates wave patterns for binary numbers.

In commercially available encoders, the output is expressed either in
binary coded decimals (BCD), or in Gray code, in which successive coded
characters differ in only one bit. Most encoders have word lengths of 4, 8,
12, or 16 bits. The word length and the accuracy of the encoder are determined by the internal structure of an encoder. The width of the segments
on the disc specifies the number of the segments, and thus the resolution.
If the width of the segments in a given disc is decreased, the accuracy of
the encoder readings can be improved. The accuracy of the encoder output
as a position indicator is considerably better than that of an analog device,
such as the aforementioned potentiometer. Moreover, it can be enhanced
considerably by means of gears.

The output of an encoder represents the angular position of the joint shaft.
If the angular velocity of the shaft is needed, for example for a controller, it
can be formed from the encoder outputs by means of hardware or software.
Thus, additional operations are needed to obtain the angular velocity (and
possibly the acceleration) from the encoder readings.

A *tachometer* is a transducer whose analog output voltage is directly proportional to the angular velocity of its shaft. A typical tachometer can be considered as a DC generator with constant excitation field. Its output, e_g, is directly determined by the angular velocity, that is, $e_g = K_{tach}\dot{\theta}$. Other types of tachometers, for example AC tachometers, are also used. A tachometer is a common component in the feedback loop of many industrial speed-control applications, in which it is often used to reduce the oscillations in the system response.

1.3.4   FORCE/TORQUE MEASURING DEVICES

The force and torque measurements provide information about the contact of the end-effector with the environment. They are needed for the proper function of a manipulator, particularly in dexterous tasks requiring skillful and precise handling of objects. Typical examples of such tasks can be found in material handling and assembly in which grasping of a fragile object, insertions, gear meshing, and push and twist actions are necessary [5]. These tasks often require the control of small movements.

A simplest type of force sensing is based on *binary contact* (microswitch); if a contact is detected, the manipulator is quickly stopped. The same idea can be generalized in several directions. The motion of the manipulator is stopped in the direction of the observed contact, while it can continue to the other directions. This type of force sensing can be used, for example, when unexpected obstacles are encountered in the path of a slowly moving manipulator.

In order to use force/torque measurements in a feedback control mode, more accurate information is needed. Several methods for measuring forces and torques are available at present [2,4]. We will discuss next two types of transducers used commonly in manipulators for force/torque sensing.

To measure forces exerted by a manipulator arm on the environment, elastic elements can be used for the measurements. When the dimensions of such an element change due to the force applied, changes in some physical property of the element may be observed. If the element is a spring, its length changes due to the force applied. If it is a *strain gauge* of properly chosen material, its electrical resistance changes. The gauges are often made of good conducting material bonded to the object. With the deformation of the object, the strain gauge is subject to strain, causing its electrical resistance to change. These changes correspond to the applied forces.

A strain gauge can be made very small. The sizes of the available strain gauges vary from about 20 mm^2 to 6–7 cm^2. They are often very thin; foil gauges can be 0.02 mm thick, or even less. The resistance of the strain gauges vary: 50–100Ω is a typical range. Strain gauges can be made of different materials. Silicon is common in semiconductor strain gauges. They can also be made of piezoelectric material, in which the deformation generates an electric signal proportional to the applied force. Typical strain gauge designs are shown in Figure 1.14a [4].

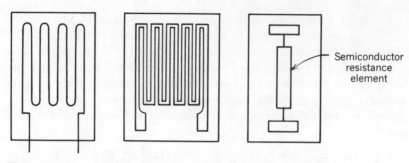

FIGURE 1.14*a*
Strain gauge designs.

The resistance of the strain gauges is sensitive to temperature variations. It is a common practice to use temperature compensation for resistance strain gauges. This is accomplished, for example, by means of Wheatstone bridge circuit, which is common in many applications. The resistances associated with electrically opposite elements in the bridge are usually installed physically at the opposite positions. The installation (attachment) of strain gauges to an object should be performed with extreme care for accurate measurements.

The strain gauges are also used in wrist-force sensors [4]. Figure 1.14*b* shows a transducer whose output vector has three components of the force and three components of the torque, (moment). The output is determined by the displacement of the strain gauges installed at the precise locations on the device. The wrist-force sensor measures the force and torque exerted by the end-effector on the environment.

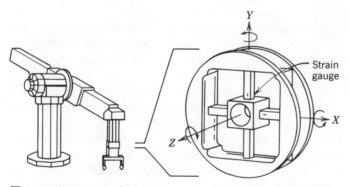

FIGURE 1.14*b*
A wrist-force sensor ("Maltese cross").

1.4 SENSORY-BASED ROBOT MANIPULATORS

It is highly desirable to make robot manipulators autonomous and intelligent so that they can function independently and adapt their operations to changing environmental circumstances [6,7]. Sensory information is essential to achieve this behavior in a manipulator. Force/torque sensors, and position/velocity transducers discussed in the previous sections give basic sensory information to the computer of a robot system. A force/torque sensor, which is usually mounted to the wrist of a manipulator, provides a computer with composite measurements about the contact forces and torques. Additionally, tactile sensors can be installed on the gripper plates, or used as separate entities to convey information about pressure on contact surfaces to a computer. These measurements are typically in the form of an array having, for example, 10×16 elements. The tactile measurements can be used also to determine contact forces or to identify object shapes. Tactile sensors do not require ideal environmental conditions to function properly. The integration of force/torque measurements with tactile information can enhance the capabilities of manipulators, particularly in assembly operations, which are expected to be the major growth area of robotics and industrial manufacturing [8].

If the environment is clean and lighting conditions are appropriate, a camera can give a computer very detailed information about the environment. The computer images can be used to determine, for example, the locations and the shapes of environmental objects. Since the pictures are in the form of large pixel (picture cell) arrays, usually of size 512×512, the processing and extraction of essential information in an appropriate form from such a large data set is very time consuming, even in large digital computers. Efficient on-line use of image information for the motion control of manipulators is presently a rapidly progressing research area.

Although computer images can be used to determine depth information, for example, to locate objects, direct measurements using range sensors is quite common. In fact, range sensors can determine directly distances between the gripper of a manipulator and objects or obstacles in the environment. Thus, both visual and range sensors can give similar information without contacting the environment. The range sensors can advantageously be used also in obstacle avoidance problems and in improving the safety aspects of robot manipulator operations. The fusion of visual, range, tactile, and force/torque information in a robot system for specific tasks is a very attractive idea. In fact, it should be an essential ingredient in the realization of intelligent control schemes [7] for robot manipulators.

Another attraction for researchers on robotics is to use auditory information for the command and control of robot manipulators. One of the main problems is the small vocabulary that can be stored in the computers of the present. Fast algorithms, neuro computing, and possible artificial intelligence methods for quick interpretation are needed before robots using speech recognition and synthesis will become practical for applications.

In the sensory-based robot systems, necessary information must be acquired, stored, and extracted in an appropriate form. The presentation of this information must be suitable for control algorithms, or vice versa. The integration of the acquired sensory information and control schemes is essential in designing intelligent robot systems, which can function successfully in unknown or partially known changing environments. Sophisticated control schemes such as those relying on adaptive controllers are needed in order to incorporate continuously acquired sensory information into control algorithms. Our objectives in this book will be to present fundamentals needed to design technically advanced controllers for such robot manipulator systems.

1.5 SUMMARY

The reader is exposed in this chapter to the basic concepts of robotic systems and to the components connected in a meaningful manner to make them function together as a system.

We have discussed the definition of the word *robot* and manipulator applications emphasizing the flexibility in their operations. This is achieved when a robot manipulator is reprogrammable and multifunctional. Common features characterizing a manipulator arm include workspace, deadspace, and reach and stroke of the arm. The shape of the workspace depends on the structure of a manipulator and the motions of links (joints), which can be translational or prismatic. Robot manipulator arms may be classified on the basis of the coordinate system used to describe the position of the end-effector: rectangular, cylindrical, or spherical.

Actuators, which represent essential components of a robotic system, move the arm; they can be electric, hydraulic or pneumatic. We discussed the characteristics of the electric and hydraulic motors. The interfacing of the components requires the use of A/D and D/A converters, whose functions we outlined. For the purpose of implementing feedback controls in a manipulator, encoders and tachometers are needed to measure the actual positions and velocities of the joints. For advanced functions of manipulators, forces and torques exerted by the end-effector must be measured. We discussed transducers that can provide this sensory information for a computer and controller.

In order to make robot manipulators function intelligently, sensory information about the environment must be obtained. Force/torque and position/velocity measurements can be complemented with tactile, visual, and auditory information that can be incorporated in the decision making, command, and control of manipulator motion. Different sensory measurements need to be fused with control schemes to achieve independence and intelligence in the function of robot manipulator systems. Such systems represent future trends in robotics.

REFERENCES

[1] "The Robot Revolution," *Time*, pp. 72–83, December 8, 1980.

[2] E. KAFRISSEN AND M. STEPHANS, *Industrial Robots and Robotics*, Reston Publishing Company, Inc., Reston, VA, 1984.

[3] W. SNYDER, *Industrial Robots: Computer Interfacing and Control*, Prentice-Hall, Englewood Cliffs, NJ, 1985.

[4] A.J. CRITCHLOW, *Introduction to Robotics*, Macmillan Publishing Company, New York, 1985.

[5] P.G. RANKY AND C.Y. HO, *Robot Modelling*, IFS Ltd., U.K., Springer-Verlag, 1985.

[6] J.F. ENGELBERGER, "Robotics in Practice," *1980 AMACOM*, A Division of The American Management Association, New York, 1980.

[7] G. S. SARIDIS, "Intelligent Robotic Control," *IEEE Transactions on Automatic Control*, Vol. AC-28, No. 5, pp. 547–557, May 1983.

[8] A.L. PORTER AND F.A. ROSSINI, "Robotics in the Year 2000," *Robotics Today*, June 1987.

PROBLEMS

1.1 For the robot shown in Figure 1.15, specify the horizontal and vertical stroke, and the horizontal and vertical reach by writing them on the figure.

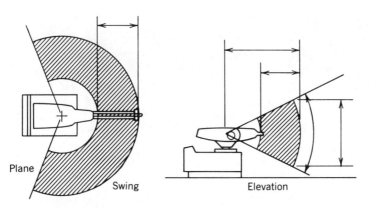

FIGURE 1.15
Workspace of a cylindrical robot manipulator.

1.2 *a.* Identify the joints and links in the human arm. Describe
           the envelop of the workspace of a point on the elbow.

     b. Approximate the ranges of the angles that the shoulder
           can cover.

     c. Explain two DOF on the upper arm (shoulder joint).
           What provides another DOF?

     d. Explain which motions the wrist can do. Approximate
           the ranges of these motions.

1.3 In the cylindrical manipulator shown in Figure 1.1*a*, the
     ranges of the joint variables are limited. Give physical rea-
     sons for these restrictions.

1.4 *a.* If a gear transmission is considered lossless, determine
           the motor torque T_m needed to cancel the load torque
           T_L, when the gear ratio $\dot{\theta}/\dot{\theta}_L = 100$. Compare this value
           with the torque needed if the coupling is direct.

     b. If the moment of inertia due to a load is 50 kgm², and
           the gear ratio is the same as in part *(a)*, what is the
           reflected moment of inertia due to the load sensed on
           the shaft of the motor?

1.5 A planar manipulator has two revolute joints and two links
     (an inverted pendulum). The lengths of the links are: $\ell_1 = 50$
     cm, and $\ell_2 = 25$ cm. Suppose that the ranges of the joint
     variables are restricted as follows: $0 \le \theta_1 \le \pi$ rad, and
     $-3\pi/4$ rad $\le \theta_2 \le 3\pi/4$ rad.

     a. Graph schematically the envelope of the workspace and
           indicate the total workspace by shading it.

     b. Determine the essential points and distances that specify
           the workspace.

1.6 Show that the dimension of β in Equation 1.3.9 is in time-
     units. Assume that the dimension of I denoted by $[I]$ is
     kgm²; $[B] = 1$ Nms; $[R_a] = 1\ \Omega$; $[K_b] = 0.5$ Vs/rad; $[K_m] = 1$
     Nm/A $= 1$ kgm²/(As²), where s $=$ seconds; 1N $= 1$ Newton
     $= 1$ kgm/s²; A $=$ Ampere.

1.7 Draw the characteristics of the frictional force versus speed for

     a. Viscous friction

     b. Coulomb friction and static friction.

1.8 In an electromechanical system driven by a separately excited
     DC motor a load is connected to the motor shaft through
     a gear train. The numerical values of the load are $I_L = 10$ kgm²,

$B_L = 4$ Nms, $T_L = 0$. The gear is assumed to be lossless with gear ratio $\dot{\theta}_m/\dot{\theta}_L = 100/1$. The motor constants are $K_m = 5$ Nm/A, $K_b = 5$ Vs, $R_a = 0.5$ Ω, $B_m = 0.5$ Nms, and $I_m = 1$ kgm².

a. Determine the time constant of the system specified by the transfer function $G(s)$ in Equation 1.3.9.

b. Suppose that the load is directly connected to the shaft of the motor by removing the gear train. Determine the time constant of the electromechanical system in this case. Compare with part (a).

1.9 A resolution of 0.05% is needed for an application [resolution $= 1/(2^n - 1)$ where n is the number of bits in a binary number]. The encoder used has a disc shown in Figure 1.13a. The width of a segment at the edge of the disc is 0.1 mm.

a. Suppose that the range of the encoder reading is 0–360°; thus, there is no sign bit used. Determine the number of bits needed to obtain the required resolution.

b. Repeat part (a) by assuming that the most significant bit in the binary numbers is a sign bit, that is, the range of the encoder readings is now from 0 to ±180°.

c. Determine the diameter of the disc in the encoder to obtain the required resolution for parts (a) and (b).

1.10 Suppose the diameter of a disc in an encoder is 3.26 cm and the width of a segment at the edge of the disc is 0.1 mm.

a. Determine the number of segments on a single circumference.

b. Calculate the minimum number of bits needed to represent the binary numbers in the encoder.

CHAPTER TWO

KINEMATICS FOR MANIPULATOR JOINTS, LINKS, AND GRIPPER

The arm of a robotic manipulator consists of links connected by joints. The links usually form an open kinematic chain when they are connected in series; parallel connections of the links are also possible. Each link is moved by an actuator. The motion of the links is an angular rotation or rectilinear translation. Definite relationships can be established between the joint positions, links, and the position and orientation of the end-effector (gripper). These mathematical relations depend on the coordinate systems chosen. They play an important role in the kinematics, dynamics, and control of the manipulators, which are the focus in our presentation.

It is a common practice for the analysis of a manipulator system to define a world coordinate system and a local coordinate system attached to each link (joint). This latter coordinate frame moves with the link. The world coordinate system may be a Cartesian coordinate system whose origin is fixed to the base, for example, at the shoulder of the manipulator. Alternatively, a cylindrical or spherical coordinate system can equally well be chosen. Another commonly used coordinate system consists of roll-pitch-yaw (RPY) angles, which appear in aeronautical applications. It is often desirable to describe the position of the end-effector, that is, the location of the centerpoint of the gripper as well as its orientation for grasping position. It is then useful to form a composite coordinate system. For example, a Cartesian coordinate system attached to the base combined with the roll-pitch-yaw angles will form a six-dimensional coordinate system, which can advantageously be used to specify the location and the orientation of the end-effector in the workspace of the manipulator.

The position and orientation of the end-effector depend on the positions of the joints, which can be expressed relative to the link coordinate frames. The relative position of a joint can be described in the link coordinate frame, which moves with the link in the fixed world (base) coordinate system. For the analysis, the designer first defines the origins and the axes of the coordinate systems and determines the transformations that express the relationships between the defined coordinates. Then the position and orientation of the end-effector can be specified in terms of manipulator parameters.

The relationships between different coordinate frames are described by functions that are static in the sense that there are no variable derivatives (or variable differences) of time explicitly in the equations. If the variables in these relations depend implicitly on time, all variables are evaluated at the same specific time; it implies that the time is "frozen" at the particular instance. Kinematic equations describe the relationships between the coordinate frames. *Kinematics* is the branch of mechanics that deals with the motion of rigid bodies without reference to their masses or forces producing the motion. Thus, the transformations between the coordinate frames are examples of kinematic relations.

We will first consider an example of a planar manipulator. It demonstrates the use of a transformation between two coordinate frames. This transformation can also result from the successive operations of the rotation and translation of the coordinate frames. The systematic selection of rectangular link coordinate frames will be discussed by following the approach proposed by Denavit and Hartenberg [1]. Having defined the local coordinate frames, the transformation relating these coordinate frames will be presented. Another possible choice for the coordinate systems will then be introduced: the angles defining roll, pitch, and yaw relative to the world coordinate system will be discussed.

2.1 COORDINATE FRAMES AND TRANSFORMATIONS FOR A PLANAR MANIPULATOR

A simple manipulator shown in Figure 2.1 can move on the plane of the paper by rotating the first joint about the base (the zeroth link), and by changing the length of the link by means of the second joint, which is prismatic. It represents a cylindrical manipulator in which the elevation coordinate is held constant. A Cartesian coordinate system $x_0 \, y_0 \, z_0$ is defined so that the z_0-axis coincides with the axis of the rotation of the first joint; the x_0- and y_0-axes are chosen on the plane of the motion so that the $x_0 y_0 z_0$-coordinate system will obey the right-hand rule (i.e., the x_0-axis is represented by the thumb, the y_0-axis by the index finger, and the z_0-axis corresponds to the direction of the middle finger). A moving local coordinate frame $x_1 y_1 z_1$ will then be defined so that the z_1-axis will be aligned with the

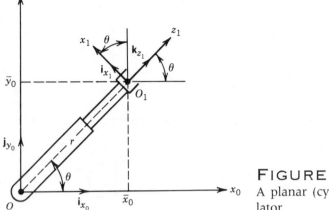

FIGURE 2.1
A planar (cylindrical) manipulator.

axis of the motion of the translational (second) joint. The x_1-axis will be
specified so that its direction is perpendicular to the plane formed by the z_0-
and z_1-axes. Finally, the direction of the y_1-axis is perpendicular to the $x_1 z_1$-
plane so that the coordinate frame $x_1 y_1 z_1$ will also obey the right-hand rule.
The origin of the $x_0 y_0 z_0$-coordinate frame is fixed to the point where the axis
of the first joint meets the plane, and that of the $x_1 y_1 z_1$-coordinate frame is
attached to the end-effector. We should observe that in this example the axes
z_0 and y_1 come upward from the paper. A Cartesian world coordinate system
in this particular example may be chosen so that it coincides with the $x_0 y_0 z_0$-
system; or it could be defined so that its origin is located away from the base
of the manipulator, for example, at a distinct point in the working space of
the manipulator.

   Having defined the coordinate systems for the planar manipulator in Figure 2.1, the next task for the kinematic analysis is to establish the transformation matrices between the two coordinate systems. That is, we want to
determine the transformation that relates the coordinates of a point expressed
in the $x_1 y_1 z_1$-coordinate system to the coordinates of the *same point* expressed
in the $x_0 y_0 z_0$-coordinate system.

   For convenience, let the coordinates of the origin O_1 of the $x_1 y_1 z_1$-
coordinate frame be denoted by $\bar{x}_0$, $\bar{y}_0$, $\bar{z}_0$ in the $x_0 y_0 z_0$-coordinate system,
that is,

$$\bar{x}_0 = r\cos\theta$$

$$\bar{y}_0 = r\sin\theta \qquad (2.1.1)$$

$$\bar{z}_0 = \bar{z}_0$$

where r and θ are defined in Figure 2.1. They may change with time when
the manipulator moves. In this case, the z_0-axis and the y_1-axis are parallel,
and we could select $\bar{z}_0 = 0$.

Let us consider a point P: $(p_{x_1}, p_{y_1}, p_{z_1})$ in the $x_1 y_1 z_1$-coordinate frame. The coordinates of point P in the $x_0 y_0 z_0$-coordinate system are obtained by projecting the local coordinates of P to the x_0-, y_0- and z_0-axes:

$$p_{x_0} = \bar{x}_0 - p_{x_1} \sin \theta + p_{z_1} \cos \theta$$

$$p_{y_0} = \bar{y}_0 + p_{x_1} \cos \theta + p_{z_1} \sin \theta \tag{2.1.2}$$

$$p_{z_0} = \bar{z}_0 + p_{y_1}$$

The equations in 2.1.2 can be expressed using vector matrix notations as follows:

$$\begin{bmatrix} p_{x_0} \\ p_{y_0} \\ p_{z_0} \end{bmatrix} = \begin{bmatrix} -\sin \theta & 0 & \cos \theta \\ \cos \theta & 0 & \sin \theta \\ 0 & 1 & 0 \end{bmatrix} \begin{bmatrix} p_{x_1} \\ p_{y_1} \\ p_{z_1} \end{bmatrix} + \begin{bmatrix} \bar{x}_0 \\ \bar{y}_0 \\ \bar{z}_0 \end{bmatrix} \tag{2.1.3}$$

The right side of Equation 2.1.3 may be combined to the product of a matrix and a vector. Indeed, the vector Equation 2.1.3 may be rewritten more concisely after introducing the notations $\sin\theta = s$, $\cos\theta = c$, and the four-dimensional vectors:

$$\begin{bmatrix} p_{x_0} \\ p_{y_0} \\ p_{z_0} \\ 1 \end{bmatrix} = \begin{bmatrix} -s & 0 & c & \bar{x}_0 \\ c & 0 & s & \bar{y}_0 \\ 0 & 1 & 0 & \bar{z}_0 \\ 0 & 0 & 0 & 1 \end{bmatrix} \begin{bmatrix} p_{x_1} \\ p_{y_1} \\ p_{z_1} \\ 1 \end{bmatrix} \tag{2.1.4}$$

where $\bar{x}_0 = rc$, and $\bar{y}_0 = rs$. A fourth component has been added to the position vector in Equation 2.1.4 to account for the translation vector $[\bar{x}_0, \bar{y}_0, \bar{z}_0]'$, where the superscript prime signifies the transposition. It can be used as a scaling factor. Moreover, the transformation matrix has been expanded to a square matrix by adding the fourth row; it can serve as a perspective transformation in vision-related applications. The introduction of the enlarged matrix in Equation 2.1.4 makes the concise representation of Equation 2.1.3 possible. Moreover, when expressions similar to the right-hand side of Equation 2.1.3 must be multiplied, these operations are reduced to only matrix multiplications by means of the enlarged matrix notation of Equation 2.1.4 [2].

Equation 2.1.4 may be expressed in a concise form by defining two four-dimensional vectors $p_0 = [p_{x_0}, p_{y_0}, p_{z_0}, 1]'$ and $p_1 = [p_{x_1}, p_{y_1}, p_{z_1}, 1]'$:

$$p_0 = A_0^1 p_1 \tag{2.1.5}$$

where the 4×4 matrix A_0^1 is specified by Equation 2.1.4 as

$$A_0^1 = \begin{bmatrix} -s & 0 & c & \bar{x}_0 \\ c & 0 & s & \bar{y}_0 \\ 0 & 1 & 0 & \bar{z}_0 \\ 0 & 0 & 0 & 1 \end{bmatrix} \tag{2.1.6}$$

Matrix A_0^1 is called a *transformation matrix*. It converts the coordinates of a point expressed in the first $(x_1 y_1 z_1)$ coordinate frame to the coordinates of the same point described in the zeroth $(x_0 y_0 z_0)$ coordinate system. It can be partitioned as follows:

$$A_0^1 = \begin{bmatrix} \text{Rotation} & \text{Translation} \\ \text{submatrix} & \text{vector} \\ \text{Perspective vector} & \text{Scaling vector} \end{bmatrix} \qquad (2.1.7)$$

The last column of the transformation matrix A_0^1 determines vector $[\bar{x}_0, \bar{y}_0, \bar{z}_0, 1]'$ describing the translation between the origins of the $x_0 y_0 z_0$- and $x_1 y_1 z_1$-coordinate systems. The rotation matrix is a 3×3 submatrix specifying the rotation of the $x_1 y_1 z_1$-coordinate axes relative to the $x_0 y_0 z_0$-coordinate system. The scaling factor can be used to adjust the desired scale for the components of the translation vector. The perspective vector may be applied to determine the position (size) of an object image using the focal length of a camera [3].

2.1.1 ROTATION MATRIX IN TERMS OF INNER PRODUCTS

The (pure) rotation matrix can often be calculated conveniently by means of inner products. To illustrate this, we may first define the unit vectors in the positive directions of the x_0, y_0, z_0-axes as $\mathbf{i}_{x_0}, \mathbf{j}_{y_0}, \mathbf{k}_{z_0}$, and for the x_1, y_1, z_1-axes as $\mathbf{i}_{x_1}, \mathbf{j}_{y_1}$, and $\mathbf{k}_{z_1}$. By assuming that the translation vector is zero, it follows that the coordinates of point P can be expressed in two ways:

$$\mathbf{p}_1 = p_{x_1} \mathbf{i}_{x_1} + p_{y_1} \mathbf{j}_{y_1} + p_{z_1} \mathbf{k}_{z_1} \qquad (2.1.8)$$

$$\mathbf{p}_0 = p_{x_0} \mathbf{i}_{x_0} + p_{y_0} \mathbf{j}_{y_0} + p_{z_0} \mathbf{k}_{z_0} \qquad (2.1.9)$$

The projection of $\mathbf{p}_1$ on the axis of $\mathbf{i}_{x_0}$ is determined by forming the inner product of $\mathbf{p}_1$ and $\mathbf{i}_{x_0}$. Thus, the component of $\mathbf{p}_1$ on the x_0-axis is $p_{x_0} = \mathbf{p}_1 \cdot \mathbf{i}_{x_0}$. Similarly, the projections of vector $\mathbf{p}_1$ on the y_0- and z_0-axes can be specified. Hence

$$p_{x_0} = (\mathbf{i}_{x_1} \cdot \mathbf{i}_{x_0}) p_{x_1} + (\mathbf{j}_{y_1} \cdot \mathbf{i}_{x_0}) p_{y_1} + (\mathbf{k}_{z_1} \cdot \mathbf{i}_{x_0}) p_{z_1} \qquad (2.1.10)$$

$$p_{y_0} = (\mathbf{i}_{x_1} \cdot \mathbf{j}_{y_0}) p_{x_1} + (\mathbf{j}_{y_1} \cdot \mathbf{j}_{y_0}) p_{y_1} + (\mathbf{k}_{z_1} \cdot \mathbf{j}_{y_0}) p_{z_1} \qquad (2.1.11)$$

$$p_{z_0} = (\mathbf{i}_{x_1} \cdot \mathbf{k}_{z_0}) p_{x_1} + (\mathbf{j}_{y_1} \cdot \mathbf{k}_{z_0}) p_{y_1} + (\mathbf{k}_{z_1} \cdot \mathbf{k}_{z_0}) p_{z_1} \qquad (2.1.12)$$

Equations 2.1.10–2.1.12 may be expressed with vector notations in the form similar to Equation 2.1.4:

$$\begin{bmatrix} p_{x_0} \\ p_{y_0} \\ p_{z_0} \\ 1 \end{bmatrix} = \begin{bmatrix} \mathbf{i}_{x_1} \cdot \mathbf{i}_{x_0} & \mathbf{j}_{y_1} \cdot \mathbf{i}_{x_0} & \mathbf{k}_{z_1} \cdot \mathbf{i}_{x_0} & 0 \\ \mathbf{i}_{x_1} \cdot \mathbf{j}_{y_0} & \mathbf{j}_{y_1} \cdot \mathbf{j}_{y_0} & \mathbf{k}_{z_1} \cdot \mathbf{j}_{y_0} & 0 \\ \mathbf{i}_{x_1} \cdot \mathbf{k}_{z_0} & \mathbf{j}_{y_1} \cdot \mathbf{k}_{z_0} & \mathbf{k}_{z_1} \cdot \mathbf{k}_{z_0} & 0 \\ 0 & 0 & 0 & 1 \end{bmatrix} \begin{bmatrix} p_{x_1} \\ p_{y_1} \\ p_{z_1} \\ 1 \end{bmatrix} \qquad (2.1.13)$$

Equation 2.1.13 specifies the rotation matrix, and the translation vector is a zero vector. If the origins of the two coordinate systems in Figure 2.1 coincide (i.e., $\bar{x}_0 = 0$, $\bar{y}_0 = 0$, $\bar{z}_0 = 0$ in Equation 2.1.4), then the matrix in Equation 2.1.13 represents the A_0^1-matrix of Equation 2.1.6. The inner products in the matrix of Equation 2.1.13 can be determined by inspection on the basis of Figure 2.1. For example, the first row of A_0^1 is $\mathbf{i}_{x_1} \cdot \mathbf{i}_{x_0} = \cos(90° + \theta) = -s$, which is the $(1,1)$ entry of the A_0^1-matrix in Equation 2.1.6; moreover, $\mathbf{j}_y \cdot \mathbf{i}_{x_0} = 0$, and $\mathbf{k}_{z_1} \cdot \mathbf{i}_{x_0} = c$. The other rows of the rotation submatrix of A_0^1 are calculated similarly using Equation 2.1.13. The use of the inner products of the unit vectors provides a straightforward method for determining the entries of the rotation submatrix even in the analysis of more general and complicated manipulators.

2.1.2 PRODUCT OF TRANSLATION MATRIX AND ROTATION MATRIX EQUALS TRANSFORMATION MATRIX

Equation 2.1.7 represents the transformation matrix that contains both the rotation submatrix and translation vector. Equation 2.1.6 specifies the partitioned parts of the A_0^1-matrix for the planar manipulator shown in Figure 2.1.

The origin of the first coordinate frame $x_1 y_1 z_1$ is expressed in the zeroth coordinate system $x_0 y_0 z_0$ by means of vector $[\bar{x}_0, \bar{y}_0, \bar{z}_0, 1]'$. When the directions of the coordinate axes remain unchanged, this *translation* can be described by means of the following matrix:

$$\text{Trans}(\bar{x}_0, \bar{y}_0, \bar{z}_0) = \begin{bmatrix} 1 & 0 & 0 & \bar{x}_0 \\ 0 & 1 & 0 & \bar{y}_0 \\ 0 & 0 & 1 & \bar{z}_0 \\ 0 & 0 & 0 & 1 \end{bmatrix} \qquad (2.1.14)$$

The rotation submatrix of the right side in Equation 2.1.14 is represented by an identity matrix, which implies that the directions of the coordinate axes are not changed.

The matrix for a pure rotation about the y_1-axis for the planar manipulator is next determined. Thus, the translation vector $[\bar{x}_0, \bar{y}_0, \bar{z}_0, 1]'$ in Equation 2.1.6 for matrix A_0^1 is taken as a zero vector for a moment. This means that the origin of the $x_1 y_1 z_1$-coordinate frame is assumed to coincide with the origin of the $x_0 y_0 z_0$-coordinate system in Figure 2.1 for this consideration. The transformation matrix $\text{Rot}(y_1, \theta)$ for the *pure rotation* about the y_1-axis can be written from Equation 2.6 as follows:

$$\text{Rot}(y_1, \theta) = \begin{bmatrix} -s & 0 & c & 0 \\ c & 0 & s & 0 \\ 0 & 1 & 0 & 0 \\ 0 & 0 & 0 & 1 \end{bmatrix} \qquad (2.1.15)$$

The second column of $\text{Rot}(y_1, \theta)$ indicates that the rotation is about the y_1-axis, and that the projection of $\mathbf{k}_{z_0}$ on the y_1-axis is one. The rotation matrix can also be obtained by determining the projections of the unit vectors of the x_0-, and y_0-axes on the x_1- and z_1-axes.

Having introduced the pure translation matrix $\text{Trans}(\bar{x}_0, \bar{y}_0, \bar{z}_0)$ and the pure rotation matrix $\text{Rot}(y_1, \theta)$ about the y_1-axis, the transformation matrix A_0^1 can now be expressed as the product of two matrices, that is,

$$A_0^1 = \text{Trans}(\bar{x}_0, \bar{y}_0, \bar{z}_0)\,\text{Rot}(y_1, \theta) \qquad (2.1.16)$$

which can be referred to as the polar decomposition of matrix A_0^1.

Equation 2.1.16 decomposes the transformation matrix into two parts: the rotation matrix $\text{Rot}(y_1, \theta)$ converts the three components p_{x_1}, p_{y_1}, and p_{z_1} of a point P given along x_1-, y_1- and z_1-axes, respectively, to the components in a new coordinate frame, whose axes are parallel to those of the $x_0 y_0 z_0$-coordinate frame, and whose origin coincides with the origin of the $x_1 y_1 z_1$-coordinate frame. Then the premultiplying translation matrix Trans $(\bar{x}_0, \bar{y}_0, \bar{z}_0)$ converts the resulting vector to vector $p_{x_0}\mathbf{i}_{x_0} + p_{y_0}\mathbf{j}_{y_0} + p_{z_0}\mathbf{k}_{z_0}$ of the $x_0 y_0 z_0$-coordinate system.

The basic kinematic equations for a planar manipulator consist of Equations 2.1.5 and 2.1.6, or 2.1.16. These equations are based on the coordinate systems (frames) defined in Figure 2.1. The generalization of the approach is next applied to an N-joint serial link manipulator.

2.2 DETERMINATION OF RECTANGULAR COORDINATE FRAMES FOR MANIPULATOR LINKS (JOINTS)

A concrete representation of a point (the tip of a vector) in a vector space can be obtained after a coordinate system for the space has been defined. A coordinate system is composed of a set of linearly independent basis vectors such that every other vector in the space can be described as a linear combination of these basis vectors. The number of the basis vectors is the same as the dimension of the space. These concepts of a vector space will next be used to define appropriate coordinate systems for the links (joints) of a manipulator. Then, relationships between the coordinate frames will be established.

A systematic method to define a local coordinate system for each link (joint) of a manipulator is first discussed as proposed by Denavit and Hartenberg (D-H) [1]. The variables in these coordinate frames describe the D-H representation of the points on the manipulator. The transformations relating the defined coordinate frames in general, and specifically in a Stanford/JPL arm are then presented. The relationships between the local coordinate frames and a world base coordinate system will be derived.

2.2.1   D-H (DENAVIT—HARTENBERG) COORDINATE FRAMES

The problem to be addressed is to assign rectangular coordinate frames for the links (joints) in a serial link N-joint manipulator. This manipulator has $N + 1$ links, and one degree of freedom (DOF) per joint. The base of the manipulator is considered as the zeroth link; thus, the first link is attached to the base through the first joint.

As a first step, a Cartesian (rectangular right-handed) coordinate system $(x_0 y_0 z_0)$ is assigned to the base of the manipulator so that the z_0-axis will represent the rotational axis of the motion for the first link (joint 1). The x_0- and y_0-axes are chosen perpendicular to the z_0-axis according to the right-hand rule. The origin of this coordinate system is placed here on the shoulder of the manipulator, that is, at the intersection of the rotational axes of links one and two. Alternatively, the origin can be placed on the root of the base of the manipulator. Thus, the $x_0 y_0 z_0$-coordinate system is now specified.

For the next link $i + 1$, the coordinate axes are chosen systematically as follows [1,2,4], $i = 1, \ldots, N$:

Determination of the coordinate frames

1. The z_i-axis is assigned along the axis for the motion of link $i + 1$. Thus, the z_i-axis coincides with the axis about which the rotation of a revolute link takes place. Or it is aligned with the axis along which a prismatic link moves. The positive z_i-direction for a revolute link is such that the positive rotation of θ_{i+1} is counterclockwise.

2. If the z_i- and z_{i-1}-axes intersect, the x_i-axis has a direction determined by the cross-product $\pm(\mathbf{k}_{z_{i-1}} \times \mathbf{k}_{z_i})$ where $\mathbf{k}_{z_i}$ is the unit vector in the positive direction of the z_i-axis, and $\mathbf{k}_{z_{i-1}}$ is defined similarly. If $\mathbf{k}_{z-1}$ and $\mathbf{k}_{z_i}$ are parallel, then the direction of the x_i-axis is along the common normal of the $\mathbf{k}_{z_{i-1}}$- and $\mathbf{k}_{z_i}$- vectors. (Note: If these vectors together define a plane, then the common normal is also on this plane.)

3. The y_i-axis is determined by means of the right-hand rule so as to complete the Cartesian $x_i y_i z_i$-coordinate system. In other words, $\mathbf{j}_{y_i} = \mathbf{k}_{z_i} \times \mathbf{i}_{x_i}$, where $\mathbf{j}_{y_i}$ and $\mathbf{i}_{x_i}$ are the unit vectors in the directions of the y_i- and x_i-axes, respectively.

4. The origin of the $x_i y_i z_i$-coordinate system for link i is placed on the intersection of the z_{i-1}- and z_i-axes, or on the intersection of the z_i-axis and the common normal between the z_i and z_{i-1} axes.

5. Steps 1 through 4 are repeated for each i. Thus, the coordinate frames for the links (joints) will be determined.

Steps 1-5 provide guidelines to define the local coordinate frames for the links (joints) systematically, although other choices of the coordinate systems are possible.

In order to establish the transformation matrix that relates a vector expressed in the $x_i\, y_i\, z_i$-coordinate system to an expression in the $x_{i-1}\, y_{i-1}\, z_{i-1}$-coordinate frame, it is necessary to define certain parameters for the chosen coordinate frames. An example is presented to illustrate the choice of these parameters.

EXAMPLE 2.2.1

A straight line L shown in Figure 2.2a in the three-dimensional $x_0\, y_0\, z_0$-space is assumed to represent a rotational axis. A local coordinate frame is to be assigned for the system.

The rotational axis, line L, specifies the z_1-axis. If the origin is placed at the point O', then the x_1-axis is aligned with the common perpendicular to the z_0- and z_1- axis. The y_1-axis completes the right-hand coordinate system.

The parameters will next be chosen to describe the relative positions of the successive pair of the axes in these two coordinate frames. Since a straight line such as L in Figure 2.2a can be described in terms of (the minimum of) four parameters, we may select to use $\theta_1, a_1, d_1,$ and α_1, shown in Figure 2.2a for the aforementioned parameters. In Figure 2.2a, plane $OH_1O'H_2$ can be drawn so that it is perpendicular to the x_0y_0-plane and contains the z_0-axis. It defines angle θ_1. By placing a plane parallel to the x_0y_0-plane through the point O', the point H_1 on the z_0-axis becomes specified. The length of the line segment $O'H_1$ that is perpendicular to both the z_0-axis

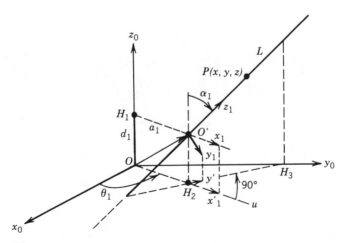

FIGURE 2.2a
Illustration of choosing the structural parameters.

and line L (the z_1-axis) equals a_1, that is, parameter a_1 is the perpendicular distance between the z_0- and z_1-axes. The distance of point H_1 from the origin O along the z_0-axis is designated as d_1, which is also the perpendicular distance between the x_0- and x_1-axes. In a plane drawn so that it is perpendicular to the x_0y_0-plane and contains line L the z_1-axis and the extension of H_2O' form angle α_1. This parameter is measured from line H_2O', which is parallel with the z_0-axis. Parameters d_1, a_1, α_1 and θ_1 can conveniently be used to express the transformation between the two coordinate systems.

In a multiple joint serial link manipulator, the parameters introduced in Example 2.2.1 can be defined for each link coordinate frame. The relative positions of two adjacent coordinate frames associated with link i and link $i + 1$ in Figure 2.2b [3] can be characterized by the following parameters: length a_i, the twist angle α_i, distance d_i, and angle θ_i, between the links. These parameters for link i are indicated in Figure 2.2b. Since the foregoing parameters depend on the structure of the given manipulator, they will be called the *structural kinematic parameters* of the manipulator.

After the coordinate frames to all links (joints) in the manipulator have been assigned, the aforementioned structural parameters are identified. They are determined for the ith link, $i = 1, \ldots , N$ as follows:

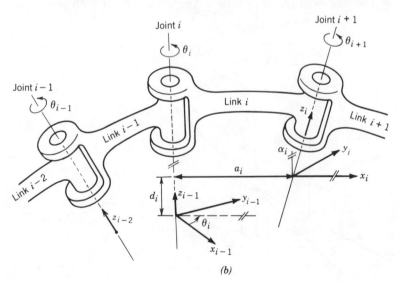

(b)

FIGURE 2.2b
Structural kinematic parameters for a general link i.

Determination of structural kinematic parameters

1. a_i: The distance from the origin of the ith coordinate frame to the intersection of the z_{i-1}- and the x_i-axis along the x_i-axis is specified as a_i; in other words, parameter a_i is the perpendicular distance between the z_i- and z_{i-1}-axes. The distance a_i is commonly called the length of the link.

2. α_i: The angle of rotation about the positive x_i-axis is measured from the positive z_{i-1}-axis (or its parallel projection) to the positive z_i-axis, and designated as angle α_i, where the positive direction is counterclockwise. Angle α_i is called the offset (twist) angle of link i.

3. θ_i: The angle of rotation about the positive z_{i-1}-axis is measured from the positive x_{i-1}-axis to the positive x_i-axis (or its parallel projection), and denoted by θ_i. It is positive in the counterclockwise direction.

4. d_i: The distance from the origin of the $(i-1)$st coordinate frame to the intersection of the z_{i-1}-axis, and the x_i-axis along the z_{i-1}-axis is d_i; if the z_{i-1}-axis and the x_i-axis do not intersect, then it is the perpendicular distance between the x_i- and x_{i-1}-axis.

Some of the structural parameters may serve as joint variables that may vary with time assuming different values in a moving manipulator, for example, angle θ_i in a revolute link, or distance d_i for a prismatic link.

Having established the coordinate frames and parameters d_i, a_i, α_i, and θ_i, $i = 1, \ldots, N$, for the links (joints) of a particular manipulator, the transformation matrix A_{i-1}^i between the adjacent coordinate frames may be written. If vector p_i is known in the ith coordinate frame, then it can be expressed in the $(i-1)$st coordinate frame as p_{i-1}, that is

$$p_{i-1} = A_{i-1}^i p_i \qquad (2.2.1)$$

where matrix A_{i-1}^i can be expressed in terms of the structural parameters associated with the ith link. The determination of the transformation matrix A_{i-1}^i for the coordinate frames in Figure 2.2a is illustrated next.

EXAMPLE 2.2.2

Using the structural kinematic parameters defined in Example 2.2.1, the transformation matrix relating the coordinate frames in Figure 2.2a is to be determined. Thus, the matrix A_0^1 in Equation 2.2.1 with $i = 1$ must be specified in terms of the structural parameters a_1, α_1, θ_1, and d_1.

The transformation matrix A_0^1 can be obtained from Figure 2.2a on the basis of Equations 2.1.7 and 2.1.13. The fourth column of matrix A_0^1 is the translation vector, that is, the coordinates of the origin O' in the base coordinate system. The projection of point O' on the $x_0 y_0$-plane is H_2, and OH_2 equals a_1. Therefore, the x_0- and y_0-coordinates of point O' are $a_1 \cos \theta_1 = a_1 c_1$ and $a_1 \sin \theta_1 = a_1 s_1$, respectively; furthermore, the z_0-coordinate is d_1. Thus, the translation vector is specified: $[a_1 \cos \theta_1, a_1 \sin \theta_1, d_1, 1]'$.

To determine the submatrix of rotation, the inner products of the unit vectors shown in Equation 2.1.13 must be evaluated. In order to facilitate the determination of these inner products, the unit vectors of the $x_1 y_1 z_1$-coordinate frame are decomposed into two components: the one is parallel to the z_0-axis, and the other parallel to the $x_0 y_0$-plane. We notice from Figure 2.2a that the x_1-axis is parallel to the $x_0 y_0$-plane and perpendicular to the z_0-axis. To specify the inner products in the first column of the matrix A_0^1, the unit vector $\mathbf{i}_{x_1}$ of the x_1-axis is first projected onto the $x_0 y_0$-plane. The inner products of the first column in A_0^1 can now be formed: $\mathbf{i}_{x_1} \cdot \mathbf{i}_{x_0} = \cos \theta_1$; $\mathbf{i}_{x_1} \cdot \mathbf{j}_{y_0} = \sin \theta_1$; and $\mathbf{i}_{x_1} \cdot \mathbf{k}_{z_0} = 0$.

In order to calculate the entries in the second column of A_0^1, the unit vector of the y_1-axis is projected onto the $x_0 y_0$-plane to obtain $|\mathbf{j}_{y_1}| \sin(90° - \alpha_1)$. It follows that the inner products are: $\mathbf{j}_{y_1} \cdot \mathbf{i}_{x_0} = \sin(90° - \alpha_1) \cos(\theta_1 + 90°) = -\cos \alpha_1 \sin \theta_1$; $\mathbf{j}_{y_1} \cdot \mathbf{j}_{y_0} = \sin(90° - \alpha_1) \sin(\theta_1 + 90°) = \cos \alpha_1 \cos \theta_1$; and $\mathbf{j}_{y_1} \cdot \mathbf{k}_{z_0} = \cos(90° - \alpha_1) = \sin \alpha_1$. Thus, the second column of A_0^1 is now specified.

To determine the third column of the A_0^1-matrix, the unit vector of the z_1-axis is projected onto the $x_0 y_0$-plane to obtain $|\mathbf{k}_{z_1}| \sin(-\alpha_1)$, and onto the z_0-axis to obtain $|\mathbf{k}_{z_1}| \cos(-\alpha_1)$. The following inner products determine the third column: $\mathbf{k}_{z_1} \cdot \mathbf{i}_{x_0} = \sin(-\alpha_1) \cos(\theta_1 + 90°) = \sin \alpha_1 \sin \theta_1$; $\mathbf{k}_{z_1} \cdot \mathbf{j}_{y_0} = \sin(-\alpha_1) \cos \theta_1$; and $\mathbf{k}_{z_1} \cdot \mathbf{k}_{z_0} = \cos \alpha_1$. The transformation matrix can now be written:

$$A_0^1 = \begin{bmatrix} \cos \theta_1 & -\cos \alpha_1 \sin \theta_1 & \sin \alpha_1 \sin \theta_1 & a_1 \cos \theta_1 \\ \sin \theta_1 & \cos \alpha_1 \sin \theta_1 & -\sin \alpha_1 \cos \theta_1 & a_1 \sin \theta_1 \\ 0 & \sin \alpha_1 & \cos \alpha_1 & d_1 \\ 0 & 0 & 0 & 1 \end{bmatrix} \qquad (2.2.2)$$

An explicit expression for the general matrix A_{i-1}^i in Equation 2.2.1 can be written by using the structural parameters for a *revolute* link and Equation 2.1.13. In fact, let us consider two adjacent coordinate frames $(i - 1)$, and i. The structural parameters are d_i, a_i, α_i and θ_i. In order to write the coordinate transformation matrix A_{i-1}^i, the approach presented in Example 2.2.2 is generalized using Figure 2.2b.

In the translation vector (the fourth column) of the matrix A_{i-1}^i, the first two components are specified by the inner products of vector $a_i \mathbf{i}_{x_i}$ with the unit vectors $\mathbf{i}_{x_{i-1}}$ and $\mathbf{j}_{y_{i-1}}$, and the third component equals d_i by the definition. Thus, the translation vector is $[a_i \cos \theta_i, a_i \sin \theta_i, d_i, 1]'$, since the angle θ_i is measured counterclockwise from the positive x_{i-1}-axis.

The first column of the rotation submatrix in A_{i-1}^i is then determined by the inner products of the unit vector $\mathbf{i}_{x_i}$ with the unit vectors $\mathbf{i}_{x_{i-1}}$, $\mathbf{j}_{y_{i-1}}$, and $\mathbf{k}_{z_{i-1}}$. Thus, the first column of A_{i-1}^i equals $[\cos\theta_i, \sin\theta_i, 0, 0]'$.

To write down the second and third columns of the transformation matrix A_{i-1}^i, we notice that $\mathbf{j}_{y_i}$ has a projection $|\mathbf{j}_{y_i}|\cos\alpha_i$ on the $x_{i-1}y_{i-1}$-plane; similarly, the projection of $\mathbf{k}_{z_i}$ on the same plane is $|\mathbf{k}_{z_i}|\sin(-\alpha_i)$ (see Figure 2.2a). Then, the inner products can be formed to obtain the second and third column of A_{i-1}^i in Equation 2.1.13. The entire transformation matrix assumes now the following general form:

$$A_{i-1}^i = \begin{bmatrix} \cos\theta_i & -\cos\alpha_i\sin\theta_i & \sin\alpha_i\sin\theta_i & a_i\cos\theta_i \\ \sin\theta_i & \cos\alpha_i\sin\theta_i & -\sin\alpha_i\cos\theta_i & a_i\sin\theta_i \\ 0 & \sin\alpha_i & \cos\alpha_i & d_i \\ 0 & 0 & 0 & 1 \end{bmatrix} \qquad (2.2.3)$$

Equation 2.2.3 specifies the transformation matrix between two coordinate frames i and $(i-1)$ attached to links i and $i-1$ which are connected by a *rotational* joint.

If link i is prismatic, then the fourth column of the transformation matrix A_{i-1}^i assumes a different expression. Since the fourth column represents a translation, the transformation between the coordinate frames i and $(i-1)$ has the following form ($a_i = 0$) for a *prismatic* link:

$$A_{i-1}^i = \begin{bmatrix} \cos\theta_i & -\cos\alpha_i\sin\theta_i & \sin\alpha_i\sin\theta_i & 0 \\ \sin\theta_i & \cos\alpha_i\cos\theta_i & -\sin\alpha_i\cos\theta_i & 0 \\ 0 & \sin\alpha_i & \cos\alpha_i & d_i \\ 0 & 0 & 0 & 1 \end{bmatrix} \qquad (2.2.4)$$

The type of the ith link determines whether the transformation matrix is specified by Equation 2.2.3 or 2.2.4.

Equation 2.2.3 or 2.2.4 determines A_{i-1}^i when the coordinate frames have been chosen and the structural parameters of a manipulator are known. Knowing A_{i-1}^i, the relationship between vectors p_i in the ith coordinate frame and p_{i-1} in the $(i-1)$st coordinate frame in Equation 2.2.1 is specified.

Equation 2.2.1 represents a recursive relation, which allows us to express vector p_6 of the sixth (gripper) coordinate frame as a vector p_0 of the base (zeroth) coordinate system. By writing Equation 2.2.1 successively we have

$$p_0 = A_0^1 A_1^2 A_2^3 A_3^4 A_4^5 A_5^6 p_6$$

$$= A_0^6 p_6 \qquad (2.2.5)$$

where $A_0^6 = A_0^1 A_1^2 A_2^3 A_3^4 A_4^5 A_5^6$. Equation 2.2.5 expresses a point in the sixth (gripper) coordinate frame as a point in the base (zeroth) coordinate system. For example, the origin $p_6 = [0, 0, 0, 1]'$ of the gripper coordinate frame can be expressed as $p_0 = [p_{0_x}, p_{0_y}, p_{0_z}, 1]'$ in the base coordinates by means of Equation 2.2.5. Matrix A_0^6 in Equation 2.2.5 is a function of the joint variables. When these values are known, the manipulator has a specific configuration. Then, the position of the end-effector can be calculated by

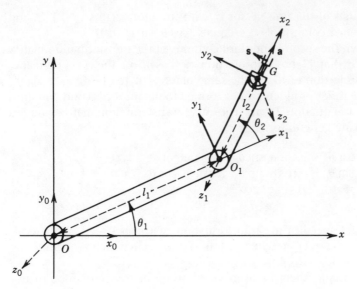

FIGURE 2.3
A planar revolute manipulator.

Equation 2.2.5. The determination of this position from the values of the joint variables is referred to as solving the forward kinematic equations.

EXAMPLE 2.2.3

A planar manipulator with three revolute joints and a gripper is shown in Figure 2.3. The third joint is assumed to be locked in the position in which the third link is a direct extension of the second link. It is required to assign coordinate frames for the manipulator using the D-H procedure.

The coordinate frames will be defined by following the steps 1–5 of Section 2.2.1 for $i = 1$ and 2. The base coordinate frame ($i = 0$) is first determined. Indeed, the rotational axis of link (joint) one is perpendicular to the plane of the paper. Hence, the z_0-axis is chosen upward from the plane of the paper. The x_0-, and y_0-axes are determined so that the $(x_0 y_0 z_0)$ coordinates form a rectangular right-hand (Cartesian) coordinate system. By a similar reasoning, the z_1-axis is chosen upward and normal to the plane of the paper. Since the z_0- and z_1-axes are parallel, the x_1-axis is along the common normal to the z_0- and z_1-axes, and on the plane determined by these two axes. The y_1-axis completes the right-hand rectangular coordinate frame. The z_2-axis is also upward normal to the paper plane. The x_2-axis is again chosen to be along the common normal to the z_1- and z_2-axes and on the plane containing the z_1- and z_2-axes. Thus, the coordinate frames shown in Figure 2.3 become specified.

The structural kinematic parameters are then determined following the guidelines 1–4 presented in Section 2.2.1. The resulting values are given in Table 2.1. The variables θ_1 and θ_2 serve as the joint variables.

TABLE 2.1

link $i =$	d_i	a_i	α_i	θ_i
1	0	ℓ_1	0	θ_1
2	0	ℓ_2	0	θ_2

The transformation matrix A^i_{i-1} relating the coordinates of the ith frame to those of the $(i-1)$st frame $(i = 1, 2)$ can be written on the basis of Equation 2.2.3. Thus, the following matrices are obtained:

$$A^1_0 = \begin{bmatrix} c_1 & -s_1 & 0 & \ell_1 c_1 \\ s_1 & c_1 & 0 & \ell_1 s_1 \\ 0 & 0 & 1 & 0 \\ 0 & 0 & 0 & 1 \end{bmatrix}$$

$$A^2_1 = \begin{bmatrix} c_2 & -s_2 & 0 & \ell_2 c_2 \\ s_2 & c_2 & 0 & \ell_2 s_2 \\ 0 & 0 & 1 & 0 \\ 0 & 0 & 0 & 1 \end{bmatrix} \qquad (2.2.6)$$

where $s_i = \sin \theta_i$, $c_i = \cos \theta_i$, and $i = 1, 2$. The transformation matrix A^2_0 can also be computed:

$$A^2_0 = A^1_0 A^2_1 = \begin{bmatrix} c_{12} & -s_{12} & 0 & \ell_1 c_1 + \ell_2 c_{12} \\ s_{12} & c_{12} & 0 & \ell_1 s_1 + \ell_2 s_{12} \\ 0 & 0 & 1 & 0 \\ 0 & 0 & 0 & 1 \end{bmatrix} \qquad (2.2.7)$$

The matrix in Equation 2.2.7 has been simplified using the identities $c_1 c_2 - s_1 s_2 = c_{12}$, and $s_1 c_2 + c_1 s_2 = s_{12}$ where $s_{12} = \sin(\theta_1 + \theta_2)$ and $c_{12} = \cos(\theta_1 + \theta_2)$.

The matrix A^2_0 in Equation 2.2.7 relates the gripper coordinates directly to the base coordinates. Equations 2.2.6 and 2.2.7 represent the transformation matrices that relate the variables of the defined coordinate frames. Moreover, the position of the end-effector in the base coordinate system can be obtained from Equations 2.2.6 and 2.2.7 when the values of the joint angles are known. Thus, the kinematic equations can be solved in the forward direction.

2.2.2 D-H COORDINATE FRAMES FOR STANFORD/JPL MANIPULATOR

One of the early manipulators that has been studied in great detail [2] is the Stanford/JPL arm designed by V. Scheinman. A schematic picture of a Stanford/JPL manipulator is shown in Figure 2.4. It has six joints:

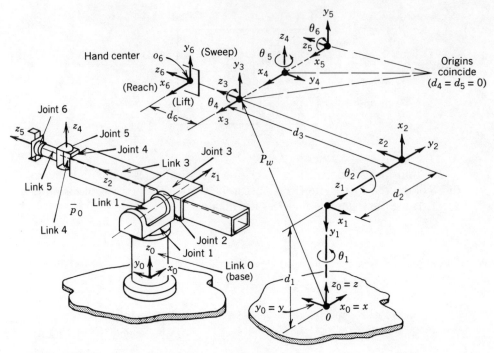

FIGURE 2.4
Stanford/JPL robot manipulator.

five of the links are revolute (rotational), and one link, the third one, is prismatic (translational). The problem is to assign rectangular right-hand coordinate frames for the links. Then, the transformation matrices relating the coordinate frames are to be written.

The coordinate frames for the joints of the manipulator in Figure 2.4 are assigned by following steps 1–5 described in Section 2.2.1:

1. The first joint makes the first link rotate on the supporting base of the manipulator about the vertical axis, which is aligned with the physical axis of the cylindrical base (link zero). Thus, the z_0-axis must coincide with this vertical axis. The second link is rotated by the second joint; the axis of this rotation is horizontal. Thus, the z_1-axis is aligned with the axis of the rotation of link two. The positive directions of the z_0- and z_1-axes are chosen in accordance with the the right-hand screw rule leading to the positive directions of the angular positions θ_1, and θ_2, respectively. They are shown in Figure 2.4. The unit vectors $\mathbf{k}_{z_0}$ and $\mathbf{k}_{z_1}$ are defined along the positive directions of the z_0- and z_1-axes, respectively. The x_0- and

y_0-axes are selected according to the right-hand rule to complete the zeroth coordinate system.

2. Since the z_0- and z_1-axes intersect, the x_1-axis is determined by the cross-product $\pm(\mathbf{k}_{z_0} \times \mathbf{k}_{z_1})$, to which the right-hand rule applies. The positive direction of the x_1-axis is chosen as shown in Figure 2.4.

3. The direction of the y_1-axis is specified so that the $x_1 y_1 z_1$-coordinate frame follows the right-hand rule. Thus, the $x_1 y_1 z_1$-coordinate frame is defined with the exception of its origin.

4. The origin of the $x_0 y_0 z_0$-coordinate system can be placed to the root of the supporting base. The origin of the $x_1 y_1 z_1$-coordinate frame can be made to coincide with the point of the intersection of the z_0- and z_1-axes, that is, at the top of the base (the shoulder) of the manipulator.

5. The same steps will next be repeated for the coordinate frame of link three. Since link (joint) three is prismatic, the direction of the motion specifies the z_2-axis. The right-hand rule of the unit vectors $\mathbf{k}_{z_1}$ and $\mathbf{k}_{z_2}$ defined on the z_1- and z_2-axes determines the direction of the x_2-axis; its positive direction is shown in Figure 2.4. The coordinate frame $x_2 y_2 z_2$ is completed by defining the y_2-axis according to the right-hand rule. The origin of the coordinate frame is placed at the intersection of the z_1- and z_2-axes. Variable $\theta_3 = d_3$ represents the joint variable of the prismatic link.

    For the $x_3 y_3 z_3$-coordinate frame, the z_3-axis representing the axis of rotation for the fourth joint (link) is aligned with the z_2-axis. The $x_3 y_3$-plane is perpendicular to the z_3-axis (as well as to the z_2-axis). For convenience, the x_3-axis is chosen to lie on the $z_2 y_2$-plane. The y_3-axis is specified by the right-hand rule. The origin of this coordinate frame is placed on the z_2-axis at the intersection of the z_3-axis and the z_4-axis which is the rotational axis of the fifth joint (link).

    In the coordinate frame of the fifth link, the x_4-axis is defined so as to align it with the x_3-axis according to the rule $\pm(\mathbf{k}_{z_{i-1}} \times \mathbf{k}_{z_i}), i = 4$. Again the y_4-axis is determined by the right-hand rule. The origin coincides with that of the third coordinate frame. The positive directions for the rotational angles θ_4 and θ_5 are indicated in Figure 2.4. The fifth $(x_5 y_5 z_5)$ and sixth $(x_6 y_6 z_6)$ coordinate frames are similarly defined; they are shown in Figure 2.4. Thus, a coordinate frame for each joint has been defined.

In order to determine the transformation matrices between the coordinate frames, the structural kinematic parameters of the manipulator must next be determined. By following rules 1–4 of Section 2.2.1, the values of d_i, a_i, α_i, and θ_i can be obtained. With reference to Figure 2.4 we have for $i = 1$: (1) a_1 is the distance from the origin of the first coordinate frame to the intersection of the z_0- and x_1-axes along the x_1-axis; thus $a_1 = 0$; (2) α_1 is the angle measured from the z_0-axis to the z_1-axis about the x_1-axis, which is $-90°$; (3) the rotational angle θ_1 about the z_0-axis is measured from the x_0-axis in the counterclockwise (positive) direction; (4) d_1 is the distance from the origin of the zeroth (base) coordinate frame to the intersection of the z_0-axis with the x_1-axis along the z_0-axis (Figure 2.4). Steps 1–4 are repeated for each joint i, $i = 2, \ldots, 6$; the reader is encouraged to follow similar reasoning to obtain the parameters of the other links.

The structural parameters are presented in Table 2.2 for the Stanford/JPL manipulator. The manipulator in Figure 2.4 is represented in the resting (parking) position; hence, each angular position θ_i, $i = 1, 2, 4, 5, 6$ and d_3 assume definite values in the park configuration. Some of the parameters in Table 2.2 are constant regardless of the arm configuration; indeed, these parameters, and their typical values are: $d_1 = 41.250$ cm, $d_2 = 15.367$ cm, and $d_6 = 26.290$ cm.

The structural kinematic parameters of the Stanford/JPL arm can be substituted into Equations 2.2.3 and 2.2.4 to obtain the coordinate transformation A_{i-1}^i for $i = 1, 2, 3, 4, 5, 6$. For example, the application of the values on the last row in Table 2.2 for link six to Equation 2.2.3 with $i = 6$ gives matrix A_5^6 in Table 2.3a. The other matrices are also given in Table 2.3. The variables that can be changed by the actuators are θ_i $i = 1, 2, 4, 5, 6$ and d_3 in the Stanford/JPL manipulator, and they serve as the joint variables. The transformation matrices are written here by first determining the structural parameters for the chosen coordinate frames and then substituting these values into Equation 2.2.3 or 2.2.4. After some practice, the designer may

TABLE 2.2

Structural Kinematic Parameters for Stanford/JPL Arm
Joint Variables: θ_i, $i = 1, 2, 4, 5, 6$ and d_3

link $i =$	d_i	a_i	α_i	θ_i
1	d_1	0	$-90°$	θ_1
2	d_2	0	$90°$	θ_2
3	d_3	0	0	$-90°$
4	0	0	$-90°$	θ_4
5	0	0	$90°$	θ_5
6	d_6	0	0	θ_6

TABLE 2.3

a. Transformation Matrices for Stanford/JPL Arm Based on D-H Coordinate Frames

$$
A_5^6 = \begin{bmatrix}
\cos\theta_6 & -\sin\theta_6 & 0 & 0 \\
\sin\theta_6 & \cos\theta_6 & 0 & 0 \\
0 & 0 & 1 & d_6 \\
0 & 0 & 0 & 1
\end{bmatrix}
$$

$$
A_4^5 = \begin{bmatrix}
\cos\theta_5 & 0 & \sin\theta_5 & 0 \\
\sin\theta_5 & 0 & -\cos\theta_5 & 0 \\
0 & 1 & 0 & 0 \\
0 & 0 & 0 & 1
\end{bmatrix}
$$

$$
A_3^4 = \begin{bmatrix}
\cos\theta_4 & 0 & -\sin\theta_4 & 0 \\
\sin\theta_4 & 0 & \cos\theta_4 & 0 \\
0 & -1 & 0 & 0 \\
0 & 0 & 0 & 1
\end{bmatrix}
$$

$$
A_2^3 = \begin{bmatrix}
0 & 1 & 0 & 0 \\
-1 & 0 & 0 & 0 \\
0 & 0 & 1 & d_3 \\
0 & 0 & 0 & 1
\end{bmatrix}
$$

$$
A_1^2 = \begin{bmatrix}
\cos\theta_2 & 0 & \sin\theta_2 & 0 \\
\sin\theta_2 & 0 & -\cos\theta_2 & 0 \\
0 & 1 & 0 & d_2 \\
0 & 0 & 0 & 1
\end{bmatrix}
$$

$$
A_0^1 = \begin{bmatrix}
\cos\theta_1 & 0 & -\sin\theta_1 & 0 \\
\sin\theta_1 & 0 & \cos\theta_1 & 0 \\
0 & -1 & 0 & d_1 \\
0 & 0 & 0 & 1
\end{bmatrix}
$$

b. Relationships between Vectors in Local Coordinate Frame and Base Coordinate System

$$
p_{01} = A_0^1 p_1
$$
$$
p_{02} = A_0^1 A_1^2 p_2
$$
$$
p_{03} = A_0^1 A_1^2 A_2^3 p_3
$$
$$
p_{04} = A_0^1 A_1^2 A_2^3 A_3^4 p_4
$$
$$
p_{05} = A_0^1 A_1^2 A_2^3 A_3^4 A_4^5 p_5
$$
$$
p_{06} = A_0^1 A_1^2 A_2^3 A_3^4 A_4^5 A_5^6 p_6
$$

(continued)

TABLE 2.3b *(Continued)*

$$A_0^1 A_1^2 = \begin{bmatrix} c_1 c_2 & -s_1 & c_1 s_2 & -d_2 s_1 \\ s_1 c_2 & c_1 & s_1 s_2 & d_2 c_1 \\ -s_2 & 0 & c_2 & d_1 \\ 0 & 0 & 0 & 1 \end{bmatrix}$$

$$A_0^1 A_1^2 A_2^3 = \begin{bmatrix} s_1 & c_1 c_2 & c_1 s_2 & d_3 c_1 s_2 - d_2 s_1 \\ -c_1 & s_1 c_2 & s_1 s_2 & d_3 s_1 s_2 + d_2 c_1 \\ 0 & -s_2 & c_2 & d_3 c_2 + d_1 \\ 0 & 0 & 0 & 1 \end{bmatrix}$$

$$A_0^1 A_1^2 A_2^3 A_3^4 = \begin{bmatrix} s_1 c_4 + c_1 c_2 s_4 & -c_1 s_2 & -s_1 s_4 + c_1 c_2 c_4 & d_3 c_1 s_2 - d_2 s_1 \\ -c_1 c_4 + s_1 c_2 s_4 & -s_1 s_2 & +c_1 s_4 + s_1 c_2 c_4 & d_3 s_1 s_2 + d_2 c_1 \\ -s_2 s_4 & -c_2 & -s_2 c_4 & d_3 c_2 + d_1 \\ 0 & 0 & 0 & 1 \end{bmatrix}$$

$$A_0^1 A_1^2 A_2^3 A_3^4 A_4^5 = \begin{bmatrix} c_5(s_1 c_4 + c_1 c_2 s_4) - c_1 s_2 s_5 & -s_1 s_4 + c_1 c_2 c_4 & s_5(s_1 c_4 + c_1 c_2 s_4) + c_1 s_2 c_5 & d_3 c_1 s_2 - d_2 s_1 \\ c_5(-c_1 c_4 + s_1 c_2 s_4) - s_1 s_2 s_5 & +c_1 s_4 + s_1 c_2 c_4 & s_5(-c_1 c_4 + s_1 c_2 s_4) + s_1 s_2 c_5 & d_3 s_1 s_2 + d_2 c_1 \\ -c_2 s_5 - s_2 s_4 c_5 & -s_2 c_4 & c_2 c_5 - s_2 s_4 s_5 & d_3 c_2 + d_1 \\ 0 & 0 & 0 & 1 \end{bmatrix}$$

$$A_0^1 A_1^2 A_2^3 A_3^4 A_4^5 A_5^6 = \begin{bmatrix} c_6[c_5(s_1 c_4 + c_1 c_2 s_4) - c_1 s_2 s_5] + s_6[-s_1 s_4 + c_1 c_2 c_4] & -s_6[c_5(s_1 c_4 + c_1 c_2 s_4) - c_1 s_2 s_5] + c_6[-s_1 s_4 + c_1 c_2 c_4] & s_5(s_1 c_4 + c_1 c_2 s_4) + c_1 s_2 c_5 & d_6[s_5(s_1 c_4 + c_1 c_2 s_4) + c_1 s_2 c_5] + d_3 c_1 s_2 - d_2 s_1 \\ c_6[c_5(-c_1 c_4 + s_1 c_2 s_4) - s_1 s_2 s_5] + s_6[+c_1 s_4 + s_1 c_2 c_4] & -s_6[c_5(-c_1 c_4 + s_1 c_2 s_4) - s_1 s_2 s_5] + c_6[+c_1 s_4 + s_1 c_2 c_4] & s_5(-c_1 c_4 + s_1 c_2 s_4) + s_1 s_2 c_5 & d_6[s_5(-c_1 c_4 + s_1 c_2 s_4) + s_1 s_2 c_5] + d_3 s_1 s_2 + d_2 c_1 \\ -c_2 s_5 c_6 - s_2(s_4 c_5 c_6 + c_4 s_6) & c_2 s_5 s_6 + s_2 s_4 c_5 s_6 - c_4 s_2 c_6 & c_2 c_5 - s_2 s_4 s_5 & d_6(c_2 c_5 - s_2 s_4 s_5) + d_3 c_2 + d_1 \\ 0 & 0 & 0 & 1 \end{bmatrix}$$

$$A_1^2 A_2^3 A_3^4 A_4^5 A_5^6 = \begin{bmatrix} c_2(s_4 c_5 c_6 + c_4 s_6) - s_2 s_5 c_6 & c_2(-s_4 c_5 s_6 + c_4 c_6) + s_2 s_5 s_6 & c_2 s_4 s_5 + s_2 c_5 & d_6 c_2 s_4 s_5 + d_3 s_2 + d_6 s_2 c_5 \\ s_2(s_4 c_5 c_6 + c_4 s_6) + c_2 s_5 c_6 & s_2(-s_4 c_5 s_6 + c_4 c_6) - c_2 s_5 s_6 & s_2 s_4 s_5 - c_2 c_5 & d_6 s_2 s_4 s_5 - d_6 c_2 c_5 - d_3 c_2 \\ -c_4 c_5 c_6 + s_4 s_6 & c_4 c_5 s_6 + s_4 c_6 & -c_4 s_5 & -d_6 c_4 s_5 + d_2 \\ 0 & 0 & 0 & 1 \end{bmatrix}$$

$$A_2^3 A_3^4 A_4^5 A_5^6 = \begin{bmatrix} s_4 c_5 c_6 + c_4 s_6 & -s_4 c_5 s_6 + c_4 c_6 & s_4 s_5 & d_6 s_4 s_5 \\ -c_4 c_5 c_6 + s_4 s_6 & c_4 c_5 s_6 + s_4 c_6 & -c_4 s_5 & -d_6 c_4 s_5 \\ -s_5 c_6 & s_5 s_6 & c_5 & d_6 c_5 + d_3 \\ 0 & 0 & 0 & 1 \end{bmatrix}$$

$$A_3^4 A_4^5 A_5^6 = \begin{bmatrix} c_4 c_5 c_6 - s_4 s_6 & -c_4 c_5 s_6 - s_4 c_6 & c_4 s_5 & d_6 c_4 s_5 \\ s_4 c_5 c_6 + c_4 s_6 & -s_4 c_5 s_6 + c_4 c_6 & s_4 s_5 & d_6 s_4 s_5 \\ -s_5 c_6 & s_5^6 & c_5 & d_6 c_5 \\ 0 & 0 & 0 & 1 \end{bmatrix}$$

$$A_4^5 A_5^6 = \begin{bmatrix} c_5 c_6 & -c_5 c_6 & s_5 & d_6 s_5 \\ s_5 c_6 & -s_5 s_6 & -c_5 & -d_6 c_5 \\ s_6 & c_6 & 0 & 0 \\ 0 & 0 & 0 & 1 \end{bmatrix}$$

$$T = A_0^1 A_1^2 A_2^3 A_3^4 A_4^5 A_5^6 = \begin{bmatrix} c_1[c_2(s_4 c_5 c_6 + c_4 s_6) - s_2 s_5 c_6] - s_1[-c_4 c_5 c_6 + s_4 s_6] & c_1[c_2(-s_4 c_5 s_6 + c_4 c_6) + s_2 s_5 s_6] - s_1[c_4 c_5 c_6 + s_4 c_6] & c_1(c_2 s_4 s_5 + s_2 c_5) + s_1 c_4 s_5 & c_1[d_6(c_2 s_4 s_5 + s_2 c_5) + d_3 s_2] - s_1[-d_6 c_4 s_5 + d_2] \\ s_1[c_2(s_4 c_5 c_6 + c_4 s_6) - s_2 s_5 c_6] + c_1[-c_4 c_5 c_6 + s_4 s_6] & s_1[c_2(-s_4 c_5 s_6 + c_4 c_6) + s_2 s_5 s_6] + c_1[c_4 c_5 s_6 + s_4 c_6] & s_1(c_2 s_4 s_5 + s_2 c_5) - c_1 c_4 s_5 & s_1[d_3 s_2 + d_6(c_2 s_4 s_5 + s_2 c_5)] + c_1[-d_6 c_4 s_5 + d_2] \\ -s_2(s_4 c_5 c_6 + c_4 s_6) - c_2 s_5 c_6 & -s_2(-s_4 c_5 s_6 + c_4 c_6) + c_2 s_5 s_6 & -s_2 s_4 s_5 + c_2 c_5 & d_6(-s_2 s_4 s_5 + c_2 c_5) + d_3 c_2 + d_1 \\ 0 & 0 & 0 & 1 \end{bmatrix}$$

55

start writing the matrices in question directly for the specified coordinate systems.

The expressions in Tables 2.3a and 2.3b can be used to determine the location of any point on the Stanford/JPL manipulator relative to the Cartesian base coordinate system when the values of the joint variables are given. Specifically, if the values of the joint variables measured for a given configuration of the arm are substituted into the p_{06}-equation of Table 2.3b, the position of a point in the end-effector coordinate frame is obtained *uniquely* in the base coordinate system. The calculations involve several trigonometric expressions, but they are straightforward.

The forward kinematic equations presented in Table 2.3b for the Stanford/JPL manipulator can similarly be obtained for other serial link multijoint manipulators. For example, the corresponding kinematic equations for the revolute manipulator PUMA 600 are described in [5, 6]. Indeed, the coordinate frames are first chosen, and then the structural kinematic parameters for the manipulator are determined. Consequently, the transformation matrices relating the adjacent coordinate frames become specified. The forward kinematic equations are thus established for the manipulator. These equations can be used to determine the position of any point on the manipulator in the Cartesian base coordinate system when the values of the joint variables of the manipulator are known.

2.2.3 TRANSFORMATION MATRIX EQUALS THE PRODUCT OF ROTATION AND TRANSLATION MATRICES

The general transformation A_{i-1}^{i} is determined in the previous sections by geometrical considerations. It can also be obtained by performing successive rotations and translations of the axes of the $(i-1)$st coordinate frame on the basis of the definitions of the structural parameters. The goal with these operations is to align the axes of the $(i-1)$st coordinate frame with the corresponding axes of the ith coordinate frame. It can be achieved by means of the general translation and rotation matrices used in a proper order.

A pure translation matrix is introduced in Equation 2.1.14. In this matrix, the submatrix representing rotation is the identity matrix, and the last column specifies the translation vector. Thus, the transformation matrix describing the pure translation by the amount of p_x in the x-direction, p_y in the y-direction, and p_z in the z-direction assumes the following form:

$$\text{Trans}(p_x, p_y, p_z) = \begin{bmatrix} 1 & 0 & 0 & p_x \\ 0 & 1 & 0 & p_y \\ 0 & 0 & 1 & p_z \\ 0 & 0 & 0 & 1 \end{bmatrix} \qquad (2.2.8)$$

The rotation of a plane perpendicular to the rotational axis by a specified angle θ can be performed about any one of the three basis vectors in the rect-

angular coordinate system. The corresponding transformations can be determined by applying Equation 2.1.13 to the specific case. Thus, the pure rota-
tion matrices $\text{Rot}(x, \theta_x)$, $\text{Rot}(y, \theta_y)$, $\text{Rot}(z, \theta_z)$ about the x-axis, y-axis, and
z-axis, respectively, are as follows:

$$\text{Rot}(x, \theta_x) = \begin{bmatrix} 1 & 0 & 0 & 0 \\ 0 & \cos\theta_x & -\sin\theta_x & 0 \\ 0 & \sin\theta_x & \cos\theta_x & 0 \\ 0 & 0 & 0 & 1 \end{bmatrix} \tag{2.2.9}$$

$$\text{Rot}(y, \theta_y) = \begin{bmatrix} \cos\theta_y & 0 & \sin\theta_y & 0 \\ 0 & 1 & 0 & 0 \\ -\sin\theta_y & 0 & \cos\theta_y & 0 \\ 0 & 0 & 0 & 1 \end{bmatrix} \tag{2.2.10}$$

$$\text{Rot}(z, \theta_z) = \begin{bmatrix} \cos\theta_z & -\sin\theta_z & 0 & 0 \\ \sin\theta_z & \cos\theta_z & 0 & 0 \\ 0 & 0 & 1 & 0 \\ 0 & 0 & 0 & 1 \end{bmatrix} \tag{2.2.11}$$

The axis of rotation is represented in the rotation submatrix by the column
in which there is only one 1 and the remaining elements are zeros; for
example, matrix $\text{Rot}(y, \theta_y)$ in Equation 2.2.10 has 1 only in the second
column of the rotation submatrix, and the other entries of this column are
zeros; hence, the rotation takes place about the y-axis.

The general transformation matrix A_{i-1}^i can be decomposed into a set
of successive pure rotations and pure translations. Thus, the transformation
matrix A_{i-1}^i can result from the following transformations:

$$A_{i-1}^i = \text{Trans}(0, 0, d_i)\,\text{Rot}(z_{i-1}, \theta_i)\,\text{Rot}(x_i, \alpha_i)\,\text{Trans}(a_i, 0, 0)$$

$$= \begin{bmatrix} 1 & 0 & 0 & 0 \\ 0 & 1 & 0 & 0 \\ 0 & 0 & 1 & d_i \\ 0 & 0 & 0 & 1 \end{bmatrix} \begin{bmatrix} \cos\theta_i & -\sin\theta_i & 0 & 0 \\ \sin\theta_i & \cos\theta_i & 0 & 0 \\ 0 & 0 & 1 & 0 \\ 0 & 0 & 0 & 1 \end{bmatrix} \begin{bmatrix} 1 & 0 & 0 & 0 \\ 0 & \cos\alpha_i & -\sin\alpha_i & 0 \\ 0 & \sin\alpha_i & \cos\alpha_i & 0 \\ 0 & 0 & 0 & 1 \end{bmatrix}$$

$$\begin{bmatrix} 1 & 0 & 0 & a_i \\ 0 & 1 & 0 & 0 \\ 0 & 0 & 1 & 0 \\ 0 & 0 & 0 & 1 \end{bmatrix} = \begin{bmatrix} \cos\theta_i & -\cos\alpha_i\sin\theta_i & \sin\alpha_i\sin\theta_i & a_i\cos\theta_i \\ \sin\theta_i & \cos\alpha_i\cos\theta_i & -\sin\alpha_i\cos\theta_i & a_i\sin\theta_i \\ 0 & \sin\alpha_i & \cos\alpha_i & d_i \\ 0 & 0 & 0 & 1 \end{bmatrix} \tag{2.2.12}$$

The decomposition of A_{i-1}^i into translational and rotational operations as
indicated in Equation 2.2.12 may be viewed as a sequence of basic operations
on the coordinate frames. For example, when a transform is *postmultiplied*
by another transform, the described operation is performed relative to the
coordinate frame resulting from the previous transformation. To illustrate
this, the translation and rotation operations in Equation 2.2.12 may be

applied with $i = 1$ to Figure 2.2a. The first translation operation Trans$(0,0,d_i)$ on the right side of Equation 2.2.12 moves the origin of a frame that coincides with the zeroth coordinate system to point H_1. The resulting coordinate frame is rotated in accordance with the postmultiplication in Equation 2.2.12 about the z_0-axis counterclockwise by the amount of θ_1, which makes the x_0-axis to align with the x_1-axis. The coordinate frame is then rotated about the x_1-axis by the angle of α_1 counterclockwise. As the result, the coordinate axes of the resulting frame are parallel with those of the (x_1, y_1, z_1)-coordinate frame. These two coordinate systems can be made to coincide by the last translational operation, Trans$(a_1, 0, 0)$ in Equation 2.2.12. Thus, by considering matrix A_{i-1}^i as a sequence postmultiplying translational and rotational operations, the axes of the two coordinate systems are made to coincide. Since the translation and rotation matrices can usually be written by inspection, the transformation matrix A_{i-1}^i can alternatively be determined by forming and calculating the product of these matrices. In this case, special attention should be paid to the proper order of these basic operations.

Translation and rotation operations are basically described in the definitions of the structural kinematic parameters of a manipulator given in Section 2.2.1. The decomposition of matrix A_{i-1}^i into the translational and rotational operations offers a simple and interesting geometric interpretation for the transformation matrix.

2.3 ORIENTATION OF END-EFFECTOR

In many applications, the end-effector of a manipulator is guided to the object that is to be grasped and moved by the manipulator. The location of the end-effector in the base coordinates is then controlled so that the gripper is moved to the object, while its orientation assumes an appropriate position for grasping the object. The manipulator has six degrees of freedom: the position of the centerpoint of the gripper is represented by a three-dimensional vector, and the orientation of the gripper is described by another three-dimensional vector.

We will next discuss how to express the orientation of the end-effector in certain commonly used coordinate frames. Specifically, the orientation of the gripper will be described by means of vectors s, a, n and roll, pitch, and yaw (RPY) angles.

2.3.1 COORDINATES s, a, n FOR END-EFFECTOR

It has become a common practice to use specific notations in the definitions of a rectangular coordinate frame for the end-effector. Such a coordinate system consists of the unit vectors **s**, **a**, and **n** emanating from the center of the gripper, which serves as the origin of the coordinate frame. This coordinate frame is displayed in Figure 2.5. The sliding vector **s** is defined

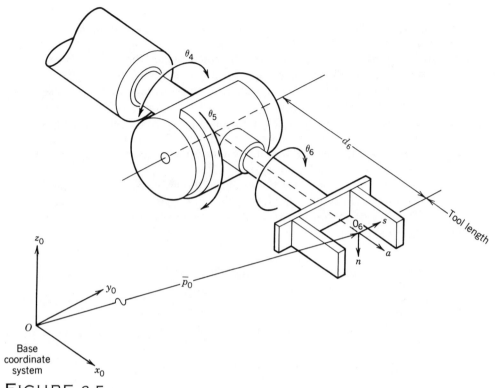

FIGURE 2.5

Local end-effector coordinate frame. Position vector $p_0 = [p_{x_0}, p_{y_0}, p_{z_0}, 1]'$

so that it is aligned with the axis along which the grasping surfaces (plates) move. The approach vector **a** coincides with the axis about which the gripper rotates; that is, the rotational axis of the sixth joint. The direction of the third vector **n** is specified by the right-hand rule so as to complete the local rectangular coordinate frame. Vectors **s**, **a** and **n** represent unit vectors, which are perpendicular to one another. These vectors define a coordinate frame for the end-effector, which can also be considered as the sixth coordinate frame in the D-H representation.

The *san*-coordinate frame may be compared with the sixth (end-effector) coordinate frame of the D-H representation shown in Figure 2.4 for the Stanford/JPL manipulator. By introducing the standard unit vectors for the gripper coordinate frame of Figure 2.4, a straightforward comparison gives in this case: $i_{x_6} = n$, $j_{y_6} = s$, and $k_{z_6} = a$.

Vectors **n**, **s**, and **a** can be expressed in the component form as $n = [n_{x_0} \quad n_{y_0} \quad n_{z_0} \quad 1]'$, $s = [s_{x_0} \quad s_{y_0} \quad s_{z_0} \quad 1]'$ and $a = [a_{x_0} \quad a_{y_0} \quad a_{z_0} \quad 1]'$, which also represent the unit vectors of the $x_6 y_6 z_6$-coordinate frame in the base $x_0 y_0 z_0$-coordinate system. Moreover, the vector describing origin O_6 in the sixth coordinate frame is $p_6 = [0, 0, 0, 1]'$. When p_6 is substituted into Equation 2.2.5, the tip of vector $\bar{p}_0 = [\bar{p}_{x_0}, \bar{p}_{y_0}, \bar{p}_{z_0}, 1]'$ will specify point O_6 in the base coordinate system. This vector $\bar{p}_0$ is also the translation vector (the last column) in the transformation matrix A_0^6, which relates the sixth (gripper) coordinate frame and the zeroth (base) coordinate system.

Having defined the rectangular coordinate frame for the gripper, the relationships between the variables of this coordinate frame, those of the base coordinate system, and the variables of the joints need to be established. First, the specific transformation that relates the variables of the gripper coordinate frame to those of the base coordinate system will be discussed. It will be expressed in terms of the joint variables.

If $\bar{p}_0 = [\bar{p}_{x_0}, \bar{p}_{y_0}, \bar{p}_{z_0}, 1]'$ specifies the origin of the gripper coordinate frame in the base coordinate system, the tips of the unit vectors **n**, **s**, and **a** in the base coordinate system can be expressed as the sum of vectors, $\bar{p}_0$ and the respective unit vectors:

$$\begin{bmatrix} p_{0x}^n \\ p_{0y}^n \\ p_{0z}^n \\ 1 \end{bmatrix} = \begin{bmatrix} n_{x_0} + \bar{p}_{x_0} \\ n_{y_0} + \bar{p}_{y_0} \\ n_{z_0} + \bar{p}_{z_0} \\ 1 \end{bmatrix} \tag{2.3.1}$$

$$\begin{bmatrix} p_{0x}^s \\ p_{0y}^s \\ p_{0z}^s \\ 1 \end{bmatrix} = \begin{bmatrix} s_{x_0} + \bar{p}_{x_0} \\ s_{y_0} + \bar{p}_{y_0} \\ s_{z_0} + \bar{p}_{z_0} \\ 1 \end{bmatrix} \tag{2.3.2}$$

$$\begin{bmatrix} p_{0x}^a \\ p_{0y}^a \\ p_{0z}^a \\ 1 \end{bmatrix} = \begin{bmatrix} a_{x_0} + \bar{p}_{x_0} \\ a_{y_0} + \bar{p}_{y_0} \\ a_{z_0} + \bar{p}_{z_0} \\ 1 \end{bmatrix} \tag{2.3.3}$$

where all components are described in the base coordinate system. As an example, Equation 2.3.3 represents the position vector from the origin of the base coordinate to the tip of the **a** vector.

Using the structural kinematic parameters and the joint variables, the transformation matrix A_0^6 in Equation 2.2.5 is obtained. It can be used to convert a vector of the sixth coordinate frame to a vector in the base coordinate frame. We will next look for another transformation matrix T_0^6 that plays a role similar to that of A_0^6 but which is expressed in terms of the components of vectors n, s, a, and $\bar{p}_0$.

Let T_0^6 be a 4×4 matrix that relates the gripper coordinate frame to the base coordinate system:

$$T_0^6 = \begin{bmatrix} t_{11} & t_{12} & t_{13} & t_{14} \\ t_{21} & t_{22} & t_{23} & t_{24} \\ t_{31} & t_{32} & t_{33} & t_{34} \\ t_{41} & t_{42} & t_{43} & t_{44} \end{bmatrix} \tag{2.3.4}$$

where entries t_{ij}, $i, j = 1, \ldots, 4$ are to be determined. The transformation matrix T_0^6 maps a vector in the end-effector coordinate frame to a vector in the base coordinate system.

The origin of the sixth (gripper) coordinate frame is described in the base coordinate system as vector $\bar{p}_0 = [\bar{p}_{x_0}, \bar{p}_{y_0}, \bar{p}_{z_0}, 1]'$. The same point in the sixth coordinate frame is $p_6 = [0, 0, 0, 1]'$. Transformation T_0^6 converts vector p_6 into expression $T_0^6 p_6$ in the base coordinate system, which must be equal to $\bar{p}_0$. Thus

$$T_0^6 \begin{bmatrix} 0 \\ 0 \\ 0 \\ 1 \end{bmatrix} = \begin{bmatrix} t_{14} \\ t_{24} \\ t_{34} \\ t_{44} \end{bmatrix} = \begin{bmatrix} \bar{p}_{x_0} \\ \bar{p}_{y_0} \\ \bar{p}_{z_0} \\ 1 \end{bmatrix} \qquad (2.3.5)$$

Moreover, the unit vector $\mathbf{n} = \mathbf{i}_{x_6}$, can be described in the end-effector coordinate frame as $n_6 = [1, 0, 0, 1]$. When T_0^6 operates on vector n_6, the resulting expression reads:

$$T_0^6 \begin{bmatrix} 1 \\ 0 \\ 0 \\ 1 \end{bmatrix} = \begin{bmatrix} t_{11} + t_{14} \\ t_{21} + t_{24} \\ t_{31} + t_{34} \\ t_{41} + t_{44} \end{bmatrix} \qquad (2.3.6)$$

Equation 2.3.6 describes the n-vector in the base coordinate system; it is also given by Equation 2.3.1.

Equation 2.3.5 and the equation resulting from combining Equations 2.3.6 and 2.3.1 are expressed in the component form:

$$
\begin{aligned}
t_{14} &= \bar{p}_{x_0} \\[4pt]
t_{24} &= \bar{p}_{y_0} \\[4pt]
t_{34} &= \bar{p}_{z_0} \\[4pt]
t_{44} &= 1 \\[4pt]
t_{11} + t_{14} &= n_{x_0} + \bar{p}_{x_0} \\[4pt]
t_{21} + t_{24} &= n_{y_0} + \bar{p}_{y_0} \\[4pt]
t_{31} + t_{34} &= n_{z_0} + \bar{p}_{z_0} \\[4pt]
t_{41} + t_{44} &= 1
\end{aligned}
\qquad (2.3.7)
$$

Equation 2.3.7 is solved for t_{11}, t_{21}, t_{31}, and t_{41}. Thus, the first and fourth columns of matrix T_0^6 are specified.

When transformation T_0^6 is applied to the other unit vectors $s_6 = [0, 1, 0, 1]'$ and $a_6 = [0, 0, 1, 1]'$ expressed in the end-effector coordinate frame, the comparison of the resulting expressions with Equations 2.3.2 and 2.3.3, respectively, gives:

$$t_{12} + t_{14} = s_{x_0} + \bar{p}_{x_0}$$

$$t_{22} + t_{24} = s_{y_0} + \bar{p}_{y_0}$$

$$t_{32} + t_{34} = s_{z_0} + \bar{p}_{z_0}$$

$$t_{42} + t_{44} = 1$$

$$t_{13} + t_{14} = a_{x_0} + \bar{p}_{x_0} \qquad (2.3.8)$$

$$t_{23} + t_{24} = a_{y_0} + \bar{p}_{x_0}$$

$$t_{33} + t_{34} = a_{z_0} + \bar{p}_{z_0}$$

$$t_{43} + t_{44} = 1$$

Equation 2.3.8 determines the elements in the second and third column of matrix T_0^6 in Equation 2.3.4. Indeed, the solutions to Equations 2.3.7 and 2.3.8 specify the T_0^6-matrix. It can now be expressed in terms of the components of the orientation and position vectors as follows:

$$T_0^6 = \begin{bmatrix} n_{x_0} & s_{x_0} & a_{x_0} & \bar{p}_{x_0} \\ n_{y_0} & s_{y_0} & a_{y_0} & \bar{p}_{y_0} \\ n_{z_0} & s_{z_0} & a_{z_0} & \bar{p}_{z_0} \\ 0 & 0 & 0 & 1 \end{bmatrix}$$

$$= \begin{bmatrix} n, & s, & a, & \bar{p}_0 \\ 0 & 0 & 0 & 1 \end{bmatrix} \qquad (2.3.9)$$

Matrix T_0^6 in Equation 2.3.9 is a concrete representation of the transformation relating the end-effector coordinates to the base coordinates. In fact, the transformation matrix T_0^6 in Equation 2.3.9 plays the same role as the transformation matrix A_0^6 in Equation 2.2.5. Matrix T_0^6 is specified in terms of the components of the unit coordinate vectors n, s, a, and the translation vector $\bar{p}_0$, all of which are expressed in the base coordinate system. On the other hand, matrix $A_0^6 = A_0^1 A_1^2 A_2^3 A_3^4 A_4^5 A_5^6$ is written in terms of the structural kinematic parameters and joint variables as shown in Equations 2.2.3 and 2.2.4.

Since the transformation matrices T_0^6 and A_0^6 represent the same relationship, these matrices must be equal. Thus

$$T_0^6 = A_0^6 = A_0^1 A_1^2 A_2^3 A_3^4 A_4^5 A_5^6 \qquad (2.3.10)$$

By equating the corresponding entries in the matrix Equation 2.3.10, the components of the vectors n, s, a, and $\bar{p}_0$ can be written in terms of the joint positions and the structural kinematic parameters. They are determined by Equation 2.3.10 and Table 2.3b for the Stanford/JPL manipulator, and presented in Table 2.4.

If the positions of the joints for a Stanford/JPL manipulator are known relative to chosen reference axes, the corresponding components in the base coordinate system can be calculated from the expressions of Table 2.4. The solution to these kinematic equations in the forward direction ($\{\theta_i\} \rightarrow \{n, s, a, \bar{p}_0\}$)

TABLE 2.4

Equations Relating End-Effector Position and Orientation in Base Coordinates to Joint Variables in Stanford/JPL Arm (Figure 2.4)

$$\bar{p}_{x_0} = d_6 [s_5 (s_1 c_4 + c_1 c_2 s_4) + c_1 s_2 c_5] + d_3 c_1 s_2 - d_2 s_1$$

$$\bar{p}_{y_0} = d_6 [s_5 (-c_1 c_4 + s_1 c_2 s_4) + s_1 s_2 c_5] + d_3 s_1 s_2 + d_2 c_1$$

$$\bar{p}_{z_0} = d_6 (c_2 c_5 - s_2 s_4 s_5) + d_3 c_2 + d_1$$

$$a_{x_0} = s_5 (s_1 c_4 + c_1 c_2 s_4) + c_1 s_2 c_5$$

$$a_{y_0} = s_5 [-c_1 c_4 + s_1 c_2 s_4] + s_1 s_2 c_5$$

$$a_{z_0} = c_2 c_5 - s_2 s_4 s_5$$

$$s_{x_0} = -s_6 [c_5 (s_1 c_4 + c_1 c_2 s_4) - c_1 s_2 s_5] + c_6 [-s_1 s_4 + c_1 c_2 c_4]$$

$$s_{y_0} = -s_6 [c_5 (-c_1 c_4 + s_1 c_2 s_4) - s_1 s_2 s_5] + c_6 [c_1 s_4 + s_1 c_2 c_4]$$

$$s_{z_0} = c_2 s_5 s_6 + s_2 s_4 c_5 s_6 - c_4 s_2 c_6$$

$$n_{x_0} = c_6 [c_5 (s_1 c_4 + c_1 c_2 s_4) - c_1 s_2 s_5] + s_6 [-s_1 s_4 + c_1 c_2 c_4]$$

$$n_{y_0} = c_6 [c_5 (-c_1 c_4 + s_1 c_2 s_4) - s_1 s_2 s_5] + s_6 [c_1 s_4 + s_1 c_2 c_4]$$

$$n_{z_0} = -c_2 s_5 c_6 - s_2 (s_4 c_5 c_6 + c_4 s_6)$$

is unique. That is, the position and orientation of the gripper are uniquely determined by the positions of the manipulator joints.

EXAMPLE 2.3.1

The configuration of a Stanford/JPL manipulator is such that the joint variables have the following values with respect to the reference positions: $\theta_1 = \pi/4$ rad, $\theta_2 = 0$ rad, $d_3 = 20$ cm, $\theta_4 = \pi/2$ rad, $\theta_5 = \pi/3$ rad, $\theta_6 = 0$ rad. The problem is to determine the position and orientation of the end-effector in the Cartesian base coordinate system.

The solution is determined on the basis of the equality $T_0^6 = A_0^6$ given in Equations 2.3.10 and 2.3.9, which are rewritten here:

$$T_0^6 = \begin{bmatrix} n_{x_0} & s_{x_0} & a_{x_0} & \bar{p}_{x_0} \\ n_{y_0} & s_{y_0} & a_{y_0} & \bar{p}_{y_0} \\ n_{z_0} & s_{z_0} & a_{z_0} & \bar{p}_{z_0} \\ 0 & 0 & 0 & 1 \end{bmatrix} \qquad (2.3.11)$$

$$A_0^6 = A_0^1 A_1^2 A_2^3 A_3^4 A_4^5 A_5^6 \qquad (2.3.12)$$

The right side expressions of Equation 2.3.12 are given in Table 2.3b and Table 2.4. Substituting the values of θ_i, $i = 1, \ldots, 6$ into these expressions gives:

$$T_0^6 = \begin{bmatrix} \dfrac{1}{2\sqrt{2}} & \dfrac{-1}{\sqrt{2}} & \dfrac{\sqrt{3}}{2\sqrt{2}} & \dfrac{d_6\sqrt{3}}{2\sqrt{2}} - \dfrac{d_2}{\sqrt{2}} \\[3mm] \dfrac{1}{2\sqrt{2}} & \dfrac{1}{\sqrt{2}} & \dfrac{\sqrt{3}}{2\sqrt{2}} & \dfrac{d_6\sqrt{3}}{2\sqrt{2}} + \dfrac{d_2}{\sqrt{2}} \\[3mm] -\dfrac{\sqrt{3}}{2} & 0 & \dfrac{1}{2} & \dfrac{d_6}{2} + d_3 + d_1 \\[3mm] 0 & 0 & 0 & 1 \end{bmatrix}$$ (2.3.13)

where d_1, d_2, and d_6 assume specific numerical values. Thus, the position and orientation of the end-effector are given by the following vectors:

$$\bar{p}_0 = \left[\dfrac{d_6\sqrt{3}}{2\sqrt{2}} - \dfrac{d_2}{\sqrt{2}} \quad \dfrac{d_6\sqrt{3}}{2\sqrt{2}} + \dfrac{d_2}{\sqrt{2}} \quad \dfrac{d_6}{2} + d_3 + d_1 \quad 1 \right]'$$

$$n = \left[\dfrac{1}{2\sqrt{2}} \quad \dfrac{1}{2\sqrt{2}} \quad -\dfrac{\sqrt{3}}{2} \quad 1 \right]'$$

$$s = \left[-\dfrac{1}{\sqrt{2}} \quad \dfrac{1}{\sqrt{2}} \quad 0 \quad 1 \right]'$$

$$a = \left[\dfrac{\sqrt{3}}{2\sqrt{2}} \quad \dfrac{\sqrt{3}}{2\sqrt{2}} \quad \dfrac{1}{2} \quad 1 \right]'$$

The foregoing vectors determine the position and orientation of the gripper in Cartesian base coordinate systems.

The orientation of the end-effector should be specified by three independent variables. In the transformation matrix T_0^6 of Equation 2.3.9, vectors s, a, and n are represented by nine components in the Cartesian base coordinate system that describe the rotation submatrix for the orientation. However, two of these three unit vectors determine the third unit vector, because these vectors represent a right-hand rectangular coordinate frame. Moreover, the orthogonality and the lengths of the two remaining vectors impose three additional relations reducing the number of the independent variables that describe the orientation to three. Thus, three independent variables in matrix T_0^6 do specify the orientation of the gripper.

The computation of the components of the unit vectors s, a, and n from the values of the joint variables is a straightforward application of the kinematic equations. On the other hand, if the vectors n, s, a, and $\bar{p}_0$ are known, the positions of the joint variables are usually not uniquely determined without additional constraints. This problem of solving the kinematic equations in the backward direction will be discussed in the next chapter.

2.3.2 ROLL-PITCH-YAW (RPY) COORDINATES

It is common practice, particularly in aerospace applications, that the position of a rigid body with six DOF be described in the three-dimensional space by giving the positional coordinates of the center of mass and the orientation of the object with respect to a reference coordinate system. The angular position of an object, such as an end-effector, is specified relative to some coordinate frame. An attractive set of such coordinate variables can be established by defining the orientation angles relative to a rectangular coordinate frame. Roll, pitch, and yaw (RPY) can be defined as the angles of rotations about the x-, y-, and z-axes, respectively, relative to a reference coordinate system. Figure 2.6a shows these angles and their rotational axes. The motions representing roll, pitch, and yaw are illustrated in Figure 2.6b.

The coordinate transformation RPY is expressed in terms of the angles of roll (ψ_z), pitch (ψ_y), and yaw (ψ_x) defined with respect to a reference coordinate frame whose axes are parallel to the respective axes of the base coordinate system. It can be described by means of successive rotational operations:

$$\text{RPY}(\psi_z, \psi_y, \psi_x) = \text{Rot}(z, \psi_z)\,\text{Rot}(y, \psi_y)\,\text{Rot}(x, \psi_x)$$

$$= \begin{bmatrix} \cos\psi_z & -\sin\psi_z & 0 & 0 \\ \sin\psi_z & \cos\psi_z & 0 & 0 \\ 0 & 0 & 1 & 0 \\ 0 & 0 & 0 & 1 \end{bmatrix} \begin{bmatrix} \cos\psi_y & 0 & \sin\psi_y & 0 \\ 0 & 1 & 0 & 0 \\ -\sin\psi_y & 0 & \cos\psi_y & 0 \\ 0 & 0 & 0 & 1 \end{bmatrix}$$

$$\begin{bmatrix} 1 & 0 & 0 & 0 \\ 0 & \cos\psi_x & -\sin\psi_x & 0 \\ 0 & \sin\psi_x & \cos\psi_x & 0 \\ 0 & 0 & 0 & 1 \end{bmatrix} \quad (2.3.14)$$

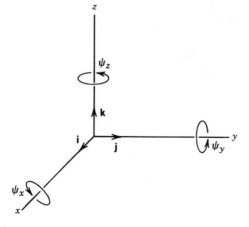

FIGURE 2.6a
Roll ψ_z, pitch ψ_y, and yaw ψ_x angles.

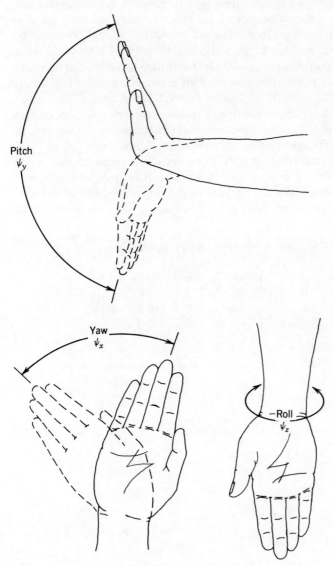

FIGURE 2.6b
Illustration of roll, pitch, and yaw.

where the rotation is first performed about the x-axis, and then about the y- and z-axes.

The multiplication of the matrices leads to:

$$\mathrm{RPY}(\psi_z, \psi_y, \psi_x)$$

$$=\begin{bmatrix} \cos\psi_z\cos\psi_y & \cos\psi_z\sin\psi_y\sin\psi_x - \sin\psi_z\cos\psi_x \\ \sin\psi_z\cos\psi_y & \sin\psi_z\sin\psi_y\sin\psi_x + \cos\psi_z\cos\psi_x \\ -\sin\psi_y & \cos\psi_y\sin\psi_x \\ 0 & 0 \end{bmatrix}$$

$$\begin{bmatrix} \cos\psi_z\sin\psi_y\cos\psi_x + \sin\psi_z\sin\psi_x & 0 \\ \sin\psi_z\sin\psi_y\cos\psi_x - \cos\psi_z\sin\psi_x & 0 \\ \cos\psi_y\cos\psi_x & 0 \\ 0 & 1 \end{bmatrix} \quad (2.3.15)$$

The rotational submatrix of the transformation matrix T according to Equation 2.3.9 has the following form in the xyz-coordinate frame, whose origin coincides with that of the RPY-coordinate frame:

$$T = \begin{bmatrix} n_x & s_x & a_x & 0 \\ n_y & s_y & a_y & 0 \\ n_z & s_z & a_z & 0 \\ 0 & 0 & 0 & 1 \end{bmatrix} \quad (2.3.16)$$

A particular orientation represented in these two different coordinate frames can be expressed in the base coordinate system by means of Equations 2.3.15 and 2.3.16. The relationship between these two coordinate frames is established by equating the corresponding entries in Equations 2.3.15 and 2.3.16. These equations are displayed in Table 2.5. If the roll, pitch, and yaw angles are known, the equations in Table 2.5 can be applied to determine the components of the rectangular coordinates n, s, and a in the base coordinate system.

EXAMPLE 2.3.2

The orientation of the end-effector is given in a Cartesian coordinate system by angles $\psi_z = 0$ rad, $\psi_y = \pi/6$ rad, and $\psi_x = \pi/2$ rad representing the roll, pitch, and yaw, respectively. The problem is to determine the orientation of the gripper in the Cartesian coordinates by specifying vectors n, s, and a.

Equations 2.3.15 and 2.3.16 establish the required relationship (Table 2.5). The substitution of the given angles into the equations of Table 2.5 gives

$$\begin{bmatrix} n_{x_0} & s_{x_0} & a_{x_0} & 0 \\ n_{y_0} & s_{y_0} & a_{y_0} & 0 \\ n_{z_0} & s_{z_0} & a_{z_0} & 0 \\ 0 & 0 & 0 & 1 \end{bmatrix} = \begin{bmatrix} \dfrac{\sqrt{3}}{2} & \dfrac{1}{2} & 0 & 0 \\ 0 & 0 & -1 & 0 \\ -\dfrac{1}{2} & \dfrac{\sqrt{3}}{2} & 0 & 0 \\ 0 & 0 & 0 & 1 \end{bmatrix} \quad (2.3.17)$$

TABLE 2.5

Relationships between Rectangular Coordinates and RPY-Variables

$$n_x = \cos \psi_z \cos \psi_y$$

$$n_y = \sin \psi_z \cos \psi_y$$

$$n_z = - \sin \psi_y$$

$$s_x = \cos \psi_z \sin \psi_y \sin \psi_x - \sin \psi_z \cos \psi_x$$

$$s_y = \sin \psi_z \sin \psi_y \sin \psi_x + \cos \psi_z \cos \psi_x$$

$$s_z = \cos \psi_y \sin \psi_x$$

$$a_x = \cos \psi_z \sin \psi_y \cos \psi_x + \sin \psi_z \sin \psi_x$$

$$a_y = \sin \psi_z \sin \psi_y \cos \psi_x - \cos \psi_z \sin \psi_x$$

$$a_z = \cos \psi_y \cos \psi_x$$

The rotation submatrix in the matrix of Equation 2.3.17 specifies the orientation of the end-effector in the base coordinate system.

The roll, pitch, and yaw angles used here to describe the orientation of an end-effector are measured relative to a reference coordinate frame. The origin of this coordinate frame can conveniently be placed to coincide with the origin of the hand (sixth) coordinate frame, and its axes must be parallel with those of the Cartesian base coordinate system for all configurations of the arm. The orientation of the end-effector can then be described by expressing, for example, the position of the approach vector that is fixed to the hand, and thus changing with the hand orientation by means of the roll, pitch, and yaw angles at any time. Three independent variables are needed to describe the orientation.

The orientation of the gripper in a manipulator is described in this chapter by means of the unit vectors n, s, a, and the roll, pitch, and yaw angles ψ_z, ψ_y, ψ_x, although other alternative descriptions are possible [3]. The roll, pitch, and yaw angles are determined from the joint variables indirectly by first computing the components of vectors n, s, and a, and then using the expressions of Table 2.5. Indeed, these calculations require more mathematical operations than those for computing only the unit vectors from the measurements of the encoders. When the orientation of the gripper needs to be described, for example, to plan a desired trajectory for manipulator motion, the desired changes in the orientation can usually be visualized and described readily by means of the approach, and slide vectors. The specifications of the changes in the roll, pitch, and yaw angles for a particular task can usually be visualized only by an experienced eye. The desired values of the orientation and position variables are often used to calculate the corresponding desired values of the joint variables, if the joint variables are controlled. In this case, the orientation of the gripper may conveniently be described in terms of the vectors n, s, and a. On the other hand, if the control of the gripper orientation is performed directly in the Cartesian base coordinate system, the roll, pitch, and yaw angles, ψ_z, ψ_y, ψ_x, should preferably be used. The choice of the variables for the orientation depends also on the particular task that the manipulator must perform.

2.4 SUMMARY

The configuration of a manipulator is determined by the values of the joint variables and the structural kinematic parameters of the manipulator. These values determine, in particular, the position and orientation of the end-effector. To establish relationships between these variables, appropriate coordinate frames are defined for the links of the manipulator.

We have presented a systematic method for defining coordinate systems for the links (joints) of a serial link manipulator in the framework of Denavit–Hartenberg (D-H) procedure. The structural parameters for the given manipulator are determined. The transformation matrix relating two adjacent coordinate systems can then be obtained. Using this transformation sequentially in a serial link manipulator, the position and orientation of the end-effector in the base coordinate system can be calculated from the measurements of the joint variables. These computations provide the solution to the forward kinematic equations.

The orientation of the gripper is described by means of three independent variables. They may be selected conveniently from the vector set s, a, n representing the unit vectors in the end-effector coordinate frame, or they can be chosen as the roll, pitch, and yaw angles. The choice of the variables representing the orientation of the gripper often depends on the particular task that the manipulator should perform.

The coordinate transformations establish the forward kinematic equations. They determine the position and orientation of the gripper uniquely. In controlling the motion of a manipulator, the values of the joint variables must often be calculated from the values of the position and orientation of the end-effector expressed in the base coordinate system. Thus, the inverse solution to the kinematic equations is determined. It is discussed in the next chapter.

REFERENCES

[1] J. DENAVIT AND R. S. HARTENBERG, "A Kinematic Notation for Lower-Pair Mechanisms Based on Matrices," *J. of Applied Mechanics*, pp. 215–221, June 1955.

[2] A. K. BEJCZY, "Robot Arm Dynamics and Control," Technical Memo 33-669, Jet Propulsion Lab., February 1974.

[3] R.P. PAUL, *Robot Manipulators: Mathematics, Programming and Control*, The MIT Press, Cambridge, MA, 1981.

[4] C. S. LEE, "Robotic Arm Kinematics, Dynamics and Control," *Computer*, IEEE, pp. 62–80, December 1982.

[5] A. BAZERGHI, A. A. Goldenberg, J. Apkarian, "An Exact Kinematic Model of PUMA 600 Manipulator," *IEEE Transactions on Systems, Man, and Cybernetics*, Vol. SMC-14, No. 3, pp. 483–487, May/June 1984.

[6] D.E. WHITNEY AND J.S. SHAMMA, "Comments on 'Exact Kinematic Model of the PUMA 600 Manipulator'," *IEEE Transactions on Systems, Man and Cybernetics (Corresp.)*, Vol. SMC-16, No. 1, pp. 182–184, January/February 1986.

PROBLEMS

2.1 Write the equation for the straight line L in Figure 2.2a.

    a. How many parameters do you need to describe the straight line L?

    b. Write the equation for the projection of L on the yz-plane, and on the xz-plane. Use these equations to describe the straight line L. How many parameters are needed?

2.2 Are the coordinate systems in Figure 2.1 chosen according to the D-H guidelines? Explain.

2.3 Figure 2.7 shows three joints of a simple manipulator with the rotational axes.

    a. Select the coordinate frames for joints 4, 5, and 6 following the guidelines of the D-H representation, when

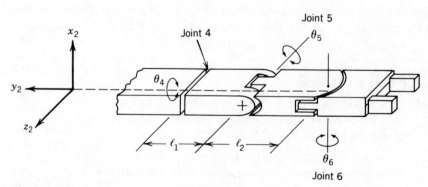

FIGURE 2.7

Three joints of a simple manipulator for Problem 2.3.

the coordinate frame for joint three is given, as shown in the figure.

b. Complete the table below for the structural kinematic parameters when the hand is in the resting position shown

link i	θ_i	a_i	α_i	d_i
4				
5				
6				

c. Write the homogeneous transformations A_4^5 and A_5^6 between the coordinate frames selected.

d. Select the positive direction of the z_4-axis opposite to that used in (a). Repeat parts (a), (b), and (c) for this choice. How do the results differ?

2.4 Given the transformation A_0^1 between the first and the base coordinate frames:

$$A_0^1 = \begin{bmatrix} c_1 & 0 & -s_1 & 2 \\ s_1 & 0 & c_1 & 3 \\ 0 & -1 & 0 & 4 \\ 0 & 0 & 0 & 1 \end{bmatrix}$$

A point P_1 in the first coordinate frame is specified as $p_1 = [1, 2, -3, 1]'$. If $\theta_1 = 30°$, determine the position of this point in the base coordinates.

2.5 The following transformation matrix is given:

$$T = \begin{bmatrix} 0.500 & 0 & -0.866 & 1.5 \\ 0 & 1 & 0 & 0 \\ 0.866 & 0 & 0.500 & 2.1 \\ 0 & 0 & 0 & 1 \end{bmatrix}$$

a. What is the axis of rotation, and what is the angle of rotation specified by the matrix T?

b. Specify the position and orientation of the gripper if T represents the transformation matrix in the base coordinate system; that is, give expressions for vectors $\bar{p}_0, n, s,$ and a.

2.6 Suppose that the joint variables of a Stanford/JPL arm have values $\theta_1 = \pi/3$ rad; $\theta_2 = -\pi/4$ rad; $d_3 = 0$ cm; $\theta_4 = 0$ rad;

$\theta_5 = 3\pi/2$ rad; $\theta_6 = \pi/2$ rad relative to certain reference axes. Determine the position and orientation of the gripper by calculating the components of the n, s, a and $\bar{p}_0$ vectors.

2.7 The configuration of PUMA 600 with the coordinate frames is shown at the *rest* position in Figure 2.8. Have the D-H guidelines been followed? Determine the structural kinematic parameters by completing the table below for the manipulator (see also [5]).

Link i	θ_i	a_i	α_i	d_i
1				
2				
3				
4				
5				
6				

2.8 a. Show that position vector p_w of the wrist in a Stanford/JPL arm expressed in the base coordinate system of Figure 2.4 is completely specified by the positional variables of the first three joints.

b. Express the wrist position p_w in terms of variables θ_1, θ_2, and d_3. (Note: For the wrist position, $d_6 = 0$ and parameters d_1 and d_2 are given constants.)

2.9 The general transformation matrix A^i_{i-1} is given.

a. Show that the rotation submatrix $(A^i_{i-1})_R$ for a rotational joint is invertible, that is, determine its determinant.

b. Does the inverse of A^i_{i-1} always exist? Provide evidence for your answer.

c. Determine the inverse $(A^i_{i-1})^{-1}$ of the transformation matrix A^i_{i-1}.

2.10 Show that the following expressions are, or are not, equalities:

a. $\text{Trans}(1,0,1)\text{Rot}(z,90°) \overset{?}{=} \text{Rot}(z,-90°)\text{Trans}(-1,0,1)$

b. $\text{Trans}(5,0,0)\text{Rot}(z,90°) \overset{?}{=} \text{Rot}(z,90°)\text{Trans}(5,0,0)$

c. $[\text{Trans}(1,0,1)\text{Rot}(z,90°)][\text{Rot}(z,-90°)\text{Trans}(-1,0,1)] \overset{?}{=} I$

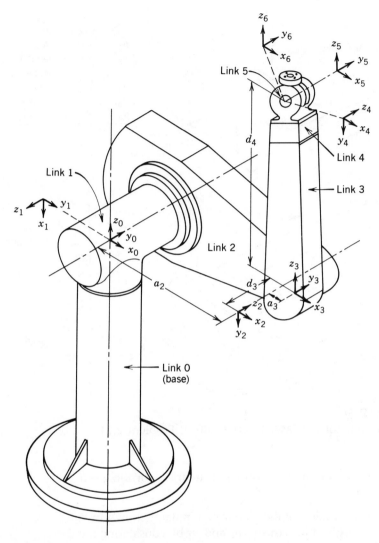

FIGURE 2.8
PUMA 600 at rest (parking) configuration for Problem 2.7.

2.11 A six joint manipulator (UNIMATE Series 2000) is shown in Figure 2.9.

    a. The coordinate systems for the system have been chosen as shown in the figure. Check if they have been chosen by following the D-H guidelines.

    b. For the resting position shown, determine the offset angles α_i relative to the chosen coordinate frames.

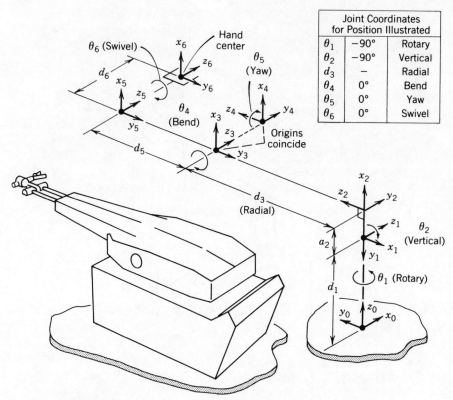

Joint Coordinates for Position Illustrated		
θ_1	$-90°$	Rotary
θ_2	$-90°$	Vertical
d_3	—	Radial
θ_4	$0°$	Bend
θ_5	$0°$	Yaw
θ_6	$0°$	Swivel

FIGURE 2.9

Six joint manipulator (UNIMATE series 2000); Problem 2.11.

c.   Determine the transformation matrices for the given
     coordinate frames.

2.12 The configuration of the planar manipulator shown in Fig-
     ure 2.3 is given in terms of θ_1 and θ_2 by considering the
     last link to be an extension of the second link. The lengths
     of the links are: $\ell_1 = 0.75$m and $\ell_2 = 0.5$m.

a.   If $\theta_1 = \pi/2$ rad and $\theta_2 = -\pi/3$ rad, determine the posi-
     tion of the axis of the third joint in the base coordinate
     system.

b.   If θ_3 denotes the position of the third joint variable,
     determine the general expression for the end-effector
     position. The length of the third link is denoted by ℓ_3.

c.   If $\theta_1 = \pi/4$ rad, $\theta_3 = -\pi/4$ rad, $\theta_3 = -\pi/2$ rad, and $\ell_3 =$
     0.2 m, determine the end-effector position. The lengths
     of the links are given above.

2.13 The crank shown in Figure 2.10 has a rotational joint at the pivot and at the end of the crank. Coordinate frames have been assigned to the pivot, the end of the crank, and the handle tip, as shown. If the handle is grasped and moved in a clockwise direction through an angle θ, the crank will turn and the joints at the pivot and at the handle will rotate. The transformation matrix T relating the position of the tip of the handle to the base coordinate frame is defined as: $T =$ pivot $\cdot$ Rot$(\cdot , \cdot) \cdot$ crank $\cdot$ Rot$(\cdot , \cdot) \cdot$ handle, where the transformations are pivot $=$ Trans$(\cdot , \cdot , \cdot)$, crank $=$ Trans$(\cdot , \cdot , \cdot)$, and handle $=$ Trans$(\cdot , \cdot , \cdot)$.

By inspection of Figure 2.10, determine the transformation matrices for the translations and rotations.

COMPUTER PROBLEM

2.C.1 The joint positions of a Stanford manipulator at a specific time t_1 are as follows: $\theta_1 = -.711$ rad; $\theta_2 = 1.857$ rad; $d_3 = 8.633$ cm; $\theta_4 = .013$ rad; $\theta_5 = 1.393$ rad; and $\theta_6 = -.726$ rad. The structural kinematic parameters are: $d_1 = 41.250$ cm, $d_2 = 15.367$ cm, and $d_6 = 26.290$ cm.

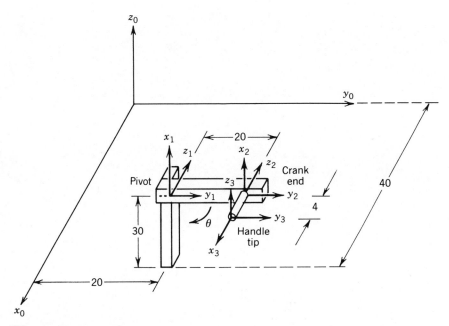

FIGURE 2.10
Configuration of a crank; Problem 2.13.

a. Compute the homogeneous transformation matrices A^i_{i-1}, $i = 1,2,\ldots,6$, when the coordinate frames defined in Figure 2.4 are used.

b. Specify the position and orientation of the end-effector in this configuration.

c. Suppose that the (constant) velocities of the joints at the time t_1 are $\dot{\theta}_1 = -1.034$ rad/s; $\dot{\theta}_2 = .127$ rad/s; $\dot{d}_3 = 12.399$ in./s; $\dot{\theta}_4 = 1.475$ rad/s; $\dot{\theta}_5 = 2.905$ rad/s; and $\dot{\theta}_6 = -3.129$ rad/s. Determine the joint positions at time $t_1 + 0.5$ s.

d. Compute the homogeneous transformations A^i_{i-1}, $i = 1,\ldots,6$, at the time $t_1 + 0.5$ s.

e. Determine the new position and orientation of the end-effector. Specify the change in the position and orientation of the end-effector that occurred.

CHAPTER THREE

INVERSE SOLUTION TO KINEMATIC EQUATIONS

In the previous chapter, the kinematic equations between the variables of two rectangular coordinate frames and between the variables of the hand coordinate frame and joint variables are established. A functional relationship between the joint variables of a manipulator arm and the components of the unit vectors s, a, n, and $\bar{p}_0$ in the base coordinate system represents a transformation between the joint space and Cartesian base coordinate variables. As a typical example, the equations in Table 2.4 describe the orientation and position vectors of the end-effector of a Stanford/JPL arm in terms of the structural kinematic parameters and the joint variables. If the values of the joint variables are known or measured, say by encoders attached to the joints, the position and orientation of the end-effector in the Cartesian base coordinate system can be uniquely determined. The kinematic equations are used in the forward direction; thus, the forward kinematic equations are solved.

For a given task, the end-effector of a manipulator is often controlled so that it should follow a desired path (trajectory) determined by the planning in terms of the Cartesian base coordinates. These values need to be converted to the values of the joint variables when the control of the manipulator is performed in the joint space, as is usually the case. In particular, when position $\bar{p}_0$ and orientation s, a, n of the end-effector are specified in the base coordinate system, the values of the joint variables that give rise to the specified location of the end-effector in the workspace must often be determined for control purposes. This is the inverse kinematic problem. The kinematic equations are used in the backward direction.

The determination of the inverse of the transformation matrix is first presented. The solution to the kinematic equations in the backward direction (i.e., the inverse kinematic solution) is then discussed. Particularly, the inverse solution to the equations relating the roll, pitch, and yaw angles to the rectangular coordinates is given. Then, the inverse solution to the kinematic equations of the Stanford/JPL manipulator is presented.

3.1 CALCULATION OF INVERSE TRANSFORMATION BETWEEN COORDINATE SYSTEMS

To determine the appropriate values of the joint variables that correspond to a given location of the end-effector, the kinematic equations must be solved in the backward direction; that is, the transformation relating the position and orientation of the end-effector to the joint variables and the structural kinematic parameters needs to be inverted. The inverse transformation will be determined after some helpful properties of the transformation matrix are first established.

The transformation matrix T_0^6 in Equation 2.3.9 contains the unit vectors of the rectangular right-hand coordinate frame attached to the gripper expressed in the base coordinate system. The rotation submatrix has the following properties:

1. $n'n = 1$, $s's = 1$, and $a'a = 1$. Thus, vectors n, s, and a are of unit length.

2. $n's = 0$, $n'a = 0$, and $s'a = 0$. That is, the unit vectors are orthogonal.

3. $\mathbf{n} = \mathbf{s} \times \mathbf{a}$, $\mathbf{s} = \mathbf{a} \times \mathbf{n}$, and $\mathbf{a} = \mathbf{n} \times \mathbf{s}$. Thus, the unit vectors obey the right-hand rule and are orthogonal.

4. The determinant of the rotation submatrix equals one, that is, $\begin{vmatrix} n & s & a \end{vmatrix} = 1$.

Property 4 can be shown by using the components of the unit vectors in calculating the determinant while making use of properties 1–3.

A matrix in which the columns of the rotation submatrix possess the foregoing properties 1–4 is called a *homogeneous transformation*. This transformation matrix represents an orthogonal transformation. The transformations T_0^6 and A_{i-1}^i specified by Equations 2.2.3, 2.2.4, and 2.3.9 satisfy conditions 1–4, and hence they qualify as homogeneous transformation matrices.

The inverse of a rotation submatrix, and that of the homogeneous transformation matrix, can now efficiently be determined.

3.1.1 INVERSE OF PURE ROTATION MATRIX

A vector p_i in the ith coordinate frame can be expressed as a vector p_{i-1} in the $(i-1)$st coordinate frame as indicated by Equation 2.2.1:

$$p_{i-1} = A_{i-1}^i p_i \qquad (3.1.1)$$

where A_{i-1}^i is a homogeneous transformation matrix. Equation 3.1.1 can also be expressed as

$$p_i = A_i^{i-1} p_{i-1} \qquad (3.1.2)$$

where $A_i^{i-1} = (A_{i-1}^i)^{-1}$ represents the inverse of the homogeneous transformation matrix A_{i-1}^i.

If matrix A_{i-1}^i is a pure rotation matrix, it can be expressed in terms of the unit vectors of the foregoing coordinate frames (see Equation 2.1.13):

$$A_{i-1}^i = \begin{bmatrix} \mathbf{i}_{x_i} \cdot \mathbf{i}_{x_{i-1}} & \mathbf{j}_{y_i} \cdot \mathbf{i}_{x_{i-1}} & \mathbf{k}_{z_i} \cdot \mathbf{i}_{x_{i-1}} & 0 \\ \mathbf{i}_{x_i} \cdot \mathbf{j}_{y_{i-1}} & \mathbf{j}_{y_i} \cdot \mathbf{j}_{y_{i-1}} & \mathbf{k}_{z_i} \cdot \mathbf{j}_{y_{i-1}} & 0 \\ \mathbf{i}_{x_i} \cdot \mathbf{k}_{z_{i-1}} & \mathbf{j}_{y_i} \cdot \mathbf{k}_{z_{i-1}} & \mathbf{k}_{z_i} \cdot \mathbf{k}_{z_{i-1}} & 0 \\ 0 & 0 & 0 & 1 \end{bmatrix} \qquad (3.1.3)$$

Equation 3.1.3 can also be written for matrix A_i^{i-1} using the unit vectors by properly changing the subscripts:

$$A_i^{i-1} = \begin{bmatrix} \mathbf{i}_{x_{i-1}} \cdot \mathbf{i}_{x_i} & \mathbf{j}_{y_{i-1}} \cdot \mathbf{i}_{x_i} & \mathbf{k}_{z_{i-1}} \cdot \mathbf{i}_{x_i} & 0 \\ \mathbf{i}_{x_{i-1}} \cdot \mathbf{j}_{y_i} & \mathbf{j}_{y_{i-1}} \cdot \mathbf{j}_{y_i} & \mathbf{k}_{z_{i-1}} \cdot \mathbf{j}_{y_i} & 0 \\ \mathbf{i}_{x_{i-1}} \cdot \mathbf{k}_{z_i} & \mathbf{j}_{y_{i-1}} \cdot \mathbf{k}_{z_i} & \mathbf{k}_{z_{i-1}} \cdot \mathbf{k}_{z_i} & 0 \\ 0 & 0 & 0 & 1 \end{bmatrix} \qquad (3.1.4)$$

The comparison of Equations 3.1.3 and 3.1.4 reveals that the A_i^{i-1}-matrix is the transpose of the A_{i-1}^i-matrix, since the order of the vectors in an inner product can be interchanged. Thus, the pure rotation matrices A_{i-1}^i and A_i^{i-1} in Equations 3.1.3 and 3.1.4 are orthonormal transformations.

When a general homogeneous transformation matrix is decomposed as shown in Equation 2.1.7, the inverse of the rotation submatrix can be obtained by a straightforward transposition operation. Indeed, the rotation submatrices $(A_{i-1}^i)_R$ and $(A_i^{i-1})_R$ of the general homogeneous transformation matrices A_{i-1}^i and A_i^{i-1} satisfy

$$(A_i^{i-1})_R = (A_{i-1}^i)_R^t = (A_{i-1}^i)_R^{-1} \qquad (3.1.5)$$

Thus, the rotation submatrix describes a *proper orthogonal* transformation, that is, the inverse of the rotation submatrix in the homogeneous transformation matrix A_{i-1}^i is equal to the transpose of the same rotation submatrix.

3.1.2 INVERSE OF HOMOGENEOUS TRANSFORMATION MATRIX

The homogeneous transformation matrix T_0^6 in Equation 2.3.9 is specified by expressing the unit vectors n, s, a and the position vector $\bar{p}_0$ of the gripper coordinate frame in the cartesian base coordinate system:

$$T_0^6 = \begin{bmatrix} n & s & a & \bar{p}_0 \\ 0 & 0 & 0 & 1 \end{bmatrix} = \begin{bmatrix} n_{x_0} & s_{x_0} & a_{x_0} & \bar{p}_{x_0} \\ n_{y_0} & s_{y_0} & a_{y_0} & \bar{p}_{y_0} \\ n_{z_0} & s_{z_0} & a_{z_0} & \bar{p}_{z_0} \\ 0 & 0 & 0 & 1 \end{bmatrix} \qquad (3.1.6)$$

The inverse of a homogeneous transformation matrix is often needed, for example, to calculate the position and orientation of an object relative to some specified coordinate system (as in problem 3.10). The aforementioned properties of the homogeneous transformation matrix make the determination of the inverse matrix fast and straightforward.

The inverse of matrix T_0^6 in Equation 3.1.6 denoted by $(T_0^6)^{-1} = T_6^0$ can be calculated by a standard procedure of the matrix inversion. To demonstrate typical calculations, the elements (1,1) and (2,1) of the inverse matrix T_6^0 are determined. First, the determinant of matrix T_0^6 is obtained:

$$|T_0^6| = n_{x_0}(s_{y_0}a_{z_0} - a_{y_0}s_{z_0}) - n_{y_0}(s_{x_0}a_{z_0} - a_{x_0}s_{z_0}) + n_{z_0}(s_{x_0}a_{y_0} - a_{x_0}s_{y_0}) = 1 \qquad (3.1.7)$$

since $\mathbf{n} = \mathbf{s} \times \mathbf{a}$ by property 3, and $n'n = 1$ by property 1 of the homogeneous transformation matrix. Element (1,1) in the inverse matrix T_6^0 is equal to $(s_{y_0}a_{z_0} - a_{y_0}s_{z_0}) = n_{x_0}$ and element (2,1) in T_6^0 is $-(n_{y_0}a_{z_0} - a_{y_0}n_{z_0}) = s_{x_0}$. By similar calculations, the remaining elements of the inverse matrix $(T_0^6)^{-1} = T_6^0$ become specified. Thus,

$$T_6^0 = \begin{bmatrix} n_{x_0} & n_{y_0} & n_{z_0} & -\bar{p}'_0 n \\ s_{x_0} & s_{y_0} & s_{z_0} & -\bar{p}'_0 s \\ a_{x_0} & a_{y_0} & a_{z_0} & -\bar{p}'_0 a \\ 0 & 0 & 0 & 1 \end{bmatrix} \qquad (3.1.8)$$

where $\bar{p}_0 = [\bar{p}_{x_0}, \bar{p}_{y_0}, \bar{p}_{z_0}]'$. The subscripts in the entries of the matrix emphasize that the variables are expressed in the base coordinate system.

By inspecting Equation 3.1.8, we observe that the inverse of matrix T_0^6 in Equation 3.1.6 is obtained by (1) transposing the submatrix of the rotation, and (2) forming the dot-products for the last column of $(T_0^6)^{-1}$, that is, the components of the translation vector are $-\bar{p}'_0 n$, $-\bar{p}'_0 s$, $-\bar{p}'_0 a$, and 1. Thus, the inverse $(T_0^6)^{-1} = T_6^0$ in Equation 3.1.8 can be written directly from the expression in Equation 3.1.6.

The homogeneous transformation matrix T_0^6 can be used to express the coordinates of a point p_6 known in the end-effector coordinate frame in terms of the base coordinates, p_0, for example, $p_0 = T_0^6 p_6$. Similarly, the homogeneous transformation matrix T_6^0 relates the coordinates of a point known in the base coordinate system to the coordinates of the end-effector coordinates, that is, $p_6 = (T_0^6)^{-1} p_0 = T_6^0 p_0$.

Three examples are presented to illustrate the determination of inverse transformations.

EXAMPLE 3.1.1

The homogeneous transformation matrix T_0^6 for a manipulator is given as

$$T_0^6 = \begin{bmatrix} 0 & 1/\sqrt{2} & 1/\sqrt{2} & 7 \\ 1 & 0 & 0 & 3 \\ 0 & -1/\sqrt{2} & 1/\sqrt{2} & -2 \\ 0 & 0 & 0 & 1 \end{bmatrix} \quad (3.1.9)$$

The inverse $T_6^0 = (T_0^6)^{-1}$ of the homogeneous transformation matrix T_0^6 is to be determined.

The answer is obtained by using Equation 3.1.8. The submatrix of T_6^0 representing rotation can be written immediately. The components of the translational vector in the fourth column are obtained by evaluating the inner products. For example, the (1,4) entry is specified by $-\bar{p}_0'n = -[7,3,-2,1][0,1,0,0]' = -3$. The resulting matrix is

$$T_6^0 = (T_0^6)^{-1} = \begin{bmatrix} 0 & 1 & 0 & -3 \\ 1/\sqrt{2} & 0 & -1/\sqrt{2} & -9/\sqrt{2} \\ 1/\sqrt{2} & 0 & 1/\sqrt{2} & -5/\sqrt{2} \\ 0 & 0 & 0 & 1 \end{bmatrix} \quad (3.1.10)$$

Equation 3.1.10 is the inverse of matrix T_0^6 given in Equation 3.1.9. It is readily verified by multiplying the two matrices to obtain the identity matrix.

EXAMPLE 3.1.2

A planar manipulator shown in Figure 3.1 has three revolute joints, and the third (wrist) joint is locked in the position so that links 2 and 3 function as one link. Three coordinate frames for the system are assigned in Example 2.2.3 following the D-H procedure. The homogeneous transformation matrices A_0^1, A_1^2, and A_0^2 between the three coordinate frames are given by Equations 2.2.6, and 2.2.7, respectively. The transformation matrix A_0^2 relating the end-effector coordinate frame to the base (zeroth) coordinate system is rewritten here:

$$A_0^2 = \begin{bmatrix} c_{12} & -s_{12} & 0 & \ell_1 c_1 + \ell_2 c_{12} \\ s_{12} & c_{12} & 0 & \ell_1 s_1 + \ell_2 s_{12} \\ 0 & 0 & 1 & 0 \\ 0 & 0 & 0 & 1 \end{bmatrix} \quad (3.1.11)$$

where the first column specifies the n-vector, the second column the s-vector, and the third column the a-vector in the base coordinate system. The problem here is to determine the transformation matrix $(A_0^2)^{-1} = A_2^0$.

To obtain $A_2^0 = (A_0^2)^{-1}$, Equation 3.1.8 is directly applied. Indeed, the translation vector in A_2^0 has the following components: $-n'p_0 = -\ell_1 c_2 - \ell_2$, $-s'p_0 = \ell_1 s_2$, and $-a'p_0 = 0$. The rotation submatrix is equal to the transpose of the corresponding submatrix $(A_0^2)_R$ in A_0^2. The inverse of the A_0^2-matrix is therefore

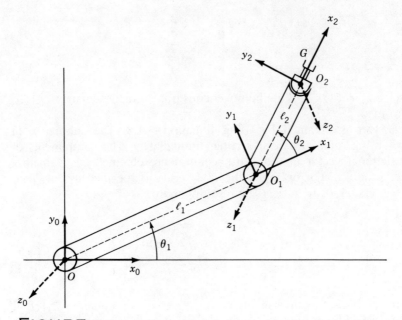

FIGURE 3.1
A planar revolute manipulator.

$$(A_0^2)^{-1} = A_2^0 = \begin{bmatrix} c_{12} & s_{12} & 0 & -\ell_1 c_2 - \ell_2 \\ -s_{12} & c_{12} & 0 & \ell_1 s_2 \\ 0 & 0 & 1 & 0 \\ 0 & 0 & 0 & 1 \end{bmatrix} \qquad (3.1.12)$$

Equation 3.1.12 is the requested transformation matrix.

Any point $p_2 = [p_{x_2}, p_{y_2}, p_{z_2}, 1]'$ in the second coordinate frame can be described in the base coordinate system as $p_0 = [p_{x_0}, p_{y_0}, p_{z_0}, 1]'$ using the following relation

$$p_0 = A_0^2 p_2 \qquad (3.1.13)$$

As an illustration, the origin O_2 in the second coordinate frame is specified by the vector $\bar{p}_2 = [0, 0, 0, 1]'$. The same point in the base coordinate system is $A_0^2 \bar{p}_2 = \bar{p}_{02} = [\ell_1 c_1 + \ell_2 c_{12}, \ell_1 s_1 + \ell_2 s_{12}, 0, 1]'$, which can also be verified directly from Figure 3.1.

When Equation 3.1.13 is solved for p_2, matrix A_2^0 appears:

$$p_2 = (A_0^2)^{-1} p_0 = A_2^0 p_0 \qquad (3.1.14)$$

Equation 3.1.14 can now be used, for instance, to express the origin O of the base coordinate system described as $\bar{p}_0 = [0, 0, 0, 1]$ in terms of the coordinates $(x_2 y_2 z_2)$ of the second coordinate frame: $A_2^0 \bar{p}_0 = \bar{p}_{20} = [-\ell_1 c_2 - \ell_2, \ell_1 s_2, 0, 1]'$. These components can also be verified directly in Figure 3.1

by projecting the vector from O_2 to O to the axes of the second coordinate frame and reading the components of $\bar{p}_{20}$ from the x_2-, y_2-, and z_2-axes.

EXAMPLE 3.1.3

Suppose that the position $\bar{p}_0 = [\bar{p}_{x_0}, \bar{p}_{y_0}, \bar{p}_{z_0}, 1]'$ of the end-effector of the planar manipulator discussed in Example 3.1.2 is known in the base coordinate system $(x_0 y_0 z_0)$. The problem is to determine the values of the joint variables θ_1 and θ_2 that give rise to the given location of the end-effector. This problem is the inverse kinematic problem for the planar manipulator.

The equations for solving the problem can be obtained by comparing the homogeneous transformation matrices A_0^2 given in Equation 2.2.7 and T given in Equation 2.3.9. The former equation gives matrix A_0^2 in terms of the joint variables, and the latter specifies T in terms of the position and orientation vectors expressed relative to the base coordinates. The unknown variables θ_1 and θ_2 can then be solved from the equations resulting from the comparison (see problem 3.5). Alternatively, the unknown joint variables θ_1 and θ_2 can be solved by using a general approach that is applicable to the inverse kinematic problem of other multijoint manipulators. This solution is discussed next.

The relationship between the homogeneous transformations is:

$$T = A_0^1 A_1^2 \tag{3.1.15}$$

The left-hand side of Equation 3.1.15 contains the components of vectors n, s, a, and $\bar{p}_0$, whereas the right-hand side is in terms of the joint variables. In order to solve for the joint variable θ_1 from Equation 3.1.15, the procedure is to isolate the unknown variable to the left-hand side only, and then manipulate the right side so as to eliminate the other unknown variable θ_2. This can be accomplished by premultiplying both sides of Equation 3.1.15 by $(A_0^1)^{-1}$ to obtain

$$(A_0^1)^{-1}T = A_1^2 \tag{3.1.16}$$

The right side of Equation 3.1.16 is determined in Example 2.2.3 and given by Equation 2.2.6, which is repeated here:

$$A_1^2 = \begin{bmatrix} c_2 & -s_2 & 0 & \ell_2 c_2 \\ s_2 & c_2 & 0 & \ell_2 s_2 \\ 0 & 0 & 1 & 0 \\ 0 & 0 & 0 & 1 \end{bmatrix} \tag{3.1.17}$$

The matrix on the left-hand side of Equation 3.1.16 is computed by using the inverse of matrix A_0^1 that is given in Equation 2.2.6:

$$(A_0^1)^{-1}T = \begin{bmatrix} c_1 & s_1 & 0 & -\ell_1 \\ -s_1 & c_1 & 0 & 0 \\ 0 & 0 & 1 & 0 \\ 0 & 0 & 0 & 1 \end{bmatrix} \begin{bmatrix} n_{x_0} & s_{x_0} & a_{x_0} & \bar{p}_{x_0} \\ n_{y_0} & s_{y_0} & a_{y_0} & \bar{p}_{y_0} \\ n_{z_0} & s_{z_0} & a_{z_0} & \bar{p}_{z_0} \\ 0 & 0 & 0 & 1 \end{bmatrix}$$

$$\begin{bmatrix} c_1 n_{x_0} + s_1 n_{y_0} & c_1 s_{x_0} + s_1 s_{y_0} & c_1 a_{x_0} + s_1 a_{y_0} & c_1 \bar{p}_{x_0} + s_1 \bar{p}_{y_0} - \ell_1 \\ -s_1 n_{x_0} + c_1 n_{y_0} & -s_1 s_{x_0} + c_1 s_{y_0} & -s_1 a_{x_0} + c_1 a_{y_0} & -s_1 \bar{p}_{x_0} + c_1 \bar{p}_{y_0} \\ n_{z_0} & s_{z_0} & a_{z_0} & \bar{p}_{z_0} \\ 0 & 0 & 0 & 1 \end{bmatrix}$$

$$(3.1.18)$$

The corresponding entries in the matrix Equation 3.1.16 or in Equations 3.1.17 and 3.1.18 are set equal. In this case, it is convenient to choose the entries (1,4) and (2,4) on both sides of Equation 3.1.16. Using Equation 3.1.18, and matrix A_1^2 in Equation 3.1.17, these elements give the following equations:

$$c_1 \bar{p}_{x_0} + s_1 \bar{p}_{y_0} - \ell_1 = \ell_2 c_2 \qquad (3.1.19)$$

$$-s_1 \bar{p}_{x_0} + c_1 \bar{p}_{y_0} = \ell_2 s_2 \qquad (3.1.20)$$

Equations 3.1.19 and 3.1.20 must then be solved for θ_1 and θ_2.

By forming the square on both sides of Equations 3.1.19 and 3.1.20, and then adding the respective sides, the unknown θ_2 is eliminated. The result can be written as

$$(2\ell_1 \bar{p}_{x_0})c_1 + (2\ell_1 \bar{p}_{y_0})s_1 = \bar{p}_{x_0}^2 + \bar{p}_{y_0}^2 + \ell_1^2 - \ell_2^2 \qquad (3.1.21)$$

Equation 3.1.21 is solved for θ_1 by the standard method: Let $2\ell_1 \bar{p}_{x_0} = r\cos\rho$, and $2\ell_1 \bar{p}_{y_0} = r\sin\rho$, where $r = 2\ell_1\sqrt{(\bar{p}_{x_0}^2 + \bar{p}_{y_0}^2)}$, and $\rho = \tan^{-1}(\bar{p}_{y_0}/\bar{p}_{x_0})$. Equation 3.1.21 can then be expressed as:

$$r\cos(\theta_1 - \rho) = \bar{p}_{x_0}^2 + \bar{p}_{y_0}^2 + \ell_1^2 - \ell_2^2 \qquad (3.1.22)$$

where the equality $c_1 \cos\rho + s_1 \sin\rho = \cos(\theta_1 - \rho)$ has been employed. Equation 3.1.22 can be rewritten in terms of a tan-function to avoid possible numerical problems if small numbers appear in the denominator of division. The resulting equation is solved for θ_1:

$$\theta_1 = \tan^{-1}\left(\frac{\bar{p}_{y_0}}{\bar{p}_{x_0}}\right) + \tan^{-1}\left[\frac{\pm\sqrt{4\ell_1^2(\bar{p}_{x_0}^2 + \bar{p}_{y_0}^2) - (\bar{p}_{x_0}^2 + \bar{p}_{y_0}^2 + \ell_1^2 - \ell_2^2)^2}}{(\bar{p}_{x_0}^2 + \bar{p}_{y_0}^2 + \ell_1^2 - \ell_2^2)}\right] + k\pi$$

$$(3.1.23)$$

where $k = 0, \pm 1, \ldots$ In many cases, the primary branch, or $k = \pm 1$ is used for determining the joint angle. Knowing the joint variable θ_1, the other unknown variable θ_2 can be solved by dividing Equation 3.1.20 by Equation 3.1.19:

$$\theta_2 = \tan^{-1}\left(\frac{-\bar{p}_{x_0}s_1 + \bar{p}_{y_0}c_1}{\bar{p}_{x_0}c_1 + \bar{p}_{y_0}s_1 - \ell_1}\right) + k\pi \qquad (3.1.24)$$

Equations 3.1.23 and 3.1.24 establish the inverse solution to the kinematic equations. They are obtained by solving the kinematic equations in the backward direction.

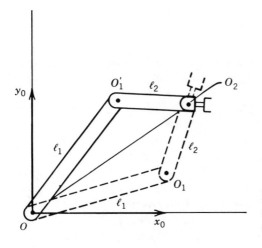

FIGURE 3.2
Location of point O_2 is the same when joint 2 is in the down-position.

Equations 3.1.23 and 3.1.24 reveal that the inverse solution to the kinematic equations is not unique. When this occurs, the inverse solution to the kinematic equations is said to be *degenerate* (e.g., 1). Thus, the same position of the end-effector can be obtained by having joint 2 in the down- or up-position (point O, or O'_1, respectively), which is determined in Figure 3.1 by the position of point O_1 or O'_1 relative to the vector from O to O_2. It is demonstrated schematically in Figure 3.2. In the sequel $k\pi$ indicates degeneracy in the joint angle solution.

The procedure described here can be applied to determine the inverse solution to the kinematic equations of multiple joint manipulators, such as the Stanford/JPL arm. It will be presented in Section 3.3. First we will present a simpler case: the inverse kinematic equations for the roll, pitch, and yaw angles.

---

3. 2 AN INVERSE KINEMATIC SOLUTION: ROLL-PITCH-YAW ANGLES FROM RECTANGULAR COORDINATES

A Cartesian $x_0 y_0 z_0$-coordinate system for a manipulator may be chosen so that the origin is located at the base of the manipulator. A local rectangular coordinate frame of the end-effector can be placed so that its origin is at the center of the gripper. In a six-joint manipulator, this hand coordinate frame is the sixth coordinate frame with the unit vectors $\mathbf{n} = \mathbf{i}_{x6}$, $\mathbf{s} = \mathbf{j}_{y6}$, and $\mathbf{a} = \mathbf{k}_{z6}$. The selection of these two coordinate systems is illustrated for a Stanford/JPL arm in Figure 2.4, in which all coordinates frames are shown. The components of the unit vectors in the sixth coordinate frame can be projected to the axes of the base coordinate system; for example, the n-vector has components n_{x_0}, n_{y_0}, and n_{z_0} on the x_0-, y_0-, and z_0-axes, respectively.

The problem to be addressed is to determine the roll, pitch, and yaw angles (i.e., ψ_z, ψ_y, and ψ_x of Section 2.3.2) for the orientation of the gripper when the components of the unit vectors n, s, and a are given in the $x\,y\,z$-coordinate frame. For convenience, the origin of the sixth coordinate frame is assumed to coincide with that of the $x\,y\,z$-coordinate system. The axes of the $x\,y\,z$-coordinates are parallel with the corresponding axes of the base coordinate system. The RPY-coordinates result after the roll, pitch, and yaw operations have been performed with respect to the $x\,y\,z$-coordinates. The posed problem implies that the equations given in Table 2.5 are to be solved for ψ_x, ψ_y, and ψ_z when the components of the n-, s-, and a-vectors are known. In solving these trigonometric equations, numerical difficulties should be anticipated, if $\sin(\,\cdot\,)$- and $\cos(\,\cdot\,)$-expressions are used. Therefore, the solutions for the angles should be expressed in terms $\tan^{-1}(\,\cdot\,)$ to avoid possible problems that division by a very small number could cause.

To solve the posed inverse kinematic problem, matrix RPY given in Equation 2.3.15 is equated with the homogeneous transformation matrix T given in Equation 2.3.9. The resulting equations are rewritten here as a matrix equation:

$$
\begin{bmatrix}
\cos\psi_z\cos\psi_y & \cos\psi_z\sin\psi_y\sin\psi_x - \sin\psi_z\cos\psi_x & \cos\psi_z\sin\psi_y\cos\psi_x + \sin\psi_z\sin\psi_x & 0 \\
\sin\psi_z\cos\psi_y & \sin\psi_z\sin\psi_y\sin\psi_x + \cos\psi_z\cos\psi_x & \sin\psi_z\sin\psi_y\cos\psi_x - \cos\psi_z\sin\psi_x & 0 \\
-\sin\psi_y & \cos\psi_y\sin\psi_x & \cos\psi_y\cos\psi_x & 0 \\
0 & 0 & 0 & 1
\end{bmatrix}
=
\begin{bmatrix}
n_x & s_x & a_x & p_x \\
n_y & s_y & a_y & p_y \\
n_z & s_z & a_z & p_z \\
0 & 0 & 0 & 1
\end{bmatrix}
\quad (3.2.1)
$$

Since the corresponding elements on both sides must be equal, Equation 3.2.1 contains 12 equalities, when the fourth row of obvious identities is not counted. Six of these equations are independent, and three of them determine the position (translation) vector to be zero. Thus, three independent equations will specify the orientation.

The solution for ψ_x, ψ_y, and ψ_z in terms of the components of n, s, and a will be obtained by equating the corresponding elements in Equation 3.2.1. To solve for ψ_x, the respective elements (3,2) and (3,3) on both sides of Equation 3.2.1 are set equal to obtain the following equations (also contained in the equations of Table 2.5):

$$
s_z = \cos\psi_y\sin\psi_x
$$

$$
a_z = \cos\psi_y\cos\psi_x \quad (3.2.2)
$$

By dividing both sides of Equations 3.2.2, angle ψ_x can be calculated:

$$\psi_x = \tan^{-1}\left(\frac{s_z}{a_z}\right) + k\pi \tag{3.2.3}$$

Equation 3.2.3 determines the yaw angle.

To solve for the pitch angle ψ_y, the respective elements (1,1) and (2,1) in Equation 3.2.1 are set equal. By squaring both sides of these equations and adding them, one obtains an equation for ψ_y; it is augmented by an additional equation resulting from equating elements (3,1) in Equation 3.2.1:

$$n_x^2 + n_y^2 = \cos^2 \psi_y \tag{3.2.4}$$

$$n_z = -\sin \psi_y \tag{3.2.5}$$

By taking the square root of both sides in Equation 3.2.4, and dividing the resulting expression by Equation 3.2.5, the following equation is obtained:

$$\psi_y = \tan^{-1}\left[\frac{-n_z}{\pm\sqrt{n_x^2 + n_y^2}}\right] + k\pi \tag{3.2.6}$$

Equation 3.2.6 specifies the pitch angle.

Finally, elements (1,1) and (2,1) in Equation 3.2.1 are equated to obtain equations for n_x and n_y that can be solved for ψ_z:

$$\psi_z = \tan^{-1}\left[\frac{n_y}{n_x}\right] + k\pi \tag{3.2.7}$$

Equation 3.2.7 determines the roll angle.

Equations 3.2.3, 3.2.6, and 3.2.7 determine the angles for roll ψ_z, pitch ψ_y, and yaw ψ_x corresponding to the given components of the unit vectors n, s, and a in the xyz-coordinate system in a closed form. The degeneracy of the solution indicated in Equations 3.2.3, 3.2.6, and 3.2.7 corresponds to a flip or no-flip orientation of the wrist [1,2].

The roll, pitch, and yaw angles can play an important role in specifying the orientation of the end-effector. When the homogeneous transformation T_0^6 specified by Equation 2.3.9 is applied to a vector $p_6 = [p_{x_6}\, p_{y_6}\, p_{z_6}\, 1]'$ given in the end-effector (sixth) coordinate frame, the resulting vector p_0 is expressed in the zeroth (base) coordinate system as follows:

$$p_0 = T_0^6 p_6 = \begin{bmatrix} n_{x_0} & s_{x_0} & a_{x_0} & \bar{p}_{x_0} \\ n_{y_0} & s_{y_0} & a_{y_0} & \bar{p}_{y_0} \\ n_{z_0} & s_{z_0} & a_{z_0} & \bar{p}_{z_0} \\ 0 & 0 & 0 & 1 \end{bmatrix} \begin{bmatrix} p_{x_6} \\ p_{y_6} \\ p_{z_6} \\ 1 \end{bmatrix} \tag{3.2.8}$$

For example, the approach vector p_6^a at the parking position becomes p_0^a in the base coordinate system. It can serve as the reference vector when the orientation is determined at any later time during the motion.

EXAMPLE 3.2.1

The positions of the joints in a Stanford/JPL arm relative to the reference positions are as follows: $\theta_1 = \pi/6$ rad, $\theta_2 = \pi/4$ rad, $d_3 = 0$ cm, $\theta_4 = \pi/2$ rad, $\theta_5 = \pi/2$ rad, and $\theta_6 = \pi/2$ rad. The problem is to specify the orientation of the end-effector by calculating the roll, pitch, and yaw angles for the gripper, that is, ψ_z, ψ_y, and ψ_x.

Using the expressions in Table 2.4 and the given values of the joint variables, the rotation submatrix in the homogeneous transformation matrix T_0^6 can be determined:

$$
T_0^6 = \begin{bmatrix}
-s_1 & c_1 s_2 & c_1 c_2 & \bar{p}_{x_0} \\
c_1 & s_1 s_2 & s_1 c_2 & \bar{p}_{y_0} \\
0 & c_2 & -s_2 & \bar{p}_{z_0} \\
0 & 0 & 0 & 1
\end{bmatrix}
$$

$$
= \begin{bmatrix}
-0.500 & 0.612 & 0.612 & \bar{p}_{x_0} \\
0.867 & 0.353 & 0.353 & \bar{p}_{y_0} \\
0 & 0.707 & -0.707 & \bar{p}_{x_0} \\
0 & 0 & 0 & 1
\end{bmatrix} \tag{3.2.9}
$$

Equation 3.2.9 specifies the components of vectors n, s, a, and $\bar{p}_0$, which form the columns of matrix T_0^6 in Equation 2.3.11.

The roll, pitch, and yaw angles can now be computed on the basis of Equations 3.2.7, 3.2.6, and 3.2.3, respectively. They are given in radians by the following equations:

$$
\psi_x = \tan^{-1}\left[\frac{0.707}{-0.707}\right] + k\pi = \frac{3}{4}\pi + k\pi
$$

$$
\psi_y = \tan^{-1}\left[\frac{0}{\pm\sqrt{(-0.500)^2 + (0.867)^2}}\right] + k\pi = 0 + k\pi \tag{3.2.10}
$$

$$
\psi_z = \tan^{-1}\left[\frac{0.867}{-0.500}\right] + k\pi = \frac{2}{3}\pi + k\pi
$$

The expressions in Equation 3.2.10 specify the angles of the roll (ψ_z), pitch (ψ_y), and yaw (ψ_x) for the orientation of the end-effector relative to the $x\,y\,z$-coordinate frame whose axes are parallel with those of the base coordinate system.

3.3 INVERSE SOLUTION TO KINEMATIC EQUATIONS OF STANFORD/JPL MANIPULATOR

When the end-effector of a manipulator is to be placed at a certain location specified in the base coordinate system, it is accomplished by making the joint variables assume the values that result in the desired position of the end-effector. In order to compute these values, the coordinates of the end-effector expressed in the base coordinate system are substituted into the kinematic equations. Then, the mathematical problem is to solve for the values of the joint variables in the equations. The determination of explicit expressions for the joint variables in terms of the position and orientation of the end-effector given in the base coordinate system will be discussed for a Stanford/JPL manipulator, that is, the solution to the backward kinematic equations for the six-joint manipulator will be given.

A Stanford/JPL manipulator with the D-H-coordinate frames is shown schematically in Figure 2.4. The homogeneous coordinate transformations A_{i-1}^i, $i = 1, \ldots, 5, 6$, between two adjacent coordinate frames for the system in Figure 2.3 are given in Table 2.3. The basic equation to be used in obtaining the inverse solution to the kinematic equations results from combining Equations 2.3.11 and 2.3.12:

$$T_0^6 = \begin{bmatrix} n_x & s_x & a_x & p_x \\ n_y & s_y & a_y & p_y \\ n_z & s_z & a_z & p_z \\ 0 & 0 & 0 & 1 \end{bmatrix}$$

$$= A_0^1 A_1^2 A_2^3 A_3^4 A_4^5 A_5^6 \tag{3.3.1}$$

where the terms on the right-hand side of Equation 3.3.1 are specified in terms of the structural kinematic parameters and the joint variables. They are given in Tables 2.4a and 2.4b. The entries of the T_0^6-matrix in the case under consideration are known.

To solve for the joint positions θ_i appearing in the transformation matrices, Equation 3.3.1 will be used repeatedly to isolate the unknown variable to be determined to the left side of the equation, as demonstrated previously in Example 3.1.3. After both sides of Equation 3.3.1 are premultiplied by the inverse of the matrix appearing as the first term in the product on the right side of Equation 3.3.1, such an equation is obtained. This approach isolates the unknown joint variable to the left side of the equation, and the right side is manipulated so that it is expressed in terms of known variables [1–3].

The first matrix equation is obtained by premultiplying both sides of Equation 3.3.1 by $(A_0^1)^{-1}$. It leads to

$$(A_0^1)^{-1} T_0^6 = A_1^2 A_2^3 A_3^4 A_4^5 A_5^6 \tag{3.3.2}$$

The inverse matrix $(A_0^1)^{-1}$ is written by applying the procedure described in Section 3.1.2 to expression A_0^1 given in Table 2.3. Hence, Equation 3.1.8 gives

$$(A_0^1)^{-1} = \begin{bmatrix} c_1 & s_1 & 0 & 0 \\ 0 & 0 & -1 & d_1 \\ -s_1 & c_1 & 0 & 0 \\ 0 & 0 & 0 & 1 \end{bmatrix} \tag{3.3.3}$$

The product on the left-hand side of Equation 3.3.1 can be determined:

$$(A_0^1)^{-1} T_0^6 = \begin{bmatrix} n_x c_1 + n_y s_1 & s_x c_1 + s_y s_1 & a_x c_1 + a_y s_1 & p_x c_1 + p_y s_1 \\ -n_z & -s_z & -a_z & -p_z + d_1 \\ -n_x s_1 + n_y c_1 & -s_x s_1 + s_y c_1 & -a_x s_1 + a_y c_1 & -p_x s_1 + p_y c_1 \\ 0 & 0 & 0 & 1 \end{bmatrix} \tag{3.3.4}$$

where the components of vectors n, s and a are expressed in the Cartesian base coordinate system. The only unknown variable in Equation 3.3.4 is the joint angle θ_1.

The matrix product occurring on the right side of Equation 3.3.2 is given in Table 2.3b. By setting the corresponding entries (3,4) and (3,3) equal on both sides of Equation 3.3.2, one obtains

$$-p_x s_1 + p_y c_1 = d_2 - d_6 c_4 s_5 \tag{3.3.5}$$

$$-a_x s_1 + a_y c_1 = -c_4 s_5 \tag{3.3.6}$$

Substituting $c_4 s_5$ from Equation 3.3.6 into Equation 3.3.5 and regrouping the terms gives

$$(a_x d_6 - p_x)s_1 - (a_y d_6 - p_y)c_1 = d_2 \tag{3.3.7}$$

Equation 3.3.7 is solved for θ_1 by the standard procedure: if $a_x d_6 - p_x = r \cos \gamma$, $a_y d_6 - p_y = r \sin \gamma$, then

$$r = \sqrt{(a_x d_6 - p_x)^2 + (a_y d_6 - p_y)^2} \qquad \gamma = \tan^{-1}\left[\frac{a_y d_6 - p_y}{a_x d_6 - p_x}\right] \tag{3.3.8}$$

It follows that Equation 3.3.7 can be expressed as

$$r \sin(\theta_1 - \gamma) = d_2 \tag{3.3.9}$$

A right triangle with a hypotenuse equals r, leg d_2, and an acute angle $\theta_1 - \gamma$, gives the expression for $\tan(\theta_1 - \gamma)$ that is solved for θ_1:

$$\theta_1 = \gamma + \tan^{-1}\left|\frac{d_2}{\pm\sqrt{r^2 - d_2^2}}\right| + k\pi$$

$$\tag{3.3.10}$$

$$= \tan^{-1}\left(\frac{a_y d_6 - p_y}{a_x d_6 - p_x}\right) + \tan^{-1}\left|\frac{d_2}{\pm\sqrt{r^2 - d_2^2}}\right| + k\pi$$

Equation 3.3.10 specifies the angular position θ_1 of joint (link) 1.

To determine θ_2, the respective entries (1,4) and (1,3) on both sides of Equation 3.3.2 are successively equated, again using the expressions in Table 2.3b:

$$p_x c_1 + p_y s_1 = d_3 s_2 + d_6(s_2 c_5 + c_2 s_4 s_5) \tag{3.3.11}$$

$$a_x c_1 + a_y s_1 = s_2 c_5 + c_2 s_4 s_5 \tag{3.3.12}$$

Similarly, the respective elements (2,3) and (2,4) of the matrices in Equation 3.3.2 give

$$a_z = c_2 c_5 - s_2 s_4 s_5 \tag{3.3.13}$$

$$p_z - d_1 = d_3 c_2 + d_6(c_2 c_5 - s_2 s_4 s_5) \tag{3.3.14}$$

Equation 3.3.12 is used to substitute $s_2 c_5 + c_2 s_4 s_5$ into Equation 3.3.11, and Equation 3.3.13 to substitute for $c_2 c_5 - s_2 s_4 s_5$ in Equation 3.3.14. The resulting equations for s_2 and c_2 can be written as

$$d_3 s_2 = (p_y - a_y d_6)s_1 + (p_x - a_x d_6)c_1 \tag{3.3.15}$$

$$d_3 c_2 = p_z - d_1 - a_z d_6 \tag{3.3.16}$$

Since θ_1 is already known, Equations 3.3.15 and 3.3.16 can now be solved for θ_2 after forming $\tan \theta_2$, and for d_3 after multiplying Equation 3.3.15 by s_2 and Equation 3.3.16 by c_2:

$$\theta_2 = \tan^{-1}\left[\frac{(p_y - a_y d_6)s_1 + (p_x - a_x d_6)c_1}{p_z - d_1 - a_z d_6}\right] + k\pi \tag{3.3.17}$$

$$d_3 = \left[(p_y - a_y d_6)s_1 + (p_x - a_x d_6)c_1\right]s_2 + (p_z - d_1 - a_z d_6)c_2 \tag{3.3.18}$$

Equation 3.3.17 specifies the angular position θ_2 of joint (link) 2, and Equation 3.3.18 gives position d_3 of the prismatic link 3.

To determine θ_4, Equation 3.3.2 is premultiplied by $(A_1^2)^{-1}$:

$$(A_1^2)^{-1}(A_0^1)^{-1}T_0^6 = A_2^3 A_3^4 A_4^5 A_5^6 \tag{3.3.19}$$

where $(A_1^2)^{-1}(A_0^1)^{-1}T_0^6 = (A_0^1 A_1^2)^{-1}T_0^6$, and the inverse of A_1^2 equals

$$(A_1^2)^{-1} = \begin{bmatrix} c_2 & s_2 & 0 & 0 \\ 0 & 0 & 1 & -d_2 \\ s_2 & -c_2 & 0 & 0 \\ 0 & 0 & 0 & 1 \end{bmatrix} \tag{3.3.20}$$

The left-hand side of Equation 3.3.19 can be calculated by premultiplying Equation 3.3.4 by $(A_1^2)^{-1}$:

$$(A_1^2)^{-1}\left[(A_0^1)^{-1}T_0^6\right] = \begin{bmatrix} c_2(n_x c_1 + n_y s_1) - n_z s_2 & c_2(s_x c_1 + s_y s_1) - s_z s_2 \\ -n_x s_1 + n_y c_1 & -s_x s_1 + s_y c_1 \\ s_2(n_x c_1 + n_y s_1) + n_z c_2 & s_2(s_x c_1 + s_y s_1) + s_z c_2 \\ 0 & 0 \end{bmatrix}$$

$$\begin{bmatrix} c_2(a_x c_1 + a_y s_1) - a_z s_2 & c_2(p_x c_1 + p_y s_1) + s_2(-p_z + d_1) \\ -a_x s_1 + a_y c_1 & -p_x s_1 + p_y c_1 - d_2 \\ s_2(a_x c_1 + a_y s_1) + a_z c_2 & s_2(p_x c_1 + p_y s_1) + c_2(p_z - d_1) \\ 0 & 1 \end{bmatrix} \quad (3.3.21)$$

The matrix product on the right side of Equation 3.3.19 is given in Table 2.3b. In this matrix, elements (1,3) and (2,3) are equated with the corresponding entries of the matrix product on the right side of Equation 3.3.21. Thus

$$-a_z s_2 + (a_x c_1 + a_y s_1) c_2 = s_4 s_5 \quad (3.3.22)$$

$$a_x s_1 - a_y c_1 = c_4 s_5 \quad (3.3.23)$$

Equations 3.3.22 and 3.3.23 can be solved for $\tan \theta_4$; thus, we have

$$\theta_4 = \tan^{-1}\left[\frac{-a_z s_2 + (a_x c_1 + a_y s_1) c_2}{a_x s_1 - a_y c_1}\right] + k\pi \quad (3.3.24)$$

Equation 3.3.24 specifies the angular position θ_4 of joint 4.

To determine the joint angle θ_5 corresponding to the given position and orientation of the gripper, Equation 3.3.19 is premultiplied successively by $(A_2^3)^{-1}$ and $(A_3^4)^{-1}$, that is

$$(A_3^4)^{-1}[(A_2^3)^{-1}(A_1^2)^{-1}(A_0^1)^{-1}T_0^6| = A_4^5 A_5^6 \quad (3.3.24)$$

where $(A_3^4)^{-1}(A_2^3)^{-1}(A_1^2)^{-1}(A_0^1)^{-1} = (A_0^1 A_1^2 A_2^3 A_3^4)^{-1}$ and

$$(A_2^3)^{-1} = \begin{bmatrix} 0 & -1 & 0 & 0 \\ 1 & 0 & 0 & 0 \\ 0 & 0 & 1 & -d_3 \\ 0 & 0 & 0 & 1 \end{bmatrix} \quad (A_3^4)^{-1} = \begin{bmatrix} c_4 & s_4 & 0 & 0 \\ 0 & 0 & -1 & 0 \\ -s_4 & c_4 & 0 & 0 \\ 0 & 0 & 0 & 0 \end{bmatrix}$$

$$(3.3.26)$$

Equating elements (1,3) and (2,3) of Equation 3.3.25 with those of $A_4^5 A_5^6$ given in Table 2.3b leads to

$$(a_x s_1 - a_y c_1) c_4 + \left[(a_x c_1 + a_y s_1) c_2 - a_z s_2\right] s_4 = s_5 \quad (3.3.27)$$

$$(a_x c_1 + a_y s_1) s_2 + a_z c_2 = c_5 \quad (3.3.28)$$

Equations 3.3.27 and 3.3.28 can be solved for θ_5:

$$\theta_5 = \tan^{-1}\left[\frac{(a_x s_1 - a_y c_1) c_4 + [(a_x c_1 + a_y s_1) c_2 - a_z s_2] s_4}{(a_x c_1 + a_y s_1) s_2 + a_z c_2}\right] + k\pi \quad (3.3.29)$$

Equation 3.3.29 specifies the joint angle θ_5.

To determine the sixth joint angle θ_6, Equation 3.3.25 is written as follows:

$$(A_4^5)^{-1}[(A_3^4)^{-1}(A_2^3)^{-1}(A_1^2)^{-1}(A_0^1)^{-1}T_0^6| = A_5^6 \quad (3.3.30)$$

where $(A_3^4)^{-1}(A_2^3)^{-1}(A_1^2)^{-1}(A_0^1)^{-1} = (A_0^1 A_1^2 A_2^3 A_3^4)^{-1}$ and

$$(A_4^5)^{-1} = \begin{bmatrix} c_5 & s_5 & 0 & 0 \\ 0 & 0 & 1 & 0 \\ s_5 & -c_5 & 0 & 0 \\ 0 & 0 & 0 & 1 \end{bmatrix}$$ (3.3.31)

The respective entries (1,2) and (2,2) on both sides of Equation 3.3.30 are set equal; these equations are

$$s_6 = s_5 [s_2(s_x c_1 + s_y s_1) + s_z c_2] - c_5 \{c_4(s_x s_1 - s_y c_1) + s_4[c_2(s_x c_1 + s_y s_1) - s_z s_2]\} = R_s$$ (3.3.32)

$$c_6 = s_4(-s_x s_1 + s_y c_1) + c_4[c_2(s_x c_1 + s_y s_1) - s_z s_2] = R_c$$ (3.3.33)

where the variables R_s and R_c are introduced for the notational convenience. Equations 3.3.32 and 3.3.33 can be solved for θ_6; thus

$$\theta_6 = \tan^{-1}(\frac{R_s}{R_c}) + k\pi$$ (3.3.34)

where R_s and R_c are defined in Equations 3.3.32 and 3.3.33. The expression in 3.3.34 specifies the angular position θ_6 of the sixth joint.

The inverse solution for $\theta_1, \ldots, \theta_6$ of the kinematic equations in Table 2.4 for the Stanford/JPL arm is determined explicitly by Equations 3.3.10, 3.3.17, 3.3.18, 3.3.24, 3.3.29, and 3.3.34, when the position and orientation of the end-effector are known in the base coordinates.

The foregoing inverse solution is degenerate. Indeed, it is not unique in general unless additional constraints are imposed, for example, on the ranges of the positional angles. As discussed in Example 3.1.3, the same position and orientation of the end-effector can be achieved for two or more values of the joint angles. By comparing the geometry of the manipulator configuration with that of a human, a shoulder, an elbow, and a wrist may be identified. Consequently, we may consider the configuration of a robot arm to be righty or lefty, which bears resemblance to the right or left arm in a human. The elbow can be up or down, depending on its position relative to the line from the elbow to the wrist. Moreover, the wrist can assume a flip or a no-flip orientation, which is determined by its bending. Depending on the application, a systematic approach to determine the desired configuration can be used [1,2]. In specific problems such as in collision avoidance and path planning, degenerate solutions may offer attractive alternatives as acceptable solutions.

A numerical example is next presented to illustrate the determination of the inverse kinematic solution discussed.

EXAMPLE 3.3.1

The initial position $p_{06}(t_1)$ of the end-effector of a Stanford/JPL arm is given at t_1 in the Cartesian base coordinate system and expressed in inches: $p_{06}(t_1) = [42.5 \quad 2.4 \quad 23.2 \quad 1]'$. The orientation of the gripper is given in radians as RPY($\psi_{z_0}, \psi_{y_0}, \psi_{x_0}$) = RPY(0 0 0), that is, $\psi_{06}(t_1) = [0\,0\,0]'$, relative to a coordinate frame whose axes are parallel in the corresponding axes of the base coordinate system. The gripper moves to the new position $p_{06}(t_2) = [-12.3 \quad -10.5 \quad 20.0]'$ in time $t_2 - t_1$. The orientation of the gripper rotates $\pi/2$ rad about the axis parallel with the z_0-axis during the given time, and there is no change in the rotational angles about the axes parallel with the x_0- and y_0-axes. The problem is to determine the values of the joint variables at times t_1 and t_2.

In order to apply the Equations of Section 3.3 for the inverse kinematics, the homogeneous transformation matrix $T_0^6(t)$ is first determined at times t_1 and t_2. The position vectors $p_{06}(t_1)$ and $p_{06}(t_2)$ specify the fourth columns in the two matrices. The elements in the rotation submatrices are calculated on the basis of Equation 2.3.15 or Table 2.5 by substituting the angles of roll, pitch, and yaw into the RPY-matrix. Then, the homogeneous transformation matrices at t_1 and t_2 become

$$T_0^6(t_1) = \begin{bmatrix} 1 & 0 & 0 & 42.5 \\ 0 & 1 & 0 & 2.4 \\ 0 & 0 & 1 & 23.2 \\ 0 & 0 & 0 & 1 \end{bmatrix} \qquad T_0^6(t_2) = \begin{bmatrix} 0 & -1 & 0 & -12.3 \\ 1 & 0 & 0 & -10.5 \\ 0 & 0 & 1 & 20.0 \\ 0 & 0 & 0 & 1 \end{bmatrix} \quad (3.3.35)$$

The values of the joint variables $\theta_i(t_1)$ in degrees, $d_3(t_1)$ in inches at time t_1, and similarly $\theta_i(t_2)$ and $d_3(t_2)$ at time t_2, $i = 1, 2, 4, 5, 6$ can now be calculated from Equations 3.3.10, 3.3.17, 3.3.18, 3.3.24, 3.3.29, and 3.3.34. The results at the respective times are:

$$\begin{bmatrix} \theta_1 & \theta_2 & d_3 & \theta_4 & \theta_5 & \theta_6 \end{bmatrix}_{t_1}$$
$$= \begin{bmatrix} +23.838 & -41.152 & -60.548 & +90.000 & +41.152 & -38.602 \end{bmatrix} \quad (3.3.36)$$
$$\begin{bmatrix} \theta_1 & \theta_2 & d_3 & \theta_4 & \theta_5 & \theta_6 \end{bmatrix}_{t_2}$$
$$= \begin{bmatrix} -26.844 & +7.280 & -49.187 & -90.000 & +7.280 & -62.738 \end{bmatrix} \quad (3.3.37)$$

Thus, the values of the joint variables at the given times have been determined.

We have demonstrated in this section the determination of the explicit inverse solutions to the kinematic equations of the Stanford/JPL manipulator. The basic approach described can be applied to a certain class of other serial link multijoint manipulators such as PUMA 560 and 600 [1,2,3]. If an explicit inverse solution to the kinematic equations cannot be obtained, numerical solutions by means of iterative algorithms for solving nonlinear equations may be applied [4]. Alternatively, the kinematic model of a manipulator may be decomposed into two subsystems so that an iterative numerical method can be applied to determine some of the joint variables while the other variables are determined in closed forms [5].

The principal uses of the inverse solutions to the kinematic equations occur in the motion planning of a manipulator when the desired Cartesian coordinates of the end-effector must be expressed in terms of the joint variables.

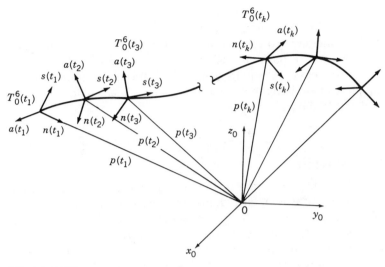

FIGURE 3.3

Sequence $\{ T_0^6(t_k) \}$ determines a path for the end-effector.

This is necessary when the motion of a manipulator is controlled in the joint space. Usually, the calculations are performed off-line, that is, before the motion is executed. However, if the desired motion of the end-effector is not known in advance, these calculations may have to be performed on-line (in real time), in fact, possibly at each sampling instant. This represents a considerable computational burden to a computer.

Since the homogeneous transformation matrix T_0^6 specifies the position and orientation of the end-effector, it can be used to describe the sequence of desired trajectory points for the end-effector to design control strategies. In particular, suppose that the homogeneous transformation matrix $T_0^6(t_k)$ given at time $t_k, k = 1, 2, \ldots$, is used to specify the desired trajectory for the end-effector of a manipulator. It is illustrated in Figure 3.3. It follows that the difference $\Delta T_0^6(t_k)$ in the homogeneous transformation matrices at times t_{k+1} and t_k is $\Delta T_0^6(t_k) = T_0^6(t_{k+1}) - T_0^6(t_k)$. This difference matrix $\Delta T_0^6(t_k)$ contains the changes in the position $\Delta p(t_k) = p(t_{k+1}) - p(t_k)$, and in the orientation that can be expressed in terms of vector $\psi(t_k)$ composed of the RPY-angles, that is, $\Delta \psi(t_k) = \psi(t_{k+1}) - \psi(t_k)$. When the time interval $\Delta t_k = t_{k+1} - t_k$ is chosen sufficiently small, the specified changes in the positions and orientations may be used to determine the constant velocities at which the end-effector should be driven over this time interval; in particular, the positional velocity $\dot{p}(t_k) = \Delta p(t_k)/\Delta t_k$, and the orientational velocity $\dot{\psi}(t_k) = \Delta \psi(t_k)/\Delta t_k$. Thus, the constant velocities that lead to the specified changes in the position and orientation during the time interval Δt_k are determined. They can serve as the control variables that can be realized by generating appropriate velocities in the joint variables. We should observe that the described kinematic control scheme does not take into account the dynamics of the manipulator. Therefore, it may be applied only at the slow or moderate values of the speeds and accelerations of the manipulator.

The foregoing kinematic control requires that the actual Cartesian velocities corresponding to the position and orientation of the end-effector in a manipulator be converted into joint velocities at the given time instants. The exact equations describing the relationships for such calculations will be discussed in detail in the next chapter.

3.4 SUMMARY

When the position and orientation of the end-effector are specified in the base coordinate system, it is often required to determine the values of the joint variables that give rise to the given coordinates of the end-effector. We have discussed in this chapter the solution to this problem. We have presented a straightforward procedure to calculate the inverse of the homogeneous transformation matrix. Then, we have discussed the relationship between the roll, pitch, and yaw angles and the variables of a rectangular coordinate system. We have given an explicit inverse solution to the kinematic equations for the Stanford/JPL arm. It demonstrates that this solution is not unique unless additional constraints are imposed on the variables. Moreover, the calculations of the inverse solution to the kinematic equations can be time consuming, particularly in medium-sized computers. If the inverse solution must be computed at each sampling instant, special hardware such as VLSI-design can be constructed for the mathematical operations. If the calculations are performed off-line, the time for calculations is not usually of great importance. Although the inverse solution to the kinematic equations are usually needed in the generation of the desired trajectory in the path planning, it can also be used to determine a crude control for manipulator motion at slow or moderate speeds and accelerations. However, the dynamics of the manipulator should be taken into account if high performance in the motion control is to be achieved.

REFERENCES

[1] R. FEATHERSTONE, "Position and Velocity Transformations Between Robot End Effector Coordinates and Joint Angles," *Intl. J. Robot Research*, Vol. 3, No. 2, pp. 35–45, 1983.

[2] S. ELGAZZAR, "Efficient Kinematic Transformations for the PUMA 560 Robot," *IEEE Journal of Robotics and Automation*, Vol. RA-1, No. 3, pp. 142–151, September 1985.

[3] R. P. PAUL, B. SHIMANO, and G. E. MAYER, "Kinematic Control Equations for Simple Manipulators," *IEEE Trans. on Systems, Man, and Cybernetics* (Corresp.), Vol. SMC-11, No. 6, pp. 449–455, June 1981.

[4] A. A. GOLDENBERG, B. BENHABIB, and R.G. FENTON, "A Complete Generalized Solution to the Inverse Kinematics of Robots," *IEEE Journal of Robotics and Automation*, Vol. RA-1, No. 1, pp. 14–20, March 1985.

[5] V. J. LUMELSKY, "Iterative Coordinate Transformation Procedure for One Class of Robots," *IEEE Trans. on Systems, Man, and Cybernetics* (Corresp.), Vol. SMC-14, No. 3, pp. 500–505, May/June 1984.

PROBLEMS

3.1   Show that matrix A^i_{i-1} in Equations 2.2.3 and 2.2.4 is a homogeneous transformation.

3.2   The following matrix is given:

$$T^6_0 = \begin{bmatrix} 0.6 & 0 & -0.8 & 1.2 \\ 0 & 1 & 0 & 1 \\ 0.8 & 0 & 0.6 & 2 \\ 0 & 0 & 0 & 1 \end{bmatrix}$$

    a.   Verify that matrix T^6_0 is a homogeneous transformation matrix.

    b.   Determine the inverse of T^6_0, that is, $(T^6_0)^{-1} = T^0_6$.

3.3   A matrix is given

$$A = \begin{bmatrix} 0 & -0.6 & 0.8 & 2.0 \\ 1.0 & 0 & 0 & 1.5 \\ 0 & 0.8 & 0.5 & 0.8 \\ 0 & 0 & 0 & 1 \end{bmatrix}$$

    a.   Check if matrix A is a homogeneous transformation matrix. If not, give reasons.

    b.   Determine the inverse of matrix A.

3.4   *a.*   Show that matrix T^0_6 in Equation 3.1.8 is or is not a homogeneous transformation matrix.

    b.   Using the expressions of Equations 3.1.6 and 3.1.8, prove that $T^6_0 T^0_6 = I$.

3.5   A planar manipulator with two revolute joints and the end-effector joint is shown schematically in Figure 3.4. Assume that the gripper plates cannot rotate but their opening can be changed.

    a.   Choose a coordinate frame for the joints by following the D-H guidelines.

b.   Write down the homogeneous transformation matrices $A_0^1 A_1^2$, and $A_0^2 = A_0^1 A_1^2$.

c.   Express the position of the end-effector in terms of the joint angles θ_1 and θ_2. For specified coordinates of the end-effector, determine the corresponding expressions for θ_1 and θ_2. This will give you the inverse solution to the kinematic equations.

d.   Suppose that $\ell_1 = \ell_2 = 0.5$ m. The gripper G is to be at the position $\bar{p}_0 = [60 \text{ cm}, 0 \text{ cm}]'$. Determine the angles θ_1 and θ_2 so that the gripper G is at the position $\bar{p}_0$.

e.   Can you determine two different values for θ_1 and θ_2 that result in the same position of the end-effector? Explain the different configurations by drawing the links of the manipulator at the appropriate angles for this degenerate case.

f.   Repeat part *(d)* when $\bar{p}_0 = [0.6 \text{ m}, 0.8 \text{ m}]'$.

3.6   A planar manipulator is shown in Figure 3.4. Assume that the gripper can rotate about the axis that coincides with the axis of the third link.

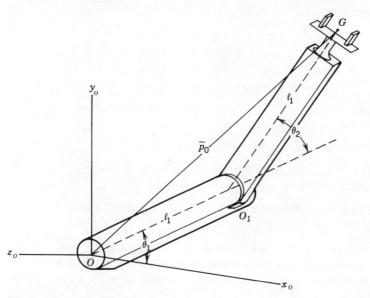

FIGURE 3.4
A planar manipulator for Problem 3.5.

a. Define a coordinate frame for the last joint. Also, define the *sna*-frame at the gripper.

b. Determine the homogeneous transformation matrices A_2^3, A_0^2, and A_0^3.

c. Determine the forward kinematic equations using vectors s, a, n, and $\bar{p}_0$ defined in the base coordinate system.

d. Determine the inverse solution to the kinematic equations, that is, establish the backward kinematic equations by solving for θ_1, θ_2, and θ_3 in terms of the components of vectors s, a, n, and $\bar{p}_0$. Is the solution unique? Explain.

e. How many DOF does this manipulator possess? Discuss.

3.7 The joint variables in a Stanford/JPL (Scheinman) manipulator are: $\theta_1 = -\pi/4$ rad; $\theta_2 = \pi/3$ rad; $d_3 = 10$ cm; $\theta_4 = 0$ rad; $\theta_5 = -\pi/6$ rad; and $\theta_6 = 0$ rad relative to the reference (resting) position.

a. Determine position ($\bar{p}_0$) and orientation (n, s, a) of the end-effector in the base coordinates.

b. Calculate the roll (ψ_z), pitch (ψ_y), and yaw (ψ_x) angles for the position of the gripper in the base coordinates specified by the values given in *(a)* for the joint variables. Thus, specify RPY (ψ_z, ψ_y, ψ_x) for the gripper.

3.8 The kinematic equations for a cylindrical robot are

$$x = r\cos\theta$$

$$y = r\sin\theta$$

$$z = z$$

a. Determine the inverse solution, that is, solve the kinematic equations in the backward direction.

b. Assume that the variables vary with time. Determine $v_x = \dot{x}$, $v_y = \dot{y}$, and $v_z = \dot{z}$. Form the magnitude $|v| = (v_x^2 + v_y^2 + v_z^2)^{1/2}$ of the velocity vector.

c. Express the answer of *(b)* in the form $v = J_p\dot{q}$ where $v = [v_x v_y v_z]'$, and $\dot{q} = [\dot{r}\dot{\theta}\dot{z}]'$, and J_p is a 3×3 matrix.

d. Determine $\dot{q} = J_p^{-1}v$ by calculating the inverse of the matrix.

3.9   A student (engineer) solved the kinematic equations of the planar manipulator in Example 3.1.3 and obtained

$$\theta_2 = \tan^{-1} \left[\frac{\sqrt{4\ell_1^2\ell_2^2 - [(p_x^d)^2 + (p_y^d)^2 - \ell_1^2 - \ell_2^2]^2}}{\pm[(p_x^d)^2 + (p_y^d)^2 - \ell_1^2 - \ell_2^2]} \right] \pm \pi$$

$$\theta_1 = \tan^{-1} \left[\frac{p_y^d}{p_x^d} \right] - \tan^{-1} \left[\frac{\ell_2 \sin \theta_2}{\ell_1 + \ell_2 \cos \theta_2} \right] \pm \pi$$

Compare these expressions with those in Equations 3.1.23 and 3.1.24. Which expressions are the correct solutions? Explain.

3.10   A workstation has a table, a robot, and a camera at fixed locations as shown in Figure 3.5. A Cartesian base coordinate system is defined with the origin at the corner of the table and at the base of the manipulator. A local Cartesian coordinate frame $(x_c \, y_c \, z_c)$ is attached to the stationary camera. Another local Cartesian coordinate frame $(x_p y_p z_p)$ is defined for the part (object) that has been placed on the table as shown in Figure 3.5.

The transformation matrices between the coordinate frames are specified as follows:

$$T_{camera}^{part} = \begin{bmatrix} n_1, \, s_1, \, a_1, \, \bar{p}_1 \\ 0, \, 0, \, 0, \, 1 \end{bmatrix}$$

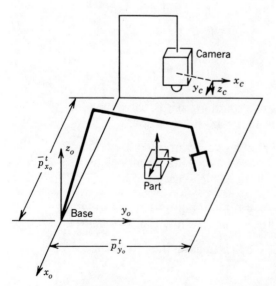

FIGURE 3.5
Manipulator, part, and camera on worktable.

where $s_1 = [1, 0, 0]'$, $a_1 = [0, 0, -1]'$, $p_1 = [1, 10, 9]'$ and

$$T^{\text{base}}_{\text{camera}} = \begin{bmatrix} n_2, & s_2, & a_2, & \bar{p}_2 \\ 0, & 0, & 0, & 1 \end{bmatrix}$$

where $n_2 = [1, 0, 0]'$, $s_2 = [0, -1, 0]'$, $p_2 = [-10, 20, 10]'$. The coordinate frame indicated by the superscript in the transformation matrix T is described in the coordinate system shown as the subscript.

a.   Determine the orientation and position of the part in the base coordinate system by specifying the transformation matrix $T^{\text{part}}_{\text{base}}$.

b.   What are the location and orientation of the coordinate frame on the part (object) in the base coordinate system?

3.11   A planar manipulator with three joints and three links is shown in Figure 3.6. The last joint can be used for determining the orientation of the gripper on the plane. The lengths of the links are: $\ell_1 = 0.5$ m, $\ell_2 = 0.5$ m, and $\ell_3 = 0.25$ m. The end-effector of the manipulator is in the position (0.7 m 0.5 m).

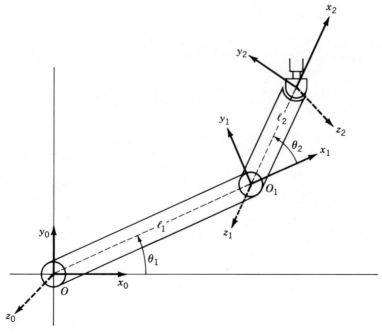

FIGURE 3.6
A three-link planar manipulator.

a.   Determine the values of the joint angles θ_1, θ_2, and
θ_3 that result in the given position of the end-effector.
The orientation is to be kept constant for horizontal
grasping.

b.   Is the general configuration of this manipulator degenerate? Discuss.

COMPUTER PROBLEM

3.C.1   For a Stanford/JPL arm, the position of the end-effector
in the base coordinate system is to be moved from
point P_1 (80.6, 6.1, 41.5) to point $P_2(-24.8, -22.0, 41.2)$
expressed in cm. The orientation of the gripper rotates
$\pi/4$ rad about the x_0-axis, $3\pi/4$ rad about the z_0-axis,
and there is no rotation about the y_0-axis relative to a
resting position use Equation 2.3.14. Assume that initial orientation of the gripper serves as the zero reference, that is, initially $n = [1,0,0]'$, $s = [0,1,0]'$, and
$a = [0,0,1]'$ relative to the base coordinate system.
The structural kinematic parameters have the same
numerical values as those in Computer Problem 2.C.1.

a.   Compute the positions of the joint variables at the
initial point P_1 as well as at the final point P_2.

b.   Assume that the gripper is to move from P_1 to P_2
along a straight line in 8 s. Determine the maximum
magnitude of the velocity vector $v = [v_x \quad v_y \quad v_z]'$ in
the base coordinates needed for this motion under
the assumption that the profile of the velocity has the
shape of an equilateral trapezoid in which the time for
the acceleration lasts 2 s. (Draw the velocity profile.)

CHAPTER FOUR

GENERALIZED VELOCITY AND TORQUE RELATIONS IN JOINT AND BASE COORDINATES

The kinematics of serial link manipulators have been discussed in Chapters 2 and 3. We have described the relationships between the joint variables and the positions of the points on the links, when the robot manipulator is stationary at a specific configuration. When a manipulator is moving to perform a task, the velocities of the points on the links and in particular the speed of the end-effector are of utmost importance, since they often determine the time needed for the completion of an assigned task.

Since the positions, velocities, and accelerations of the joint variables in a manipulator often change with time, the new positions, velocities, and accelerations of the points on the manipulator, and in particular those of the end-effector (hand), must be determined. In this chapter, the velocities (accelerations) of the joint variables will be related to the rotational and translational velocities (accelerations) of the serially connected links in a manipulator. These relations are expressed by means of a Jacobian matrix. Also, a relation between the generalized joint torque and the external generalized force exerted by the end-effector on the environment will be established.

4.1 VELOCITY AND ACCELERATION IN FIXED AND ROTATING COORDINATE SYSTEMS

When the joints of a manipulator move, the points on the links will move at a certain translational (linear) and rotational velocity. These velocities can concretely be described after the coordinate frames have been defined for the manipulator. In the sequel, we assume that necessary coordinate systems have been assigned using the D-H method discussed in Section 2.2. The world coordinate system is assumed to be fixed to the base. The local coordinate frames are attached to the links and move relative to the fixed base coordinate system.

To determine the equations that relate the translational and rotational velocities of the links to the velocities of the joint variables, the kinematic equations can be differentiated with respect to time. The resulting equations indeed describe the sought relationships between the aforementioned velocities. By differentiating the velocity equations again with respect to time, the relations for the corresponding accelerations can be established. Although the approach is straightforward, the resulting equations are usually very lengthy and tedious to obtain for many manipulators. On the other hand, these relationships can be described *recursively* by considering the positions, velocities, and accelerations of the links in two adjacent local coordinate frames that move in the fixed base coordinate system. This formulation leads to equations that are computationally attractive because discrete equations lend themselves to recursive calculations efficiently on a digital computer. We will discuss the approach in the following sections.

4.1.1   TIME DERIVATIVES OF A VECTOR IN ROTATING COORDINATE FRAME

A fixed coordinate system $(x_0 y_0 z_0)$, and a rotating ith coordinate frame $(x_i y_i z_i)$ are assumed to be defined so that their *origins coincide*. When the latter frame rotates about the z_0-axis, the angular velocity of the moving coordinate frame is represented by vector $\boldsymbol{\omega}_{z_0}$ on the z_0-axis of the fixed coordinate system as shown in Figure 4.1a. The position of an arbitrary point P can be expressed in the ith coordinate frame as follows:

$$\mathbf{p}_i = p_{x_i}\mathbf{i}_{x_i} + p_{y_i}\mathbf{j}_{y_i} + p_{z_i}\mathbf{k}_{z_i} \tag{4.1.1}$$

where $\mathbf{i}_{x_i}$, $\mathbf{j}_{y_i}$ and $\mathbf{k}_{z_i}$ are the unit vectors on the x_i-, y_i-, and z_i-axes, respectively. The position of vector $\mathbf{p}_i$ is assumed to change with time, that is, $\mathbf{p}_i = \mathbf{p}_i(t)$.

The time derivative of the $\mathbf{p}_i(t)$-vector is

$$\frac{d\mathbf{p}_i(t)}{dt} = \dot{p}_{x_i}\mathbf{i}_{x_i} + \dot{p}_{y_i}\mathbf{j}_{y_i} + \dot{p}_{z_i}\mathbf{k}_{z_i} + p_{x_i}\frac{d\mathbf{i}_{x_i}}{dt} + p_{y_i}\frac{d\mathbf{j}_{y_i}}{dt} + p_z\frac{d\mathbf{k}_{z_i}}{dt} \tag{4.1.2}$$

where the dot refers to the time derivative of the variable; for example, $\dot{p}_{x_i} = dp_{x_i}/dt$.

In order to evaluate the time derivatives of the unit vectors, let us consider an arbitrary vector $\boldsymbol{\rho}(t)$ that rotates about the z-axis at an angular velocity ω_z, as shown in Figure 4.1b. The definition of the derivative of vector $\boldsymbol{\rho}(t)$ is

$$\frac{d\boldsymbol{\rho}(t)}{dt} = \lim_{\Delta t \to 0} \frac{\boldsymbol{\rho}(t + \Delta t) - \boldsymbol{\rho}(t)}{\Delta t} \tag{4.1.3}$$

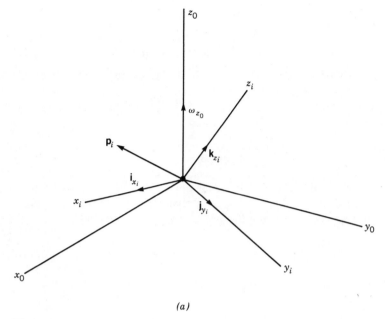

(a)

FIGURE 4.1*a*

Rotating coordinate frame $(x_i\, y_i\, z_i)$ in fixed coordinate system $(x_0\, y_0\, z_0)$.

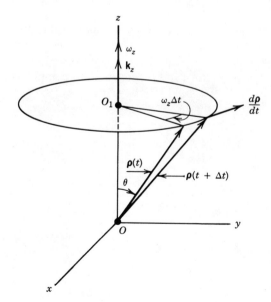

FIGURE 4.1*b*

Vector $\rho\,(\Delta t)$ at times t and $t + \Delta t$ and $d\rho(t)/dt$ in fixed coordinate system.

Figure 4.1b shows that the following relation holds for a sufficiently small time increment Δt:

$$|\boldsymbol{\rho}(t + \Delta t) - \boldsymbol{\rho}(t)| = |(\omega_z \Delta t) \boldsymbol{\rho}(t) \sin \theta| \tag{4.1.4}$$

where θ is the angle between vectors $\boldsymbol{\omega}_z$ and $\boldsymbol{\rho}(t)$. Furthermore, the direction of the vector $[\boldsymbol{\rho}(t + \Delta t) - \boldsymbol{\rho}(t)]$ is perpendicular to vectors $\boldsymbol{\omega}_z$ and $\boldsymbol{\rho}(t)$, as shown in Figure 4.1b. Hence, the derivative of vector $\boldsymbol{\rho}(t)$ can be expressed as the cross-product of vectors $\boldsymbol{\omega}_z$ and $\boldsymbol{\rho}$ as follows:

$$\frac{d\boldsymbol{\rho}(t)}{dt} = \boldsymbol{\omega}_z \times \boldsymbol{\rho} \tag{4.1.5}$$

By applying Equation 4.1.5 to the time derivatives of the unit vectors appearing in Equation 4.1.2, and recalling that the ith coordinate frame rotates about the z_0-axis at the angular velocity $\boldsymbol{\omega}_{z_0}$, one obtains

$$\frac{d\mathbf{i}_{x_i}}{dt} = \boldsymbol{\omega}_{z_0} \times \mathbf{i}_{x_i} \qquad \frac{d\mathbf{j}_{y_i}}{dt} = \boldsymbol{\omega}_{z_0} \times \mathbf{j}_{y_i} \qquad \frac{d\mathbf{k}_{z_i}}{dt} = \boldsymbol{\omega}_{z_0} \times \mathbf{k}_{z_i} \tag{4.1.6}$$

The substitution of the expressions of Equation 4.1.6 into Equation 4.1.2 leads to

$$\frac{d\mathbf{p}_i(t)}{dt} = \frac{d^*\mathbf{p}_i(t)}{dt} + p_{x_i}(\boldsymbol{\omega}_{z_0} \times \mathbf{i}_{x_i}) + p_{y_i}(\boldsymbol{\omega}_{z_0} \times \mathbf{j}_{y_i}) + p_{z_i}(\boldsymbol{\omega}_{z_0} \times \mathbf{k}_{z_i}) \tag{4.1.7}$$

where the star in $d^*\mathbf{p}_i(t)/dt$ signifies the time derivative of $\mathbf{p}_i(t)$ in the moving ith coordinate frame with the unit vectors fixed. This coordinate frame will be referred to as the starred coordinate system. The components of the $\mathbf{p}_i$-vector can be combined with the unit vectors in Equation 4.1.7 to obtain

$$\frac{d\mathbf{p}_i(t)}{dt} = \frac{d^*\mathbf{p}_i(t)}{dt} + \boldsymbol{\omega}_{z_0} \times \mathbf{p}_i \tag{4.1.8}$$

Thus, the time derivative of the $\mathbf{p}_i$-vector in the fixed coordinate system is equal to the time derivative of $\mathbf{p}_i(t)$ in the starred coordinate frame plus the unstarred derivative it would have if it were at rest in the starred system.

The acceleration $d^2\mathbf{p}_i(t)/dt^2$ of the $\mathbf{p}_i$-vector can be determined by applying Equation 4.1.8 to $d\mathbf{p}_i/dt$:

$$
\begin{aligned}
\frac{d^2\mathbf{p}_i(t)}{dt^2} &= \frac{d}{dt}\left(\frac{d\mathbf{p}_i(t)}{dt}\right) \\
&= \frac{d}{dt}\left(\frac{d^*\mathbf{p}_i(t)}{dt}\right) + \boldsymbol{\omega}_{z_0} \times \frac{d\mathbf{p}_i(t)}{dt} + \frac{d\boldsymbol{\omega}_{z_0}}{dt} \times \mathbf{p}_i(t) \\
&= \frac{d^{*2}\mathbf{p}_i(t)}{dt^2} + \boldsymbol{\omega}_{z_0} \times \frac{d^*\mathbf{p}_i(t)}{dt} + \boldsymbol{\omega}_{z_0} \times \left(\frac{d^*\mathbf{p}_i(t)}{dt} + \boldsymbol{\omega}_{z_0} \times \mathbf{p}_i\right) \\
&\quad + \frac{d\boldsymbol{\omega}_{z_0}}{dt} \times \mathbf{p}_i(t) \\
&= \frac{d^{*2}\mathbf{p}_i(t)}{dt^2} + 2\boldsymbol{\omega}_{z_0} \times \frac{d^*\mathbf{p}_i}{dt} + \frac{d\boldsymbol{\omega}_{z_0}}{dt} \times \mathbf{p}_i + \boldsymbol{\omega}_{z_0} \times (\boldsymbol{\omega}_{z_0} \times \mathbf{p}_i) \quad (4.1.9)
\end{aligned}
$$

The first term on the right side of Equation 4.1.9 represents the acceleration relative to the starred coordinate frame. The second term signifies the acceleration of Coriolis. Thus, the Coriolis-effect is dependent on the vector product of the angular velocity $\boldsymbol{\omega}_{z_0}$ and the relative velocity $d^*\mathbf{p}_i/dt$ with respect to the starred (rotating) coordinate frame. The third term determines the acceleration in the direction that is perpendicular to vectors $d\boldsymbol{\omega}_{z_0}/dt$ and $\mathbf{p}_i$. The last term on the right of Equation 4.1.9 describes the centripetal acceleration directed toward the center of the curvature of the motion. It forms a right angle with the $\boldsymbol{\omega}_{z_0}$-vector; moreover, its magnitude equals $\omega_{z_0}^2 p_i \sin \theta_i$, where θ_i is the angle between the $\boldsymbol{\omega}_{z_0}$- and $\mathbf{p}_i$-vectors.

Equations 4.1.8 and 4.1.9 are next applied to determine equations for the angular and translational (linear) velocities of a rotating manipulator arm.

EXAMPLE 4.1.1

In a spherical $(r\theta\psi)$ manipulator shown in Figure 4.2, the arm rotates at a constant speed of 0.30 rad/s about the (vertical) axis of the base. Simultaneously, the arm is being raised with a constant angular velocity of 0.50 rad/s relative to the horizontal line. The length of the arm is equal to 40 cm, and it is kept constant. The task is to determine (a) the actual trans-

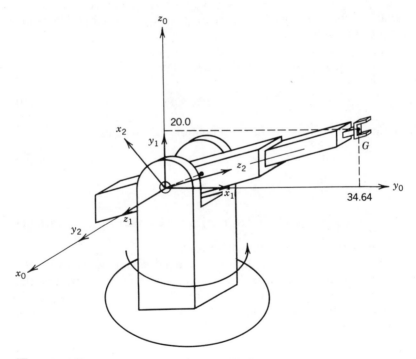

FIGURE 4.2
Spherical manipulator; moving coordinate frames $(x_1 y_1 z_1)$ and $(x_2 y_2 z_2)$ in fixed coordinate system $(x_0 y_0 z_0)$ with origin removed from base axis.

lational (Cartesian) velocity, (b) acceleration of the end-effector, and (c) the actual (Cartesian) angular velocity and acceleration of the arm in the base coordinate system.

To solve the problem, a Cartesian base coordinate system is defined so that the origin is at the top of the base, that is, at the shoulder. This coordinate system $(x_0 y_0 z_0)$ is fixed in the space. Local coordinate frames $(x_1 y_1 z_1)$ and $(x_2 y_2 z_2)$ are next defined following the rules of the D-H method. The coordinate frame $(x_1 y_1 z_1)$ rotates with the arm about the z_0-axis; it represents a moving coordinate frame. The second local coordinate frame $(x_2 y_2 z_2)$ is attached to the arm, and it moves with respect to the $(x_1 y_1 z_1)$-coordinate system. The origins of the first and second coordinate frames coincide with that of the base coordinate system.

The given angular velocity vectors can now be expressed explicitly in rad/s: $\boldsymbol{\omega}_{z_0} = 0.30 \mathbf{k}_{z_0}$ and $\boldsymbol{\omega}_{z_1} = 0.50 \mathbf{k}_{z_1}$. The former velocity specifies the angular speed of the first moving coordinate frame in the base coordinate system. The latter velocity gives the relative angular speed of the second coordinate frame and that of the arm in the $(x_1 y_1 z_1)$-coordinate system.

Suppose that the position $\mathbf{p}_0$ of the gripper G at the time of our consideration is given in the base coordinate system $\mathbf{p}_0 = p_{0x} \mathbf{i}_{x_0} + p_{0y} \mathbf{j}_{y_0} + p_{0z} \mathbf{k}_{z_0} = 34.64 \mathbf{j}_{y_0} + 20.0 \mathbf{k}_{z_0}$ expressed in cm. The same point in the first coordinate frame is $\mathbf{p}_1 = p_{1x} \mathbf{i}_{x_1} + p_{1y} \mathbf{j}_{y_1} + p_{1z} \mathbf{k}_{z_1} = 34.64 \mathbf{i}_{x_1} + 20.0 \mathbf{j}_{y_1}$ in cm. The dimensions of the variables will be dropped in the sequel. Now the posed problems (a), (b), and (c) can be solved.

(a) The actual translational velocity of the gripper in the base coordinate system is determined by using Equation 4.1.8. For $i = 0$,

$$\frac{d \mathbf{p}_0}{dt} = \frac{d^* \mathbf{p}_0}{dt} + \boldsymbol{\omega}_{z_0} \times \mathbf{p}_0 \qquad (4.1.10)$$

The evaluation of the cross-product in Equation 4.1.10 gives

$$\boldsymbol{\omega}_{z_0} \times \mathbf{p}_0 = \begin{vmatrix} \mathbf{i}_{x_0} & \mathbf{j}_{y_0} & \mathbf{k}_{z_0} \\ 0 & 0 & \omega_{z_0} \\ p_{0x} & p_{0y} & p_{0z} \end{vmatrix}$$

$$= (-\omega_{z_0} p_{0y}) \mathbf{i}_{x_0} + (\omega_{z_0} p_{0x}) \mathbf{j}_{y_0} = -10.392 \mathbf{i}_{x_0} \qquad (4.1.11)$$

where ω_{z_0} represents the magnitude of vector $\boldsymbol{\omega}_{z_0}$.

In Equation 4.1.10, the starred derivative $d^* \mathbf{p}_0(t)/dt$ can also be expressed as $d \mathbf{p}_1/dt$, since the tip of vector $\mathbf{p}_0$ is the same as that of vector $\mathbf{p}_1$, and the derivative is taken with respect to the moving first coordinate frame. Since the arm rotates in this frame at the speed of $\boldsymbol{\omega}_{z_1}$, and the length of the arm is constant, it follows that

$$\frac{d \mathbf{p}_1(t)}{dt} = \boldsymbol{\omega}_{z_1} \times \mathbf{p}_1$$

$$= (-\omega_{z_1} p_{1y}) \mathbf{i}_{x_1} + (\omega_{z_1} p_{1x}) \mathbf{j}_{y1}$$

$$= -10.0 \mathbf{i}_{x_1} + 17.320 \mathbf{j}_{y_1} \qquad (4.1.12)$$

The terms in Equation 4.1.10 are now determined by Equations 4.1.11 and 4.1.12; however, these expressions contain unit vectors of two different coordinate systems. In order to combine Equations 4.1.11 and 4.1.12, it is convenient to express the unit vectors in Equation 4.1.12 in terms of the unit vectors of the base coordinate system used in Equation 4.1.11. Since $\mathbf{i}_{x_1} = \mathbf{j}_{y_0}$, and $\mathbf{j}_{y_1} = \mathbf{k}_{z_0}$ by Figure 4.2, Equation 4.1.12 becomes

$$\frac{d\mathbf{p}_1(t)}{dt} = -10.0\mathbf{i}_{x_1} + 17.320\mathbf{j}_{y_1}$$

$$= -10.0\mathbf{j}_{y_0} + 17.320\mathbf{k}_{z_0} \qquad (4.1.13)$$

The substitution of Equations 4.1.11 and 4.1.13 into Equation 4.1.10 leads to

$$\frac{d\mathbf{p}_0(t)}{dt} = -10.392\mathbf{i}_{x_0} - 10.0\mathbf{j}_{y_0} + 17.320\mathbf{k}_{z_0} \qquad (4.1.14)$$

Equation 4.1.14 specifies the actual velocity $\dot{\mathbf{p}}_0$ of the gripper G in the base coordinate system. The speed of gripper G relative to the moving first coordinate frame is given by Equation 4.1.12.

Equation 4.1.14 can also be expressed as vector

$$\dot{p}_0 = \begin{vmatrix} -10.392 & -10.0 & 17.320 \end{vmatrix}' \qquad (4.1.15)$$

(b) The actual value of the translational acceleration of gripper G in the base coordinate system is determined by Equation 4.1.9. In this particular case, Equation 4.1.9 gives for $i = 0$ and constant $\boldsymbol{\omega}_{z_0}$:

$$\frac{d^2\mathbf{p}_0}{dt^2} = \frac{d^{*2}\mathbf{p}_0(t)}{dt^2} + 2\boldsymbol{\omega}_{z_0} \times \frac{d^*\mathbf{p}_0(t)}{dt} + \boldsymbol{\omega}_{z_0} \times (\boldsymbol{\omega}_{z_0} \times \mathbf{p}_0) \qquad (4.1.16)$$

where $d^*\mathbf{p}_0(t)/dt$ is specified by Equation 4.1.12.

The cross-products in Expression 4.1.16 are first determined:

$$\boldsymbol{\omega}_{z_0} \times (\boldsymbol{\omega}_{z_0} \times \mathbf{p}_0) = \boldsymbol{\omega}_{z_0} \times \begin{bmatrix} (-\omega_{z_0}p_{0y})\mathbf{i}_{x_0} + (\omega_{z_0}p_{0x})\mathbf{j}_{y_0} + 0\mathbf{k}_{z_0} \end{bmatrix}$$

$$= -3.118\mathbf{j}_{y_0} \qquad (4.1.17)$$

$$2\boldsymbol{\omega}_{z_0} \times \frac{d^*\mathbf{p}_0}{dt} = 2\boldsymbol{\omega}_{z_0} \times (\boldsymbol{\omega}_{z_1} \times \mathbf{p}_1)$$

$$= (0.60\mathbf{k}_{z_0}) \times (-10.0\mathbf{i}_{x_1} + 17.320\mathbf{j}_{y_1}) \qquad (4.1.18)$$

Although the cross-products in Equation 4.1.18 can be evaluated in the form given, it is convenient and advisable in view of the general case to express the vectors in Equation 4.1.18 in the base coordinates. Thus,

$$2\boldsymbol{\omega}_{z_0} \times \frac{d^*\mathbf{p}_0}{dt} = (0.60\mathbf{k}_{z_0}) \times (-10.0\mathbf{j}_{y_0} + 17.320\mathbf{k}_{z_0})$$

$$= 6.0\mathbf{i}_{x_0} \qquad (4.1.19)$$

The last two terms on the right side of Equation 4.1.16 are now determined. The first term on the right side of Equation 4.1.16 involving

$\mathbf{p}_0$ represents the derivative of the gripper position vector $\mathbf{p}_0(t)$ with respect to the moving first coordinate frame. The position of gripper G can also be expressed as vector $\mathbf{p}_1(t)$ in the first coordinate frame; its time derivative is specified by Equation 4.1.12. It follows that

$$\frac{d^{*2}\mathbf{p}_1}{dt^2} = \boldsymbol{\omega}_{z_1} \times (\boldsymbol{\omega}_{z_1} \times \mathbf{p}_1)$$

$$= -8.660\mathbf{i}_{x_1} - 5.0\mathbf{j}_{y_1} \qquad (4.1.20)$$

In Equation 4.1.16, the terms on the right side are now given by Equations 4.1.17, 4.1.19, and 4.1.20, and they must be combined. Therefore, Equation 4.1.20 is written in terms of the unit vectors of the base coordinate system; indeed, Figure 4.2 shows that $\mathbf{i}_{x_1} = \mathbf{j}_{y_0}$ and $\mathbf{j}_{y_1} = \mathbf{k}_{z_0}$, as was observed earlier. Although the relationship between the unit vectors is established here by inspection, we should notice that the rotation submatrix in the homogeneous transformation establishes the same relationship. By combining the terms in Equation 4.1.16, the translational Cartesian acceleration in cm/s^2 is obtained:

$$\frac{d^2\mathbf{p}_0}{dt^2} = (-3.118\mathbf{j}_{y_0}) + (6.0\mathbf{i}_{x_0}) + (-8.660\mathbf{j}_{y_0} - 5.0\mathbf{k}_{z_0})$$

$$= 6.0\mathbf{i}_{x_0} - 11.778\mathbf{j}_{y_0} - 5.0\mathbf{k}_{z_0} \qquad (4.1.21)$$

Equation 4.1.21 describes the actual translational acceleration of gripper G in the base coordinate systems.

(c) The angular velocity $\boldsymbol{\omega}_G$ of gripper G is

$$\boldsymbol{\omega}_G = \boldsymbol{\omega}_{z_0} + \boldsymbol{\omega}_{z_1} \qquad (4.1.22)$$

where $\boldsymbol{\omega}_{z_0} = 0.30\mathbf{k}_{z_0}$, and $\boldsymbol{\omega}_{z_1} = 0.50\mathbf{k}_{z_1} = 0.50\mathbf{i}_{x_0}$ in rad/s. Hence,

$$\boldsymbol{\omega}_G = 0.50\mathbf{i}_{x_0} + 0.30\mathbf{k}_{z_0} \qquad (4.1.23)$$

The angular acceleration is, by Equation 4.1.22

$$\dot{\boldsymbol{\omega}}_G = \dot{\boldsymbol{\omega}}_{z_0} + \dot{\boldsymbol{\omega}}_{z_1} \qquad (4.1.24)$$

Since $\boldsymbol{\omega}_{z_0}$ is constant in this example, it follows that $\dot{\boldsymbol{\omega}}_{z_0} = 0$. Vector $\boldsymbol{\omega}_{z_1}$ rotates with the z_1-axis at the angular speed of $\boldsymbol{\omega}_{z_0}$ about the z_0-axis. Its time derivative $\dot{\boldsymbol{\omega}}_{z_1}$ is determined by Equation 4.1.8 with the starred derivative being equal to zero. Thus, $\dot{\boldsymbol{\omega}}_{z_1}$ is obtained by replacing the $\mathbf{p}_i$-vector in Equation 4.1.8 by the vector $\boldsymbol{\omega}_{z_1}$. Therefore, the angular velocity $\dot{\boldsymbol{\omega}}_G$ is in rad/s^2:

$$\dot{\boldsymbol{\omega}}_G = \boldsymbol{\omega}_{z_0} \times \boldsymbol{\omega}_{z_1}$$

$$= 0.15\mathbf{j}_{y_0} \qquad (4.1.25)$$

Equation 4.1.25 specifies the angular acceleration $\dot{\boldsymbol{\omega}}_G$ of the gripper in the base coordinate system.

We have discussed the determination of the velocity and acceleration vectors in the fixed and rotating coordinate systems with a common origin using the cross-products, which are commonly used in the dynamics. Since many engineers are familiar with the matrix-vector manipulations, we like to note that the general Equations 4.1.8 and 4.1.9 can also be presented in matrix-vector notations.

Suppose that the rotation for the $(i + 1)$st coordinate frame relative to the ith coordinate system is specified by the angular velocity vector $\boldsymbol{\omega}_i$:

$$\boldsymbol{\omega}_i = \omega_{x_i}\, \mathbf{i}_{x_i} + \omega_{y_i}\, \mathbf{j}_{y_i} + \omega_{z_i}\, \mathbf{k}_{z_i} \tag{4.1.26}$$

We want to determine the velocity of point P. It is specified by vector p_i in the ith coordinate frame, and moving relative to the local coordinate systems. Thus, the velocity of point P is determined by Equation 4.1.8:

$$\frac{d\mathbf{p}_i}{dt} = \frac{d^*\mathbf{p}_i}{dt} + \boldsymbol{\omega}_i \times \mathbf{p}_i \tag{4.1.27}$$

where the star refers to the $(i + 1)$st coordinate frame. By expanding the cross-product, we can see that the components are the same as in the product of a velocity matrix Ω_i and vector p_i; thus, Equation 4.1.27 can equivalently be written as follows:

$$\frac{d}{dt}\begin{bmatrix} p_{x_i} \\ p_{y_i} \\ p_{z_i} \\ 1 \end{bmatrix} = \frac{d^*}{dt}\begin{bmatrix} p_{x_i} \\ p_{y_i} \\ p_{z_i} \\ 1 \end{bmatrix} + \begin{bmatrix} 0 & -\omega_{z_i} & \omega_{y_i} & 0 \\ \omega_{z_i} & 0 & -\omega_{x_i} & 0 \\ -\omega_{y_i} & \omega_{x_i} & 0 & 0 \\ 0 & 0 & 0 & 0 \end{bmatrix}\begin{bmatrix} p_{x_i} \\ p_{y_i} \\ p_{z_i} \\ 1 \end{bmatrix} \tag{4.1.28}$$

Or in a concise form

$$\frac{dp_i}{dt} = \frac{d^*p_i}{dt} + \Omega_i p_i \tag{4.1.29}$$

where $d^*p_i/dt = (A_i^{i+1})_R dp_{i+1}/dt$ and Ω_i is the velocity matrix containing the components of the angular velocity of the $(i + 1)$st coordinate frame relative to the ith coordinate system. The velocity matrix Ω_i in Equation 4.1.28 is skew-symmetric, that is, the (i, j) element ω_{ij} is equal to the negative of the (j, i) element $-\omega_{ij}, j = 1, \ldots, 4$. To illustrate the use of the matrix-vector manipulations, the solution to the part (a) of Example 4.1.1 is now presented.

EXAMPLE 4.1.2

A spherical (polar) manipulator has the same specifications as in Example 4.1.1. The problem is to determine the actual translational Cartesian velocity of the end-effector G in the component form using the matrix-vector notations.

The solution to the problem is obtained by applying Equation 4.1.29 to the case under consideration. Indeed for $i = 0$, Equation 4.1.29 gives

$$\frac{dp_0}{dt} = \frac{d^*p_0}{dt} + \Omega_0 p_0 \tag{4.1.30}$$

where the star refers to the moving first coordinate frame, and the velocity matrix Ω_0 by Equation 4.1.28 equals

$$
\Omega_0 = \begin{bmatrix} 0 & -\omega_{z_0} & \omega_{y_0} & 0 \\ \omega_{z_0} & 0 & -\omega_{x_0} & 0 \\ -\omega_{y_0} & \omega_{x_0} & 0 & 0 \\ 0 & 0 & 0 & 0 \end{bmatrix} \tag{4.1.31}
$$

In Equation 4.1.30, the starred derivative d^*p_0/dt is evaluated by first determining dp_1/dt on the basis of Equation 4.1.29 when $i = 1$:

$$
\frac{dp_1}{dt} = \frac{d^*p_1}{dt} + \Omega_1 p_1 \tag{4.1.32}
$$

where the starred derivative is taken relative to the second coordinate frame; hence, it is zero. The velocity matrix Ω_1 contains the angular velocities of the second rotating coordinate frame expressed with respect to the first coordinate system. Thus

$$
\Omega_1 = \begin{bmatrix} 0 & -\omega_{z_1} & \omega_{y_1} & 0 \\ \omega_{z_1} & 0 & -\omega_{x_1} & 0 \\ -\omega_{y_1} & \omega_{x_1} & 0 & 0 \\ 0 & 0 & 0 & 0 \end{bmatrix} \tag{4.1.33}
$$

According to Equation 2.2.1

$$
p_i = A_i^{i+1} p_{i+1} \tag{4.1.34}
$$

where A_i^{i+1} represents the homogeneous transformation matrix between the $(i + 1)$st and ith coordinate systems. In general, the rotation submatrix $(A_i^{i+1})_R$ in the transformation matrix A_i^{i+1} causes the rotation of a vector of the $(i + 1)$st coordinate frame so that the components are obtained in the ith coordinate system; an example of this operation is shown in Equation 4.1.13. The rotation submatrix $(A_0^1)_R$ is applied to vector dp_1/dt to obtain from Equation 4.1.30

$$
\frac{dp_0}{dt} = (A_0^1)_R \frac{dp_1}{dt} + \Omega_0 p_0 \tag{4.1.35}
$$

The rotation matrix $(A_0^1)_R$ is written here as a (4×4) matrix with zero translation vector. It is obtained from the homogeneous transformation matrix A_0^1 that can be written on the basis of Figure 4.2. Thus

$$
(A_0^1)_R = \begin{bmatrix} 0 & 0 & 1 & 0 \\ 1 & 0 & 0 & 0 \\ 0 & 1 & 0 & 0 \\ 0 & 0 & 0 & 1 \end{bmatrix} \tag{4.1.36}
$$

The actual velocity of gripper G in the base coordinate system is determined by Equations 4.1.30, 4.1.32, and 4.1.35:

$$\frac{dp_0}{dt} = (A_0^1)_R \Omega_1 p_1 + \Omega_0 p_0 \qquad (4.1.37)$$

In this particular example, $\omega_{z_0} = 0.30$ rad/s, $\omega_{x_0} = \omega_{y_0} = 0$; $\omega_{z_1} = 0.50$ rad/s, $\omega_{x_1} = \omega_{y_1} = 0$; $p_0 = [0, 34.64, 20.0, 1]'$ cm, and $p_1 = [34.64, 20.0, 0, 1]'$ cm. When these numerical values are substituted into Equation 4.1.37, the actual translational velocity vector of the gripper in the base coordinate system is obtained; it is naturally the same as that given by Equation 4.1.15.

4.2 JACOBIAN MATRIX RELATES CARTESIAN VELOCITIES AND JOINT VELOCITIES

In a serial link N-joint manipulator, let us consider the ith and $(i + 1)$st coordinate frames attached to links i and $(i + 1)$, respectively, $i = 0, \ldots, N - 1$. Each coordinate system moves with the link to which it is attached. The coordinate frames are shown schematically in Figure 4.3a relative to the base coordinate system. The origin O_i of the ith coordinate frame is specified by vector $\mathbf{p}_{0i}$ in the base coordinate system; similarly, vector $\mathbf{p}_{0(i+1)}$ determines the origin O_{i+1} of the $(i + 1)$st coordinate frame. The rotational and translational velocities of the ith coordinate frame relative to the base coordinate system are $\omega_{0i} = [\omega_{0ix}, \omega_{0iy}, \omega_{0iz}]'$, and $dp_{0i}/dt = v_{0i} = [v_{0ix}, v_{0iy}, v_{0iz}]'$, respectively. Similar expressions apply to the motion of the $(i + 1)$st coordinate frame. The rotational motion of the $(i + 1)$st link about the z_i-axis is specified by $\dot{\theta}_{i+1}\mathbf{k}_{z_i}$, where $\mathbf{k}_{z_i}$ is the unit vector in the positive z_i-direction. Figure 4.3b illustrates the foregoing local coordinate frames ($i = 3$) defined in the fixed base coordinate system on a PUMA manipulator.

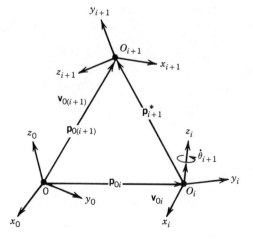

FIGURE 4.3a

Local coordinate frames ($x_i y_i z_i$) and ($x_{i+1} y_{i+1} z_{i+1}$) in fixed coordinate system ($x_0 y_0 z_0$).

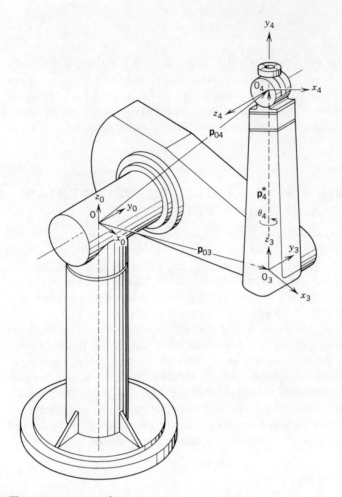

FIGURE 4.3*b*

Two moving coordinate frames ($i = 3, 4$) in base coordinate system on PUMA manipulator.

If joint i is revolute, the rotational velocity vector $\boldsymbol{\omega}_{0(i+1)}$ of the coordinate frame ($i + 1$) is the sum of the angular velocity $\boldsymbol{\omega}_{0i}$ of point O_i and the rotational velocity $\dot{\theta}_{i+1}\mathbf{k}_{z_i}$ of link ($i + 1$) relative to the ith coordinate frame:

$$\boldsymbol{\omega}_{0(i+1)} = \boldsymbol{\omega}_{0i} + \dot{\theta}_{i+1}\mathbf{k}_{z_i} \qquad (4.2.1)$$

If joint i is prismatic, then the angular velocity of link ($i + 1$) is the same as that of the ith coordinate frame expressed in the base coordinate system:

$$\boldsymbol{\omega}_{0(i+1)} = \boldsymbol{\omega}_{0i} \qquad (4.2.2)$$

The translational velocity vector $\mathbf{v}_{0(i+1)}$ of the end point of link $(i + 1)$ can be written on the basis of Equation 4.1.8. Indeed, the relative speed equals $d[\mathbf{p}_{0(i+1)} - \mathbf{p}_{0i}]/dt = d\mathbf{p}_{0(i+1)}^*/dt = \mathbf{v}_{0(i+1)} - \mathbf{v}_{0i}$ expressed in the base coordinate system where vector $\mathbf{p}_{0(i+1)}^* = \mathbf{p}_{0(i+1)} - \mathbf{p}_{0i}$. Thus

$$\mathbf{v}_{0(i+1)} - \mathbf{v}_{0i} = \frac{d^*\mathbf{p}_{i+1}^*}{dt} + \boldsymbol{\omega}_{0i} \times \mathbf{p}_{i+1}^* \tag{4.2.3}$$

where the time derivative of $\mathbf{p}_{i+1}^*$ in the starred coordinate frame equals to $(\mathbf{k}_{z_i}\dot{\theta}_{i+1}) \times \mathbf{p}_{i+1}^*$. Using Equation 4.2.1 for a revolute joint, the terms in Equation 4.2.3 can be combined to obtain

$$\mathbf{v}_{0(i+1)} = \mathbf{v}_{0i} + \boldsymbol{\omega}_{0(i+1)} \times \mathbf{p}_{i+1}^* \tag{4.2.4}$$

where vector $\mathbf{p}_{i+1}^*$ is expressed relative to the ith coordinate frame.

The translational velocity vector for a prismatic joint is also obtained on the basis of Equation 4.1.8:

$$\mathbf{v}_{0(i+1)} - \mathbf{v}_{0i} = \mathbf{k}_{z_i}\dot{\theta}_{i+1} + \boldsymbol{\omega}_{0i} \times \mathbf{p}_{i+1}^* \tag{4.2.5}$$

Since by Equation 4.2.2 $\boldsymbol{\omega}_{0i} = \boldsymbol{\omega}_{0(i+1)}$ for a prismatic link, Equation 4.2.5 can be rewritten in the form similar to Equation 4.2.4:

$$\mathbf{v}_{0(i+1)} = \mathbf{v}_{0i} + \dot{\theta}_{i+1}\mathbf{k}_{z_i} + \boldsymbol{\omega}_{0(i+1)} \times \mathbf{p}_{i+1}^* \tag{4.2.6}$$

where $\dot{\theta}_{i+1}$ represents the translational velocity for the prismatic link.

Equations 4.2.1, 4.2.4, 4.2.2, and 4.2.6 specify the rotational and translational velocities for the links of the manipulator arm. These equations are summarized below after substituting $\mathbf{v}_{0i} = \dot{\mathbf{p}}_{0i}$ [2,3]:

(1) Rotational velocities:

$$\boldsymbol{\omega}_{0(i+1)} = \boldsymbol{\omega}_{0i} + \dot{\theta}_{i+1}\mathbf{k}_{z_i} \qquad [\text{link}(i + 1)\text{revolute}] \tag{4.2.1}$$

$$\boldsymbol{\omega}_{0(i+1)} = \boldsymbol{\omega}_{0i} \qquad [\text{link}(i + 1)\text{prismatic}] \tag{4.2.2}$$

(2) Translational (linear) velocities:

$$\dot{\mathbf{p}}_{0(i+1)} = \dot{\mathbf{p}}_{0i} + \boldsymbol{\omega}_{0(i+1)} \times \mathbf{p}_{i+1}^* \qquad [\text{link}(i + 1)\text{revolute}] \tag{4.2.4}$$

$$\dot{\mathbf{p}}_{0(i+1)} = \dot{\mathbf{p}}_{0i} + \dot{\theta}_{i+1}\mathbf{k}_{z_i} + \boldsymbol{\omega}_{0(i+1)} \times \mathbf{p}_{i+1}^* \quad [\text{link}(i + 1)\text{ prismatic}] \tag{4.2.5}$$

$$\mathbf{p}_{i+1}^* = \mathbf{p}_{0(i+1)} - \mathbf{p}_{0i} \tag{4.2.7}$$

where $i = 0, 1, \ldots, N - 1$, and N is the number of the moving links (joints) in the manipulator.

Equations 4.2.1, 4.2.2, 4.2.4, and 4.2.5 can also be written in terms of matrix-vector notations. As an example, Equation 4.2.1 is considered. The terms $\boldsymbol{\omega}_{0(i+1)}$ and $\boldsymbol{\omega}_{0i}$ have components on the base coordinates. In order to write the last term of Equation 4.2.1 in the component form without the unit vectors, the components of vector $\mathbf{k}_{z_i}$ in the base coordinates must be determined. They may be obtained by applying the (3×3) rotation submatrix $(A_0^i)_R$ of the homogeneous transformation matrix A_0^i to vector $k_{z_i} = [0, 0, 1]$. Thus, Equation 4.2.1 can be written in the component form by denoting $\boldsymbol{\omega}_{0i} = [\omega_{0ix}, \omega_{0iy}, \omega_{0iz}]'$:

$$\begin{bmatrix} \omega_{0(i+1)x} \\ \omega_{0(i+1)y} \\ \omega_{0(i+1)z} \end{bmatrix} = \begin{bmatrix} \omega_{0ix} \\ \omega_{0iy} \\ \omega_{0iz} \end{bmatrix} + (A_0^i)_R \begin{bmatrix} 0 \\ 0 \\ 1 \end{bmatrix} \dot{\theta}_{i+1} \qquad (4.2.8)$$

where all terms are expressed relative to the base coordinate system.

By following similar arguments, Equation 4.2.4 can be expressed in the form of column vectors as follows:

$$\begin{bmatrix} \dot{p}_{0(i+1)x} \\ \dot{p}_{0(i+1)y} \\ \dot{p}_{0(i+1)z} \\ 0 \end{bmatrix} = \begin{bmatrix} \dot{p}_{0ix} \\ \dot{p}_{0iy} \\ \dot{p}_{0iz} \\ 0 \end{bmatrix}$$

$$+ \begin{bmatrix} 0 & -\omega_{0(i+1)z} & \omega_{0(i+1)y} & 0 \\ \omega_{0(i+1)z} & 0 & -\omega_{0(i+1)x} & 0 \\ -\omega_{0(i+1)y} & \omega_{0(i+1)x} & 0 & 0 \\ 0 & 0 & 0 & 0 \end{bmatrix} \begin{bmatrix} p_{0(i+1)x} - p_{0ix} \\ p_{0(i+1)y} - p_{0iy} \\ p_{0(i+1)z} - p_{0iz} \\ 1 \end{bmatrix} \qquad (4.2.9)$$

where the last term on the right shows the product of the velocity matrix $\Omega_{0(i+1)}$ and the relative position vector.

Equations 4.2.1, 4.2.2, 4.2.4, 4.2.5, and 4.2.7 can be considered as vector difference equations with respect to i, where i refers to the ith link of the manipulator. If all joints of a manipulator are revolute, as in PUMA 560(600) manipulator, then the rotational and translational velocities of the nth frame can be determined by solving the difference Equations 4.2.1 and 4.2.4 recursively for zero initial conditions, that is, $\boldsymbol{\omega}_{00} = 0$, and $\dot{\mathbf{p}}_{00} = 0$, since the base coordinate system does not move. Thus, for a *revolute* manipulator

$$\boldsymbol{\omega}_{0n} = \dot{\theta}_1 \mathbf{k}_{z_0} + \dot{\theta}_2 \mathbf{k}_{z_1} + \ldots + \dot{\theta}_n \mathbf{k}_{z_{n-1}}$$

$$= \sum_{j=0}^{n-1} \dot{\theta}_{j+1} \mathbf{k}_{z_j} \qquad (4.2.10)$$

$$\dot{\mathbf{p}}_{0n} = \boldsymbol{\omega}_{01} \times (\mathbf{p}_{01} - \mathbf{p}_{00}) + \boldsymbol{\omega}_{02} \times (\mathbf{p}_{02} - \mathbf{p}_{01}) + \ldots + \boldsymbol{\omega}_{0n} \times (\mathbf{p}_{0n} - \mathbf{p}_{0(n-1)})$$

$$= \sum_{j=0}^{n-1} \boldsymbol{\omega}_{0(j+1)} \times (\mathbf{p}_{0(j+1)} - \mathbf{p}_{0j}) \qquad (4.2.11)$$

where $1 < n \le N$. The angular velocity $\dot{\theta}_{j+1}$ of link $(j + 1)$ represents the rotational motion about the positive z_j-axis. The vector $\mathbf{p}_{0j}$ signifies the position vector of the origin of the jth coordinate frame, and $\boldsymbol{\omega}_{0(j+1)}$ is the rotational velocity vector of the $(j + 1)$st coordinate frame both expressed in the base coordinate system.

A common task in the path planning and control of a revolute manipulator is to determine the rotational and translational velocities of the links in the base coordinate system in terms of the angular velocities of the joints. These relationships can be derived by substituting Equation 4.2.10 into Equation 4.2.11:

$$\dot{\mathbf{p}}_{0n} = \sum_{j=0}^{n-1} \left[\sum_{i=0}^{j} \dot{\theta}_{i+1} \mathbf{k}_{z_i} \right] \times (\mathbf{p}_{0(j+1)} - \mathbf{p}_{0j})$$

$$(4.2.12)$$

$$= \sum_{j=0}^{n-1} \dot{\theta}_{j+1} \mathbf{k}_{z_j} \times (\mathbf{p}_{0n} - \mathbf{p}_{0j})$$

Equation 4.2.12 determines the translational velocity $\dot{\mathbf{p}}_{0n}$ of the nth coordinate frame in the base coordinate system in terms of the angular velocities of the joints.

Equations 4.2.10 and 4.2.11 may be combined by defining a generalized velocity vector V_{0n} describing the translational and rotational velocities of link n in the base coordinate system:

$$V_{0n} = \begin{bmatrix} \dot{\mathbf{p}}_{0n} \\ \omega_{0n} \end{bmatrix}$$

$$(4.2.13)$$

where $\dot{\mathbf{p}}_{0n}$ and ω_{0n} represent three-dimensional column vectors. The velocity vector V_{0n} is dependent upon the angular velocities as shown in Equations 4.2.10 and 4.2.12. These equations can conveniently be expressed in the vector form for a revolute manipulator by combining Equations 4.2.12 and 4.2.10:

$$\mathbf{V}_{0n} = \begin{bmatrix} \dot{\mathbf{p}}_{0n} \\ \omega_{0n} \end{bmatrix} = \mathbf{J}_n \begin{bmatrix} \dot{\theta}_1 \\ \dot{\theta}_2 \\ \vdots \\ \dot{\theta}_{n-1} \\ \dot{\theta}_n \end{bmatrix}$$

$$(4.2.14)$$

where the vector form $\mathbf{J}_n$ of the Jacobian matrix for *revolute* joints (links) is written on the basis of Equations 4.2.12 and 4.2.10:

$$\mathbf{J}_n = \begin{bmatrix} \mathbf{J}_{n1} & \mathbf{J}_{n2} \dots \mathbf{J}_{nn} \end{bmatrix}$$

$$= \begin{bmatrix} \mathbf{k}_{z_0} \times (\mathbf{p}_{0n} - \mathbf{p}_{00}) & \mathbf{k}_{z_1} \times (\mathbf{p}_{0n} - \mathbf{p}_{01}) & \dots & \mathbf{k}_{z(n-1)} \times (\mathbf{p}_{0n} - \mathbf{p}_{0(n-1)}) \\ \mathbf{k}_{z_0} & \mathbf{k}_{z_1} & \dots & \mathbf{k}_{z(n-1)} \end{bmatrix}$$

$$(4.2.15)$$

The cross-product terms in the matrix of Equation 4.2.15 contain vectors that are specified in various coordinate frames. Since the position vector $(\mathbf{p}_{0n} - \mathbf{p}_{0i})$, $i = 0, \dots, (n-1)$ is expressed in the base coordinate system, it is convenient and often advisable to determine the components of the unit vector $\mathbf{k}_{z_i}$ also in the base coordinate system. It can be accomplished by applying the (3×3) rotation submatrix $(A_0^i)_R$ of the homogeneous transformation matrix A_0^i to the unit vector $k_{z_i} = [0, 0, 1]'$ to obtain the components of this vector in the base coordinate system. Moreover, the calculations in the cross-products may be converted to the calculations of the elements in the product of a matrix and a vector, as discussed in Example 4.1.2.

When a manipulator consists of only prismatic links (a Cartesian manipulator), Equations 4.2.2 and 4.2.5 can be solved for $i = 0, 1, \ldots, n - 1$ with the initial condition $\dot{p}_{00} = 0$, and $\omega_{00} = 0$ to obtain

$$\boldsymbol{\omega}_{0n} = 0 \qquad (4.2.16)$$

$$\dot{\mathbf{p}}_{0n} = \sum_{i=0}^{n-1} \dot{\theta}_{i+1} \mathbf{k}_{z_i} \qquad (4.2.17)$$

where $\dot{\theta}_{i+1} \mathbf{k}_{z_i}$ represents the translational velocity of the prismatic link $(i + 1)$. Equation 4.2.17 for a prismatic link corresponds to Equation 4.2.12 of a revolute link.

If the $(i + 1)$st link of a manipulator is prismatic and the other links revolute, then the $(i + 1)$st column in the Jacobian matrix J_n for the *prismatic* link (joint) is

$$\mathbf{J}_{n(i+1)} = \begin{bmatrix} \mathbf{k}_{z_i} \\ 0 \end{bmatrix} \qquad \text{[link } (i + 1) \text{ prismatic]} \qquad (4.2.18)$$

The other columns of the Jacobian matrix $\mathbf{J}_n$ are the same as in Equation 4.2.15.

The relationship between the velocities of the joints and the generalized Cartesian velocities may now be summarized for a manipulator containing revolute and prismatic links. The generalized velocity vector for the nth link is

$$\begin{bmatrix} \dot{p}_{0n} \\ \omega_{0n} \end{bmatrix} = J_n \dot{\theta} \qquad (4.2.19)$$

where $\dot{\theta} = [\dot{\theta}_1 \quad \dot{\theta}_2 \ldots \dot{\theta}_n]'$. The Jacobian matrix may be specified in the vector form

$$\mathbf{J}_n = [\mathbf{J}_{n1} \mathbf{J}_{n2} \cdots \mathbf{J}_{nn}] \qquad (4.2.20)$$

The column vectors in Equation 4.2.20 are calculated as follows:

$$\mathbf{J}_{n(i+1)} = \begin{bmatrix} \mathbf{k}_{z_i} \times (\mathbf{p}_{0n} - \mathbf{p}_{0i}) \\ \mathbf{k}_{z_i} \end{bmatrix} \qquad \text{[link } (i + 1) \text{ is revolute]} \qquad (4.2.21)$$

$$\mathbf{J}_{n(i+1)} = \begin{bmatrix} \mathbf{k}_{z_i} \\ 0 \end{bmatrix} \qquad \text{[link } (i + 1) \text{ is prismatic]} \qquad (4.2.22)$$

where $i = 0, 1, \ldots, n - 1$. The conversion of the vector form Jacobian matrix $\mathbf{J}_n$ to the matrix form J_n requires that all entries in $\mathbf{J}_n$ are described in the same coordinate system.

Examples are next presented to illustrate the use of Equations 4.2.19–4.2.22.

EXAMPLE 4.2.1

A planar manipulator with two revolute joints is shown in the schematic Figure 4.4. The lengths of the two links are ℓ_1 and ℓ_2. The problem is to determine the translational ($\dot{p}_{02}$) and rotational (ω_{02}) velocities of the end-effector, that is, of point O_2 in terms of the angular velocities $\dot{\theta}_1$, and $\dot{\theta}_2$ of joints one and two, respectively.

The translational velocity of the end-effector is denoted by $\dot{p}_{02}$ = $[v_{02x}, v_{02y}, v_{02z}]'$, and the angular velocity by ω_{02} = $[\omega_{02x}, \omega_{02y}, \omega_{02z}]'$. The relationship between $\dot{p}_{02}$, ω_{02}, and $\dot{\theta}_1$, $\dot{\theta}_2$ is specified by Equation 4.2.14. It assumes now the following form:

$$\mathbf{V}_2 = \begin{bmatrix} \dot{\mathbf{p}}_{02} \\ \boldsymbol{\omega}_{02} \end{bmatrix} = \mathbf{J}_2 \begin{bmatrix} \dot{\theta}_1 \\ \dot{\theta}_2 \end{bmatrix} \qquad (4.2.23)$$

where $\mathbf{V}_2$ is the generalized velocity vector.

The vector form $\mathbf{J}_2$ of the Jacobian matrix in this case is given by Equation 4.2.15 with $n = 2$:

$$\mathbf{J}_2 = \begin{bmatrix} \mathbf{J}_{21}, \mathbf{J}_{22} \end{bmatrix} = \begin{bmatrix} \mathbf{k}_{z_0} \times (\mathbf{p}_{02} - \mathbf{p}_{00}) & \mathbf{k}_{z_1} \times (\mathbf{p}_{02} - \mathbf{p}_{01}) \\ \mathbf{k}_{z_0} & \mathbf{k}_{z_1} \end{bmatrix} \qquad (4.2.24)$$

where the unit vector $k_{z_0} = [0 \quad 0 \quad 1]'$ is expressed in the $(x_0 y_0 z_0)$ coordinate system, and $k_{z_1} = [0 \quad 0 \quad 1]'$ in the $(x_1 y_1 z_1)$ coordinate frame. Moreover, vectors $\mathbf{k}_{z_0}$ and $\mathbf{k}_{z_1}$ are parallel and of equal length. Hence, a vector on the z_0-axis has an equal value on the z_1-axis, and the rotation matrix $(A_0^1)_R$ need not be used in this particular case. Vector $\mathbf{p}_{00} = 0$. Vector $\mathbf{p}_{01}$ emanates from the origin O of the base coordinate system to the origin O_1 of the first coordinate frame. Vector $\mathbf{p}_{02}$ is directed from O to the origin O_2 of the $(x_2 y_2 z_2)$ coordinate frame.

In order to calculate the entries in matrix $\mathbf{J}_2$, vectors $(\mathbf{p}_{02} - \mathbf{p}_{00})$ and $(\mathbf{p}_{02} - \mathbf{p}_{01})$ are expressed in the base coordinate system directly on the basis of Figure 4.4:

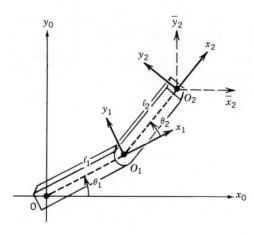

FIGURE 4.4
Revolute planar manipulator in fixed coordinate system $(x_0 y_0 z_0)$.

$$\mathbf{p}_{02} - \mathbf{p}_{00} = \begin{bmatrix} \ell_1 c_1 + \ell_2 c_{12} \\ \ell_1 s_1 + \ell_2 s_{12} \\ 0 \\ 1 \end{bmatrix} \quad (4.2.25)$$

$$\mathbf{p}_{02} - \mathbf{p}_{01} = \begin{bmatrix} \ell_1 c_1 + \ell_2 c_{12} - \ell_1 c_1 \\ \ell_1 s_1 + \ell_2 s_{12} - \ell_1 s_1 \\ 0 \\ 1 \end{bmatrix} \quad (4.2.26)$$

where $c_{12} = \cos(\theta_1 + \theta_2)$, and $s_{12} = \sin(\theta_1 + \theta_2)$. The cross-products in Equation 4.2.24 can then be determined:

$$\mathbf{k}_{z_0} \times (\mathbf{p}_{02} - \mathbf{p}_{00}) = \begin{vmatrix} \mathbf{i}_{x_0} & \mathbf{j}_{y_0} & \mathbf{k}_{z_0} \\ 0 & 0 & 1 \\ \ell_1 c_1 + \ell_2 c_{12} & \ell_1 s_1 + \ell_2 s_{12} & 0 \end{vmatrix}$$

$$= -(\ell_1 s_1 + \ell_2 s_{12})\mathbf{i}_{x_0} + (\ell_1 c_1 + \ell_2 c_{12})\mathbf{j}_{y_0} + 0\mathbf{k}_{z_0} \quad (4.2.27)$$

$$\mathbf{k}_{z_1} \times (\mathbf{p}_{02} - \mathbf{p}_{01}) = \begin{vmatrix} \mathbf{i}_{x_0} & \mathbf{j}_{y_0} & \mathbf{k}_{z_0} \\ 0 & 0 & 1 \\ \ell_2 c_{12} & \ell_2 s_{12} & 0 \end{vmatrix}$$

$$= -\ell_2 s_{12}\mathbf{i}_{x_0} + \ell_2 c_{12}\mathbf{j}_{y_0} + 0\mathbf{k}_{z_0} \quad (4.2.28)$$

The columns of the Jacobian matrix $\mathbf{J}_2$ in Equation 4.2.15 are now specified by Equations 4.2.27 and 4.2.28. Indeed, the Jacobian matrix in the vector form in this case is

$$\mathbf{J}_2 = \begin{bmatrix} -(\ell_1 s_1 + \ell_2 s_{12})\mathbf{i}_{x_0} + (\ell_1 c_1 + \ell_2 c_{12})\mathbf{j}_{y_0} & -\ell_2 s_{12}\mathbf{i}_{x_0} + \ell_2 c_{12}\mathbf{j}_{y_0} \\ \mathbf{k}_{z_0} & \mathbf{k}_{z_1} \end{bmatrix} \quad (4.2.29)$$

Using the column vector notations, the relationship between the angular velocities of the joints and the velocity components of the end-effector expressed in the base coordinate system can be written by Equation 4.2.23:

$$\begin{bmatrix} \dot{\mathbf{p}}_{02} \\ \omega_{02} \end{bmatrix} = \begin{bmatrix} -\ell_1 s_1 - \ell_2 s_{12} & -\ell_2 s_{12} \\ \ell_1 c_1 + \ell_2 c_{12} & \ell_2 c_{12} \\ 0 & 0 \\ 0 & 0 \\ 0 & 0 \\ 1 & 1 \end{bmatrix} \begin{bmatrix} \dot{\theta}_1 \\ \dot{\theta}_2 \end{bmatrix} \quad (4.2.30)$$

since vectors $\mathbf{k}_{z_0}$ and $\mathbf{k}_{z_1}$ are parallel. Equation 4.2.30 describes the translational and rotational velocities of the end-effector (point O_2) in the base coordinate system as a function of the angular velocities of the joints.

In calculating matrix $\mathbf{J}_2$ in Equation 4.2.23, vectors $(\mathbf{p}_{02} - \mathbf{p}_{01})$ and

($\mathbf{p}_{02} - \mathbf{p}_{00}$) are determined in the base coordinate system by inspection of Figure 4.4. The expressions given in Equations 4.2.25 and 4.2.26 can also be obtained using the homogeneous transformation matrices. Since this approach can systematically be applied to general multijoint manipulators, the solution to the previous problem is briefly outlined by this alternative procedure.

For the planar manipulator given, the homogeneous transformation matrices can be written

$$A_0^1 = \begin{bmatrix} c_1 & -s_1 & 0 & \ell_1 c_1 \\ s_1 & c_1 & 0 & \ell_1 s_1 \\ 0 & 0 & 1 & 0 \\ 0 & 0 & 0 & 1 \end{bmatrix} \qquad A_1^2 = \begin{bmatrix} c_2 & -s_2 & 0 & \ell_2 c_2 \\ s_2 & c_2 & 0 & \ell_2 s_2 \\ 0 & 0 & 1 & 0 \\ 0 & 0 & 0 & 1 \end{bmatrix} \qquad (4.2.31)$$

It follows that the homogeneous transformation A_0^2 for the planar manipulator is

$$A_0^2 = A_0^1 A_1^2 = \begin{bmatrix} c_{12} & -s_{12} & 0 & \ell_2 c_{12} + \ell_1 c_1 \\ s_{12} & c_{12} & 0 & \ell_2 s_{12} + \ell_1 s_1 \\ 0 & 0 & 1 & 0 \\ 0 & 0 & 0 & 1 \end{bmatrix} \qquad (4.2.32)$$

The homogeneous transformation matrices in Equations 4.2.31 and 4.2.32 can systematically be used to determine vectors $\mathbf{p}_{02}$, $\mathbf{p}_{01}$, and $\mathbf{p}_{00}$, which represent the origins of the second, first, and zeroth (base) coordinate frames in the base coordinate system. For example, the matrix A_0^2 can be applied to vector $[0, 0, 0, 1]'$, which represents the origin O_2 in the second coordinate frame. Thus, the same point O_2 in the base coordinate system is $p_{02} = [\ell_2 c_{12} + \ell_1 c_1, \ell_2 s_{12} + \ell_1 s_1, 0, 1]'$. This vector could also be read from Equation 4.2.32 as the translation vector. Similarly, the matrix A_0^1 gives $p_{01} = [\ell_1 c_1, \ell_1 s_1, 0, 1]'$. Then, the cross-products are directly formed as shown in Equations 4.2.27 and 4.2.28. Thus, the Jacobian matrix and the relationship between the Cartesian and joint velocities are determined.

The planar manipulator in Example 4.2.1 functions in the two-dimensional space. Thus, the task space has two DOF. Since the manipulator has two joints, it has also two DOF. Therefore, the translational velocity relation in Equation 4.2.30 can be reduced to a two-dimensional relation:

$$\begin{bmatrix} \dot{p}_{02x} \\ \dot{p}_{02y} \end{bmatrix} = \begin{bmatrix} -\ell_1 s_1 - \ell_2 s_{12} & -\ell_2 s_{12} \\ \ell_1 c_1 + \ell_2 c_{12} & \ell_2 c_{12} \end{bmatrix} \begin{bmatrix} \dot{\theta}_1 \\ \dot{\theta}_2 \end{bmatrix} \qquad (4.2.33)$$

If the velocities of the joint variables are known, the translational velocity of the end-effector can be calculated for a given configuration of the planar manipulator from Equation 4.2.33. When the translational velocity of the end-effector at a specific configuration is known, the velocities of the joint variables can be determined by inverting the square matrix in Equation 4.2.33.

The inverse of the Jacobian matrix may not exist at some values of the joint variables. This configuration of the manipulator is called *singular*. Equation 4.2.33 reveals that the determinant of the Jacobian matrix is zero when $\theta_2 = (0 + k\pi)$ rad, $k = 0, 1, \ldots$. This condition implies that link two is aligned with link one in Figure 4.4 at the singular configuration. The two links seem to act as one; thus, the manipulator loses one DOF at the singular point. The end-effector in this particular case ($k = 0$) is on the boundary of the manipulator workspace. The translational velocity of the end-effector is in this particular case

$$\begin{bmatrix} \dot{p}_{02x} & \dot{p}_{02y} \end{bmatrix} = \begin{bmatrix} -(\ell_1 + \ell_2)\dot{\theta}_1 s_1 - \ell_2 \dot{\theta}_2 s_1 & (\ell_1 + \ell_2)\dot{\theta}_1 c_1 + \ell_2 \dot{\theta}_2 c_1 \end{bmatrix}$$

When a control strategy is constructed for a manipulator, the designer should be aware of possible occurrences of singular points. In particular, if the control algorithm involves the inversion of the Jacobian (square) matrix, proper precautions should be taken to avoid numerical problems even in the neighborhoods of singular points.

EXAMPLE 4.2.2

The translational and rotational velocities of the end-effector in a six-joint revolute manipulator, such as a PUMA 600, can be determined by the application of Equations 4.2.14 and 4.2.15. The task here is to obtain the general expression that relates the angular velocities of the joints to the translational and rotational velocities of the end-effector expressed in the base coordinate system.

For a six-joint revolute manipulator, Equations 4.2.14 and 4.2.15 specifying the generalized velocity vector $\mathbf{V}_{06}$ assume the following form:

$$\mathbf{V}_{06} = \begin{bmatrix} \dot{\mathbf{p}}_{06} \\ \boldsymbol{\omega}_{06} \end{bmatrix} = \mathbf{J}_6^p \begin{bmatrix} \dot{\theta}_1 \\ \dot{\theta}_2 \\ \dot{\theta}_3 \\ \dot{\theta}_4 \\ \dot{\theta}_5 \\ \dot{\theta}_6 \end{bmatrix} \tag{4.2.34}$$

where the translational velocity vector $\dot{\mathbf{p}}_{06}$ and the angular velocity vector $\boldsymbol{\omega}_{06}$ are both expressed in the base coordinate system using the unit vectors. The Jacobian matrix $\mathbf{J}_6^p$ can be calculated from Equation 4.2.15:

$$\mathbf{J}_6^p = \begin{bmatrix} \mathbf{J}_{61}^p & \mathbf{J}_{62}^p & \mathbf{J}_{63}^p & \mathbf{J}_{64}^p & \mathbf{J}_{65}^p & \mathbf{J}_{66}^p \end{bmatrix}$$

$$= \begin{bmatrix} \mathbf{k}_{z_0} \times (\mathbf{p}_{06} - \mathbf{p}_{00}) & \mathbf{k}_{z_1} \times (\mathbf{p}_{06} - \mathbf{p}_{01}) & \ldots & \mathbf{k}_{z_5} \times (\mathbf{p}_{06} - \mathbf{p}_{05}) \\ \mathbf{k}_{z_0} & \mathbf{k}_{z_1} & \ldots & \mathbf{k}_{z_5} \end{bmatrix} \tag{4.2.35}$$

The six-dimensional vector $\mathbf{J}_{6i}^p \dot{\theta}_i$ obtained by multiplying the ith column

$\mathbf{J}_{6i}^p$ and the joint velocity $\dot{\theta}_i$ in Equation 4.2.34 describes the contribution of the velocity of the ith joint to the velocity of the end-effector. In fact, the first three components of $\mathbf{J}_{6i}^p\dot{\theta}_i$ represent the translational velocity directed perpendicularly to the plane of vectors $\mathbf{k}_{z_{i-1}}$ and $(\mathbf{p}_{06} - \mathbf{p}_{0(i-1)})$, and the last three components determine the rotational velocity of the end-effector about $\mathbf{k}_{z_{i-1}}$ due to the rotation of the ith joint. The translational and rotational velocities of the end-effector are the sum of the contributions of all moving joints, as Equation 4.2.34 indicates.

Equations 4.2.34 and 4.2.35 relate the angular velocities $\dot{\theta}_j, j = 1, \dots, 6$ to the generalized velocity vector $\mathbf{V}_{06}$ in a six-joint *revolute* manipulator.

EXAMPLE 4.2.3

The problem is to determine the translational and rotational velocities of the gripper in a Stanford/JPL arm relative to the base coordinate system in terms of the velocities of the joint variables in a general form.

Since the Stanford/JPL manipulator has the third link prismatic, and the remaining five links revolute, the equation relating the joint velocities to the generalized velocity of the end-effector in the base coordinate system assumes the following form by Equation 4.2.14 ($\theta_3 = d_3$):

$$\mathbf{V}_{06} = \begin{bmatrix} \dot{\mathbf{p}}_{06} \\ \boldsymbol{\omega}_{06} \end{bmatrix} = \mathbf{J}_6^s \begin{bmatrix} \dot{\theta}_1 \\ \dot{\theta}_2 \\ \dot{\theta}_3 \\ \dot{\theta}_4 \\ \dot{\theta}_5 \\ \dot{\theta}_6 \end{bmatrix} \qquad (4.2.36)$$

The vector form of the Jacobian matrix $\mathbf{J}_6^s$ is specified by Equations 4.2.21–4.2.22:

$$\mathbf{J}_6^s = \begin{bmatrix} \mathbf{J}_{61}^s & \mathbf{J}_{62}^s & \mathbf{J}_{63}^s & \mathbf{J}_{64}^s & \mathbf{J}_{65}^s & \mathbf{J}_{66}^s \end{bmatrix}$$

$$= \begin{bmatrix} \mathbf{k}_{z_0} \times (\mathbf{p}_{06} - \mathbf{p}_{00}) & \mathbf{k}_{z_1} \times (\mathbf{p}_{06} - \mathbf{p}_{01}) & \mathbf{k}_{z_2} \\ \mathbf{k}_{z_0} & \mathbf{k}_{z_1} & 0 \\ \mathbf{k}_{z_3} \times (\mathbf{p}_{06} - \mathbf{p}_{03}) & \mathbf{k}_{z_4} \times (\mathbf{p}_{06} - \mathbf{p}_{04}) & \mathbf{k}_{z_5} \times (\mathbf{p}_{06} - \mathbf{p}_{05}) \\ \mathbf{k}_{z_3} & \mathbf{k}_{z_4} & \mathbf{k}_{z_5} \end{bmatrix} \qquad (4.2.37)$$

where the third column of matrix $\mathbf{J}_6^s$ represents the prismatic link, and the remaining columns the revolute links. Equations 4.2.36 and 4.2.37 determine the actual translational and angular velocities as a function of the joint velocities in a general form.

A numerical example is next presented to illustrate the use of Equations 4.2.36 and 4.2.37 to calculate the translational and rotational velocities of the end-effector in the Cartesian coordinate system from the link (joint) velocities.

EXAMPLE 4.2.4

During the motion of a Stanford/JPL manipulator, the following measurements were obtained for the positions and the velocities of the joints at a specific time [5]:

$$\theta_1 = -.711 \text{ rad}, \quad \theta_2 = 1.857 \text{ rad}, \quad \theta_3 = d_3 = 8.633 \text{ in.},$$

$$\theta_4 = .013 \text{ rad}, \quad \theta_5 = 1.393 \text{ rad}, \quad \theta_6 = -.726 \text{ rad} \qquad (4.2.38)$$

$$\dot{\theta}_1 = -1.034 \text{ rad/s}, \quad \dot{\theta}_2 = .127 \text{ rad/s}, \quad \dot{\theta}_3 = \dot{d}_3 = 12.399 \text{ in./s},$$

$$\dot{\theta}_4 = 1.475 \text{ rad/s}, \quad \dot{\theta}_5 = 2.905 \text{ rad/s}, \quad \dot{\theta}_6 = -3.129 \text{ rad/s} \qquad (4.2.39)$$

The problem is to determine the translational and rotational velocities of the end-effector with respect to the base coordinate system.

The solution to this problem is given by Equations 4.2.36 and 4.2.37. To evaluate the Jacobian matrix, the positions of the origins of the six coordinate frames in the base coordinate system must be determined. They will be obtained by first calculating the homogeneous transformation matrices given in Table 2.3 for the coordinate frames shown in Figure 2.4.

On the basis of the given positions of the joints, the homogeneous transformation matrices $A_0^i = A_0^1 \cdots A_{i-1}^i, i = 1, \ldots, 6$ given in Table 2.3a are calculated. They assume the following values [5]:

$$A_0^1 = \begin{bmatrix} .759 & 0 & .650 & 0 \\ -.650 & 0 & .759 & 0 \\ 0 & -1 & 0 & 14.0 \\ 0 & 0 & 0 & 1 \end{bmatrix} \qquad (4.2.40)$$

$$A_0^2 = A_0^1 A_1^2 = \begin{bmatrix} -.214 & .650 & .728 & 4.147 \\ .184 & .759 & -.624 & 4.841 \\ -.959 & 0 & -.282 & 14.00 \\ 0 & 0 & 0 & 1 \end{bmatrix} \qquad (4.2.41)$$

$$A_0^3 = A_0^1 A_1^2 A_2^3 = \begin{bmatrix} -.214 & .650 & .728 & 10.429 \\ .184 & .759 & -.624 & -.540 \\ -.959 & 0 & -.282 & 11.559 \\ 0 & 0 & 0 & 1 \end{bmatrix} \qquad (4.2.42)$$

$$A_0^4 = A_0^1 A_1^2 A_2^3 A_3^4 = \begin{bmatrix} -.211 & -.728 & .651 & 10.429 \\ .188 & .624 & .758 & -.540 \\ -.959 & -.282 & .005 & 11.559 \\ 0 & 0 & 0 & 1 \end{bmatrix} \qquad (4.2.43)$$

$$A_0^5 = A_0^1 A_1^2 A_2^3 A_3^4 A_4^5 = \begin{bmatrix} -.754 & .651 & -.079 & 10.429 \\ .647 & .758 & .076 & -.540 \\ .110 & .005 & -.993 & 11.559 \\ 0 & 0 & 0 & 1 \end{bmatrix} \qquad (4.2.44)$$

The origin of the ith coordinate frame is represented by the position vector p_{0i}. It is specified by the translation vector in the last column of the matrix $A_0^i, i = 1, \ldots, 6$:

$$p_{01} = \begin{bmatrix} 0 \\ 0 \\ 14.00 \end{bmatrix} \qquad p_{02} = \begin{bmatrix} 4.187 \\ 4.841 \\ 14.00 \end{bmatrix} \qquad p_{03} = \begin{bmatrix} 10.429 \\ -.540 \\ 11.559 \end{bmatrix} \qquad (4.2.45)$$

For the Stanford/JPL manipulator in Figure 2.4, $p_{04} = p_{03}$, $p_{05} = p_{03}$. The position of the origin of the sixth (end-effector) coordinate frame is $p_{06} = [9.652, .200, 1.868]'$.

In order to evaluate the cross-products in Equation 4.2.15, the components (projections) of the unit vector $k_{z_i}, i = 0, 1, \ldots, 5$ on the base coordinates are needed. They can be obtained by applying the (3×3) rotation submatrix $(A_0^i)_R$ to k_{z_i}; it gives $k_{0i} = (A_0^i)_R k_{z_i}$, where $k_{z_i} = [0, 0, 1]'$. Thus, the components of k_{z_i} on the base coordinates become specified:

$$k_{01} = \begin{bmatrix} .759 & 0 & .650 \\ -.650 & 0 & .759 \\ 0 & -1 & 0 \end{bmatrix} \begin{bmatrix} 0 \\ 0 \\ 1 \end{bmatrix} = \begin{bmatrix} .650 \\ .759 \\ 0 \end{bmatrix} \qquad (4.2.46)$$

$$k_{02} = \begin{bmatrix} -.214 & .650 & .728 \\ .184 & .759 & -.628 \\ -.959 & 0 & -.282 \end{bmatrix} \begin{bmatrix} 0 \\ 0 \\ 1 \end{bmatrix} = \begin{bmatrix} .728 \\ -.624 \\ -.282 \end{bmatrix} \qquad (4.2.47)$$

$$k_{03} = \begin{bmatrix} -.214 & .650 & .728 \\ .184 & .759 & -.624 \\ -.959 & 0 & -.282 \end{bmatrix} \begin{bmatrix} 0 \\ 0 \\ 1 \end{bmatrix} = \begin{bmatrix} .728 \\ -.624 \\ -.282 \end{bmatrix} \qquad (4.2.48)$$

$$k_{04} = \begin{bmatrix} -.211 & -.728 & .651 \\ .188 & .624 & .758 \\ -.959 & -.282 & .005 \end{bmatrix} \begin{bmatrix} 0 \\ 0 \\ 1 \end{bmatrix} = \begin{bmatrix} .651 \\ .758 \\ .005 \end{bmatrix} \qquad (4.2.49)$$

$$k_{05} = \begin{bmatrix} -.754 & .651 & -.079 \\ .647 & .758 & .076 \\ .110 & .005 & -.993 \end{bmatrix} \begin{bmatrix} 0 \\ 0 \\ 1 \end{bmatrix} = \begin{bmatrix} -.079 \\ .076 \\ -.993 \end{bmatrix} \qquad (4.2.50)$$

Therefore, the Jacobian matrix for the Stanford/JPL arm at the given configuration can now be computed from Equation 4.2.37 by evaluating the cross-products using Equations 4.2.46–4.2.50:

$$J_6^s = \begin{bmatrix} -.200 & -9.212 & .728 & 6.256 & -7.352 & 0 \\ 9.652 & 7.892 & -.624 & 7.278 & 6.312 & 0 \\ 0 & -7.199 & -.282 & .054 & 1.072 & 0 \\ 0 & .650 & 0 & .728 & .651 & -.079 \\ 0 & .759 & 0 & -.624 & .758 & .076 \\ 1 & 0 & 0 & -.282 & .005 & -.993 \end{bmatrix} \qquad (4.2.51)$$

As a result, the velocities of the gripper at the given point can be calculated in the base coordinates by multiplying the joint velocities by the Jacobian matrix J_6^s given in Equation 4.2.51. The numerical values of the translational and rotational velocities of the gripper at the point under consideration are as follows:

$$
\begin{bmatrix} v_{0x} \\ v_{0y} \\ v_{0z} \\ -- \\ \omega_{0x} \\ \omega_{0y} \\ \omega_{0z} \end{bmatrix} =
\begin{bmatrix}
-.200 & -9.212 & .728 & 6.256 & -7.352 & 0 \\
9.652 & 7.892 & -.624 & 7.278 & 6.312 & 0 \\
0 & -7.199 & -.282 & .054 & 1.072 & 0 \\
0 & .650 & 0 & .728 & .651 & -.079 \\
0 & .759 & 0 & -.624 & .758 & .076 \\
1 & 0 & 0 & -.282 & .005 & -.993
\end{bmatrix}
$$

$$
\begin{bmatrix} -1.034 \\ .127 \\ 12.399 \\ 1.475 \\ 2.905 \\ -3.129 \end{bmatrix} \qquad (4.2.52)
$$

$$
\begin{bmatrix} \dot{p}_{06} \\ -- \\ \omega_{06} \end{bmatrix} =
\begin{bmatrix} -4.008 \\ 12.459 \\ -1.251 \\ --- \\ 3.309 \\ 1.134 \\ 1.655 \end{bmatrix} \qquad (4.2.53)
$$

where the dimensions for the angular velocities are rad/s, and for the translational velocity in./s. Equation 4.2.53 gives the components of the translational and rotational velocities of the gripper in the Cartesian base coordinate system.

It should be remarked that the computations required in Example 4.2.4 can advantageously be performed using symbolic programs. Computationally efficient methods for determining the Jacobian matrices of manipulators are described in the literature [e.g., 3,4,6].

4.3 GENERALIZED EXTERNAL FORCES AND JOINT TORQUES RELATED BY JACOBIAN MATRIX

The actuators of the joints in a manipulator generate torques and forces, which will be termed generalized joint torques in the sequel. Their effects are transferred through the links to different parts of the manipulator causing the arm to move and/or to exert some force or torque on the environment. The static forces and torques (moments) exerted on the environment will be called the external generalized forces. We will next establish a relationship between the generalized joint torques and the external generalized forces exerted by the manipulator arm in a given configuration. Such a relationship can be determined, for example, by making use of the virtual work principle.

Suppose that a base coordinate system $(x_0 y_0 z_0)$ has been defined. The static generalized external force F_e^n exerted by the nth link will have the following components:

$$F_e^n = \begin{bmatrix} F_{x_0}^n & F_{y_0}^n & F_{z_0}^n; M_{x_0}^n & M_{y_0}^n & M_{z_0}^n \end{bmatrix}' = \begin{bmatrix} F_0^n \\ M_0^n \end{bmatrix} \qquad (4.3.1)$$

where the external force vector F_e^n is composed of the force F_0^n with components $F_{x_0}^n$, $F_{y_0}^n$, and $F_{z_0}^n$, and the moment (torque) M_0^n with components $M_{x_0}^n$, $M_{y_0}^n$, and $M_{z_0}^n$ on the x_0-, y_0-, and z_0-axes, respectively. The force component $F_{x_0}^n$ attempts to cause a translational (linear) motion in the direction of the x_0-axes; a similar interpretation holds for $F_{y_0}^n$ and $F_{z_0}^n$. The moment components $M_{x_0}^n$, $M_{y_0}^n$, and $M_{z_0}^n$ try to cause rotations about the x_0-, y_0-, and z_0-axes, respectively.

The actuator of joint i generates the joint torque τ_i. The actuators of the first n joints in a serial link N-joint manipulator ($N \geq n$) produce the generalized torques that may be expressed as the torque vector Γ_n:

$$\Gamma_n = \begin{bmatrix} \tau_1 & \tau_2 & \ldots & \tau_n \end{bmatrix}' \qquad (4.3.2)$$

Some components of Γ_n may actually represent linear forces instead of torques. The task here is to determine the equation that relates the generated joint torque vector Γ_n to the external force vector F_e^n.

The sought relationship between the generalized torque and force in a given manipulator configuration is obtained by equating the expression for the virtual work performed by the joint torque Γ_n over an incremental joint variable change with the virtual work that the generalized external force F_e^n performs over an incremental distance. When joint i moves an incremental distance, $\Delta\theta_i, i = 1, \ldots, n$, the virtual work dW performed by the torques of the joints is

$$dW = \sum_{i=1}^{n} \tau_i \Delta\theta_i$$

$$= \begin{bmatrix} \tau_1 & \tau_2 \cdots \tau_n \end{bmatrix} \begin{bmatrix} \Delta\theta_1 \\ \Delta\theta_2 \\ \vdots \\ \Delta\theta_n \end{bmatrix} \qquad (4.3.3)$$

Assuming that there is no dissipation of energy, the virtual work dW in Equation 4.3.3 equals the incremental work performed by the external generalized force when the manipulator moves an incremental distance expressed in the base coordinate system:

$$dW = \begin{bmatrix} F_{x_0}^n & F_{y_0}^n & F_{z_0}^n; M_{x_0}^n & M_{y_0}^n & M_{z_0}^n \end{bmatrix} \begin{bmatrix} \Delta p_{0x}^n \\ \Delta p_{0y}^n \\ \Delta p_{0z}^n \\ -- \\ \Delta\psi_{0x}^n \\ \Delta\psi_{0y}^n \\ \Delta\psi_{0z}^n \end{bmatrix}$$

$$= (F_e^n)' \begin{bmatrix} \Delta p_0^n \\ \Delta\psi_0^n \end{bmatrix} \qquad (4.3.4)$$

where the incremental translational change $\Delta p_0^n = [\Delta p_{0x}^n \quad \Delta p_{0y}^n \quad \Delta p_{0z}^n]'$, and the incremental angular change $\Delta \psi_0^n = [\Delta \psi_{0x}^n \quad \Delta \psi_{0y}^n \quad \Delta \psi_{0z}^n]'$ of link n are expressed relative to the base coordinate system. Since there is no dissipation of energy, Expressions 4.3.3 and 4.3.4 are equal:

$$(\Gamma_n)'\Delta\theta = (F_e^n)' \begin{bmatrix} \Delta p_0^n \\ \Delta \psi_0^n \end{bmatrix} \qquad (4.3.5)$$

where $\Delta\theta = [\Delta\theta_1 \Delta\theta_2 \ldots \Delta\theta_n]'$. The incremental changes in the rotational and translational positions can be described by means of the corresponding velocities and an incremental time Δt:

$$\Delta\theta = \dot{\theta}\Delta t \qquad \Delta p_0^n = \dot{p}_0^n \Delta t \quad \Delta \psi_0^n = \omega_0^n \Delta t \qquad (4.3.6)$$

Equation 4.3.6 can then be substituted into Equation 4.3.5. Both sides of the resulting equation are divided by Δt to obtain

$$(\Gamma_n)'\dot{\theta} = (F_e^n)' \begin{bmatrix} \dot{p}_0^n \\ \omega_0^n \end{bmatrix} \qquad (4.3.7)$$

The expressions on both sides in Equation 4.3.7 represent the power associated with the motion. Thus, Equation 4.3.7 could have been written directly by equating the powers associated with the joint moment Γ_n and the external force F_e^n.

The velocities of the joints are related to the translational and rotational velocities of the nth link by the Jacobian matrix J_n as shown in Equations 4.2.14 and 4.2.15. The substitution of Equation 4.2.14 into Equation 4.3.7 gives, after transposing both sides

$$\Gamma_n = J_n'(F_e^n) \qquad (4.3.8)$$

where matrix J_n' is the transpose of the Jacobian matrix J_n in Equation 4.2.15, and

$$\Gamma_n = \begin{bmatrix} \tau_1 \ldots \tau_n \end{bmatrix}'$$

$$F_e^n = \begin{bmatrix} F_{x_0}^n & F_{y_0}^n & F_{z_0}^n; M_{x_0}^n & M_{y_0}^n & M_{z_0}^n \end{bmatrix}' \qquad (4.3.9)$$

The entries of the Jacobian matrix depend on the values of the joint variables, and they are usually expressed relative to the base coordinate system. In this case, the components of the external generalized force vector Γ_n must also be expressed relative to the base coordinates.

Equation 4.3.8 expresses a vector relationship between the generalized joint torque Γ_n and the external generalized force F_e^n exerted by the first n links on the environment in a specific configuration of the manipulator.

An example is next presented to illustrate the basic approach described for determining the relationship between the joint torques and the external generalized forces.

EXAMPLE 4.3.1

A planar manipulator in which two joints are rotational and the third (wrist) joint is locked is shown schematically in Figure 4.5. The end-effector at O_2 is pressing against a flat horizontal surface. As a consequence, it is subject to static external (reaction) forces specified by F_{x_2}, F_{y_2} expressed in the end-effector $(x_2 y_2 z_2)$ frame. The problem is to determine the joint torques (τ_1, τ_2) that will balance the external forces F_{x_2} and F_{y_2} $(F_{z_2} = 0)$. The external moment $M_0^2 = 0$.

To solve the problem, the relationship in Equation 4.3.8 between the external generalized forces and the joint torques needs to be evaluated:

$$\Gamma_2 = J_2'(F_e) \tag{4.3.10}$$

where the joint torque vector $\Gamma_2 = [\tau_1, \tau_2]'$. The external static generalized force vector is denoted by F_e where superscript $n = N = 2$ of Equation 4.3.9 is dropped for convenience. It is a six-dimensional vector in which the first three components are the forces, and the last three components are the moments. The elements of the Jacobian (2×6) matrix J_2' depends on the configuration of the manipulator.

The Jacobian matrix J_2 and the generalized force F_e are represented relative to the base coordinate system. The Jacobian matrix is determined by Equation 4.2.15. It is calculated in Example 4.2.1 and given by Equation 4.2.29. The transpose of the Jacobian matrix is

$$J_2' = \begin{bmatrix} -\ell_1 s_1 - \ell_2 s_{12} & \ell_1 c_1 + \ell_2 c_{12} & 0 & ; & 0 & 0 & 1 \\ -\ell_2 s_{12} & \ell_2 c_{12} & 0 & ; & 0 & 0 & 1 \end{bmatrix} \tag{4.3.11}$$

The Jacobian matrix J_2 is expressed relative to the base coordinate system. The external generalized force vector F_e is given in the $(x_2\, y_2\, z_2)$-coordinate

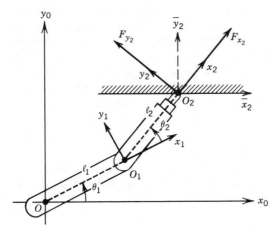

FIGURE 4.5

The end-effector (at O_2) of planar manipulator exerting force on the environment.

frame. Its components in the base coordinates are obtained by applying the rotation submatrix $(A_0^2)_R$ in Equation 4.2.32 to the generalized force vector $F_2 = [F_{x_2} \ F_{y_2} \ F_{z_2}]'$.

$$F_e = \begin{bmatrix} F_0^2 \\ M_0^2 \end{bmatrix} = \begin{bmatrix} (A_0^2)_R F_2 \\ 0 \end{bmatrix} \qquad (A_0^2)_R = \begin{bmatrix} c_{12} & -s_{12} & 0 \\ s_{12} & c_{12} & 0 \\ 0 & 0 & 1 \end{bmatrix} \qquad (4.3.12)$$

where $M_0^2 = 0$ is a three-dimensional null vector. The substitution of Equation 4.3.12 into Equation 4.3.10 leads to

$$\Gamma_2 = J_2' \begin{bmatrix} (A_0^2)_R F_2 \\ 0 \end{bmatrix} \qquad (4.3.13)$$

The matrices in Equation 4.3.13 are multiplied to obtain

$$\begin{bmatrix} \tau_1 \\ \tau_2 \end{bmatrix} = \begin{bmatrix} \ell_1 s_2 & \ell_1 c_2 + \ell_2 & 0 \ ; \ 0 & 0 & 1 \\ 0 & \ell_2 & 0 \ ; \ 0 & 0 & 1 \end{bmatrix} \begin{bmatrix} F_2 \\ 0 \end{bmatrix} \qquad (4.3.14)$$

where the force vector F_2 is expressed in the end-effector coordinate frame. The joint torques τ_1 and τ_2 given by Equation 4.3.14 balance the reaction forces F_{x_2} and F_{y_2} exerted externally on the end-effector in the given configuration of the manipulator.

Since the planar manipulator under consideration has two DOF, Equation 4.3.14 may be expressed in a reduced form:

$$\begin{bmatrix} \tau_1 \\ \tau_2 \end{bmatrix} = \begin{bmatrix} \ell_1 s_2 & \ell_1 c_2 + \ell_2 \\ 0 & \ell_2 \end{bmatrix} \begin{bmatrix} F_{x_2} \\ F_{y_2} \end{bmatrix} \qquad (4.3.15)$$

which can also be verified directly from Figure 4.5. Equation 4.3.15 determines the relationship between the externally applied forces and the joint torques.

If the joint torques τ_1 and τ_2 are known, the external force applied to the environmental object by the end-effector can be calculated by inverting the matrix in Equation 4.3.15. The inverse of the transposed Jacobian matrix in Equation 4.3.15 exists except at $\theta_2 = (0 + k\pi)$ rad, $k = 0, 1, \ldots$, which represents the singular condition. At the singular configuration ($k = 0$), link two functions as an extension of link one. In this particular case, Equation 4.3.15 gives: $\tau_1 = (\ell_1 + \ell_2) F_{y_2}$ and $\tau_2 = \ell_2 F_{y_2}$. Thus, the force component F_{x_2} at the singular configuration does not influence the joint torques.

When the manipulator is moving, possible singular configurations occur at discrete time instances (in a set of measure zero). Special precautions should be incorporated in the control strategy for possible singular points in designing force controllers, if the inverse Jacobian is used in the control scheme.

4.4 ROTATIONAL AND TRANSLATIONAL ACCELERATIONS IN BASE COORDINATE SYSTEMS

When the links of a manipulator move, they are usually subject to inertial forces and torques. These generalized forces depend on the translational and rotational accelerations of the links. We will present next recursive expressions for the translational and rotational accelerations of the links in a serial link manipulator in terms of the joint variables [2,7].

To describe mathematically the motion of the manipulator joints and links, the basic laws of Newton and Euler can be applied. The expressions for the translational and rotational accelerations are needed when manipulator dynamics are studied in Newton–Euler's formulation. The accelerations of link $(i + 1)$ are next determined in the form that can sequentially be applied to obtain the accelerations of all links in a serial link manipulator. They are derived from the equations that describe the link velocities presented in Section 4.1.

The determination of the translational and rotational accelerations of link $(i + 1)$ is based on Equations 4.1.8, 4.1.9, and 4.2.1–4.2.7. The rotational acceleration $\boldsymbol{\omega}_{0(i+1)}$ of the link $(i + 1)$ results from differentiating both sides of Equations 4.2.1 and 4.2.2 with respect to time:

$$\dot{\boldsymbol{\omega}}_{0(i+1)} = \dot{\boldsymbol{\omega}}_{0i} + \ddot{\theta}_{i+1}\mathbf{k}_{z_i} + \boldsymbol{\omega}_{0i} \times (\dot{\theta}_{i+1}\mathbf{k}_{z_i}) \quad [\text{link}(i + 1)\text{revolute}] \quad (4.4.1)$$

$$\dot{\boldsymbol{\omega}}_{0(i+1)} = \dot{\boldsymbol{\omega}}_{0i} \quad [\text{link}(i + 1)\text{prismatic}] \quad (4.4.2)$$

where $i = 0, 1, \ldots, N - 1$ in an N-joint manipulator.

The translational (linear) acceleration $\mathbf{p}_{0(i+1)}$ of link $(i + 1)$ is derived by starting from Equations 4.2.4 and 4.2.5. The differentiation of both sides of Equation 4.2.4 gives for the revolute link $(i + 1)$:

$$\ddot{\mathbf{p}}_{0(i+1)} = \ddot{\mathbf{p}}_{0i} + \dot{\boldsymbol{\omega}}_{0(i+1)} \times \mathbf{p}_{i+1}^{*}$$

$$+ \boldsymbol{\omega}_{0(i+1)} \times (\boldsymbol{\omega}_{0(i+1)} \times \mathbf{p}_{i+1}^{*}) \quad [\text{link}(i + 1)\text{revolute}] \quad (4.4.3)$$

where the vector $p_{i+1}^{*} = \mathbf{p}_{0(i+1)} - \mathbf{p}_{0i}$ (see Figure 4.3).

When link $(i + 1)$ is prismatic, the translational acceleration is determined by differentiating both sides of Equation 4.2.5. The acceleration in the translational motion of link $(i + 1)$ is thus obtained:

$$\ddot{\mathbf{p}}_{0(i+1)} = \ddot{\mathbf{p}}_{0i} + \ddot{\theta}_{0(i+1)}\mathbf{k}_{z_i} + 2\boldsymbol{\omega}_{0(i+1)} \times (\dot{\theta}_{i+1}\mathbf{k}_{z_i}) + \dot{\boldsymbol{\omega}}_{0(i+1)} \times \mathbf{p}_{i+1}^{*}$$

$$+ \boldsymbol{\omega}_{0(i+1)} \times (\boldsymbol{\omega}_{0(i+1)} \times \mathbf{p}_{i+1}^{*}) \quad [\text{link } (i + 1) \text{ prismatic}] \quad (4.4.4)$$

where $\dot{\theta}_{i+1}$ refers to the translational velocity. Equations 4.4.1–4.4.4 determine the rotational and translational accelerations of link $(i + 1)$ in an N-

joint manipulator. The accelerations for all links in a manipulator can be calculated by setting sequentially $i = 0, 1, \ldots, N - 1$. Thus, the difference Equations 4.4.1–4.4.4 with respect to i are solved in the forward (i increases) direction.

EXAMPLE 4.4.1

The angular and translational accelerations of the links in the planar manipulator described in Example 4.2.1 are to be determined. The planar manipulator is shown schematically in Figure 4.4.

The problem is solved by applying Equations 4.4.1 and 4.4.3 to the planar manipulator. The angular acceleration of link one is determined by Equation 4.4.1 when $i = 0$:

$$\dot{\boldsymbol{\omega}}_{01} = \dot{\boldsymbol{\omega}}_{00} + \ddot{\theta}_1 \mathbf{k}_{z_0} + \boldsymbol{\omega}_{00} \times (\dot{\theta}_1 \mathbf{k}_{z_0}) \tag{4.4.5}$$

Since the base (zeroth) coordinate system is fixed, it follows that $\boldsymbol{\omega}_{00} = 0$, $\dot{\boldsymbol{\omega}}_{00} = 0$. By letting $\dot{\boldsymbol{\omega}}_{01} = [\dot{\omega}_{01x}, \dot{\omega}_{01y}, \dot{\omega}_{01z}]'$, one obtains

$$[\dot{\omega}_{01x}, \dot{\omega}_{01y}, \dot{\omega}_{01z}]' = [0, 0, \ddot{\theta}_1]' \tag{4.4.6}$$

Equation 4.4.6 describes the rotational acceleration of link one.

The angular acceleration for the second link is obtained also from Equation 4.4.1. For $i = 1$, Equation 4.4.1 gives:

$$\dot{\boldsymbol{\omega}}_{02} = \dot{\boldsymbol{\omega}}_{01} + \ddot{\theta}_2 \mathbf{k}_{z_1} + \boldsymbol{\omega}_{01} \times (\dot{\theta}_2 \mathbf{k}_{z_1}) \tag{4.4.7}$$

Since the vectors $\mathbf{k}_{z_0}$ and $\mathbf{k}_{z_1}$ are parallel, Equation 4.4.7 can be simplified to

$$\begin{bmatrix} \dot{\omega}_{02x} \\ \dot{\omega}_{02y} \\ \dot{\omega}_{02z} \end{bmatrix} = \begin{bmatrix} \omega_{01y}\dot{\theta}_2 \\ \omega_{01x}\dot{\theta}_2 \\ \ddot{\theta}_1 + \ddot{\theta}_2 \end{bmatrix} = \begin{bmatrix} 0 \\ 0 \\ \ddot{\theta}_1 + \ddot{\theta}_2 \end{bmatrix} \tag{4.4.8}$$

Equation 4.4.8 specifies the rotational acceleration of link two in the base coordinate system.

The translational acceleration of link one is specified by Equation 4.4.3, when $i = 0$:

$$\ddot{\mathbf{p}}_{01} = \ddot{\mathbf{p}}_{00} + \dot{\boldsymbol{\omega}}_{01} \times (\mathbf{p}_{01} - \mathbf{p}_{00}) + \boldsymbol{\omega}_{01} \times [\boldsymbol{\omega}_{01} \times (\mathbf{p}_{01} - \mathbf{p}_{00})] \tag{4.4.9}$$

where $p_{00} = [0 \ \ 0 \ \ 0]'$; $p_{01} = [\ell_1 c_1 \ \ \ell_1 s_1 \ \ 0]'$; $\dot{\omega}_{01} = [0 \ \ 0 \ \ \ddot{\theta}_1]'$; $\omega_{01} = [0 \ \ 0 \ \ \dot{\theta}_1]'$. Equation 4.4.9 can now be rewritten in the form of column vectors after forming the cross-products of the vectors:

$$\begin{bmatrix} \ddot{p}_{01x} \\ \ddot{p}_{01y} \\ \ddot{p}_{01z} \end{bmatrix} = \begin{bmatrix} -\ell_1 s_1 \\ \ell_1 c_1 \\ 0 \end{bmatrix} \ddot{\theta}_1 - \begin{bmatrix} \ell_1 c_1 \\ \ell_1 s_1 \\ 0 \end{bmatrix} \dot{\theta}_1^2 = \begin{bmatrix} -\ddot{\theta}_1 \ell_1 s_1 - \dot{\theta}_1^2 \ell_1 c_1 \\ \ddot{\theta}_1 \ell_1 c_1 - \dot{\theta}_1^2 \ell_1 s_1 \\ 0 \end{bmatrix} \tag{4.4.10}$$

because $\omega_{01z} = \dot{\theta}_1$, and $\dot{\omega}_{01z} = \ddot{\theta}_1$ by Equation 4.4.6. Equation 4.4.10 describes $\ddot{p}_{01}$, the translational acceleration of link one.

The translational acceleration $\ddot{p}_{02}$ of link two is given similarly by Equation 4.4.3 when $i = 1$:

$$\ddot{\mathbf{p}}_{02} = \ddot{\mathbf{p}}_{01} + \dot{\boldsymbol{\omega}}_{02} \times (\mathbf{p}_{02} - \mathbf{p}_{01}) + \boldsymbol{\omega}_{02} \times \left[\boldsymbol{\omega}_{02} \times (\mathbf{p}_{02} - \mathbf{p}_{01}) \right] \qquad (4.4.11)$$

where $\ddot{p}_{01}$ is given by Equation 4.4.10, $\dot{\omega}_{02}$ by Equation 4.4.8, $p_{02} - p_{01} = [\ell_2 c_{12} \ \ \ell_2 s_{12} \ \ 0]'$ by Equation 4.2.26, and $\omega_{02} = [0, 0, \dot{\theta}_1 + \dot{\theta}_2]'$ by Equation 4.2.30. By forming the vector cross-products in Equation 4.4.11, one obtains

$$
\begin{bmatrix} \ddot{p}_{02x} \\ \ddot{p}_{02y} \\ \ddot{p}_{02z} \end{bmatrix}
=
\begin{bmatrix} -\ddot{\theta}_1 \ell_1 s_1 - \dot{\theta}_1^2 \ell_1 c_1 \\ \ddot{\theta}_1 \ell_1 c_1 - \dot{\theta}_1^2 \ell_1 s_1 \\ 0 \end{bmatrix}
+
\begin{bmatrix} -(\ddot{\theta}_1 + \ddot{\theta}_2)\ell_2 s_{12} \\ (\ddot{\theta}_1 + \ddot{\theta}_2)\ell_2 c_{12} \\ 0 \end{bmatrix}
+
\begin{bmatrix} -(\dot{\theta}_1 + \dot{\theta}_2)^2 \ell_2 c_{12} \\ -(\dot{\theta}_1 + \dot{\theta}_2)^2 \ell_2 s_{12} \\ 0 \end{bmatrix}
$$

$$(4.4.12)$$

For the planar manipulator, Equation 4.4.12 may be rewritten as

$$
\begin{bmatrix} \ddot{p}_{02x} \\ \ddot{p}_{02y} \end{bmatrix}
=
\begin{bmatrix} -\ell_1 s_1 & -\ell_2 s_{12} \\ \ell_1 c_1 & \ell_2 c_{12} \end{bmatrix}
\begin{bmatrix} \ddot{\theta}_1 \\ \ddot{\theta}_1 + \ddot{\theta}_2 \end{bmatrix}
-
\begin{bmatrix} \ell_1 c_1 & \ell_2 c_{12} \\ \ell_1 s_1 & \ell_2 s_{12} \end{bmatrix}
\begin{bmatrix} \dot{\theta}_1^2 \\ (\dot{\theta}_1 + \dot{\theta}_2)^2 \end{bmatrix}
$$

$$(4.4.13)$$

Equation 4.4.13 expresses the translational acceleration of link two in the base coordinate system. The first term on the right side of Equation 4.4.13 describes the components of the acceleration vector that is in the direction of the tangent to the trajectory of the motion on the base coordinates. Similarly, the second term specifies the corresponding components of the acceleration vector normal to the trajectory.

In this particular case, Equation 4.4.10 can also be obtained in a straightforward manner by differentiating $p_{01} = [\ell_1 c_1 \ \ \ell_1 s_1 \ \ 0]'$ twice with respect to time. Similarly, Equation 4.4.12 results when the second derivative of $p_{02} = [\ell_2 c_{12} + \ell_1 c_1 \ \ \ell_2 s_{12} + \ell_1 s_1 \ \ 0]'$ with respect to time is formed. Alternatively, the differentiation of Equation 4.2.30 with respect to time gives Equations 4.4.8 and 4.4.12.

Example 4.4.1 illustrates the use of Equations 4.4.3 and 4.4.1 in determining explicitly the translational and rotational accelerations of the links in a planar manipulator. In a multijoint manipulator with N serially connected links, the translational and rotational velocities and accelerations of the links can be calculated numerically by Equations 4.2.1–4.2.7 and 4.4.1–4.4.3, respectively. In fact, these equations may be regarded as forward difference equations with respect to integer i, $i = 0, \ldots, N - 1$. They can be solved sequentially by starting with $i = 0$ to calculate the velocities and accelerations of link one. Then, by setting $i = 1$, the velocities and accelerations of link two can be determined, and so on. Thus, the translational and rotational velocities and accelerations can be computed numerically by starting from the base (the zeroth link) of the manipulator and proceeding toward the end-effector. The calculated values of the velocities and accelerations can be used in the dynamical model of the manipulator that is expressed in Newton–Euler's formulation and in designing control strategies for the manipulator motion.

4.5 SUMMARY

The motion of the end-effector in a robot manipulator is caused by the actuators of the joints. The positions, velocities, and accelerations of the joint variables are often measured, or they are otherwise available. The position, velocity, and acceleration of the end-effector, or any other point on the links, are related to the corresponding variables of the joints. The kinematic equations discussed in Chapters 2 and 3 establish the relationships between the positions expressed in the Cartesian coordinate systems and the joint variables. The corresponding relations between the velocities and accelerations are presented in this chapter.

The translational and rotational velocities of the links of a manipulator expressed in the Cartesian base coordinate system are related to the velocities of the joint variables by means of the Jacobian matrix, which is evaluated for a specific configuration of the manipulator. Its elements are expressed with respect to the chosen coordinate frames. A method for determining the Jacobian matrix for serial link manipulators is described.

The Jacobian matrix (transpose) is also used to relate an externally applied static generalized force to the generalized joint torques at the specific configuration of a manipulator. Since the Jacobian matrix changes with the joint variables, the aforementioned generalized force/torque relation needs to be calculated often at each sampling instant. If the inverse of the Jacobian matrix is used in the control algorithm, special precautions should be taken to avoid possible singular points, at which the inverse of the Jacobian matrix ceases to exist.

The translational and rotational velocities and accelerations of the links in a serial link manipulator can be calculated recursively using the forward difference equations by starting from the base (the zeroth link) and proceeding successively link by link toward the end-effector. By this procedure, the numerical values of the translational and rotational velocities and accelerations of the links can effectively be computed, when the velocities and accelerations of the joint variables are known. These recursive equations form the basis in determining the dynamical model for the motion of a manipulator in the framework of Newton–Euler's formulation.

REFERENCES

[1]   H. GOLDSTEIN, *Classical Mechanics*, Addison-Wesley Publishers, Reading, MA, 1980.

[2]   J. Y. S. LUH, M. W. WALKER, and R. P. PAUL, "On-line
      Computational Scheme for Mechanical Manipulators," *Jour-
      nal of Dynamic Systems, Measurement, and Control*, Vol. 102,
      pp. 69–76, June 1980.

[3]   R. FEATHERSTONE, "Position and Velocity Transformations
      Between Robot End-Effector Coordinates and Joint Angles,"
      The International Journal of Robotics, Vol. 2, No. 2, pp. 35–45,
      Summer 1983.

[4]   D. ORIN and W. SCHRADER, "Efficient Jacobian Determi-
      nation for Robot Manipulators," *Robotics Research: The First
      International Symposium*, edited by M. Brady and R. P. Paul,
      MIT Press, Cambridge, Massachusetts, 1984.

[5]   S. ZAREI, "Adaptive Controller Design for Gross Motion of a
      Manipulator," M. S. Thesis, School of Electrical Engineering,
      Purdue University, 1985.

[6]   M. B. LEAHY, L. M. NUGENT, G. N. SARIDIS, and K. P.
      VALAVANIS, "Efficient PUMA Manipulator Jacobian Calcu-
      lation and Inversion," *Journal of Robotic Systems*, No. 2, pp.
      185–197, April 1987.

[7]   J. M. HOLLERBACH and G. SAHAR, "Wrist-Partitioned,
      Inverse Kinematic Accelerations and Manipulator Dynam-
      ics," *The International Journal of Robotics*, Vol. 2, No. 4, pp.
      61–76, 1983.

PROBLEMS

4.1   A point P on a link is specified by the position vector p_i in
      the base coordinate system. The link rotates at the angular
      velocity ω_z relative to the base coordinate system. Show
      that

$$|\boldsymbol{\omega}_z \times \mathbf{p}_i| = \omega_z p_i \sin \theta$$

      where θ is the angle between vectors $\mathbf{p}_i$ and $\boldsymbol{\omega}_z$ (see Figure
      4.1b).

4.2   A moving coordinate frame $(x_1\, y_1\, z_1)$ coincides at $t = 0$ with
      a fixed coordinate system $(x_0\, y_0\, z_0)$. A point P at $t = 0$ is
      given in the moving coordinate frame: $\mathbf{p}_P = 10\mathbf{i}_{x_1} + 0\mathbf{j}_{y_1} +$
      $0\mathbf{k}_{z_1}$. The moving coordinate frame rotates at the angular
      speed of $\omega_0 = \pi/4$ rad/s about the z_0-axis. Point P rotates at
      the angular speed of $3\pi/4$ rad/s about the z_1-axis relative
      to the moving coordinate frame.

  a. Determine the position of the moving coordinate frame at $t = 1$s in the fixed coordinate system.

  b. Determine the location of the point P in the fixed coordinate system at $t = 1$s.

  c. Sketch the position of the moving coordinate frame and that of point P in the fixed coordinate system at $t = 1$s.

4.3 A fixed world coordinate frame $(x_0\ y_0\ z_0)$ has its origin at point O. A (local) coordinate frame $(x_1\ y_1\ z_1)$ is located at time $t = 0$ so that its origin O_1 is at point $\mathbf{p}_{o1} = 5\mathbf{i}_{x_0} + 0\mathbf{j}_{y_0} + 0\mathbf{k}_{z_0}$. The positive x_1-axis coincides with that of the x_0-axis, and the axes z_0 and z_1 are parallel. The coordinate frame rotates at the speed of $\omega_0 = \pi/4$ rad/s about the z_0-axis. A point P is given in the moving coordinate frame: $\mathbf{p}_P = 10\mathbf{i}_{x_1} + 0\mathbf{j}_{y_1} + 0\mathbf{k}_{z_1}$. The angular speed of point P relative to the moving coordinate frame is $\omega_1 = 3\pi/4$ rad/s about the z_1-axis.

  a. At time $t_1 = 1$ s, origin O_1 will be at position O_1' and point P at P'. Determine the coordinates of points O_1' and P'. Sketch the new position of the coordinate frame $(x_1 y_1 z_1)$, and show the new position P' of point P.

  b. Determine the translational velocity of point P in the fixed coordinate system.

4.4 For a spherical (polar) manipulator the local coordinate frames have been assigned as shown in Figure 4.6.

  a. Determine the position vector $p = [p_x\ \ p_y\ \ p_z]'$ of the end-effector in terms of the joint variables $[\theta\ \ \psi\ \ r]' = [\theta_1\ \ \theta_2\ \ r]'$.

  b. Differentiate the position vector p with respect to time to obtain $\dot{p}$ in terms of the joint velocities $\dot{\theta} = [\dot{\theta}_1\ \ \dot{\theta}_2\ \ \dot{r}]'$. Express the answer in the form $\dot{p} = J_f \dot{\theta}$. How is J_f different from the Jacobian matrix given in Equation 4.2.15?

4.5 For a planar manipulator discussed in Example 4.2.1, the position vectors are given: $p_{o1} = [\ell_1 c_1, \ell_1 s_1, 0]'$ and $p_{o2} = [\ell_2 c_{12} + \ell_1 c_1, \ell_2 s_{12} + \ell_1 s_1, 0]'$ where p_{oi} is the position of the origin of the ith coordinate frame in the base coordinate system, and $i = 1, 2$. The orientation of the end-effector does not change.

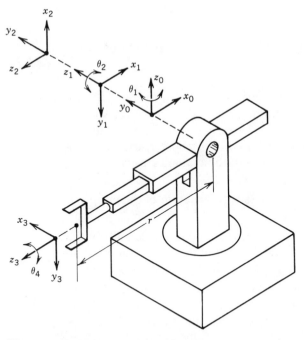

FIGURE 4.6
Spherical manipulator; Problem 4.12.

a. Derive the translational velocity vectors $\dot{p}_{01}$, and $\dot{p}_{02}$ by differentiating p_{01}, and p_{02} with respect to time. Determine the Jacobian matrix from the resulting equations.

b. Derive the translational acceleration vectors $\ddot{p}_{01}$, and $\ddot{p}_{02}$ by differentiating $\dot{p}_{01}$ and $\dot{p}_{02}$.

c. Determine the conditions for the occurrences of a singular point. Sketch the configuration of the manipulator in this situation.

4.6 For the spherical manipulator shown in Figure 4.2, the angular velocities of the joints defined in Example 4.1.1 are given: $\dot{\theta} = \omega_{z_0} = 0.30$ rad/s, $\dot{\psi} = \omega_{z_1} = 0.50$ rad/s, and $\dot{r} = 0$. The orientation of the end-effector is unchanged.

a. Determine the matrix that relates the joint velocities to the Cartesian velocity of the gripper.

b. Use Equation 4.2.14 with $[\dot{\theta}_1 \quad \dot{\theta}_2 \quad \dot{\theta}_3]' = [\dot{\theta} \quad \dot{\psi} \quad \dot{r}]'$ to determine the translational (linear) and the rotational velocities of the gripper G in the base coordinate system.

c. Can a singularity point occur in a spherical manipulator? Explain.

4.7 Suppose that the zeroth coordinate system $(x_0 \, y_0 \, z_0)$ is defined for a spherical manipulator as in Example 4.1.1. The first coordinate frame $(x_1 \, y_1 \, z_1)$ rotates at the speed of $\dot{\theta}_1$ about the z_0-axis. The second coordinate frame $(x_2 \, y_2 \, z_2)$ rotates at the speed of $\dot{\theta}_2$ about z_1-axis. Assume that the length of the arm of the manipulator is constant.

a. Determine the Cartesian acceleration of the gripper G in the general form.

b. What is (i) the absolute value (i.e., the length) of the velocity vector and (ii) that of the acceleration vector?

c. Suppose that the arm length $r = \theta_3$ changes at the velocity $\dot{r} = \dot{\theta}_3$. Determine (i) the actual Cartesian translational velocity vector of the gripper G, (ii) the actual Cartesian translational acceleration vector of the gripper G, (iii) the rotational velocity and the rotational acceleration of the gripper G.

d. Express the gripper position $p_{0G} = [p_{0x} \quad p_{0y} \quad p_{0z}]'$ in terms r, θ, ψ. These joint variables vary with time. Differentiate the expressions to form $\dot{p}_{0G}$ and $\ddot{p}_{0G}$. Compare the results with those obtained in part (c).

4.8 Write the following equations in the form of matrix-vector notations:

a. Equation 4.2.5

b. Equation 4.2.11

c. Equation 4.2.12 by noticing that $\boldsymbol{\omega}_{z_j} = \dot{\theta}_{j+1}\mathbf{k}_{z_j}$

4.9 After combining Equations 4.2.10 and 4.2.11, show that the result can be expressed in the form of Equation 4.2.12.

4.10 Suppose Equation 4.2.14 is written in the form: $V_{0n} = J_n\dot{\theta}$, where J_n and $\dot{\theta}$ change with time.

a. Determine the expression for the generalized acceleration $\dot{V}_{0n}$ in the base coordinate system.

b. Apply the resulting equation to Equation 4.2.30 to determine the generalized acceleration of the gripper of the planar manipulator in the Cartesian base coordinate system.

4.11 For the planar manipulator of Example 4.2.1, the homogeneous transformation matrices $A_0^1, A_1^2,$ and A_0^2 are given in Equations 4.2.31 and 4.2.32. The position of the end-effector is such that $\theta_1 = -\pi/6$ rad, and $\theta_2 = \pi/4$ rad. Assume $\ell_1 = \ell_2 = 0.5$ m. A force $[F_{x_2} \quad F_{y_2} \quad F_{z_2}]' = [5\text{N} \quad 10\text{N} \quad 0]'$ (N = Newton) is applied on the gripper externally. The external moments are zero.

Determine the joint torque $\Gamma_2 = [\tau_1 \quad \tau_2]'$ which is needed to cancel the effect of the external force.

4.12 In the spherical manipulator shown in Figure 4.6, the position of the end-effector is determined by the values $\theta_1 = \pi/3$ rad, $\theta_2 = \pi/4$ rad, and $r = 1.0$ m. The manipulator should exert a force of $[5.0 \quad 0 \quad 15.0]'$ N on an object in the environment where the force is expressed in the local (nsa) coordinate frame. The external moments are zero.

Determine the torques to be generated at the joints so that the external force assumes the given values.

4.13 Repeat Problem 4.12 for the cylindrical robot manipulator studied in Problem 3.8, when $r = 0.75$ m, $\theta = \pi/4$ rad, and $z = 0.5$m relative to the zero position of the coordinate system.

4.14 The differential motion of the end-effector in a manipulator can conveniently be determined on the basis of the velocity relation 4.2.14. For a sufficiently small time increment Δt,

$$\begin{bmatrix} \Delta p(t) \\ \Delta \psi(t) \end{bmatrix} = \begin{bmatrix} \dot{p}(t) \\ \omega(t) \end{bmatrix} \Delta t$$

$$= J(\theta)\,\dot{\theta}\Delta t$$

where $\Delta p(t)$ and $\Delta \psi(t)$ represent incremental changes in the position and orientation vectors, respectively.

In the planar manipulator described in Computer Problem 4.C.1, the initial position of the end-effector in the Cartesian base coordinate system is

$$[p_x(0) \quad p_y(0)] = [0.5 \quad -0.2]$$

An incremental change $\Delta\theta(t)$ occurs in the joint variable vector $\theta(t) = [\theta_1(t) \quad \theta_2(t)]'$ over a time period $\Delta t = 50$ ms. It is given as follows: $\Delta\theta = [0.4 \quad 0.2]'$ rad. Determine the new position vector, that is $p(\Delta t) = p(0) + \Delta p$.

COMPUTER PROBLEM

4.C.1 The planar manipulator with two joints is discussed in Example 4.2.1. The lengths of the links are $\ell_1 = 1.6$ m

and $\ell_2 = 1.3$ m. The end-effector is to follow a straight line
defined in the Cartesian base coordinate system: $y_0 = 0.5$ m
starting from point $(-0.5$ m, 0.5 m) and moving in the
positive direction of the x_0-axis.

The magnitude of the Cartesian translational velocity of the end-effector has the shape of an equilateral
trapezoid. The acceleration period is $10T$ where $T = 25$ms,
until the maximum velocity of 0.5 m/s is reached. The
total duration of motion is 4 s.

a.   Determine the desired trajectories for $\theta_1^d(t)$ and $\theta_2^d(t)$
vs. time.

b.   Use the values of part *(a)* to determine the desired
trajectory on the $(x_0 y_0)$-plane.

c.   Determine the desired velocity profiles $\dot{\theta}_1^d(t)$ and
$\dot{\theta}_2^d(t)$ vs. time using the Jacobian matrix. Graph the
two profiles. Compare these graphs with the Carte-
sian velocity profile.

d.   Repeat part *(a)* when $\ell_1 = 0.6$ m and $\ell_2 = 0.4$ m.

e.   Repeat part *(a)* when the manipulator motion is constrained to the subspace defined by $x_0 \le 0.5$ m.

CHAPTER FIVE

DETERMINATION OF DYNAMICAL MODELS FOR MANIPULATORS

The branch of solid body mechanics that deals with situations in which the forces acting on a body are so arranged that the body remains at rest relative to a given coordinate frame is commonly called *statics*. Another branch of mechanics that deals with the motion of rigid bodies without reference to their masses or forces producing the motion is termed *kinematics*. We have discussed kinematic equations (relations) in Chapters 2 and 3. The part of mechanics that treats the effects of forces changing the motion of solid bodies is referred to as *kinetics*.

The motion of a physical system is described by means of a mathematical model that mimics its behavior. It can conveniently be used in studying the system responses by simulating it under various circumstances and in designing the system so that it will behave in a desirable manner. A system may be presented by means of a *dynamical* model described by differential or difference equations (or partial differential equations for manipulators with nonrigid links). A typical feature of the dynamical models is that the solution to these equations describes the evolution of the motion with *time*. The model equations have time as the independent variable. Under stable steady-state conditions, the system model may be described by *static* equations, which do not contain the variable time explicitly. The dependent variables are now governed by the parametric equations whose solution specifies the steady-state operating conditions of the system.

Dynamical models of a manipulator specify the equations of motion relative to a chosen coordinate system. In many applications, the motion of the end-effector is of primary interest. The center of the end-effector traces a path in the world coordinate system when the manipulator is moved by the actuators. Controlling this path is called the *gross motion* control. It can be determined based on the dynamical model of the manipulator.

Dynamical models or equations of motion for a manipulator can be obtained by forming Euler–Lagrange's equation on the basis of Lagrange's energy function. The resulting differential equations describe the motion in terms of the joint variables and the structural parameters of the manipulator. An alternative approach to the modeling of the manipulator dynamics is to consider each link as a free body and obtain the equations of motion for each link in succession on the basis of Newton's and Euler's laws. Thus, the recursive differential equations of motion can be determined for the entire manipulator with serial links.

We will discuss in this chapter the determination of dynamical models for the gross motion of the robot manipulators. We will present the modeling of a manipulator motion first in Lagrange's formulation using Euler–Lagrange's equation. Then, we will derive the differential equations of motion in the framework of Newton–Euler's formulation.

5.1 DIFFERENTIAL EQUATION MODEL FROM LAGRANGE'S FUNCTION

For many applications, the dynamical model can conveniently be obtained from the Lagrangian function using *Euler–Lagrange's equations*, which often are called simply Lagrange's equations. Lagrange's function is formed from the kinetic and potential energy expressions of the system.

Let N independent variables $q_1(t) \cdots q_N(t)$ be chosen to describe the motion of a manipulator with N joints. These generalized coordinates are expressed as a vector $q(t) = [q_1(t) \cdots q_N(t)]'$. They can be used to determine the kinetic $K(q, \dot{q}, t)$ and potential $P(q, t)$ energy expressions for the manipulator. Lagrange's energy function $\mathcal{L}$ can then be defined for the system:

$$\mathcal{L}(q, \dot{q}, t) = K(q, \dot{q}, t) - P(q, t) \qquad (5.1.1)$$

The equations of motion for the manipulator are obtained by means of Euler–Lagrange's equations:

$$\frac{d}{dt} \left[\frac{\partial \mathcal{L}(q, \dot{q}, t)}{\partial \dot{q}_i} \right] - \frac{\partial \mathcal{L}(q, \dot{q}, t)}{\partial q_i} = F_i \qquad (5.1.2)$$

where $i = 1, \ldots, N$, and F_i is the generalized force or torque acting in the direction of the q_i-coordinate. Frictional forces can be included in the model by inserting their q_i-directional components to the right side of Equation 5.1.2. When Lagrange's Equation 5.1.2 is formed, the generalized coordinates q_i, $i = 1, \ldots, N$ must be independent. If the variables initially chosen to represent the generalized coordinates are not independent, a set of independent variables should first be chosen before applying Lagrange's equation.

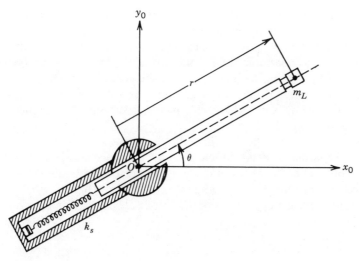

FIGURE 5.1*a*
Cylindrical robot manipulator (top view).

The derivation of Equation 5.1.2 is presented in Appendix A by starting from Hamilton's principle, which represents one of the most basic principles of mechanics.

An example is next presented to demonstrate the determination of the dynamical model using Lagrange's energy function.

EXAMPLE 5.1.1

A cylindrical robot shown schematically in Figures 5.1*a* (top view) and 5.1*b* (side view) has one revolute joint and two prismatic joints. The arm has the length ℓ, and its mass per unit length is a constant m_A/ℓ. The length of the

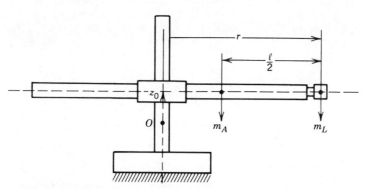

FIGURE 5.1*b*
Cylindrical robot manipulator (side view).

prismatic radial link changes when it slides through the hub. A force opposes the motion of this link. It is modeled as a spring with a constant k_s, which imposes a zero force at $r = 2\ell/3$. Another translational joint determines the height of the arm on the horizontal plane. Its motion is opposed by a force that is also modeled as a spring with constant k_z. This force is zero at $z = \ell_0$. The mass of the hub is m_h with center of gravity on the z-axis and that of the load at the end-effector is m_L. The problem is to determine a dynamical model for the motion of the manipulator.

The equations of motion will be determined by applying Euler–Lagrange's Equation 5.1.2 to Lagrange's energy function. Thus, Lagrange's energy function is first to be obtained. The position of the end-effector can be described by the independent coordinates r, θ, and z, representing the three DOF in the system. Thus, the generalized coordinates are: $r = q_1$, $\theta = q_2$, and $z = q_3$.

Lagrange's energy function is determined by writing the expression for the difference of the kinetic and potential energies in the system. The kinetic energy can be expressed using the radial and tangential velocity components on the horizontal plane, and the vertical velocity component. The energy stored in a spring is included in the expression for the potential energy. Thus, Lagrange's energy function assumes the following form:

$$\mathcal{L}(r, \theta, z; \dot{r}, \dot{\theta}, \dot{z}) = K - P \tag{5.1.3}$$

$$= \frac{1}{2}(m_A + m_L)\dot{r}^2 + \frac{1}{2}\left(m_A \frac{r}{\ell}\right)\left(\frac{1}{2}r\dot{\theta}\right)^2 + \frac{1}{2}\left(m_A \frac{\ell - r}{\ell}\right)\left(\frac{\ell - r}{2}\dot{\theta}\right)^2 + \frac{1}{2}m_L(r\dot{\theta})^2$$

$$+ \frac{1}{2}(\overline{m}_A + m_L)\dot{z}^2 + \frac{1}{2}I_\theta \dot{\theta}^2 - (\overline{m}_A + m_L)gz - \frac{1}{2}k_s\left(r - \frac{2}{3}\ell\right)^2 - \frac{1}{2}k_z(z - \ell_0)^2$$

where $\overline{m}_A = m_A + m_h$, and I_θ represents the effective moment of inertia of the rotating masses excluding m_A and m_L.

The dynamical model is now obtained by substituting Equation 5.1.3 into Equation 5.1.2 with $q_1 = r$, $q_2 = \theta$, and $q_3 = z$. Thus, we have

$$\frac{d}{dt}\left(\frac{\partial \mathcal{L}}{\partial \dot{r}}\right) - \frac{\partial \mathcal{L}}{\partial r} = F_r - B_r(\dot{r}) \tag{5.1.4}$$

$$\frac{d}{dt}\left(\frac{\partial \mathcal{L}}{\partial \dot{\theta}}\right) - \frac{\partial \mathcal{L}}{\partial \theta} = \tau_\theta - B_\theta(\dot{\theta}) \tag{5.1.5}$$

$$\frac{d}{dt}\left(\frac{\partial \mathcal{L}}{\partial \dot{z}}\right) - \frac{\partial \mathcal{L}}{\partial z} = F_z - B_z(\dot{z}) \tag{5.1.6}$$

where F_r and F_z are the translational forces acting in the directions of the r-, and z-coordinates, respectively; τ_θ is the torque causing rotation about the z-

axis. The terms $B_r(\dot{r})$, $B_\theta(\dot{\theta})$, and $B_z(\dot{z})$ represent frictional forces. They oppose the motion and act on the coordinate axes designated by the subscripts. Each frictional term can depend on the velocity indicated by the argument.

By forming the appropriate derivatives, Equations 5.1.4–5.1.6 can be rewritten as follows:

$$\frac{d}{dt}[(m_A + m_L)\dot{r}] - \frac{3}{4}\left(\frac{m_A}{\ell}r\right)\frac{r}{2}\dot{\theta}^2 + \frac{3}{4}(\ell - r)\left(\frac{m_A}{\ell}\right)\frac{\ell - r}{2}\dot{\theta}^2 - m_L r\dot{\theta}^2$$

$$+ k_s\left(r - \frac{2}{3}\ell\right) = F_r - B_r(\dot{r}) \tag{5.1.7}$$

$$\frac{d}{dt}\left[\left(\frac{m_A}{\ell}r\right)\left(\frac{r}{2}\right)^2\dot{\theta} + (\ell - r)\left(\frac{m_A}{\ell}\right)\left(\frac{\ell - r}{2}\right)^2\dot{\theta} + m_L r^2\dot{\theta} + I_\theta\dot{\theta}\right] = \tau_\theta - B_\theta(\dot{\theta}) \tag{5.1.8}$$

$$\frac{d}{dt}[(\overline{m}_A + m_L)\dot{z}] - [-(\overline{m}_A + m_L)g] + k_z(z - \ell_0) = F_z - B_z(\dot{z}) \tag{5.1.9}$$

It is often convenient to express the equations of motion in such a form that the term representing the highest order derivative is on the left side of the equation and the remaining terms on the right side. Equations 5.1.7–5.1.9 can thus be solved for the accelerations:

$$\ddot{r} = \frac{\overline{m}_e(r)r/2}{m_A + m_L}\dot{\theta}^2 - \frac{k_s(r - 2\ell/3)}{m_A + m_L} + \frac{1}{m_A + m_L}[F_r - B_r(\dot{r})] \tag{5.1.10}$$

$$\ddot{\theta} = \frac{1}{I_t(r)}\left\{-\left[\frac{3}{4}\frac{m_A}{\ell}(r^2 - (\ell - r)^2) + 2m_L r\right]\dot{r}\dot{\theta} + \tau_\theta - B_\theta(\dot{\theta})\right\} \tag{5.1.11}$$

$$\ddot{z} = -g - \frac{k_z(z - \ell_0)}{\overline{m}_A + m_L} + \frac{1}{\overline{m}_A + m_L}[F_z - B_z(\dot{z})] \tag{5.1.12}$$

where the moment of inertia $I_t(r) = [r^3 + (\ell - r)^3]m_A/(4\ell) + m_L r^2 + I_\theta$ and $\overline{m}_e(r) = 3[m_A r/\ell - m_A(\ell - r)^2/(r\ell)]/4 + 2m_L$. Equations 5.1.10 and 5.1.11 represent coupled differential equations, whereas Equation 5.1.12 is independent of the motion described by Equations 5.1.10 and 5.1.11. Equations 5.1.10–5.1.12 govern the motion of the manipulator end-effector in the cylindrical coordinate system.

Equations 5.1.10–5.1.12 can also be expressed as a second-order vector differential equation:

$$\begin{bmatrix} \ddot{r} \\ \ddot{\theta} \\ \ddot{z} \end{bmatrix} = \begin{bmatrix} \dfrac{\overline{m}_e(r)r/2}{m_A + m_L}\dot{\theta}^2 \\ -\left[\dfrac{\overline{m}_e(r)}{I_t(r)}r\right]\dot{r}\dot{\theta} \\ 0 \end{bmatrix} + \begin{bmatrix} -\dfrac{k_s(r - 2\ell/3)}{m_A + m_L} \\ 0 \\ -g - \dfrac{k_z(z - \ell_0)}{\overline{m}_A + m_L} \end{bmatrix} + \begin{bmatrix} \dfrac{F_r - B_r(\dot{r})}{m_A + m_L} \\ \dfrac{\tau_\theta - B_\theta(\dot{\theta})}{I_t(r)} \\ \dfrac{F_z - B_z(\dot{z})}{\overline{m}_A + m_L} \end{bmatrix} \tag{5.1.13}$$

In Equation 5.1.13, the first term on the right side represents the Coriolis and centripetal generalized forces, the second term combines the gravitational and the spring forces, and the last term signifies the applied generalized forces and the frictional effects.

The approach described here for a specific cylindrical manipulator will next be applied to determine a general dynamical model for serial link N-joint manipulators in Lagrange's formulation.

5.2 DYNAMICAL MANIPULATOR MODEL USING LAGRANGE FORMULATION

The equations of motion for a cylindrical manipulator are determined in the previous section on the basis of Lagrange's energy function. If the approach is applied to the incremental masses of a link, the summation of the incremental energy expressions over the entire link gives the equation for the total energy of the link. The potential and kinetic energies of all links can thus be determined for Lagrange's energy expression. The equations of motion for the entire manipulator are obtained by substituting Lagrange's energy expression into Euler–Lagrange's equation and forming the appropriate derivatives [1]. The basic equations are only summarized here. The details of the derivation of the equation are presented in Appendix B.

5.2.1   EULER–LAGRANGE'S EQUATIONS DESCRIBE TOTAL JOINT VARIABLE DYNAMICS

The equations of motion for a serial link N-joint manipulator can be determined by first writing the kinetic and potential energy expressions for Lagrange's energy function. Euler–Lagrange's equations can then be formed to describe the dynamics of the manipulator.

The kinetic energy K_i of link i in a manipulator can be determined: $K_i = \dot{p}'_{0i} I_i \dot{p}_{0i}/2$ where $\dot{p}_{0i}$ is the velocity of link i, and I_i is its inertial matrix. This energy can be expressed in terms of the joint velocities. Since $\dot{p}_{0i} = \sum_k (\partial A_0^i/\partial q_k)\dot{q}_k$, the total kinetic energy K can be written as

$$K = \sum_{i=1}^{N} K_i$$

$$= \frac{1}{2} \sum_{i=1}^{N} \mathrm{tr}\left[\sum_{k=1}^{i} \sum_{j=1}^{i} \frac{\partial A_0^i}{\partial q_k} I_i \left(\frac{\partial A_0^i}{\partial q_j} \right)' \dot{q}_j \dot{q}_k \right]$$

(5.2.1)

where tr[·] indicates the trace of matrix [·]; the homogeneous transformation matrix A_0^i relates the ith coordinate frame to the base coordinate system and it depends on the joint variables $q_1, \cdots, q_i$; $\dot{q}_k$ signifies the generalized velocity of the kth joint; and I_i is called the pseudo-inertia matrix of the ith link, that is

$$
I_i = \begin{bmatrix}
\frac{1}{2}(-I_{x_i} + I_{y_i} + I_{z_i}) & I_{x_i y_i} & I_{x_i z_i} & \bar{p}_{x_i} m_i \\
I_{x_i y_i} & \frac{1}{2}(I_{x_i} - I_{y_i} + I_{z_i}) & I_{y_i z_i} & \bar{p}_{y_i} m_i \\
I_{x_i z_i} & I_{y_i z_i} & \frac{1}{2}(I_{x_i} + I_{y_i} - I_{z_i}) & \bar{p}_{z_i} m_i \\
m_i \bar{p}_{x_i} & m_i \bar{p}_{y_i} & m_i \bar{p}_{z_i} & m_i
\end{bmatrix}
$$

$$(5.2.2)$$

The fourth column and row of the matrix in Equation 5.2.2 show the link mass m_i is concentrated on the gravity center $\bar{p}_i = [\bar{p}_{x_i}, \bar{p}_{y_i}, \bar{p}_{z_i}, 1]'$ expressed in the ith coordinate frame. The (3×3) upper left submatrix in Equation 5.2.2 contains the second moments that are determined in accordance with the subscripts, that is

$$
I_{x_i} = \int (p_{y_i}^2 + p_{z_i}^2)\, dm_i \qquad I_{y_i} = \int (p_{x_i}^2 + p_{z_i}^2)\, dm_i \qquad I_{z_i} = \int (p_{x_i}^2 + p_{y_i}^2)\, dm_i
$$

$$
I_{wv} = \int p_w p_v\, dm_i \qquad w \neq v \qquad w, v = x_i, y_i, z_i \qquad (5.2.3)
$$

where the integration is over the mass of the ith link. Terms I_{x_i}, I_{y_i}, and I_{z_i} are the second mass moments of inertia about the axis indicated by the subscript, and I_{wv} is the second mass product of inertia of the link. They are independent of the positions and velocities of the joint variables. Equation 5.2.2 reveals that I_i is a symmetric matrix.

The total potential energy of the manipulator is obtained by first determining the potential energy of the ith link, and then forming the sum over all links. The total potential energy associated with N links is

$$
P = \sum_{i=1}^{N} P_i
$$

$$(5.2.4)$$

$$
= -\sum_{i=1}^{N} m_i g' A_0^i \bar{p}_i
$$

where vector $g' = [g_{0x}, g_{0y}, g_{0z}, 0]$ describes the gravitational acceleration with components in the directions of the coordinates (x_0, y_0, z_0) of the base coordinate system.

Lagrange's energy function $\mathscr{L} = K - P$ can now be formed. It is then substituted into Euler–Lagrange's Equation 5.1.2 to obtain the equations of motion for link $n, n = 1, \cdots, N$ of the manipulator. The resulting equation can be written as follows:

$$
\sum_{i=1}^{N} \left\{ \sum_{k=1}^{i} \operatorname{tr}\left[\frac{\partial A_0^i}{\partial q_n} I_i \left(\frac{\partial A_0^i}{\partial q_k} \right)' \right] \ddot{q}_k \right.
$$

$$
\left. + \sum_{k=1}^{i} \sum_{j=1}^{i} \operatorname{tr}\left[\frac{\partial A_0^i}{\partial q_n} I_i \left(\frac{\partial^2 A_0^i}{\partial q_k \partial q_j} \right)' \right] \dot{q}_k \dot{q}_j - m_i g' \frac{\partial A_0^i}{\partial q_n} \bar{p}_i \right\} = F_n \quad (5.2.5)
$$

Equation 5.2.5 may be expressed in a more tractable form by introducing the following notations:

$$D_{nk} = \sum_{i=max(n,k)}^{N} \text{tr} \left[\frac{\partial A_0^i}{\partial q_n} I_i \left(\frac{\partial A_0^i}{\partial q_k} \right)' \right] \tag{5.2.6}$$

$$D_{nkj} = \sum_{i=max(n,k,j)}^{N} \text{tr} \left[\frac{\partial A_0^i}{\partial q_n} I_i \left(\frac{\partial A_0^i}{\partial q_k \partial q_j} \right)' \right] \tag{5.2.7}$$

$$G_n = -\sum_{i=n}^{N} m_i g' \frac{\partial A_0^i}{\partial q_n} \bar{p}_i \tag{5.2.8}$$

The dynamical model of link n in Equation 5.2.5 for the N-joint serial link manipulator can then be expressed as a second-order nonlinear differential equation:

$$\sum_{k=1}^{N} D_{nk} \ddot{q}_k + \sum_{k=1}^{N} \sum_{j=1}^{N} D_{nkj} \dot{q}_k \dot{q}_j + G_n = F_n \tag{5.2.9}$$

In Equation 5.2.9, the first term on the left describes the generalized forces due to the acceleration of the inertial masses in the system. The second term represents the Coriolis ($k \neq j$) and centripetal ($k = j$) generalized forces. The third term specifies the gravitational effect on the motion of joint n. The force of Coriolis is dependent on the product of the (generalized) velocities of two different joints describing the effect of the velocities of joints k and j on the motion of joint n. The Coriolis terms give rise to the coupling between the differential equations. When the manipulator is operated at slow speeds, the effect of the Coriolis and centripetal forces on the manipulator motion may be small. However, they can play a major role at high speeds in the dynamics of the manipulator.

Equation 5.2.9 can be rewritten as a set of second-order vector differential equations

$$D(q)\ddot{q} + C(q,\dot{q}) + G(q) = F \tag{5.2.10}$$

where vector $q = [q_1 \cdots q_N]'$ consists of the generalized coordinates q_i, $D(q)$ in the symmetric ($N \times N$) pseudo-inertia matrix, the N-dimensional vector $C(q,\dot{q})$ represents the Coriolis and centripetal terms, vector $G(q)$ signifies the gravity effects, and $F = [F_1 \cdots F_N]'$ specifies the generalized torques acting on the joint shaft. Matrix $D(q)$ can be shown to be positive definite [2,3]; hence, its inverse, which is essential for digital simulation, exists for all values of the joint variables. In some problems it may be advantageous to express $C(q,\dot{q})$ in the following form:

$$C(q,\dot{q}) = \begin{bmatrix} \dot{q}' C_1(q) \dot{q} \\ \vdots \\ \dot{q}' C_N(q) \dot{q} \end{bmatrix} \tag{5.2.11}$$

where matrix C_i, $i = 1, \cdots, N$ is symmetric, and depends on the generalized coordinates. The diagonal elements of C_i determine the centripetal (centrifugal) forces acting on link i due to the rotational velocities of the links. The off-diagonal elements are associated with the forces of Coriolis.

For a serial-link manipulator, the computation of the matrix derivatives in Equations 5.2.6–5.2.8 is very time consuming. The calculations can be made faster by first noticing that the homogeneous transformation matrix A_{k-1}^k is a function of the generalized coordinate q_k only. Moreover, the computation of derivative $\partial A_{k-1}^k/\partial q_k$ for serial link manipulators can be converted to a multiplication of matrices [1]. It can be verified by straightforward calculations (Problem 5.8) that

$$\frac{\partial A_0^i}{\partial q_k} = A_0^{k-1} \frac{\partial A_{k-1}^k}{\partial q_k} A_k^i = A_0^{k-1} Q_k A_{k-1}^k A_k^i \qquad \text{for } 1 \leq k \leq i \qquad (5.2.12)$$

$$\frac{\partial^2 A_0^i}{\partial q_k \partial q_\ell} = \begin{cases} A_0^{k-1} \dfrac{\partial A_{k-1}^k}{\partial q_k} A_k^{\ell-1} \dfrac{\partial A_{\ell-1}^\ell}{\partial q_\ell} A_\ell^i = A_0^{k-1}(Q_k A_{k-1}^k) A_k^{\ell-1}(Q_\ell A_{\ell-1}^\ell) A_\ell^i \\ \qquad\qquad \text{for } 1 \leq k < \ell \leq i \\[2mm] A_0^{k-1} \dfrac{\partial^2 A_{k-1}^k}{\partial q_k^2} A_k^i = A_0^{k-1}(Q_k^2 A_{k-1}^k) A_k^i \qquad \text{for } k = \ell \leq i \end{cases}$$

$$(5.2.13)$$

In Equations 5.2.12 and 5.2.13, matrix A_i^i for all i equals the identity matrix, and $Q_k = Q_k^{\text{rot}}$ for a rotational joint k, and $Q_k = Q_k^{\text{trans}}$ for a translational joint k, where

$$Q_k^{\text{rot}} = \begin{bmatrix} 0 & -1 & 0 & 0 \\ 1 & 0 & 0 & 0 \\ 0 & 0 & 0 & 0 \\ 0 & 0 & 0 & 0 \end{bmatrix} \qquad Q_k^{\text{trans}} = \begin{bmatrix} 0 & 0 & 0 & 0 \\ 0 & 0 & 0 & 0 \\ 0 & 0 & 0 & 1 \\ 0 & 0 & 0 & 0 \end{bmatrix} \qquad (5.2.14)$$

The derivatives in Equations 5.2.6–5.2.8 can be calculated by means of Equations 5.2.12–5.2.14, which convert the differential operations on the homogeneous transformation matrices to algebraic matrix multiplications. Thus, the calculations are considerably simplified.

EXAMPLE 5.2.1

A planar manipulator shown in Figure 5.2 can move on a vertical plane. For the coordinate frames shown, the homogeneous transformation matrices A_0^1, A_1^2, and A_0^2 are determined by Equations 2.2.6 and 2.2.7 where $\theta_1 = q_1$ and $\theta_2 = q_2$. They are repeated here ($i = 1, 2$):

$$A_{i-1}^i = \begin{bmatrix} c_i & -s_i & 0 & \ell_i c_i \\ s_i & c_i & 0 & \ell_i s_i \\ 0 & 0 & 1 & 0 \\ 0 & 0 & 0 & 1 \end{bmatrix} \qquad A_0^2 = \begin{bmatrix} c_{12} & -s_{12} & 0 & \ell_1 c_1 + \ell_2 c_{12} \\ s_{12} & c_{12} & 0 & \ell_1 s_1 + \ell_2 s_{12} \\ 0 & 0 & 1 & 0 \\ 0 & 0 & 0 & 1 \end{bmatrix}$$

$$(5.2.15)$$

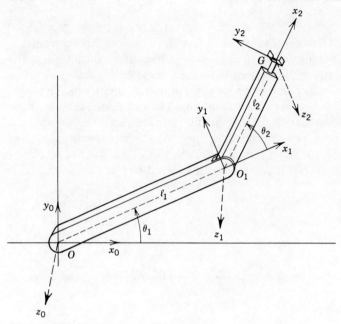

FIGURE 5.2
Planar manipulator on a vertical plane.

The problem is to determine the dynamical model for the two-link manipulator, when the masses of links one and two are m_1 and m_2, respectively. The actuator of joint i produces torque τ_i, and the viscous friction is specified by a coefficient B_i, $i = 1, 2$.

The equations of motion of the manipulator are given by Equations 5.2.5–5.2.9, where $N = 2$. To illustrate the general solution, Equation 5.2.5 is written in the expanded form for the manipulator in question for $n = 1$ and $n = 2$, respectively, by starting with the value of i in the first summation, and expanding the sums inside:

$$\text{tr}\left\{\frac{\partial A_0^1}{\partial q_1}I_1\left[\left(\frac{\partial A_0^1}{\partial q_1}\right)'\ddot{q}_1 + \left(\frac{\partial^2 A_0^1}{\partial q_1^2}\right)'\dot{q}_1^2\right]\right\} - m_1 g'\frac{\partial A_0^1}{\partial q_1}\bar{p}_1 + \text{tr}\left\{\frac{\partial A_0^2}{\partial q_1}I_2\left[\left(\frac{\partial A_0^2}{\partial q_1}\right)'\ddot{q}_1\right.\right.$$

$$\left.\left. + \left(\frac{\partial A_0^2}{\partial q_2}\right)'\ddot{q}_2 + \sum_{k=1}^{2}\left\{\left(\frac{\partial^2 A_0^2}{\partial q_k\partial q_1}\right)'\dot{q}_k\dot{q}_1 + \left(\frac{\partial^2 A_0^2}{\partial q_k\partial q_2}\right)'\dot{q}_k\dot{q}_2\right\}\right]\right\} - m_2 g'\frac{\partial A_0^2}{\partial q_1}\bar{p}_2$$

$$= \tau_1 - B_1\dot{q}_1 \quad (5.2.16)$$

$$\text{tr}\left\{\frac{\partial A_0^2}{\partial q_2}I_2\left[\left(\frac{\partial A_0^2}{\partial q_1}\right)'\ddot{q}_1 + \left(\frac{\partial A_0^2}{\partial q_2}\right)'\ddot{q}_2 + \sum_{k=1}^{2}\left\{\left(\frac{\partial^2 A_0^2}{\partial q_k\partial q_1}\right)\dot{q}_k\dot{q}_1 + \left(\frac{\partial^2 A_0^2}{\partial q_k\partial q_2}\right)'\dot{q}_k\dot{q}_2\right\}\right]\right\}$$

$$- m_2 g'\frac{\partial A_0^2}{\partial q_2}\bar{p}_2 = \tau_2 - B_2\dot{q}_2 \quad (5.2.17)$$

where $\bar{p}_i$ is the center of gravity of link i, and matrix I_i signifies the pseudo-inertia of link i both expressed relative to the ith coordinate frame, $i = 1, 2$. The gravity vector $g = [0, g_{0y}, 0, 0]'$. The generalized coordinates in this problem are $q_1 = \theta_1$ and $q_2 = \theta_2$. Matrices A_0^1, and $A_0^2 = A_0^1 A_1^2$ are given by Equation 5.2.15.

Equations 5.2.16 and 5.2.17 can be simplified by combining the terms that contain explicitly $\ddot{q}_1$, $\ddot{q}_2$, $\dot{q}_1^2$, $\dot{q}_1 \dot{q}_2$, and $\dot{q}_2^2$ to obtain

$$D_{11}\ddot{q}_1 + D_{12}\ddot{q}_2 + \sum_{k=1}^{2}[D_{1k1}\,\dot{q}_k\,\dot{q}_1 + D_{1k2}\,\dot{q}_k\,\dot{q}_2] + G_1 = \tau_1 - B_1\dot{q}_1 \quad (5.2.18)$$

$$D_{21}\ddot{q}_1 + D_{22}\ddot{q}_2 + \sum_{k=1}^{2}[D_{2k1}\,\dot{q}_k\,\dot{q}_1 + D_{2k2}\,\dot{q}_k\,\dot{q}_2] + G_2 = \tau_2 - B_2\dot{q}_2 \quad (5.2.19)$$

Equations 5.2.18 and 5.2.19 can also be obtained directly by writing Equation 5.2.9 for $n = 1$, $n = 2$, and $N = 2$.

In writing Equations 5.2.18 and 5.2.19, the notations of Equations 5.2.6–5.2.8 are applied. The derivatives are converted to matrix products according to Equations 5.2.12–5.2.14 to obtain

$$D_{11} = \sum_{i=1}^{2} \text{tr}\left[\frac{\partial A_0^i}{\partial q_1} I_i \left(\frac{\partial A_0^i}{\partial q_1}\right)'\right] = \sum_{i=1}^{2} \text{tr}\left[(Q_1 A_0^i) I_i (Q_1 A_0^i)'\right] \quad (5.2.20)$$

$$D_{22} = \text{tr}\left[\frac{\partial A_0^2}{\partial q_2} I_2 \left(\frac{\partial A_0^2}{\partial q_2}\right)'\right] = \text{tr}\left[(A_0^1 Q_2 A_1^2) I_2 (A_0^1 Q_2 A_1^2)'\right] \quad (5.2.21)$$

$$D_{12} = D_{21} = \text{tr}\left[\frac{\partial A_0^2}{\partial q_2} I_2 \left(\frac{\partial A_0^2}{\partial q_1}\right)'\right] = \text{tr}\left[(A_0^1 Q_2 A_1^2) I_2 (Q_1 A_0^2)'\right] \quad (5.2.22)$$

$$D_{111} = \sum_{i=1}^{2} \text{tr}\left[\frac{\partial A_0^i}{\partial q_1} I_i \left(\frac{\partial^2 A_0^i}{\partial q_1^2}\right)'\right] = \sum_{i=1}^{2} \text{tr}\left[(Q_1 A_0^i) I_i (Q_1^2 A_0^i)'\right] \quad (5.2.23)$$

$$D_{112} = D_{121} = \text{tr}\left[\frac{\partial A_0^2}{\partial q_1} I_2 \left(\frac{\partial^2 A_0^2}{\partial q_2 \partial q_1}\right)'\right]$$

$$= \text{tr}\left[(Q_1 A_0^2) I_2 (Q_1 A_0^1 Q_2 A_1^2)'\right] \quad (5.2.24)$$

$$D_{122} = \text{tr}\left[\frac{\partial A_0^2}{\partial q_1} I_2 \left(\frac{\partial^2 A_0^2}{\partial q_2^2}\right)'\right] = \text{tr}\left[(Q_1 A_0^2) I_2 (A_0^1 Q_2^2 A_1^2)'\right] \quad (5.2.25)$$

$$D_{211} = \text{tr}\left[\frac{\partial A_0^2}{\partial q_2} I_2\left(\frac{\partial^2 A_0^2}{\partial q_1^2}\right)'\right] = \text{tr}\left[(A_0^1 Q_2 A_1^2) I_2 (Q_1^2 A_0^2)'\right] \quad (5.2.26)$$

$$D_{212} = D_{221} = \text{tr}\left[\frac{\partial A_0^2}{\partial q_2} I_2\left(\frac{\partial^2 A_0^2}{\partial q_2 \,\partial q_1}\right)'\right]$$

$$= \text{tr}\left[(A_0^1 Q_2 A_1^2) I_2 (Q_1 A_0^1 Q_2 A_1^2)'\right] \quad (5.2.27)$$

$$D_{222} = \text{tr}\left[\frac{\partial A_0^2}{\partial q_2} I_2\left(\frac{\partial^2 A_0^2}{\partial q_2^2}\right)'\right] = \text{tr}\left[(A_0^1 Q_2^2 A_1^2) I_2 (A_0^1 Q_2^2 A_1^2)'\right] \quad (5.2.28)$$

$$G_1 = -\sum_{i=1}^{2} m_i g' \frac{\partial A_0^i}{\partial q_1} \bar{p}_i = \sum_{i=1}^{2} m_i g'(Q_1 A_0^i) \bar{p}_i \quad (5.2.29)$$

$$G_2 = -m_2 g' \frac{\partial A_0^2}{\partial q_2} \bar{p}_2 = -m_2 g'(A_0^1 Q_2^2 A_1^2) \bar{p}_2 \quad (5.2.30)$$

where $Q_k = Q_k^{\text{rot}}$, $k = 1, 2$ is given in Equation 5.2.14.

The elements of the pseudo-inertia matrix I_i are given in the general form by Equation 5.2.2 and expressed relative to the local coordinate frame i, $i = 1, 2$. Matrix I_i for the planar manipulator in Figure 5.2 is

$$I_i = \begin{bmatrix} I_{z_i}/2 & 0 & 0 & -m_i \ell_i/2 \\ 0 & I_{z_i}/2 & 0 & 0 \\ 0 & 0 & -I_{z_i}/2 & 0 \\ -m_i \ell_i/2 & 0 & 0 & m_i \end{bmatrix} \quad (5.2.31)$$

where the symmetric matrix I_{z_i} is the moment of inertia about the axis z_i of the rotation. If I_{0i} denotes the second inertial moment of link i about the center of mass, then by the parallel axis law, the second-order moment about the rotational axis becomes for $i = 1, 2$

$$I_{z_i} = I_{0i} + m_i(\ell_i/2)^2 \quad (5.2.32)$$

The parameters in Equations 5.2.18 and 5.2.19 are determined by evaluating expressions 5.2.20–5.2.30:

$$D_{11} = I_{01} + m_1 \ell_1^2/4 + I_{02} + m_2 \ell_2^2/4 + m_2 \ell_1^2 + m_2 \ell_1 \ell_2 c_2 \quad (5.2.33)$$

$$D_{12} = I_{02} + m_2 \ell_2^2/4 + m_2 \ell_1 \ell_2 c_2/2 \qquad D_{12} = D_{21} \quad (5.2.34)$$

$$D_{22} = I_{02} + m_2 \ell_2^2/4 \quad (5.2.35)$$

$$D_{111} = 0 \qquad D_{112} = -m_2 \ell_1 \ell_2 s_2/2 \qquad D_{112} = D_{121} \quad (5.2.36)$$

$$D_{122} = -m_2 \ell_1 \ell_2 s_2/2 \quad (5.2.37)$$

$$D_{211} = m_2 \ell_1 \ell_2 s_2/2 \qquad D_{212} = D_{221} = 0 \qquad D_{222} = 0 \quad (5.2.38)$$

$$G_1 = m_1 g \ell_1 c_1/2 + m_2 g \ell_1 c_1 + m_2 g \ell_2 c_{12}/2 \qquad (5.2.39)$$

$$G_2 = m_2 g \ell_2 c_{12}/2 \qquad (5.2.40)$$

Expressions D_{11} and D_{22} in Equations 5.3.33 and 5.2.35 describe the second moments of inertias about the joint axes associated with the rotational motions of joints one and two, respectively. They result when the parallel axis law is applied to I_{01} and I_{02}. The coupling inertias D_{12} and D_{21} reflect the action-reaction effects in the system. These inertia terms do not depend on q_1, the generalized coordinate of joint one. This property holds for general serial link manipulators [2]. Equations 5.2.36–5.2.38 specify the centripetal and Coriolis effects on the dynamics of the planar manipulator. Terms G_1 and G_2 in Equations 5.2.39 and 5.2.40 are due to the motion on the vertical plane (in the gravitational field).

The parameters given in Equations 5.2.33–5.2.40 are then substituted into the dynamical model Equations 5.2.18 and 5.2.19. The resulting equations describe the motion of the planar manipulator. (They are given explicitly for Computer Problem 5.C.1. in Table 5.4.)

The dynamical Equations 5.2.18 and 5.2.19 can also be expressed in the form of a second-order vector differential Equation 5.2.10:

$$\begin{bmatrix} D_{11}(q) D_{12}(q) \\ D_{21}(q) D_{22}(q) \end{bmatrix} \begin{bmatrix} \ddot{q}_1 \\ \ddot{q}_2 \end{bmatrix} + \begin{bmatrix} \dot{q}' C_1(q) \dot{q} \\ \dot{q}' C_2(q) \dot{q} \end{bmatrix} + \begin{bmatrix} G_1(q) \\ G_2(q) \end{bmatrix} = \begin{bmatrix} \tau_1 - B_1 \dot{q}_1 \\ \tau_2 - B_2 \dot{q}_2 \end{bmatrix} \qquad (5.2.41)$$

where the generalized coordinate vector $q = [\theta_1 \theta_2]'$. The inertia matrix and $C_n(q)$, $n = 1, 2$ are symmetric and given as

$$C_n(q) = \begin{bmatrix} D_{n11}(q) D_{n12}(q) \\ D_{n21}(q) D_{n22}(q) \end{bmatrix} \qquad D(q) = \begin{bmatrix} D_{11}(q) D_{12}(q) \\ D_{21}(q) D_{22}(q) \end{bmatrix} \qquad (5.2.42)$$

The classical Lagrangian formulation presented is based on the kinetic and potential energy expressions that are used in Euler–Lagrange's equations to produce the dynamical model for the manipulator motion. The kinetic energy usually gives rise to the inertial, centripetal, and Coriolis generalized forces and the potential energy to the gravitational generalized force. The resulting model is sometimes called the closed-form dynamic manipulator model [2]. It is well structured and attractive from the viewpoint of control engineers.

The equations of motions for such common manipulators as the Stanford/JPL and PUMA robots have been determined in Lagrange's formulation [1,4,5]. The expressions $D(q)$, $C(q, \dot{q})$ and $G(q)$ in Equation 5.2.10 are very lengthy and tedious to obtain by human calculations. Moreover, if their values are needed on-line for real-time control, the numerical calculations of these expressions impose severe constraints on the choice of the sampling rate in the control schemes.

If the computational complexity of an algorithm is measured on the basis of the required additions and multiplications, the numerical computational requirements of the Lagrangian model in Equation 5.2.10 is of the order of N^4, where N is the number of the joints in the manipulator [5,6]. To reduce the computational burden in calculating the dynamical coefficients of Equation 5.2.10, various alternative approaches have been sought. Recursive expressions are generally known to be computationally efficient. Lagrange's Equation 5.2.10 describing the total joint variable dynamics can be developed to a form of difference equations relative to the links. Such recursive equations speed up the calculations and are computationally economical. Using recursive Lagrange's equations, the number of total additions and multiplications is proportional to N [6]. When the desired trajectory for the manipulator motion is given, recursive Lagrange's equations can be used to determine the needed torques of the joints under ideal conditions. Thus, the inverse dynamics problem can be solved efficiently. However, the dynamical model of the manipulator represented by recursive Lagrange's equations are in the form in which the derivatives of the generalized variables do not appear explicitly. This can be a serious drawback in the controller design.

5.3 DIFFERENTIAL EQUATION MODEL IN NEWTON–EULER FORMULATION

The dynamical model of a manipulator is derived in the previous section using Lagrange's formulation, in which the equations of motion are based on the kinetic and potential energies of the entire system. The resulting equations of motion give the designer physical insight needed to understand the behavior of the overall system. The drawback in Lagrange's formulation is the computational complexity. An alternative approach is to isolate each link in succession as a free (solid) body and determine the dynamical model for one link at a time using Newton's and Euler's equations. The resulting equations of motion represent a recursive model for a link involving variables of the adjacent links. They describe the translational and rotational dynamics of the link in detail containing internal forces and torques, which do not appear in Lagrange's formulation. Such a model helps the designer to understand the dynamical behavior of each link separately, and particularly the propagation of the forces and torques through the joints and their interactions. Both formulations are important in their own right. The designer of robot manipulator systems needs to be familiar with both modeling approaches.

We will discuss next the dynamical modeling of a manipulator with serially connected links in Newton–Euler's formulation.

5.3.1 NEWTON–EULER EQUATIONS DETERMINE RECURSIVELY LINK DYNAMICS

The equations of motion for each link of a serial link manipulator can be determined using Newton–Euler's formulation. The equations of Newton and Euler that describe some of the basic principles of the classical dynamics are reviewed in Appendix C. The equations of Newton and Euler that will be applied to the free-body configurations of the links in a manipulator are, respectively

$$\frac{d(m\dot{\mathbf{r}})}{dt} = m\ddot{\mathbf{r}} = \mathbf{F} \tag{5.3.1}$$

$$\frac{d(I\boldsymbol{\omega})}{dt} = I\dot{\boldsymbol{\omega}} + \boldsymbol{\omega}x(I\boldsymbol{\omega}) = \mathbf{M} \tag{5.3.2}$$

where mass m is assumed to be concentrated at the center of gravity (centroid), and vector $\mathbf{r}$ emanates from the origin of the base coordinate system to the centroid of the rigid body. Matrix I is the inertia matrix whose elements are determined about the rotational axis, and $\boldsymbol{\omega}$ is the angular velocity vector. The left side of Equation 5.3.1 is the rate of change in the linear momentum, and that of Equation 5.3.2 the rate of change in the angular momentum (see Appendix C).

We will present next the dynamical model that results from the application of Newton's and Euler's equations to the ith link, where $i = 1, \ldots, N$ in a manipulator with N moving links.

A schematic diagram of link i is shown in Figure 5.3. The center of gravity (centroid) of the link i is at point G_i. The Cartesian base coordinate system, the ith and $(i-1)$st coordinate frames have their origins at points O, O_i, and O_{i-1}, respectively. The vectors defining the relative positions of these points are indicated in Figure 5.3. For example, vector p_{0i} determines the position of the origin of the ith coordinate frame in the zeroth (base) coordinate system.

When the ith link is isolated so as to obtain a free-body representation, appropriate forces and moments (torques) are introduced at the end points of the link to account for the interactions between the adjacent links. Thus, force $\mathbf{F}_{(i-1),i}$ exerted by link $(i-1)$ on link i at point O_{i-1} is shown to act in the positive direction. The corresponding force of equal magnitude acting in the opposite direction appears at the end of link $(i-1)$, which was separated from link i for the analysis. According to this convention, force $\mathbf{F}_{i,(i+1)}$ describes the force exerted on link i by link $(i+1)$ at point O_{i+1} in the negative direction. Similarly, torque $\mathbf{M}_{(i-1),i}$ signifies the torque exerted by link $(i-1)$ on link i in the positive direction. It tends to cause a counterclockwise rotation about point G_i. An equal torque to the opposite direction is introduced at the end of link $(i-1)$. Also, link $(i+1)$ exerts a torque $\mathbf{M}_{i,}(i+1)$ on link i; it has the tendency of causing a clockwise (the negative direction) rotation of link i about point G_i.

The equations of motion for link i are obtained by applying first Newton's Equation 5.3.1 and then Euler's Equation 5.3.2 to the homogeneous body of the ith link. When the translational (linear) force $\mathbf{F}_0^i$ is acting on link i, the balancing of the forces of link i in Figure 5.3 gives:

$$\mathbf{F}_{(i-1),i} - \mathbf{F}_{i,(i+1)} = \mathbf{F}_0^i \tag{5.3.3}$$

The external force $\mathbf{F}_0^i$ can be considered to include the acceleration term, and the gravitational force. Suppose that the gravitational acceleration vector

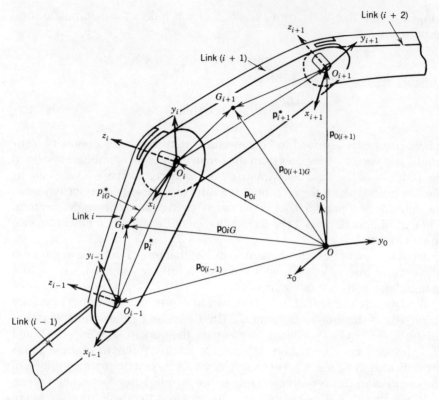

FIGURE 5.3

Coordinate frames and vectors for link i in Newton–Euler formulation.

is $\mathbf{g}_{0i}$, and the mass m_i of the ith link is concentrated at centroid G_i specified by vector $\mathbf{p}_{0iG}$ in the base coordinate system. Then, force $\mathbf{F}_0^i$ is determined by

$$m_i \ddot{\mathbf{p}}_{0iG} + m_i \mathbf{g}_{0i} = \mathbf{F}_0^i \tag{5.3.4}$$

Equations 5.3.3 and 5.3.4 represent Newton's equations for the translational motion of the ith link.

The rotational motion of link i about the centroid is governed by Euler's Equation 5.3.2. If the resultant external moment (torque) exerted on link i is denoted by $\mathbf{M}_0^i$, then Equation 5.3.2 for the ith link is

$$\frac{d}{dt}(I_{0i}\,\boldsymbol{\omega}_{0i}) = I_{0i}\,\dot{\boldsymbol{\omega}}_{0i} + \boldsymbol{\omega}_{0i} \times (I_{0i}\,\boldsymbol{\omega}_{0i}) = \mathbf{M}_0^i \tag{5.3.5}$$

where I_{0i} is the moment of inertia of link i about the rotational axis, and $\boldsymbol{\omega}_{0i}$ is the angular velocity of the ith link. The moments exerted on the ith link by link $(i-1)$ are $\mathbf{M}_{(i-1),i}$ and $(\mathbf{p}_{0(i-1)} - \mathbf{p}_{0iG}) \times \mathbf{F}_{(i-1),i}$, where $\mathbf{p}_{0(i-1)}$ specifies the origin O_{i-1} of the $(i-1)$st coordinate frame in the base coordinate system. The moments exerted on link i by link $(i+1)$ are similarly determined. By balancing the moments that have the tendency of causing the rotation of link i about centroid G_i, one obtains

$$\mathbf{M}_{(i-1),i} - \mathbf{M}_{i,(i+1)} + (\mathbf{p}_{0(i-1)} - \mathbf{p}_{0iG}) \times \mathbf{F}_{(i-1),i} - (\mathbf{p}_{0i} - \mathbf{p}_{0iG}) \times \mathbf{F}_{i,(i+1)} = \mathbf{M}_0^i$$

$$(5.3.6)$$

where the external moment $\mathbf{M}_0^i$ is determined by Equation 5.3.5. Equations 5.3.5 and 5.3.6 represent Euler's equations, which describe the rotational motion of the ith link about the centroid.

The input that produces the link motion appears in the equation for the translational or rotational motion depending on the type of the link. If the actuator at the proximal end (O_{i-1}) is causing the translational movement of the ith link, the input from the actuator is included in $\mathbf{F}_{(i-1),i}$. On the other hand, if the actuator at the same location is causing the ith link to rotate, the input is written in $\mathbf{M}_{(i-1),i}$.

Specifically for a *prismatic* link i, the generalized joint variable q_i in the coordinate frame defined according to the D-H procedure represents the motion along the z_{i-1}-axis. In this case, the z_{i-1}-component of the force-vector $F_{(i-1),i}$ in Equation 5.3.3 is

$$F'_{(i-1),i} k_{z_{i-1}} = \tau_i - B_i(\dot{q}_i) \qquad (5.3.7)$$

where $F'_{(i-1),i} k_{z_{i-1}}$ specifies the component of $F_{(i-1),i}$ on the z_{i-1}-axis, τ_i is the generalized torque (force) input, and $B_i(\dot{q}_i)$ describes the friction as a function of the joint velocity $\dot{q}_i$.

For a *rotational* link i, the generalized joint variable q_i in the D-H local coordinate frame represents the rotation about the z_{i-1}-axis. In this case, the z_{i-1}-component of the moment vector $M_{(i-1),i}$ in Equation 5.3.6 is

$$M'_{(i-1),i} k_{z_{i-1}} = \tau_i - B_i(\dot{q}_i) \qquad (5.3.8)$$

where $M'_{(i-1),i} k_{z_{i-1}}$ is the component of $M_{(i-1),i}$ on the z_{i-1}-axis, and τ_i is the generalized torque produced by the actuator of the ith joint.

Equations 5.3.3 and 5.3.4 describe the translational motion, and Equations 5.3.5 and 5.3.6 the rotational motion of link i in the Cartesian base coordinate system. These equations can be regarded as difference equations with respect to the integer i associated with the link.

The velocities and accelerations of the translational and rotational motions of a specific link are related to those of the adjacent links by the equations that are presented in Sections 4.2 and 4.4. These recursive equations are summarized in Table 5.1. They also represent difference equations with respect to the integer i of the link. They are usually needed when the dynamical model of an entire manipulator is calculated using Newton–Euler's formulation.

The use of Newton's and Euler's equations in the modeling of the motion of a manipulator is illustrated by two examples that are next presented.

EXAMPLE 5.3.1

The links of the planar manipulator with two rotating joints can move on a vertical plane, as shown in Figure 5.4a. The length of link i is denoted

TABLE 5.1
Recursive Relations for the Positions, Velocities, and Accelerations of the $(i + 1)$st Link $(i = 0,1,\ldots, N − 1)$ in the Cartesian Base Coordinate System. Serial-Link Manipulator

Revolute Link

a.1 Position Relation: $p_{0i} = A_0^i p_i$

b.1 Velocity Relations:

$$\omega_{0(i+1)} = \omega_{0i} + \dot{\theta}_{i+1}\mathbf{k}_{zi} \tag{4.2.1}$$

$$\dot{\mathbf{p}}_{0(i+1)} = \dot{\mathbf{p}}_{0i} + \omega_{0(i+1)} \times \mathbf{p}_{i+1}^* \tag{4.2.4}$$

$$\mathbf{p}_{i+1}^* = \mathbf{p}_{0(i+1)} - \mathbf{p}_{0i} \tag{4.2.7}$$

c.1 Acceleration Relations:

$$\dot{\omega}_{0(i+1)} = \dot{\omega}_{0i} + \ddot{\theta}_{i+1}\mathbf{k}_{zi} + \omega_{0i} \times (\dot{\theta}_{i+1}\mathbf{k}_{zi}) \tag{4.4.1}$$

$$\ddot{\mathbf{p}}_{0(i+1)} = \ddot{\mathbf{p}}_{0i} + \dot{\omega}_{0(i+1)} \times \mathbf{p}_{i+1}^*$$
$$+ \omega_{0(i+1)} \times (\omega_{0(i+1)} \times \mathbf{p}_{i+1}^*) \tag{4.4.3}$$

d. Centroidal Relations:

$$\mathbf{p}_{iG}^* = \mathbf{p}_{0iG} - \mathbf{p}_{0i}$$

$$\dot{\mathbf{p}}_{0iG} = \dot{\mathbf{p}}_{0i} + \omega_{0i} \times (\mathbf{p}_{0iG} - \mathbf{p}_{0i})$$

$$\ddot{\mathbf{p}}_{0iG} = \ddot{\mathbf{p}}_{0i} + \dot{\omega}_{0i} \times (\mathbf{p}_{0iG} - \mathbf{p}_{0i}) +$$
$$+ \omega_{0i} \times [\omega_{0i} \times (\mathbf{p}_{0iG} - \mathbf{p}_{0i})]$$

Prismatic Link

a.2 Position Relation: $p_{0i} = A_0^i p_i$

b.2 Velocity Relations:

$$\omega_{0(i+1)} = \omega_{0i} \tag{4.2.2}$$

$$\dot{\mathbf{p}}_{0(i+1)} = \dot{\mathbf{p}}_{0i} + \dot{\theta}_{i+1}\mathbf{k}_{zi} + \omega_{0(i+1)} \times \mathbf{p}_{i+1}^* \tag{4.2.5}$$

$$\mathbf{p}_{i+1}^* = \mathbf{p}_{0(i+1)} - \mathbf{p}_{0i} \tag{4.2.7}$$

c.2 Acceleration Relations:

$$\dot{\omega}_{0(i+1)} = \dot{\omega}_{0i} \tag{4.4.2}$$

$$\ddot{\mathbf{p}}_{0(i+1)} = \ddot{\mathbf{p}}_{0i} + \ddot{\theta}_{i+1}\mathbf{k}_{zi} + 2\omega_{0(i+1)} \times (\dot{\theta}_{i+1}\mathbf{k}_{zi})$$
$$+ \dot{\omega}_{0(i+1)} \times \mathbf{p}_{i+1}^* + \omega_{(i+1)} \times (\omega_{0(i+1)} \times \mathbf{p}_{i+1}^*) \tag{4.4.4}$$

NOTE: Subscript 0 refers to the base coordinate system, i to the ith local coordinate system; G to the centroid (center of gravity) of the link. Vector p_{0i} specifies the origin of the ith coordinate frame in the zeroth (base) coordinate system, and $p_i = [0\ 0\ 0\ 1]'$ in the ith coordinate frame. The base is the zeroth link. Hence, $\psi_{00} = 0$, $\omega_{00} = 0$, $\dot{\omega}_{00} = 0$, $p_{00} = 0$, $\dot{p}_{00} = 0$, $\ddot{p}_{00} = 0$. In the prismatic link, $\dot{\theta}_{i+1}$ and $\ddot{\theta}_{i+1}$ refer to the translational motion.

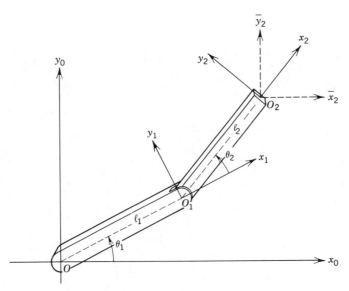

FIGURE 5.4a
Two-link revolute planar manipulator.

by ℓ_i, $i = 1, 2$. Each link is homogeneous, and mass m_i of link i is assumed to be concentrated on the centroid G_i located at the midpoint of the link. The actuator of joint i generates torque τ_i, which tends to cause a rotational movement. The problem is to determine a dynamical model (equations of motion) for each link of the manipulator using Newton–Euler's formulation. To solve the problem, the free-body representations of links one and two are first determined. Link one ($i = 1$) is affected by forces $F_{0,1}$ from the zeroth link and $F_{1,2}$ from link two, as shown in Figure 5.4b. The moments exerted on link one are $M_{0,1}$ and $M_{1,2}$. Equal forces and moments to the opposite directions are exerted by link one on the zeroth link and on the second link. Link two is also subjected to the effects of the external force $F_{2,3} = [F_{2,3x} \quad F_{2,3y} \quad 0]'$ and the external moment (torque) $M_{2,3} = [0 \quad 0 \quad M_{2,3z}]'$ from the environment.

Newton's equations for the translational motion of links one and two can now be written on the basis of Equations 5.3.3 and 5.3.4. The positions for the centroids of links one and two in the base coordinate system are denoted by $\mathbf{p}_{01G}$, and $\mathbf{p}_{02G}$, respectively. Equations 5.3.3 and 5.3.4 give

$$\mathbf{F}_{0,1} - \mathbf{F}_{1,2} = \mathbf{F}_0^1 \tag{5.3.9}$$

$$\mathbf{F}_{1,2} - \mathbf{F}_{2,3} = \mathbf{F}_0^2 \tag{5.3.10}$$

$$m_1 \, \ddot{\mathbf{p}}_{01G} + m_1 \, \mathbf{g}_{01} = \mathbf{F}_0^1 \tag{5.3.11}$$

$$m_2 \, \ddot{\mathbf{p}}_{02G} + m_2 \, \mathbf{g}_{02} = \mathbf{F}_0^2 \tag{5.3.12}$$

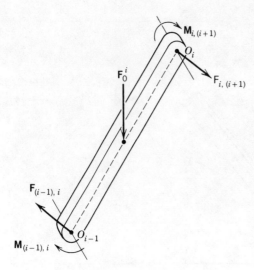

FIGURE 5.4b

Free-body configuration of link
i in the planar manipulator.

where the gravity vector is $\mathbf{g}_{01} = \mathbf{g}_{02} = g\mathbf{j}_{y_0}$. Equations 5.3.9 and 5.3.11 represent Newton's equations for link one, and Equations 5.3.10 and 5.3.12 for link two.

The dynamical equations for the rotational motion are given by Euler's equations. Indeed, Equations 5.3.5 and 5.3.6 for $i = 1, 2$ in this particular case assume the following forms:

$$I_{01}\dot{\boldsymbol{\omega}}_{01} + \boldsymbol{\omega}_{01} \times (I_{01}\boldsymbol{\omega}_{01}) = \mathbf{M}_0^1 \tag{5.3.13}$$

$$I_{02}\dot{\boldsymbol{\omega}}_{02} + \boldsymbol{\omega}_{02} \times (I_{02}\boldsymbol{\omega}_{02}) = \mathbf{M}_0^2 \tag{5.3.14}$$

$$\mathbf{M}_{0,1} - \mathbf{M}_{1,2} + (\mathbf{p}_{00} - \mathbf{p}_{01G}) \times \mathbf{F}_{0,1} - (\mathbf{p}_{01} - \mathbf{p}_{01G}) \times \mathbf{F}_{1,2} = \mathbf{M}_0^1 \tag{5.3.15}$$

$$\mathbf{M}_{1,2} - \mathbf{M}_{2,3} + (\mathbf{p}_{01} - \mathbf{p}_{02G}) \times \mathbf{F}_{1,2} - (\mathbf{p}_{02} - \mathbf{p}_{02G}) \times \mathbf{F}_{2,3} = \mathbf{M}_0^2 \tag{5.3.16}$$

where I_{0i}, $i = 1, 2$ is the moment of inertia for link i about its center of gravity and $\mathbf{p}_{00} = 0$. The external torques $\mathbf{M}_{0,1}$ and $\mathbf{M}_{1,2}$ are specified as follows: $\mathbf{M}_{0,1} = \tau_1 \mathbf{k}_{z_0} - B_1 \boldsymbol{\omega}_{01}$ and $\mathbf{M}_{1,2} = \tau_2 \mathbf{k}_{z_1} - B_2 \boldsymbol{\omega}_{12}$, where $\boldsymbol{\omega}_{12} = \dot{\theta}_2 \mathbf{k}_{z_1}$, and constants B_1 and B_2 represent the coefficients of viscous friction. Equations 5.3.13 and 5.3.15 are Euler's equation for link 1, and Equations 5.3.14 and 5.3.16 for link 2.

The dynamical model for the translational and rotational motions of the planar manipulator in the Cartesian base coordinate system is specified by Equations 5.3.9–5.3.16 in Newton–Euler's formulation. The variables describing the motion in Equations 5.3.9–5.3.16 are the translational positions $\mathbf{p}_{01G}$, and $\mathbf{p}_{02G}$ of the centroids of the two links, and their rotational velocities $\boldsymbol{\omega}_{01}$, and $\boldsymbol{\omega}_{02}$. The effects of the environment on the end-effector are specified by the external force $\mathbf{F}_{2,3}$ and torque $\mathbf{M}_{2,3}$. The input torques τ_1 and τ_2 produced by the actuators are included in terms $\mathbf{M}_{0,1}$ and $\mathbf{M}_{1,2}$, respectively. We should observe that Equations 5.3.9–5.3.16 for the motion of the manipulator can be expressed in the form of the matrices and vectors without using the unit vectors. This form is attractive if the control of the manipulator motion is performed in the Cartesian space.

EXAMPLE 5.3.2

The dynamical model for the planar manipulator of Example 5.3.1 is determined in the base coordinate system by Equations 5.3.9–5.3.16 of Newton and Euler. The task here is to develop from the aforementioned equations a dynamical model for the joint variables θ_1 and θ_2.

The problem is solved by first expressing the translational accelerations $\ddot{\mathbf{p}}_{01G}$, and $\ddot{\mathbf{p}}_{02G}$ in Equations 5.3.11 and 5.3.12 in terms of θ_1 and θ_2. It is accomplished, for example, by differentiating vectors p_{01G} and p_{02G} of the centroidal positions twice with respect to time:

$$\ddot{\mathbf{p}}_{01G} = \frac{1}{2}[-\ell_1 s_1 \ddot{\theta}_1 - \ell_1 c_1 \dot{\theta}_1^2 \quad \ell_1 c_1 \ddot{\theta}_1 - \ell_1 s_1 \dot{\theta}_1^2 \quad 0]' \tag{5.3.17}$$

$$\ddot{\mathbf{p}}_{02G} = \begin{bmatrix} -\ell_1 s_1 \ddot{\theta}_1 - \ell_1 c_1 \dot{\theta}_1^2 - \ell_2 s_{12}(\ddot{\theta}_1 + \ddot{\theta}_2)/2 - \ell_2 c_{12}(\dot{\theta}_1 + \dot{\theta}_2)^2/2 \\ \ell_1 c_1 \ddot{\theta}_1 - \ell_1 s_1 \dot{\theta}_1^2 + \ell_2 c_{12}(\ddot{\theta}_1 + \ddot{\theta}_2)/2 - \ell_2 s_{12}(\dot{\theta}_1 + \dot{\theta}_2)^2/2 \\ 0 \end{bmatrix} \tag{5.3.18}$$

Equations 5.3.17 and 5.3.18 can also be obtained by the application of the relations in Table 5.1. The rotational velocities and accelerations are also expressed in terms of the joint variables (see Equation 4.2.1):

$$\boldsymbol{\omega}_{0(i+1)} = \boldsymbol{\omega}_{0i} + \dot{\theta}_{i+1}\mathbf{k}_{z_i} \tag{5.3.19}$$

where $\boldsymbol{\omega}_{00} = 0$. It follows that

$$\boldsymbol{\omega}_{01} = \dot{\theta}_1\mathbf{k}_{z_0} \qquad \boldsymbol{\omega}_{02} = \dot{\theta}_1\mathbf{k}_{z_0} + \dot{\theta}_2\mathbf{k}_{z_1} \tag{5.3.20}$$

where the unit vectors $\mathbf{k}_{z_0}$ and $\mathbf{k}_{z_1}$ are parallel. By a similar argument, the angular accelerations are determined: $\dot{\omega}_{01} = [0, 0, \ddot{\theta}_1]'$ and $\dot{\omega}_{02} = [0, 0, \ddot{\theta}_1 + \ddot{\theta}_2]'$.

When the desired positions, velocities, and accelerations of the joint variables are known, the translational and rotational positions, velocities, and accelerations of links 1 and 2 can be determined. Since the external force $\mathbf{F}_{2,3}$ and moment $\mathbf{M}_{2,3}$ are assumed to be known, Equations 5.3.9–5.3.16 can be used to calculate first $\mathbf{F}_{1,2}$ and $\mathbf{M}_{1,2}$, and then $\mathbf{F}_{0,1}$ and $\mathbf{M}_{0,1}$. Thus, the unknown forces and torques are computed by starting from the end-effector and proceeding toward the base.

Equations 5.3.11 and 5.3.12 are first combined with Equations 5.3.9 and 5.3.10 to eliminate $\mathbf{F}_0^1$ and $\mathbf{F}_0^2$. Thus, two equations are obtained:

$$\mathbf{F}_{1,2} = \mathbf{F}_{2,3} + m_2\ddot{\mathbf{p}}_{02G} + m_2\mathbf{g}_{02} \tag{5.3.21}$$

$$\mathbf{F}_{0,1} = [\mathbf{F}_{2,3} + m_2\ddot{\mathbf{p}}_{02G} + m_2\mathbf{g}_{02}] + m_1\ddot{\mathbf{p}}_{01G} + m_1\mathbf{g}_{01} \tag{5.3.22}$$

where the translational accelerations are specified by Equations 5.3.17 and 5.3.18. Equations 5.3.21 and 5.3.22 represent Newton's equations for the translational motion of the planar manipulator. When Equations 5.3.17 and 5.3.18 are substituted into Equations 5.2.21 and 5.2.22, the equations that relate the joint variables to the input forces $\mathbf{F}_{0,1}$ and $\mathbf{F}_{1,2}$ are obtained.

Euler's equations for the rotational motion will now be described in terms of the joint variables. In Equations 5.3.15 and 5.3.16, the cross-product terms are first calculated. In these cross products, vectors $\mathbf{p}_{01}$ and $\mathbf{p}_{02}$ describe the origins of the first and the second coordinate frames, respectively, in the base (zeroth) coordinate system. Vectors $\mathbf{p}_{01G}$ and $\mathbf{p}_{02G}$ specify the positions of the centroids of links 1 and 2, respectively, also in the base coordinate system. The moments arms for Equations 5.3.15 and 5.3.16 can be written (Figure 5.4a) as follows:

$$\mathbf{p}_{00} - \mathbf{p}_{01G} = (-\frac{1}{2}\ell_1 c_1)\,\mathbf{i}_{x_0} + (-\frac{1}{2}\ell_1 s_1)\,\mathbf{j}_{y_0} + 0\,\mathbf{k}_{z_0} \tag{5.3.23}$$

$$\mathbf{p}_{01} - \mathbf{p}_{01G} = (\frac{1}{2}\ell_1 c_1)\,\mathbf{i}_{x_0} + (\frac{1}{2}\ell_1 s_1)\,\mathbf{j}_{y_0} + 0\,\mathbf{k}_{z_0} \tag{5.3.24}$$

$$\mathbf{p}_{01} - \mathbf{p}_{02G} = (-\frac{1}{2}\ell_2 c_{12})\,\mathbf{i}_{x_0} + (-\frac{1}{2}\ell_2 s_{12})\,\mathbf{j}_{y_0} + 0\,\mathbf{k}_{z_0} \tag{5.3.25}$$

$$\mathbf{p}_{02} - \mathbf{p}_{02G} = (\frac{1}{2}\ell_2 c_{12})\,\mathbf{i}_{x_0} + (\frac{1}{2}\ell_2 s_{12})\,\mathbf{j}_{y_0} + 0\,\mathbf{k}_{z_0} \tag{5.3.26}$$

The cross-product terms in Equations 5.3.15 and 5.3.16 are next developed. If the position of a vector is $\mathbf{p} = p_x\mathbf{i} + p_y\mathbf{j} + 0\mathbf{k}$, and force $\mathbf{F} = F_x\mathbf{i} + F_y\mathbf{j} + 0\mathbf{k}$, then their cross-product has the following form:

$$\mathbf{p} \times \mathbf{F} = 0\mathbf{i} + 0\mathbf{j} + (p_x F_y - p_y F_x)\mathbf{k} \tag{5.3.27}$$

Equation 5.3.27 is applied to calculate the cross-products in Equations 5.3.15 and 5.3.16.

By applying Equation 5.3.27 to Equations 5.3.15 and 5.3.16, and combining Equations 5.3.15–5.3.18 with the resulting expressions, the dynamical model for θ_1 and θ_2 is obtained. It may be written in the following form:

$$
\begin{bmatrix} M_{0,1z} - M_{2,3z} \\ M_{1,2z} - M_{2,3z} \end{bmatrix} = \begin{bmatrix} I_{11}\,I_{12} \\ I_{21}\,I_{22} \end{bmatrix} \begin{bmatrix} \ddot{\theta}_1 \\ \ddot{\theta}_1 + \ddot{\theta}_2 \end{bmatrix} + \begin{bmatrix} m_2\ell_1\ell_2 s_2/2 - m_2\ell_1\ell_2 s_2/2 \\ m_2\ell_1\ell_2 s_2/2 \quad 0 \end{bmatrix} \begin{bmatrix} \dot{\theta}_1^2 \\ (\dot{\theta}_1 + \dot{\theta}_2)^2 \end{bmatrix}
$$

$$
- \begin{bmatrix} \ell_1 c_1/2 & \ell_1 c_1 + \ell_2 c_{12}/2 \\ 0 & \ell_2 c_{12}/2 \end{bmatrix} \begin{bmatrix} m_1 g_{01} \\ m_2 g_{02} \end{bmatrix} + \begin{bmatrix} -\ell_1 s_1 - \ell_2 s_{12} & \ell_1 c_1 + \ell_2 c_{12} \\ -\ell_2 s_{12} & \ell_2 c_{12} \end{bmatrix} \begin{bmatrix} F_{2,3x} \\ F_{2,3y} \end{bmatrix}
$$

$$\tag{5.3.28}$$

where $g_{01} = g_{02} = -g$, $I_{11} = I_{01} + m_1\ell_1^2/4 + m_2\ell_1^2 + m_2\ell_1\ell_2 c_2/2$, $I_{12} = I_{02} + m_2\ell_2^2/4 + m_2\ell_1\ell_2 c_2/2$, $I_{21} = m_2\ell_1\ell_2 c_2/2$, and $I_{22} = I_{02} + m_2\ell_2^2/4$. In the dynamical model of Equation 5.3.28, the torques (moments) produced by the joint actuators are included in the expressions of moments $M_{0,1}$ and $M_{1,2}$ as the inputs. They are given by Equation 5.3.8: $M_{0,1z} = \tau_1 - B_1\dot{\theta}_1$, and $M_{1,2z} = \tau_2 - B_2\dot{\theta}_2$. Term $M_{2,3z}$ represents the external moment acting on the end-effector. The system model in Equation 5.3.28 is written for the joint angles θ_1 and $(\theta_1 + \theta_2)$. In this particular form, the inertia matrix is not symmetric, since $I_{12} \neq I_{21}$.

The second-order vector differential Equation 5.3.28 represents the equations of motion in the joint variable space for the planar manipulator. This model is derived by starting from Newton–Euler's equations.

5.4 DYNAMICAL MANIPULATOR MODEL USING NEWTON–EULER FORMULATION IN LOCAL COORDINATE FRAME

The translational and rotational positions, velocities, and accelerations of link $(i + 1)$ are presented in Section 4.4 in terms of the corresponding variables of the ith link for a serial link manipulator. These equations are recursive relative to the serially connected links. They are summarized in Table 5.1. The variables are expressed relative to the base coordinate system. Using the position, velocity, and acceleration of the origin of the local coordinate frame, the corresponding variables for the centroid (the center of gravity) of this link can also be determined. The equations of motion for the center of gravity of the link can then be obtained in Newton–Euler's formulation [7,8]. The resulting model is computationally efficient, particularly in digital simulations. The main drawback in using these model equations is that the expressions of the inertial moments depend on the configuration of the manipulator. To avoid this problem, the equations of motions can be described in the local coordinate frames.

We will next describe the determination of the equations of motion relative to moving coordinate frames in recursive forms using Newton–Euler's formulation. After summarizing the recursive equations that relate the translational and rotational variables of the serially connected links in a manipulator, we will express Newton's and Euler's equations for the motion in the local coordinate frames attached to the links.

5.4.1 POSITIONS, VELOCITIES, AND ACCELERATIONS OF LINK CENTROIDS

In the dynamical model of a link based on Newton–Euler's equations and described in Section 5.3, the second mass moment of link i appears in the model equations. Its expression in the base coordinate system is dependent on the orientation and position of this link. Since it is changing when the link moves, its determination in general becomes very cumbersome. To circumvent this difficulty, the equations of motion for each link can be expressed in the local coordinate frame moving with the link. The aforementioned moment of inertia in the moving coordinate frame remains constant during the motion. The determination of the mass moments for the links and the equations of motion for the manipulator can thus be simplified considerably.

TABLE 5.2
Recursive Relations for the Positions, Velocities, and Accelerations of Table 5.1 Referred to Local Coordinate Frame

Revolute Link	Prismatic Link
a.1 Position Relation, $i = 0, \ldots, N-1$	*a.2 Position Relation, $i = 0, \ldots, N-1$*
$^i p_{0i} = (A_i^0)_R p_{0i}$	$^i p_{0i} = (A_i^0)_R p_{0i}$
b.1 Velocity Relations, $i = 0, \ldots, N-1$	*b.2 Velocity Relations, $i = 0, \ldots, N-1$*
$^i \omega_{0i} = (A_i^0)_R \omega_{0i}$	$^i \omega_{0i} = (A_i^0)_R \omega_{0i}$
$^{i+1}\omega_{0i} = (A_{i+1}^i)_R[(A_i^0)_R \omega_{0i}]$	$^{i+1}\omega_{0i} = (A_{i+1}^i)_R[(A^0)_R \omega_{0i}]$
$^{i+1}k_{z_i} = (A_{i+1}^i)_R k_{z_i}$	$^{i+1}k_{z_i} = (A_{i+1}^i)_R k_{z_i}$
$^{i+1}\boldsymbol{\omega}_{0(i+1)} = {}^{i+1}\boldsymbol{\omega}_{0i} + \dot\theta_{i+1}({}^{i+1}\mathbf{k}_{z_i})$	$^{i+1}\boldsymbol{\omega}_{0(i+1)} = {}^{i+1}\boldsymbol{\omega}_{0i}$
$^i v_{0i} = (A_i^0)_R \dot p_{0i}$	$^i v_{0i} = (A_i^0)_R \dot p_{0i}$
$^{i+1}v_{0i} = (A_{i+1}^i)_R[(A_i^0)_R \dot p_{0i}]$	$^{i+1}v_{0i} = (A_{i+1}^i)_R[(A_i^0)_R \dot p_{0i}]$
$^{i+1}p_{i+1}^* = (A_{i+1}^0)_R p_{i+1}^* = (A_{i+1}^0)_R(p_{0(i+1)} - p_{0i})$	$^{i+1}p_{i+1}^* = (A_{i+1}^0)_R p_{i+1}^* = (A_{i+1}^0)_R(p_{0(i+1)} - p_{0i})$
$^{i+1}\mathbf{v}_{0(i+1)} = {}^{i+1}\mathbf{v}_{0i} + ({}^{i+1}\boldsymbol{\omega}_{0(i+1)}) \times ({}^{i+1}\mathbf{p}_{i+1}^*)$	$^{i+1}\mathbf{v}_{0(i+1)} = {}^{i+1}\mathbf{v}_{0i} + \dot\theta_{i+1}({}^{i+1}\mathbf{k}_{z_i})$
	$\quad + ({}^{i+1}\boldsymbol{\omega}_{0(i+1)}) \times ({}^{i+1}\mathbf{p}_{i+1}^*)$

Revolute Link	Prismatic Link
c.1 Acceleration Relations, i = 0, . . . , N − 1	*c.2 Acceleration Relations, i = 0, . . . , N − 1*

Revolute Link

$^i\alpha_{0i} = (A_i^0)_R \,\dot\omega_{0i}$

$^{i+1}\alpha_{0i} = (A_{i+1}^i)_R[(A_i^0)_R \,\dot\omega_{0i}] = (A_{i+1}^i)_R({}^i\alpha_{0i})$

$^{i+1}k_{z_i} = (A_{i+1}^i)k_{z_i}$

$^{i+1}\boldsymbol{\alpha}_{0(i+1)} = {}^{i+1}\boldsymbol{\alpha}_{0i} + \ddot\theta_{i+1}({}^{i+1}\mathbf{k}_{z_i})$
$\quad + ({}^{i+1}\boldsymbol{\omega}_{0i}) \times ({}^{i+1}\mathbf{k}_{z_i})\dot\theta_{i+1}$

$^i a_{0i} = (A_i^0)_R \,\ddot{p}_{0i}$

$^{i+1}a_{0i} = (A_{i+1}^i)_R[(A_i^0)_R \,\ddot{p}_{0i}] = (A_{i+1}^i)_R({}^i a_{0i})$

$^{i+1}a_{0(i+1)} = {}^{i+1}\mathbf{a}_{0i} + ({}^{i+1}\boldsymbol{\alpha}_{0(i+1)}) \times ({}^{i+1}\mathbf{p}_{i+1}^*)$
$\quad + ({}^{i+1}\boldsymbol{\omega}_{0(i+1)}) \times [({}^{i+1}\boldsymbol{\omega}_{0(i+1)}) \times ({}^{i+1}\mathbf{p}_{i+1}^*)]$

Prismatic Link

$^i\alpha_{0i} = (A_i^0)_R \,\dot\omega_{0i}$

$^{i+1}\alpha_{0i} = (A_{i+1}^i)_R[(A_i^0)_R \,\dot\omega_{0i}]$

$^{i+1}k_{z_i} = (A_{i+1}^i)_R k_{z_i}$

$^{i+1}\boldsymbol{\alpha}_{0(i+1)} = {}^{i+1}\boldsymbol{\alpha}_{0i}$

$^i a_{0i} = (A_i^0)_R \,\ddot{p}_{0i}$

$^{i+1}a_{0i} = (A_{i+1}^i)_R[(A_i^0)_R \,\ddot{p}_{0i}]$

$^{i+1}\mathbf{a}_{0(i+1)} = {}^{i+1}\mathbf{a}_{0i} + \ddot\theta_{i+1}({}^{i+1}\mathbf{k}_{z_i})$
$\quad + ({}^{i+1}\boldsymbol{\alpha}_{0(i+1)}) \times ({}^{i+1}\mathbf{p}_{i+1}^*)$
$\quad + 2({}^{i+1}\boldsymbol{\omega}_{0(i+1)}) \times ({}^{i+1}\mathbf{k}_{z_i})\,\dot\theta_{i+1} +$
$\quad ({}^{i+1}\boldsymbol{\omega}_{0(i+1)}) \times [({}^{i+1}\boldsymbol{\omega}_{0(i+1)}) \times ({}^{i+1}\mathbf{p}_{i+1}^*)]$

In order to determine Newton–Euler's equations for a link in a moving coordinate frame, the recursive relations for the positions, velocities, and accelerations are first determined relative to the local moving coordinate frames. Therefore, it is necessary to rotate the vectors in Table 5.1 so that their components will be along the coordinates that are parallel with those of the local coordinate frames. The rotation submatrix $(A_i^0)_R$ can be applied to the vectors in Table 5.1 to determine the desired projections. It is available in the homogeneous transformation matrix A_i^0, which maps a vector of the base (zeroth) coordinate system to a vector in the ith coordinate frame. Each vector in Table 5.1 can thus be described in a local coordinate frame while maintaining the recursive nature of the equations.

When the rotation submatrix $(A_i^0)_R$ is applied to a vector p_{0i} specifying the origin of the ith coordinate frame in the zeroth coordinate system, a vector $^ip_{0i}$ is obtained. It has the components in a new coordinate system that has its axes parallel with the corresponding axes of the ith coordinate frame, as indicated by the front superscript. Thus

$$^ip_{0i} = (A_i^0)_R p_{0i} \tag{5.4.1}$$

$$(A_i^0)_R = (A_i^{i-1})_R \cdots (A_2^1)_R (A_1^0)_R \tag{5.4.2}$$

We recall that $(A_i^0)_R = (A_0^i)_R^{-1} = (A_0^i)_R'$. When Equation 5.4.1 is applied to the relations in Table 5.1, the expressions of Table 5.2 are obtained. The vectors in Table 5.2 are denoted either as column vectors (lightface), or using the unit vectors (boldface), for example, $p_{0i} = [p_{0ix} \quad p_{0iy} \quad p_{0iz}]'$ or $\mathbf{p}_{0i} = p_{0ix}\mathbf{i}_{x_0} + p_{0iy}\mathbf{j}_{y_0} + p_{0iz}\mathbf{k}_{z_0}$. The translational velocity and acceleration vectors of the origin of the ith coordinate frame (link i), which are originally expressed in the zeroth coordinate system, are premultiplied by $(A_i^0)_R$ to obtain $^iv_{0i}$, and $^ia_{0i}$, respectively. The corresponding rotational variables are denoted by $^i\omega_{0i}$, and $^i\alpha_{0i}$, as shown in Table 5.2. The underlined equations in Table 5.2 describe the rotational and translational velocity and acceleration relations for the $(i + 1)$st link in recursive forms.

The relationships in Table 5.2 are forward difference equations relative to integer i. Thus, the components of the positions, velocities, and accelerations of link $(i + 1)$ can be calculated recursively when the corresponding values of link i are known. Since the base coordinate system is attached to the base or the shoulder of the manipulator representing the zeroth link, it follows that $\omega_{00} = 0$, $\dot{\omega}_{00} = 0$, $\dot{p}_{00} = 0$, and $\ddot{p}_{00} = 0$. To determine the position, velocity, and acceleration of link 1 in the first local coordinate frame the underlined equations in Table 5.2 are used by starting with $i = 0$. Proceeding then successively to the next link $i = 1, 2, \ldots, N-1$, the positions, velocities, and accelerations of the subsequent links in the other local coordinate frames are calculated.

The equations given in Table 5.2 can be considered as forward difference equations with respect to the links. They are applied sequentially from the base toward the end-effector to obtain the rotational and translational velocities, and accelerations for all serially connected links. The equations in Table

5.2 for the serially connected links are valid at any instant of time. The corresponding variables for the centroids of the links given in Table 5.3a can then be computed.

5.4.2 NEWTON—EULER EQUATIONS DESCRIBE MANIPULATOR DYNAMICS

After the positions, velocities and accelerations of the centroids of the links have been determined, Newton—Euler's equations are written to specify the generalized torques needed to cause the motion represented by those values. As discussed previously, the equations of motion should be expressed in the local moving coordinate frames in order to make the inertial moments of the links independent of the positions and orientations of the links. The positions, velocities, and accelerations of the centroids are described in the local moving coordinate frames as shown in Table 5.3a. The forces and torques given by Newton—Euler's Equations 5.3.3–5.3.8 are next rewritten so that the variables of each link are referenced to the local coordinate frame.

In order to express Newton—Euler's equations in terms of variables referenced to the moving coordinate frame, the rotation matrix $(A_i^0)_R$ is applied to the vectors in the dynamical equations. While manipulating the transformed vector expressions in the recursive forms, it is important to keep in mind the information available at the particular time. The positions, velocities, and accelerations of the translational and rotational motions are calculated by starting from the base and sequentially moving toward the end-effector as discussed in Section 5.4.1. Example 5.3.2 gives an indication that the generalized torques are calculated backward by starting from the last link (the end-effector) and proceeding link by link toward the base. In this way, Newton—Euler's equations can be expressed in a computationally efficient form.

The terms in Newton's Equations 5.3.3, 5.3.4 and in Euler's Equations 5.3.5, 5.3.6 can be expressed in terms of the vectors whose components are parallel to the coordinates of the ith frame. The resulting expressions for the external force and moment acting on link i are given in Table 5.3b. Before the rotation matrix is applied to the terms in Equation 5.3.6, it is advantageous to rewrite Equation 5.3.6 by substituting for $F_{(i-1),i}$ from Equation 5.3.3. The resulting expressions can be simplified by observing that by Figure 5.3

$$\mathbf{p}_{0(i-1)} - \mathbf{p}_{0i} = -\mathbf{p}_i^* \tag{5.4.3}$$

$$\mathbf{p}_{0(i-1)} - \mathbf{p}_{0iG} = -(\mathbf{p}_i^* + \mathbf{p}_{iG}^*) \tag{5.4.4}$$

where $\mathbf{p}_{iG}^*$ is a vector emanating from the origin O_i of the ith coordinate frame to the centroid G_i expressed in the base coordinate system. By making use of Equations 5.4.3, 5.4.4, and 5.3.3, Equation 5.3.6 is rewritten in a more convenient form:

$$\mathbf{M}_{(i-1),i} = \mathbf{M}_{i,(i+1)} + \mathbf{p}_i^* \times \mathbf{F}_{i,(i+1)} + (\mathbf{p}_i^* + \mathbf{p}_{iG}^*) \times \mathbf{F}_0^i + \mathbf{M}_0^i \tag{5.4.5}$$

The rotation matrix $(A_i^0)_R$ is now applied to the terms of Equation 5.4.5 to obtain the torque equation underlined in Table 5.3c for Newton—Euler's formulation. The underlined force and torque equations in Tables 5.3b and

TABLE 5.3
Recursive Relations of Link i

a. Position, Velocity, and Acceleration for the Centroid of the ith Link,

$i = 1, \ldots, N$

$^i p_{iG}^* = (A_i^0)_R p_{iG}^* = (A_i^0)_R (P_{0iG} - p_{0i})$

$^i \omega_{0i} = (A_i^0)_R \omega_{0i}$

$^i v_{0i} = (A_i^0)_R \dot{p}_{0i}$

$^i v_{0iG} = (A_i^0)_R \dot{p}_{0iG}$

$^i \mathbf{v}_{0iG} = {}^i \mathbf{v}_{0i} + ({}^i \boldsymbol{\omega}_{0i}) \times^i \mathbf{p}_{iG}^*$

$^i a_{0i} = (A_i^0)_R \ddot{p}_{0i}$

$^i a_{0iG} = (A_i^0)_R \ddot{p}_{0iG}$

$^i \alpha_{0i} = (A_i^0)_R \dot{\omega}_{0i}$

$^i \mathbf{a}_{0iG} = {}^i \mathbf{a}_{0i} + ({}^i \boldsymbol{\alpha}_{0i}) \times ({}^i \mathbf{p}_{iG}^*) + {}^i \boldsymbol{\omega}_{0i} \times [({}^i \boldsymbol{\omega}_{0i}) \times^i \mathbf{p}_{iG}^*)]$

b. External Force and Moment Acting on Link i, $i = N, \ldots, 1$

$^i F_0^i = (A_i^0)_R F_0^i$

$^i F_0^i = m_i [(A_i^0)_R (\ddot{p}_{0iG} + g_{0i})]$

$^i M_0^i = (A_i^0)_R M_0^i$

$^i I_{0i} = (A_i^0)_R I_{0i} (A_0^i)_R \quad {}^i H_{0i} = ({}^i I_{0i})({}^i \omega_{0i}) \quad {}^i M_{0i}^c = ({}^i I_{0i})({}^i \alpha_{0i})$

$^i \mathbf{M}_0^i = ({}^i \mathbf{M}_{0i}^c) + ({}^i \boldsymbol{\omega}_{0i}) \times ({}^i \mathbf{H}_{0i})$

c. Force, Torque Relations for the ith Link, $i = N, \ldots, 1$

$^i F_{(i-1),i} = (A_i^0)_R F_{(i-1),i}$

$^i F_{i,(i+1)} = (A_i^{i+1})_R ({}^{i+1} F_{i,(i+1)}) = (A_i^{i+1})_R [(A_{i+1}^0)_R F_{i,(i+1)}]$

$^i \mathbf{F}_{(i-1),i} = ({}^i \mathbf{F}_{i,(i+1)}) + {}^i \mathbf{F}_0^i$

$^i M_{(i-1),i} = (A_i^0)_R M_{(i-1),i}$

$^i M_{i,(i+1)} = (A_i^{i+1})_R [(A_{i+1}^0)_R M_{i,(i+1)}]$

$^i \mathbf{M}_{(i-1),i} = {}^i \mathbf{M}_{i,(i+1)} + ({}^i \mathbf{p}_i^*) \times ({}^i \mathbf{F}_{i,(i+1)}) + [({}^i \mathbf{p}_i^*) + ({}^i \mathbf{p}_{iG}^*)] \times ({}^i \mathbf{F}_0^i) + {}^i \mathbf{M}_0^i$

$p_i^* = p_{0i} - p_{0(i-1)} \qquad p_{0iG} - p_{0(i-1)} = (p_i^* + p_{iG}^*) \qquad {}^{i+1} p_i^* = (A_{i+1}^0)_R p_i^*$

$^i p_i^* = (A_i^{i+1})_R [(A_{i+1}^0)_R p_i^*] = (A_i^{i+1})_R ({}^{i+1} p_i^*)$

d. Generalized Torque Generated by the Actuator at Joint i

$\tau_i = [(A_i^0)_R M_{(i-1),i}]' [(A_i^{i-1})_R k_{z_{i-1}}] = ({}^i M_{(i-1),i})' [(A_i^{i-1})_R k_{z_{i-1}}]$ if link i is rotational

$\tau_i = [(A_i^0)_R F_{(i-1),i}]' [(A_i^{i-1})_R k_{z_{i-1}}] = ({}^i F_{(i-1),i})' [(A_i^{i-1})_R k_{z_{i-1}}]$ if link i is prismatic

5.3c present Newton–Euler's equations relative to the local coordinates of link i.

Input τ_i generated by the actuator of joint i acts about or along the z_{i-1}-axis. It follows that input τ_i represents the component of the generalized torque $(A_i^0)_R M_{(i-1),i}$ on the z_{i-1}-axis. This component can be determined, for example, by forming a dot product between the generalized torque vector and the unit vector $(A_i^{i-1})_R k_{z_{i-1}}$. Thus, the equations in Table 5.3d for a rotational as well as for a prismatic link are obtained.

The calculations of the dynamics for a serial link N-joint manipulator in the framework of Newton–Euler's formulation are summarized in the following.

Algorithm for Manipulator Dynamics in Newton–Euler's Formulation

A. Use the *forward* difference equations for calculating the position, velocity, and acceleration of link $j, j = 1, \ldots, N$:

1. Initialize the algorithm: $\psi_{00} = 0$, $\omega_{00} = 0$, $\dot{\omega}_{00} = 0$, $p_{00} = 0$, $\dot{p}_{00} = 0$, and $\ddot{p}_{00} = 0$. Set $i = 0$.

2. Determine the rotational and translational (a) velocities and (b) accelerations for link j where $j = i + 1$ using the equations of Table 5.2b and 5.2c. That is, compute $^j\omega_{0j} = (A_j^0)_R \omega_{0j}$, $^j\alpha_{0j} = (A_j^0)_R \dot{\omega}_{0j}$, $^jv_{0j} = (A_j^0)_R \dot{p}_{0j}$, and $^ja_{0j} = (A_j^0)_R \ddot{p}_{0j}$.

3. Compute the velocity and acceleration of the centroid $^jv_{0jG}(A_j^0)_R \dot{p}_{0jG}$, $^ia_{0jG} = (A_j^0)_R \ddot{p}_{0jG}$ for link j using the equations of Table 5.3a.

4. Calculate the external force $^jF_0^j = (A_j^0)_R F_0^j$, and moment $^jM_0^j = (A_j^0)_R M_0^j$ exerted on link j using the equations of Table 5.3b.

5. If $j < N$, increase i by one, that is, replace i by $i + 1$, and return to step 2. If $j = N$, continue.

B. Use the *backward* difference equations for computing the generalized torques for link $i = N, N - 1, \ldots, 1$.

Knowing the contact force $^NF_{N,(N+1)}$, and moment $^NM_{N,(N+1)}$, set $i = N$.

6. Compute force $^iF_{(i-1),i} = (A_i^0)_R F_{(i-1),i}$ and torque $^iM_{(i-1),i} = (A_i^0)_R M_{(i-1),i}$ acting on link i using the equations in Table 5.3c.

7. Decrease i by one, that is, replace i by $i - 1$, and repeat from step 6 on until $i = 1$.

8. Calculate the generalized torque τ_i of joint i for $i = 1$, $\ldots$, N using the equation of Table 5.3d. Stop.

The evaluation of the expressions in Newton–Euler's formulation is efficient due to the recursive nature of the equations. The computational complexity of the foregoing algorithm increases linearly with N, the number of joints in the manipulator. The complexity of calculations in Newton–Euler's formulation is proportional to N; thus, it is of the same order as in the recursive Lagrangian formulation.

The planar manipulator of Example 5.3.1 is next studied to illustrate the systematic use of Newton–Euler's formulation in the local moving coordinate frames.

EXAMPLE 5.4.1

The dynamical model for the motion of the revolute planar manipulator described in Example 5.3.1 is to be developed systematically using Newton–Euler's formulation in the local coordinate frames. The external force and torque exerted by the environment on the end-effector are $F_{2,3} = [F_{2,3x} \ F_{2,3y} \ 0]'$ and $M_{2,3} = [0 \ 0 \ M_{2,3z}]'$, respectively.

To solve the problem, the rotation submatrices are first written from the homogeneous transformation matrices associated with the coordinate frames shown in Figure 5.4a. They are contained in matrices A_0^1, A_1^2, and A_0^2 of Equations 4.2.31–4.2.32. Since the inverse of the rotation matrix is equal to its transpose, it follows that

$$(A_1^0)_R = \begin{bmatrix} c_1 & s_1 & 0 \\ -s_1 & c_1 & 0 \\ 0 & 0 & 1 \end{bmatrix} \tag{5.4.6}$$

$$(A_2^1)_R = \begin{bmatrix} c_2 & s_2 & 0 \\ -s_2 & c_2 & 0 \\ 0 & 0 & 1 \end{bmatrix} \tag{5.4.7}$$

$$(A_2^0)_R = \begin{bmatrix} c_{12} & s_{12} & 0 \\ -s_{12} & c_{12} & 0 \\ 0 & 0 & 1 \end{bmatrix} \tag{5.4.8}$$

The equations of motion will be determined by applying the equations of Tables 5.2 and 5.3 to the planar manipulator under consideration. The position, velocity, and acceleration equations are first applied in the forward direction, that is, from the base toward the end-effector. Then, the force and torque are calculated in the backward direction, that is, from the end-effector toward the base.

By applying the algorithm of Newton–Euler's formulation, the following equations from Tables 5.2 and 5.3 are obtained for the two-link revolute manipulator:

A.   *Forward Difference Equations:*

1.   Initialization of the algorithm: $\omega_{00} = 0$, $\dot{\omega}_{00} = 0$, $p_{00} = 0$, $\dot{p}_{00} = 0$, $\ddot{p}_{00} = 0$, and $p_{01} = [\ell_1 c_1 \ \ell_1 s_1 \ 0]'$.

2. (a) The underlined velocity relations in Table 5.2b for link
        1 $(i = 0)$ give

$${}^1\omega_{01} = (A_1^0)_R[(A_0^0)_R\omega_{00} + \dot{\theta}_1 k_{z_0}] \qquad (5.4.9)$$

$${}^1\mathbf{v}_{01} = {}^1\mathbf{v}_{00} + ({}^1\boldsymbol{\omega}_{01}) \times ({}^1\mathbf{p}_1^*) \qquad (5.4.10)$$

$${}^1p_1^* = (A_1^0)_R p_1^* = (A_1^0)_R(p_{01} - p_{00}) \qquad (5.4.11)$$

where ${}^1\mathbf{v}_{00} = 0$, $p_1^* = p_{01} - p_{00} = [\ell_1 c_1 \ \ \ell_1 s_1 \ \ 0]'$ and ${}^1p_1^* =$
$[\ell_1 \ 0 \ 0]'$. Since ${}^1k_{z_0} = k_{z_0}$, Equation 5.4.9 reduces to

$${}^1\omega_{01} = (A_1^0)_R\omega_{01} = [0 \ \ 0 \ \ \dot{\theta}_1]' \qquad (5.4.12)$$

Equations 5.4.10–5.4.12 can be combined to give

$${}^1\mathbf{v}_{01} = (A_1^0)_R \dot{p}_{01} = [0 \ \ \ell_1\dot{\theta}_1 \ \ 0]' \qquad (5.4.13)$$

(b) The acceleration relations (underlined) in Table 5.2c for
    link 1 $(i = 0)$ give, since ${}^1k_{z_0} = (A_1^0)_R k_{z_0} = k_{z_0}$

$${}^1\boldsymbol{\alpha}_{01} = ({}^1\boldsymbol{\alpha}_{00}) + \ddot{\theta}_1 k_{z_0} + ({}^1\boldsymbol{\omega}_{00}) \times (\dot{\theta}_1 k_{z_0}) \qquad (5.4.14)$$

$${}^1\mathbf{a}_{01} = {}^1\mathbf{a}_{00} + ({}^1\boldsymbol{\alpha}_{01}) \times ({}^1\mathbf{p}_1^*) + ({}^1\boldsymbol{\omega}_{01}) \times [({}^1\boldsymbol{\omega}_{01}) \times ({}^1\mathbf{p}_1^*)] \qquad (5.4.15)$$

Equation 5.4.14 can be simplified since ${}^1\alpha_{00} = 0$ and
${}^0\omega_{00} = 0$

$${}^1\alpha_{01} = (A_1^0)_R \dot{\omega}_{01} = [0 \ \ 0 \ \ \ddot{\theta}_1]' \qquad (5.4.16)$$

Moreover, Equations 5.4.15 can be developed by using
${}^1\alpha_{00} = 0$ to give

$${}^1a_{01} = (A_1^0)_R \ddot{p}_{01} = [-\ell_1\dot{\theta}_1^2 \ \ \ell_1\ddot{\theta}_1 \ \ 0]' \qquad (5.4.17)$$

3. The position, velocity, and acceleration of the centroid for
    link 1 are written from Table 5.3a:

$${}^1p_{1G}^* = (A_1^0)_R p_{1G}^* = (A_1^0)_R(p_{01G} - p_{01}) \qquad (5.4.18)$$

$${}^1\mathbf{v}_{01G} = {}^1\mathbf{v}_{01} + ({}^1\boldsymbol{\omega}_{01}) \times ({}^1\mathbf{p}_{1G}^*) \qquad (5.4.19)$$

$${}^1\mathbf{a}_{01G} = {}^1\mathbf{a}_{01} + ({}^1\boldsymbol{\alpha}_{01}) \times ({}^1\mathbf{p}_{1G}^*) + ({}^1\boldsymbol{\omega}_{01}) \times [({}^1\boldsymbol{\omega}_{01}) \times ({}^1\mathbf{p}_{1G}^*)] \qquad (5.4.20)$$

where $p_{01G} = [\ell_1 c_1/2 \ \ \ell_1 s_1/2 \ \ 0]'$, ${}^1\mathbf{v}_{01} = \ell_1\dot{\theta}_1 \mathbf{j}_{y_0}$ by Equation
5.4.13, ${}^1\omega_{01} = \dot{\theta}_1 k_{z_0}$ by Equation 5.4.12, ${}^1a_{01} = (-\ell_1\dot{\theta}_1^2)\mathbf{i}_{x_0} + (\ell_1\ddot{\theta}_1)\mathbf{j}_{y_0}$, and ${}^1\alpha_0 1 = \ddot{\theta}_1 k_{z_0}$. Equation 5.4.18 gives ${}^1\mathbf{p}_{1G}^* = (-\ell_1/2)\mathbf{i}_{x_0}$. Equation 5.4.19 and 5.4.20 can now be manipulated to obtain

$${}^1\mathbf{v}_{01G} = (A_1^0)_R \dot{p}_{01G} = [0 \ \ \ell_1\dot{\theta}_1/2 \ \ 0]' \qquad (5.4.21)$$

$${}^1a_{01G} = (A_1^0)_R \ddot{p}_{01G} = [-\ell_1\dot{\theta}_1^2/2 \ \ \ell_1\ddot{\theta}_1/2 \ \ 0]' \qquad (5.4.22)$$

4. The external force and torque relations for link 1 are written from Table 5.3b

$$^1F_0^1 = (A_1^0)_R F_0^1 = m_1(A_1^0)_R(\ddot{p}_{01G} + g_{01}) \tag{5.4.23}$$

$$^1I_{01} = (A_1^0)_R I_{01}(A_0^1)_R \qquad ^1H_{01} = (^1I_{01})(^1\omega_{01}) \qquad ^1M_{01}^c = (^1I_{01})(^1\alpha_{01}) \tag{5.4.24}$$

$$^1\mathbf{M}_0^1 = {}^1\mathbf{M}_{01}^c + (^1\boldsymbol{\omega}_{01}) \times (^1\mathbf{H}_{01}) \tag{5.4.25}$$

where $g_{01} = [0 \ -g \ 0]'$. Equations 5.4.23–5.4.25 can be manipulated to the following forms

$$^1F_0^1 = (A_1^0)_R F_0^1 = m_1[-\ell_1\dot{\theta}_1^2/2 - gs_1 \quad \ell_1\ddot{\theta}_1/2 - gc_1 \quad 0]' \tag{5.4.26}$$

$$^1M_0^1 = (A_1^0)_R M_0^1 = [0 \ 0 \ I_{01}\ddot{\theta}_1]' \tag{5.4.27}$$

Equations 5.4.26 and 5.4.27 describe the external force and moment acting on link 1. Steps 2, 3, and 4 of the algorithm for the first link are now completed.

5. Since $j < 2$, i is increased by one. The algorithm returns to step 2. The calculations are then repeated for link 2.

2.' (a) Velocity relations from Table 5.2b for link 2 give

$$^2\omega_{02} = (A_2^0)_R \omega_{02} = (A_2^1)_R[^1\omega_{01} + \dot{\theta}_2 k_{z_1}] \tag{5.4.28}$$

$$^2p_2^* = (A_2^0)_R p_2^* = (A_2^0)_R(p_{02} - p_{01}) \tag{5.4.29}$$

$$^2\mathbf{v}_{02} = {}^2\mathbf{v}_{01} + (^2\boldsymbol{\omega}_{02}) \times (^2\mathbf{p}_2^*) \tag{5.4.30}$$

where $^1\omega_{01}$ is given by Equation 5.4.12, $^2v_{01} = (A_2^1)_R(^1v_{01})$ by Equations 5.4.13 and 5.4.7, and $p_2^* = p_{02} - p_{01} = [\ell_2c_{12} \ \ell_2s_{12} \ 0]'$. The substitution of Equation 5.4.12 into Equation 5.4.28 gives

$$^2\omega_{02} = (A_2^0)_R \omega_{02} = [0 \ 0 \ \dot{\theta}_1 + \dot{\theta}_2]' \tag{5.4.31}$$

Equations 5.4.13, 5.4.28, and 5.4.29 are substituted into Equation 5.4.30, to obtain

$$^2v_{02} = (A_2^0)_R \dot{p}_{02} = [\ell_1\dot{\theta}_1s_2 \quad \ell_1\dot{\theta}_1c_2 + \ell_2(\dot{\theta}_1 + \dot{\theta}_2) \quad 0]' \tag{5.4.32}$$

(b) Acceleration relations from Table 5.2c for link 2 give, since $^2k_{z_1} = (A_2^1)_R k_{z_1} = k_{z_1}$

$$^2\alpha_{02} = {}^2\alpha_{01} + \ddot{\theta}_2 \mathbf{k}_{z_1} + (^2\omega_{01}) \times (\dot{\theta}_2\mathbf{k}_{z_1}) \tag{5.4.33}$$

$$^2\mathbf{a}_{02} = {}^2\mathbf{a}_{01} + (^2\boldsymbol{\alpha}_{02}) \times (^2\mathbf{p}_2^*) \tag{5.4.34}$$

where $^2\omega_{01} = (A_2^1)_R(^1\omega_{01})$, $^2\alpha_{01} = (A_2^1)_R(^1\alpha_{01})$, $^2a_{01} = (A_2^1)_R(^1a_{01})$, and $^1\alpha_{01}$ and $^1a_{01}$ are specified by Equations 5.4.16 and 5.4.17, respectively.

The substitution of Equations 5.4.16 and 5.4.12 into Equation 5.4.33 leads to

$$^2\alpha_{02} = A(^0_2)_R \dot{\omega}_{02} = [0 \; 0 \; \ddot{\theta}_1 + \ddot{\theta}_2]' \tag{5.4.35}$$

Equation 5.4.34 is developed by making use of Equations 5.4.17, 5.4.35, and 5.4.29 to obtain

$$^2a_{02} = (A^0_2)_R \ddot{p}_{02} = \begin{bmatrix} \ell_2(\dot{\theta}_1 + \dot{\theta}_2)^2 - \ell_1\dot{\theta}^2 c_2 + \ell_1\ddot{\theta}_1 s_2 \\ \ell_2(\ddot{\theta}_1 + \ddot{\theta}_2) + \ell_1\dot{\theta}^2_1 s_2 + \ell_1\ddot{\theta}_1 c_2 \\ 0 \end{bmatrix} \tag{5.4.36}$$

3.' The position, velocity, and acceleration of the centroid for link 2 are obtained from Table 5.3a:

$$^2p^*_{2G} = (A^0_2)_R p^*_{2G} = (A^0_2)_R(p_{02G} - p_{02}) \tag{5.4.37}$$

$$^2v_{02G} = {}^2v_{02} + (^2\omega_{02}) \times (^2p^*_{2G}) \tag{5.4.38}$$

$$^2a_{02G} = {}^2a_{02} + (^2\alpha_{02}) \times (^2p^*_{2G}) + (^2\omega_{02}) \times [(^2\omega_{02}) \times (^2p^*_{2G})] \tag{5.4.39}$$

where $p^*_{2G} = p_{02G} - p_{02} = [-\ell_2 c_{12}/2 \;\; -\ell_2 s_{12}/2 \; 0]'$, $^2\omega_{02}$ and $^2v_{02}$ are given in Equations 5.4.31 and 5.4.32, and $^2a_{02}$ and $^2\alpha_{02}$ in Equations 5.4.36 and 5.4.35, respectively. Equations 5.4.36–5.4.39 can be developed to the following expressions:

$$^2p^*_{2G} = (A^0_2)_R p^*_{2G} = [-\ell_2/2 \; 0 \; 0]' \tag{5.4.40}$$

$$^2v_{02G} = (A^0_2)_R \dot{p}_{02G} = [\ell_1\dot{\theta}_1 s_2 \;\; \ell_1\dot{\theta}_1 c_2 + \ell_2(\dot{\theta}_1 + \dot{\theta}_2)/2 \; 0]' \tag{5.4.41}$$

$$^2a_{02G} = (A^0_2)_R \ddot{p}_{02G} = \begin{bmatrix} -\ell_2(\dot{\theta}_1 + \dot{\theta}_2)^2/2 - \ell_1\dot{\theta}^2_1 c_2 + \ell_1\ddot{\theta}_1 s_2 \\ \ell_1\dot{\theta}^2_1 s_2 + \ell_1\ddot{\theta}_1 c_2 + \ell_2(\ddot{\theta}_1 + \ddot{\theta}_2)/2 \\ 0 \end{bmatrix} \tag{5.4.42}$$

4.' The external force and moment relation for link 2 are written from Table 5.3b

$$^2F^2_0 = (A^0_2)_R F^2_0 = m_2(A^0_2)_R(\ddot{p}_{02G} + g_{02}) \tag{5.4.43}$$

$$^2I_{02} = (A^0_2)_R I_{02}(A^0_2)_R \qquad ^2H_{02} = (^2I_{02})(^2\omega_{02}) \qquad ^2M^c_{02} = (^2I_{02})(^2\alpha_{02}) \tag{5.4.44}$$

$$^2M^2_0 = {}^2M^c_{02} + (^2\omega_{02}) \times (^2H_{02}) \tag{5.4.45}$$

where $^2\omega_{02}$ and $^2\alpha_{02}$ are determined by Equations 5.4.31 and 5.4.35, respectively, and $g_{02} = [0 \; -g \; 0]'$. When Equation 5.4.42 is substituted into Equation 5.4.43, the external force $^2F^2_0$ becomes

$$^2F_0^2 = (A_2^0)_R F_0^2 = m_2 \begin{bmatrix} \ell_1 \ddot{\theta}_1 s_2 - \ell_1 \dot{\theta}_1^2 c_2 - \ell_2(\dot{\theta}_1 + \dot{\theta}_2)^2/2 - g s_{12} \\ \ell_1 \ddot{\theta}_1 c_2 + \ell_2(\ddot{\theta}_1 + \ddot{\theta}_2)/2 + \ell_1 \dot{\theta}_1^2 s_2 - g c_{12} \\ 0 \\ 0 \end{bmatrix} \quad (5.4.46)$$

After the substitution of Equations 5.4.35 and 5.4.31 into
Equations 5.4.44 and 5.4.45, the external torque $^2M_0^2$ may
be expressed as

$$^2M_0^2 = (A_2^0)_R^2 M_0 = [0 \quad 0 \quad I_{01}(\ddot{\theta}_1 + \ddot{\theta}_2)]' \quad (5.4.47)$$

5.'   Since $j = 2$, the computations of part A of the algorithm are
complete. Indeed, the rotational and translational velocities
and accelerations for the links and their gravity centers are
determined recursively first for link 1 by Equations 5.4.12–
5.4.22, and then for link 2 by Equations 5.4.31–5.4.42. Thus,
the calculations are performed by starting from the base and
proceeding toward the end-effector.
    The algorithm then moves to part B.

B.   *Backward Difference Equations*

6.   For $i = 2$, the required force and torque at joint 2 ($N = 2$)
are determined by the equations of Table 5.3c. Since
$(A_2^3)_R(^3F_{2,3}) = (A_2^3)_R(A_3^0)_R F_{2,3} = (A_2^0)_R F_{2,3}$

$$^2F_{1,2} = (A_2^0)_R F_{1,2} = (A_2^0)_R F_{2,3} + (A_2^0)_R F_0^2 \quad (5.4.48)$$

$$^2F_0^2 = (A_2^0)_R F_0^2 \quad ^2M_{1,2} = (A_2^0)_R M_{1,2} \quad ^2M_{2,3} = (A_2^0)_R M_{2,3} \quad (5.4.49)$$

$$^2\mathbf{M}_{1,2} = {}^2\mathbf{M}_{2,3} + (^2\mathbf{p}_2^*) \times (^2\mathbf{F}_{2,3}) + $$

$$[(^2\mathbf{p}_2^*) + (^2\mathbf{p}_{2G}^*)] \times (^2\mathbf{F}_0^2) + {}^2\mathbf{M}_0^2 \quad (5.4.50)$$

where $F_{2,3} = [F_{2,3x} \quad F_{2,3y} \quad 0]'$, $M_{2,3} = [0 \quad 0 \quad M_{2,3z}]'$, and $p_2^* + p_{2G}^* = p_{02G} - p_{01} = [\ell_2 c_{12}/2 \quad \ell_2 s_{12}/2 \quad 0]'$. The substitution of
Equation 5.4.46 into Equation 5.4.48 leads to

$$^2F_{1,2} = \quad (5.4.51)$$

$$\begin{bmatrix} F_{2,3x} c_{12} + F_{2,3y} s_{12} + m_2[\ell_1 \ddot{\theta}_1 s_2 - \ell_1 \dot{\theta}_1^2 c_2 - \ell_2(\dot{\theta}_1 + \dot{\theta}_2)^2/2 - g s_{12}] \\ F_{2,3y} c_{12} - F_{2,3x} s_{12} + m_2[\ell_1 \ddot{\theta}_1 c_2 + \ell_2(\ddot{\theta}_1 + \ddot{\theta}_2)/2 + \ell_1 \dot{\theta}_1^2 s_2 - g c_{12}] \\ 0 \end{bmatrix}$$

Equation 5.4.50 can be expressed in terms of the joint variables by observing Equations 5.4.40, 5.4.46, and 5.4.49 to
obtain

$$^2M_{1,2} = (A_2^0)_R M_{1,2} = [0 \quad 0 \quad {}^2M_{1,2z}]' \quad (5.4.52)$$

where

$$^2M_{1,2z} = M_{2,3z} + \ell_2 c_{12} F_{2,3y} - \ell_2 s_{12} F_{2,3x} + \tag{5.4.53}$$

$$(m_2\ell_2/2)[(\ell_1\ddot{\theta}_1 c_2 + \ell_2(\ddot{\theta}_1 + \ddot{\theta}_2)/2 + \ell_1\dot{\theta}_1^2 s_2 - gc_{12} + I_{02}(\ddot{\theta}_1 + \ddot{\theta}_2)]$$

Equation 5.4.52 is the torque acting on link 2. It will be
used to determine the torque the actuator of joint 2 needs
to generate.

7. Since $i > 1$, integer i is reduced by 1, that is, set $i = 1$. Step
    6 is then repeated.

6.' The equations of Table 5.3c are rewritten for link 1 ($i = 1$):

$$^1F_{0,1} = (A_1^0)_R F_{0,1} = (A_1^2)_R(^2F_{1,2}) + (A_1^0)_R F_0^1 \tag{5.4.54}$$

$$^1F_{1,2} = (A_1^2)_R(^2F_{1,2}) \qquad ^1M_{0,1} = (A_1^0)_R M_{0,1} \qquad ^1M_{1,2} = (A_1^2)_R(^2M_{1,2}) \tag{5.4.55}$$

$$^1\mathbf{M}_{0,1} = {}^1\mathbf{M}_{1,2} + (^1\mathbf{p}_1^*) \times (^1\mathbf{F}_{1,2}) +$$

$$[(^1\mathbf{p}_1^*) + (^1\mathbf{p}_{1G}^*)] \times (^1\mathbf{F}_0^1) + {}^1\mathbf{M}_0^1 \tag{5.4.56}$$

where $^1p_1^*$ and $^1p_{1G}^*$ are specified by Equations 5.4.11
and 5.4.18, $^1F_0^1$ and $^1M_0^1$ by Equations 5.4.26 and 5.4.27,
respectively. Moreover, $^2F_{1,2}$ and $^2M_{1,2}$ are determined by
Equations 5.4.51 and 5.4.52. The substitution of Equations
5.4.51, and 5.4.26 into Equation 5.4.54 gives, after some
simplifications

$$^1F_{0,1} = \begin{bmatrix} ^1F_{1,2}^x \\ ^1F_{1,2}^y \\ 0 \end{bmatrix} + \begin{bmatrix} -m_1\ell_1\dot{\theta}_1^2/2 - m_1 g s_1 \\ m_1\ell_1\ddot{\theta}_1/2 - m_1 g c_1 \\ 0 \end{bmatrix} \tag{5.4.57}$$

$$^1F_{1,2}^x = F_{2,3x}c_1 + F_{2,3y}s_1 - m_2[\ell_1\dot{\theta}_1^2 + \ell_2(\dot{\theta}_1 + \dot{\theta}_2)^2 c_2/2 + \ell_2(\ddot{\theta}_1 + \ddot{\theta}_2)s_2/2 - gs_1]$$

$$^1F_{1,2}^y = -F_{2,3x}s_1 + F_{2,3y}c_1 + m_2[\ell_1\ddot{\theta}_1 - \ell_2(\dot{\theta}_1 + \dot{\theta}_2)^2 s_2/2 + \ell_2(\ddot{\theta}_1 + \ddot{\theta}_2)c_2/2 - gc_1]$$

To express Equation 5.4.56 in terms of the joint variables,
Equations 5.4.26, 5.4.27, 5.4.51, and 5.4.52 are substituted
into Equations 5.4.55 and 5.4.56. After forming the cross-
products and simplifying, torque $^1M_{0,1}$ assumes the following
expression:

$$^1M_{0,1} = (A_1^0)_R M_{0,1} = [0 \quad 0 \quad ^1M_{0,1z}]' \tag{5.4.58}$$

where

$$^1M_{0,1z} = I_{11}\ddot{\theta}_1 + I_{12}(\ddot{\theta}_1 + \ddot{\theta}_2) + (m_2\ell_1\ell_2 s_2/2)\dot{\theta}_1^2$$

$$-(m_2\ell_1\ell_2 s_2/2)(\dot{\theta}_1 + \dot{\theta}_2)^2 - m_1 g\ell_1 c_1/2 - m_2 g(\ell_1 c_1 + \ell_2 c_{12}/2)$$

$$-(\ell_1 s_1 + \ell_2 s_{12})F_{2,3x} + (\ell_1 c_1 + \ell_2 c_{12})F_{2,3y} + M_{2,3z}] \tag{5.4.59}$$

with $I_{11} = I_{01} + m_1(\ell_1/2)^2 + m_2(\ell_1^2 + \ell_1\ell_2 c_2/2)$, and $I_{12} = I_{02} + m_2\ell_1\ell_2 c_2/2 + m_2(\ell_2/2)^2$. Equation 5.4.58 is used to determine the torque that the actuator on joint 1 needs to generate.

The generalized forces and torques acting on the links are now determined. Indeed, $^2F_{1,2}$ and $^2M_{1,2}$ for link 2 are first calculated by Equations 5.4.51–5.4.53, and then $^1F_{0,1}$ and $^1M_{0,1}$ for link 1 by Equations 5.4.57–5.4.59. Thus, the determination of the generalized forces and torques is performed by starting from the link of the end-effector and proceeding toward the base.

7.'   Since $i = 1$, step 8 is performed next.

8.   Equations 5.4.58 and 5.4.52 determine the joint torques of the actuators. The torque actuator i needs to provide is given by the following expression:

    (a)   The moment needed at joint $i = 1$ is

$$\tau_1 = (^1M_{0,1})'[(A_1^0)_R k_{z_0}] - B_1\dot{\theta}_2 \qquad (5.4.60)$$

    where $(A_1^0)_R k_{z_0} = [0 \quad 0 \quad 1]'$, and the first term on the right is given by Equations 5.4.58 and 5.4.59.

    (b)   The moment needed at joint $i = 2$ is

$$\tau_2 = (^2M_{1,2})'[(A_2^1)_R k_{z_1}] - B_2\dot{\theta}_2 \qquad (5.4.61)$$

    where $(A_2^0)_R k_{z_1} = [0 \quad 0 \quad 1]'$, and the first term on the right is specified by Equations 5.4.52 and 5.4.53.

The dynamical model obtained can be developed to the same form as Equation 5.3.28 for the given revolute planar manipulator. The derivation presented here is systematic, and the resulting equations are computationally efficient. The approach is applicable to any serial link manipulator.

5.5   COMPUTATIONAL ROBOT DYNAMICS

The dynamical models for the motion of robot manipulators are discussed in this chapter in the framework of Lagrange's and Newton–Euler's formulations. Both formulations are based on fundamental physical laws of dynamics [9] and provide the designer insight in the dynamical behavior of the system. This is an important aspect in the engineering analysis and design. When Lagrange's formulation is used, the designer has the dynamical model for the whole robot system. Therefore, the interactions between the variables and the couplings between the dynamical equations of the joints are quite apparent. The general form of the model in Equation 5.2.10 is attractive to control engineers, although it contains lengthy expressions. It fits the general framework of many control design methods. When Newton–Euler's formulation is used, the effects of the external and internal forces and torques on a single link are transparent to the designer. In many cases, however, it is difficult for him/her to visualize the interrelationships between different parts of the entire robot system, because they do not appear in the model explicitly. We emphasize that the designer needs to be familiar with both models. The choice of the model used in a study is usually dependent on the specific application and task to be performed by the robot manipulator system.

The equations of motion obtained for a manipulator by either of the foregoing approaches can be used to solve the problem of the inverse dynamics, that is, to determine the generalized joint torques when the desired positions, velocities, and accelerations of the joints are specified (this is sometimes called the computed torque method). It is accomplished by an algorithm that *numerically* solves the equations of the manipulator dynamics for the generalized joint torques.

Because of the lengthy and cumbersome expressions appearing in the dynamical model for the motion of a general robot manipulator, *symbolic* modeling of robot mechanisms plays an important role in computational robot dynamics. It is described by a symbolic program that can manipulate algebraic expressions (instead of numbers) after encoding them to forms that facilitate the symbolic mathematical operations internally in the computer. A symbolic robot modeling algorithm can generate complete dynamical manipulator models from a set of manipulator specifications (given by the program user). Such a description includes the total DOF, the types of the links (joints), the structural kinematic parameters, the masses, and inertial moments of the links [10].

The symbolic modeling software packages are implemented in symbol-oriented programming languages that perform symbolic operations. For example, LISP and the symbolic program package Macsyma (or Vaxyma) are also widely used. The basic mathematical operations needed for robot modeling in a symbolic software package include matrix–matrix and matrix–vector multiplications, calculations of quadratic forms, trace operations, evaluations of vector cross- and dot-products, matrix and vector additions, and several subroutines for simplifying trigonometric expressions. When the classical Lagrangian formulation is used, symbolic differentiation is also needed. The differentiation of the homogeneous transformation matrices can be converted to matrix multiplications, as discussed in Section 5.2.1. When Newton–Euler's formulation is used, the recursive structure of the model can be preserved.

A symbolic program can be written as a customized algorithm for a specific structured manipulator (which possesses parallel and perpendicular joint axes) to automate the computations. Customized algorithms for determining robot dynamics reduce the computational burden of the algorithms written for general manipulator dynamics. They can provide faster simulation results and permit the use of shorter sampling periods in the real-time implementations of control schemes. Moreover, since the program codes are generated by a computer, manual errors are reduced, and the reliability of the calculations is increased.

A symbolic computer program called Algebraic Robot Modeler (ARM) is well documented and described in [10]. It generates complete dynamical robot models. It is implemented at two levels: (1) a composer delineates the symbolic mathematical operations to produce the dynamical model, and (2) a performer carries out the operations specified by the composer. The composer is programmed in the C-language, and the performer for symbolic processing is implemented in LISP. The program has been verified by extensive testing procedures. The computational requirements of customized algorithms and several applications are discussed in [10], which provides additional details on the subject.

For a specific manipulator, several dynamical models can be generated by means of customized symbolic algorithms. For example, a single model can be constructed to describe the total manipulator dynamics as Euler–Lagrange equations, or the recursive equations of Newton–Euler for the motion of the links. In selecting a computational robot dynamic model for general purposes, an algorithm for the recursive Newton–Euler formulation is computationally the most efficient. It is to be noted, however, that a symbolic robot modeling program that is customized for specific manipulators such as a six DOF Stanford/JPL manipulator can reduce the computational requirements of a general-purpose recursive Newton–Euler algorithm by a factor of more than two [10]. It is accomplished by removing unnecessary repetitive arithmetic operations that are inherent in the formulation or in the manipulator structure and by systematically organizing the mathematical calculations in the symbolic program.

The symbolic computer programs for generating dynamical manipulator models are becoming important tools in the analysis and design of manipulator systems. They enable the designer to study complicated robotic systems that are in their complexity beyond the serial link, open-loop kinematic chain mechanisms composed of rigid links. The symbolic software approach facilitates the development of realistic and accurate models for such manipulator systems. We can expect that symbolic computer programs will be constructed in the future for complicated systems such as closed-chain mechanisms including multiple manipulator systems and multifingered devices holding a common object.

5.6 SUMMARY

The dynamical model of a system helps us to study its behavior under various circumstances, for example, by means of digital simulations. Moreover, it usually provides us with the basic information to design controllers for the system. We have discussed in this chapter the determination of the equations of motion for serial-link manipulators in the framework of Lagrange's formulation and Newton–Euler's formulation.

In the Lagrangian formulation, the expressions for the kinetic and potential energies of *all* moving links in a manipulator are first determined. After

Lagrange's energy function is written, it is substituted into Euler–Lagrange's equation to obtain the dynamical model of the manipulator. The resulting equations of motion for the entire manipulator are lengthy and complex, involving many trigonometric expressions. These equations can be converted to recursive forms, which are computationally efficient for digital simulations. However, the equations in the recursive Lagrangian formulation have a limited use for controller design because of their specific forms.

In Newton–Euler's formulation, *one* link of a serial-link manipulator is isolated at a time. The dynamical equations resulting from the balancing of the forces and torques acting on the link are written. They involve terms reflecting the effects of the adjacent links. These equations can be considered as difference equations relative to the links, and they can be solved recursively. The translational and rotational positions, velocities, and accelerations of the links are solved in the forward direction from the base to the end-effector, and the generalized force (moment) relations are determined in the backward direction from the end-effector to the base. These calculations are computationally efficient.

In comparing the two approaches, we observe that Lagrange's formulation results in a dynamical model of the entire manipulator in terms of all joint variables. Newton–Euler's formulation gives a dynamical model for one link in a form that represents a set of difference equations with respect to the links in a serial link manipulator. They can be solved recursively to obtain the dynamical model for the entire manipulator. In calculating the dynamics of a serial link manipulator, for example in digital simulations, the complexity of computations using Newton–Euler's formulation is of order N, and using the general closed-form Equation 5.2.10 of Lagrange's formulation of order N^4. However, the Lagrangian model can be developed into a recursive form [6]. In this case, the computational complexity is of the same order in recursive Lagrange's formulation and Newton–Euler's formulation. The equations in the recursive Lagrangian formulation are not in the form that is suitable to most controller design methods.

Because of the cumbersome expressions that need to be determined for the equations of motion, symbolic programming for the computational robot dynamics offers an attractive alternative to manual manipulations. It can be applied to customize the algorithms of dynamical modeling for specific manipulators. These customized symbolic programs for computational robot dynamics improve the computational efficiency and the reliability of simulation models. Their efficiency is achieved by systematically organizing the program so as to take advantage of special structural properties and by minimizing repetitive and unnecessary calculations. Thus, the computational requirements of the customized algorithms can be reduced so as to make them even more efficient than recursive dynamical model algorithms constructed for general purposes. The recursive Newton–Euler's formulation for general purposes is at present computationally the most efficient among the dynamical models of the manipulator motions.

REFERENCES

[1] A. K. BEJCZY, "Robot Arm Dynamics and Control," Technical Memo 33-669, Jet Propulsion Lab., February 1974.

[2] V. D. TOURASSIS and C. P. NEUMAN, "Properties and Structure of Dynamic Robot Models for Control Engineering Applications," *Mechanism and Machine Theory*, Vol. 20, No. 1, pp. 27–40, 1985.

[3] S. S. MAHIL, "On the Application of Lagrange's Methods to the Description of Dynamic Systems," *IEEE Transactions on Systems, Man and Cybernetics*, Vol. SMC-12, No. 6, November/December 1982.

[4] R. PAUL, M. RONG, and H. ZHANG, "The Dynamics of the PUMA Manipulator," Technical Report TR-EE 84-19, School of Electrical Engineering, Purdue University, West Lafayette, IN, July 1984.

[5] C.P. NEUMAN and J.J. MURRAY, "The Complete Dynamic Model and Customized Algorithms of the PUMA Robot," *IEEE Transactions on Systems, Man and Cybernetics*, Vol. SMC-17, No. 4, pp. 635–644, July/August 1987.

[6] J. M. HOLLERBACH, "A Recursive Lagrangian Formulation of Manipulator Dynamics and Comparative Study of Dynamics Formulation Complexity," *IEEE Transactions on Systems, Man, and Cybernetics*, Vol. SMC-10, No. 11, pp. 730–736, November 1980.

[7] J. Y. S. LUH, M. W. WALKER, and R. P. C. PAUL, "On-line Computational Scheme for Mechanical Manipulators," *Journal of Dynamic Systems, Measurement, and Control*, Vol. 102, pp. 69–76, June 1980.

[8] M. W. WALKER and D. E. ORIN, "Efficient Dynamic Computer Simulation of Robotic Mechanisms," *Journal of Dynamical Systems, Measurements, and Control*, Vol. 104, pp. 205–211, 1982.

[9] H. GOLDSTEIN, *Classical Mechanics*, Addison-Wesley, Reading, MA, 1980.

[10] C. P. NEUMAN and J. J. MURRAY, "Symbolically Efficient Formulations for Computational Robot Dynamics," *Journal of Robotic Systems*, Vol. 4, No. 6, pp. 743–769, 1987.

PROBLEMS

5.1 The first-order differential in a real space "corresponds" to the first-order variation in a function space. Consider a real-valued function $f:\mathbf{R}^2 \to \mathbf{R}$ where

$$f(x_1,x_2) = 2x_1^2 + 4x_1x_2 - x_1 - 2x_2 + 5$$

a. Determine necessary conditions for a stationary point of $f(x_1,x_2)$ by using the partial derivatives.

b. Show that the same necessary conditions for a stationary point are obtained by setting the first-order differential equal to zero.

c. Generalize the result to a real-valued function defined in an n-dimensional real space, that is, $f = f(x)$, where $f:\mathbf{R}^n \to \mathbf{R}$.

5.2 Given a three-dimensional vector p_0, compute

a. $p_0' p_0$

b. $p_0 p_0'$

c. Show that $p_0' p_0 = \text{tr}(p_0 p_0')$

d. The following holds for the trace operation: $\text{tr}[uu' + vv'] = \text{tr}[uu'] + \text{tr}[vv']$. Prove it for three-dimensional vectors u and v.

5.3 For the cylindrical robot discussed in Example 5.1.1, assume that the origin of the base coordinate $(x_0\ y_0\ z_0)$ system is at point O. Assume that the coordinates of the end-effector are x, y, z. In the manipulator, the arm has the mass $\rho_A = m_A/\ell$ per unit length, and the mass of the load is m_L.

a. Express the kinematic and potential energies in terms of the coordinates of the base coordinate system. Form Lagrange's energy function.

b. Use the relationships between the Cartesian and cylindrical coordinates to show that Lagrange's energy function formed in (a) can be expressed as given in Equation 5.1.3.

5.4 The behavior of the planar manipulator shown in Figure 5.2 and discussed in Example 5.2.1 is described by Equations 5.2.18 and 5.2.19, in which the coefficients are specified by Equations 5.2.20–5.2.30.

a. Evaluate the coefficients in Equations 5.2.20–5.2.30 explicitly, and substitute them into Equations 5.2.18 and 5.2.19 to obtain explicit expressions for the equations of motion.

b. Determine the kinetic and potential energies for the planar manipulator of Example 5.2.1 moving on a vertical plane (without using the homogeneous transformation matrices). Form the Lagrange's energy function, and obtain the dynamical model directly.

c. Compare the approaches used in *(a)* and *(b)*.

5.5 a. Starting from Equation B.5 of Appendix B, show that the inertial moment matrix I_i can be expressed as given in Equation 5.2.2. *Hint:* Start from Equation B.5 and use the definitions given in Equation B.7.

b. Explain why I_i is called the "pseudo-inertia" matrix.

5.6 a. For general (4×4) square matrices, show that $\text{tr}(AB')$ $= \text{tr}(BA')$. Note that in general $AB' \neq BA'$.

b. If I_i is a pseudo-inertia matrix, show

$$\text{tr}\left[\frac{\partial^2 A_0^i}{\partial q_j \partial q_n} I_i \left(\frac{\partial A_0^i}{\partial q_k} \right)' \right] = \text{tr}\left[\frac{\partial A_0^i}{\partial q_k} I_i \left(\frac{\partial^2 A_0^i}{\partial q_j \partial q_n} \right)' \right]$$

c. Verify that $\partial \mathcal{L}/\partial q_n$ and $\partial \mathcal{L}/\partial \dot{q}_n$ formed on the basis of Equations B.8 and B.10 can be expressed as given by Equations B.12 and B.13, respectively.

5.7 Verify that Equation 5.2.5 can be rewritten as Equations 5.2.6–5.2.9 by appropriate change of variables in the summation. (*Hint:* Use a coordinate graph to change the variables; recall how the changes in the variables of integration change the integration limits.)

5.8 In the dynamical model of a Stanford/JPL arm, the matrix derivative $\partial A_0^i/\partial q_n$ appears in Equation 5.2.5. Define [1]

$$U_{jn} = \frac{\partial A_0^j}{\partial q_n} = A_0^1 \cdots Q_n A_{n-1}^n \cdots A_{j-1}^j$$

where $1 \leq n \leq j$, and $j = 1, \ldots, N$. For a rotational link, $q_n = \theta_n$ and $Q_n = Q_{\theta_n}$. For a translational (linear) link, $q_n = d_n$ and $Q_n = Q_{d_n}$. Specifically,

$$Q_{\theta_n} = \begin{bmatrix} 0 & -1 & 0 & 0 \\ 1 & 0 & 0 & 0 \\ 0 & 0 & 0 & 0 \\ 0 & 0 & 0 & 0 \end{bmatrix}; \quad Q_{d_n} = \begin{bmatrix} 0 & 0 & 0 & 0 \\ 0 & 0 & 0 & 0 \\ 0 & 0 & 0 & 1 \\ 0 & 0 & 0 & 0 \end{bmatrix}$$

$$A_{i-1}^i = \begin{bmatrix} c\theta_i & -c\alpha_i s\theta_i & s\alpha_i s\theta_i & a_i c\theta_i \\ s\theta_i & c\alpha_i c\theta_i & -s\alpha_i c\theta_i & a_i s\theta_i \\ 0 & s\alpha_i & c\alpha_i & d_i \\ 0 & 0 & 0 & 1 \end{bmatrix}$$

a. Apply this method to evaluate at least three derivatives of the matrices in Example 5.2.1.

b. Prove that the equation defining U_{jn} holds for U_{32} and U_{11}. Determine U_{12}.

c. Prove that the general expression given for U_{jn} is true.

d. Determine U_{21}. If $j < n$, what is Q_n in the expression U_{jn}?

5.9 The base of a cylindrical manipulator has been installed on a vertical wall; thus, Figure 5.1a represents a side view. The parameters of the mechanical system are m_A/ℓ, m_L, I_θ, ℓ, k_z, and k_s defined for the vertical manipulator in Example 5.1.1.

a. Determine the kinetic and potential energy expressions for the system. Form Lagrange's energy function.

b. Obtain the equations of motion for this manipulator in terms of the joint variables.

5.10 In the cylindrical robot manipulator shown in Figures 5.5a and 5.5b, the origin of the base coordinate system $(x_0\ y_0\ z_0)$ is located at point O. The moving cylindrical coordinate

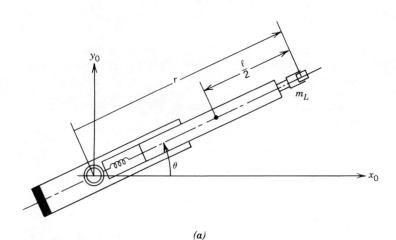

(a)

FIGURE 5.5a
Cylindrical robot manipulator; (top view).

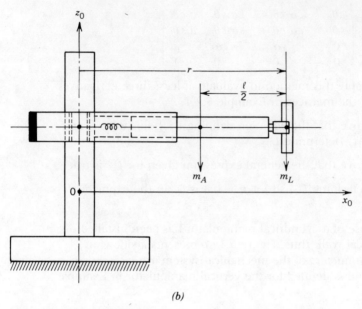

(b)

FIGURE 5.5*b*
Cylindrical robot manipulator; (side view).

frame $(r\theta z)$ is shown in the figure. The hub has a mass of
m_h with the center of gravity on the z_0-axis, and its moment
of inertia about the z_0-axis is I_θ. The mass of the prismatic
link is m_A with the center of gravity at the midpoint of
the link. The length of the link is ℓ. The mass of the load
is m_L. The springs are acting along the r- and z-axes and
described by the spring constant k_s; they are at the resting
positions (i.e., they produce zero forces) when the lengths
of the springs are $\ell/3$.

a.   Verify that Lagrange's energy function can be expressed
     as follows:

$$\mathscr{L}(r, \theta, z; \dot{r}, \dot{\theta}, \dot{z}) = K - P$$

$$= \frac{1}{2}(m_A + m_L)\dot{r}^2 + \frac{1}{2}m_A[(r - \frac{\ell}{2})\dot{\theta}]^2 + \frac{1}{2}m_L(r\dot{\theta})^2 + \frac{1}{2}(\overline{m}_A + m_L)\dot{z}^2$$

$$+ \frac{1}{2}I_\theta\,\dot{\theta}^2 - (\overline{m}_A + m_L)gz - \frac{1}{2}k_s(r - \frac{1}{3}\ell)^2 - \frac{1}{2}k_s(z - \frac{\ell}{3})^2$$

where $\overline{m}_A = m_A + m_h$.

b. Determine the equations of motion for the manipula-
    tor in terms of r, θ, and z. Verify that they can be
    written as

$$
\begin{bmatrix} \ddot{r} \\ \ddot{\theta} \\ \ddot{z} \end{bmatrix} = \begin{bmatrix} (r - \dfrac{m_A \ell/2}{m_A + m_L})\dot{\theta}^2 - \dfrac{k_s(r - \ell/3)}{m_A + m_L} + \dfrac{F_r - B_r(\dot{r})}{m_A + m_L} \\ \dfrac{1}{I_t}[m_A\ell - 2(m_A + m_L)r]\dot{r}\dot{\theta} + \dfrac{1}{I_t}[\tau_\theta - B_\theta(\dot{\theta})] \\ \dfrac{F_z - B_z(\dot{z})}{\overline{m}_A + m_L} - g - \dfrac{k_s(z - \ell/3)}{\overline{m}_A + m_L} \end{bmatrix}
$$

where the moment of inertia $I_t = I_\theta + m_A\ell^2/2 + (m_A + m_L)r^2 - m_A r\ell$.

5.11 The dynamical behavior of a planar manipulator discussed
      in Example 5.2.1. can be modeled by Equations 5.2.18–
      5.2.30 (see also Table 5.4), when the inputs consist of the
      joint torques, and the outputs are the joint variables. Each
      joint shaft is driven by a separately excited DC motor. The
      motor shafts are connected to the joint shafts through gears
      with gear ratios 1/100.

      The task is to include the dynamics of the actuators in
      the manipulator model. Thus, the inputs into the whole
      system are the voltages applied to the motors. The outputs
      are the same as in Example 5.2.1.

      a. Write the entire differential equation model. What is
          the order of the total system?

      b. Write the dynamical model described in part (a) in
          the form of Equation 5.1.13, that is, the highest order derivatives
          are on the left.

5.12 In the model of the cylindrical robot manipulator discussed
      in Example 5.1.1, the spring constants are such that $k_s = 0$,
      $k_z = 0$, and the frictions are omitted (no energy losses in
      the system).

      a. Rewrite the equations of motion given in Equation
          5.1.13 as the total time derivatives in the following
          forms:

$$\frac{d(I\omega)}{dt} = \tau_\theta(k)$$

$$\frac{d(\overline{m}\dot{z})}{dt} = F_z^t(k)$$

$$\frac{1}{2}\frac{d(I\omega^2 + m\dot{r}^2)}{dt} = \frac{d}{dt}[\theta\tau_\theta(k) + rF_r(k)]$$

      where $d\theta/dt = \omega$ is the rotational velocity, I is the
      second moment of inertia, $F_z^t = F_z - \overline{m}g$, and $\tau_\theta(k)$,

TABLE 5.4

Dynamical Model for a Vertical Revolute Two-Joint Planar Manipulator

$$\tau_1 - B_{\theta_1}\dot{\theta}_1 = \ddot{\theta}_1 \left[\frac{m_1 L_1^2}{4} + m_2 L_1^2 + \frac{m_2 L_2^2}{4} + m_2 L_1 L_2 \cos(\theta_2) + I_{01} + I_{02} \right]$$

$$+ \ddot{\theta}_2 \left[\frac{m_2 L_2^2}{4} + \frac{m_2 L_1 L_2}{2} \cos\theta_2 + I_{02} \right]$$

$$+ (\dot{\theta}_1)^2 [0] + (\dot{\theta}_2)^2 [-\frac{1}{2} m_2 L_1 L_2 \sin\theta_2] + (\dot{\theta}_1 \dot{\theta}_2)[-m_2 L_1 L_2 \sin\theta_2]$$

$$+ \left[\frac{m_1 L_1}{2} \cos\theta_1 + m_2 L_1 \cos\theta_1 + \frac{1}{2} m_2 L_2 \cos(\theta_1 + \theta_2) \right] g_y$$

$$\tau_2 - B_{\theta_2}\dot{\theta}_2 = \ddot{\theta}_1 \left[\frac{m_2 L_2^2}{4} + \frac{m_2 L_1 L_2}{2} \cos\theta_2 + I_{02} \right] + \ddot{\theta}_2 \left[\frac{m_2 L_2^2}{4} + I_{02} \right]$$

$$+ (\dot{\theta}_1)^2 \left[\frac{m_2 L_1 L_2}{2} \sin\theta_2 \right] + (\dot{\theta}_2)^2 [0] + (\dot{\theta}_1 \dot{\theta}_2)[0] + \frac{m_2 g_y L_2}{2} \cos(\theta_1 + \theta_2)$$

where each link is a solid right circular cylinder of length L_i, $i = 1, 2$; and m_1, m_2 are the link masses. The gravity vector $g = [0, g_y, 0]'$. The joint positions are θ_1 and θ_2, the corresponding velocities $\dot{\theta}_1$ and $\dot{\theta}_2$ and the accelerations $\ddot{\theta}_1$ and $\ddot{\theta}_2$. The inertias of the links about the mass centers are I_{01} and I_{02} where $I_{0i} = m_i L_i^2/12$, $i = 1, 2$. The coefficient of viscous friction of joint i is B_{θ_i}, and the external input torque to joint i is denoted by τ_i, $i = 1, 2$.

> $F_z(k)$, $F_r(k)$ are actuator inputs assumed to be constant over a sampling interval $[kT, (k + 1)T)$.

   b. Identify the linear and angular momenta in part (a) (see Appendix C), and specify I, $\overline{m}$, and m.

   c. Discuss the expressions obtained in (a) in view of the energy of the system.

5.13 A cylindrical robotic manipulator is shown schematically in Figures 5.1a and 5.1b. The end-effector of the cylindrical manipulator moves on a plane when the z-coordinate is kept constant. Thus, the manipulator can be considered to have one revolute (ψ) and one prismatic joint (link).

   a. Draw a free-body configuration for the system. Designate the forces and torques in your figure.

b.   Determine Newton's and Euler's equation for the sys-
      tem.

c.   Compare the model determined in part *(b)* with the
      equations given in Example 5.1.1.

5.14 The computations in the dynamical model given by Equa-
      tions 5.2.5 can be made faster if the equations are written
      by using only the rotational (3×3) matrix of the homoge-
      neous (4×4) transformation matrix and three-dimensional
      vector, instead of the four-dimensional ones [6]. Verify the
      following equations.
         The position vector p_{0i} of the incremental mass element
      dm_i may be expressed as

$$p_{0i} = \bar{p}_{0i} + (A_0^i)_R p_i$$

      where the (3×3) matrix $(A_0^i)_R$ accounts for the rotation,
      $\bar{p}_{0i}$ represents the three-dimensional vector specifying the
      origin of the i th coordinate frame in the base coordinates,
      and the three-dimensional vector p_i specifies the position
      of dm_i in the ith coordinate frame. The kinetic energy dK_i
      of the mass dm_i is then

$$dK_i = \frac{1}{2} \operatorname{tr}\left[(\dot{p}_{0i})(\dot{p}_{0i})'\right]dm_i$$

      The kinetic energy of link i is now obtained by
      integration. Let $\int dm_i = m_i;\ \int p_i dm_i = m_i \bar{p}_i;\ \int p_i p_i' d m_i = \bar{I}_i.$
      It follows that the total kinetic energy of an N-joint manip-
      ulator is

$$K = \frac{1}{2} \sum_{i=1}^{N} \operatorname{tr}\left[m_i(\dot{p}_{0i})(\dot{p}_{0i})' + 2(\dot{A}_0^i)_R(m_i \bar{p}_i)(\dot{p}_{0i})' + (\dot{A}_0^i)_R \bar{I}_i(\dot{A}_0^i)_R'\right]$$

      where

$$\dot{A}_0^i = \dot{A}_0^{i-1} A_{i-1}^i + A_0^{i-1}\frac{\partial A_{i-1}^i}{\partial q_i}\dot{q}_i$$

      The equations of motion may be expressed as

$$\sum_{i=n}^{N} \operatorname{tr}[m_i \frac{\partial p_{0i}}{\partial q_n}(\ddot{p}_{0i})' + \frac{\partial p_{0i}}{\partial q_n}(m_i \bar{p}_i)(\ddot{A}_0^i)_R' + \frac{\partial (A_0^i)_R'}{\partial q_n}(m_i \bar{p}_i)(\ddot{p}_{0i})'$$

$$+ \frac{\partial (A_0^i)_R}{\partial q_n}\bar{I}_i(\ddot{A}_0^i)_R'] - m_i g'\frac{\partial (A_0^i)_R}{\partial q_n}\bar{p}_i = F_n; \qquad n = 1, \ldots, N$$

$$\ddot{A}_0^i = \ddot{A}_0^{i-1}A_{i-1}^i + 2\dot{A}_0^{i-1}\frac{\partial A_{i-1}^i}{\partial q_i}\dot{q}_i + A_0^{i-1}\frac{\partial^2 A_{i-1}^i}{\partial q_i^2}\dot{q}_i^2 + A_0^{i-1}\frac{\partial A_{i-1}^i}{\partial q_i}\ddot{q}_i$$

Hint: Note that

$$\frac{\partial p_{oi}}{\partial q_n} = \frac{\partial \dot{p}_{oi}}{\partial \dot{q}_n}; \qquad \frac{\partial \dot{p}_{oi}}{\partial q_n} = \frac{d}{dt}\left[\frac{\partial \dot{p}_{oi}}{\partial \dot{q}_n}\right]$$

The resulting model is computationally more advantageous than the model given in Equation 5.2.5. However, it suffers from a drawback. What is it? Explain.

COMPUTER PROBLEMS

5.C.1   The dynamical model for the planar manipulator with two revolute links is described in Table 5.4. Assume that the following numerical values are given:

$$m_1 = 1.0 \text{ kg} \qquad g_x = 0$$

$$m_2 = 1.0 \text{ kg} \qquad g_y = -9.8 \text{ m/s}^2$$

$$L_1 = 0.5 \text{ m}$$

$$L_2 = 0.5 \text{ m}$$

a.   Write the dynamical model as a set of first-order differential equations for this (vertical) planar manipulator. (*Hint:* We need to solve the two equations for $\ddot{\theta}_1$ and $\ddot{\theta}_2$. For example, solve the first equation for $\ddot{\theta}_1$ and substitute into the second equation. Then, eliminate $\ddot{\theta}_2$ from the first equation.)

b.   In the equations obtained in part (a), introduce the following approximations for $i = 1, 2$

$$\dot{\theta}_i(t) \approx [\theta_i(t + \Delta t) - \theta_i(t)]/\Delta t$$

$$\ddot{\theta}_i(t) \approx [\dot{\theta}_i(t + \Delta t) - \dot{\theta}_i(t)]/\Delta t$$

$$\approx \frac{[\theta_i(t + 2\Delta t) - 2\theta_i(t + \Delta t) + \theta_i(t)]}{(\Delta t)^2}$$

where Δt is a sufficiently small subinterval for integration. Set $t = k(\Delta t)$ where $k = 0, 1, 2, \ldots$. Solve the approximate manipulator equations for

$$\theta_i[(k + 2)\Delta t] = f_i\{\theta[(k + 1)\Delta t], \theta(k\Delta t), u(k\Delta t)\}$$

where $i = 1, 2, \theta(k\Delta t) = [\theta_1(k\Delta t) \quad \theta_2(k\Delta t)]'$ and $u(k\Delta t) = [\tau_1(k\Delta t) \quad \tau_2(k\Delta t)]'$. Using the equations for $\theta_i[(k + 2)\Delta t]$ when $k = 0, 1, 2, \ldots, \Delta t = 10^{-4}$ (time

units), known initial conditions and the specified
inputs, the equations of motion for the manipulator
can be integrated, that is, a digital simulation of the
system can be obtained.

c.   Assume

$$\tau_1 = 0.5 \text{ Nm} \qquad\qquad \theta_1(0) = 25^0$$

$$\tau_2 = 0.5 \text{ Nm} \qquad\qquad \theta_2(0) = 30^0$$

$$B_{\theta_i} = 1.5 \text{ Nms}, i = 1, 2 \quad \dot{\theta}_1(0) = \dot{\theta}_2(0) = 0$$

(Note: Nms = Newton-meter-second)

Obtain digital simulation for the foregoing system for
$0 \le t \le 20.0$ s.

Graph (i) $\theta_1(t)$ and $\theta_2(t)$ vs. time; (ii) $\dot{\theta}_1(t)$ and $\dot{\theta}_2(t)$
vs. time.

d.   Determine the steady-state value (equilibrium state)
of the joint variables from the graphs obtained in (c).
Determine the equilibrium state also analytically.

5.C.2   The cylindrical manipulator described in Example 5.1.1
is operated so that the z-coordinate remains constant.
Assume the following numerical values: $m_A = 1.0$ kg,
$m_L = 0.125$ kg, $m_h = 0.5$ kg, $I_\theta = 0.1$ kgm^2, $\ell = 1.0$ m,
$k_s = 100$ N/m, $k_z = 0$ and the coefficients of viscous friction
$B_r = 5.0$ Ns/m, $B_z = 5.0$ Ns/m, $B_\theta = 0.001$ Nms.

a.   Write the dynamical model in Equations 5.1.10 and
5.1.11 in the state variable form.

b.   Apply the approximate expressions of the derivatives
given in Computer Problem 5.C.1b to determine the
model as a set of difference equations. Describe the
recursive calculations using the resulting difference
equations.

c.   The following inputs are applied to the actuators

$$F_r = 30 \text{ N for } 0 \le t \le 5s, \qquad r(0) = 0.70 \text{ m}$$
$$\text{and zero otherwise}$$

$$\tau_\theta = 1.0 \text{ Nm for } 0 \le t \le 5s \qquad \theta(0) = 1.57 \text{ rad}$$
$$\text{and zero otherwise} \qquad \dot{r}(0) = \dot{\theta}(0) = 0.$$

Obtain digital simulation of the foregoing system for
$0 \le t \le 35$s. Graph (i) $r(t)$ and $\theta(t)$ vs. time; (ii) $\dot{r}(t)$
and $\dot{\theta}(t)$ vs. time.

d.   Repeat (b) when the $m_A = 10.0$ kg, $m_L = 1.25$ kg,
$m_h = 5.0$ kg and $I_\theta = 1$ kgm^2.

CHAPTER SIX

STATE VARIABLE REPRESENTATION AND LINEARIZATION OF NONLINEAR MODELS

Since the mathematical models for manipulator dynamics are nonlinear, transformation methods such as Laplace and Fourier transforms cannot be applied directly to study these systems. The models can be written as a set of first-order nonlinear differential equations that are represented in a vector form as a state-variable equation. The first-order vector differential equations are commonly used when general integration software packages are written. They offer advantages over a single high-order differential equation, for example, when the derivatives are approximated for numerical integration in digital simulations.

In this chapter, we review and discuss continuous-time state variable representations [1]. We will study the determination of the equilibrium states of nonlinear continuous-time stable systems. The dynamical behavior of nonlinear systems is often studied by digital simulations, for which we review some basic algorithms. We will discuss two different approaches to linearize the set of nonlinear state variable equations. The first method results in a linearized model that is valid for small (weak) variations about a nominal trajectory. The second method makes use of feedback–feedforward loops that cause cancellation of nonlinear functions and render the system model linear.

We will present the general solution to linear time-invariant vector differential and difference equations and describe some basic properties of linear systems. Specifically, we will discuss the determination of the eigenvalues of the plant matrix for stability studies and present necessary and sufficient conditions for the complete controllability in linear time-invariant systems.

The applicability of these concepts to robot manipulator models is illustrated by examples.

6.1   STATE VARIABLE REPRESENTATIONS FOR DYNAMICAL MODELS

Most dynamical models written on the basis of physical laws are differential equations in which time appears as the independent variable. The dependent variables assume some real values at any instant of time; thus, the model is referred to as a "continuous-time model." The dependent variables usually specify the state of the system. Indeed, they serve as the state variables. The set of the state variable equations containing one or more time derivatives of the dependent variables represent the dynamic model of the system.

The *state* in a deterministic system can roughly be considered as the minimal amount of information about the past history of the system that suffices with the dynamical model to predict the effect of the past on the future. This information is contained, for example, in the set of initial conditions of the differential equation model. When the minimal number of the state variables is used to describe the system behavior, the corresponding model is called the minimal state representation. These state variables span the state space, that is, they form the basis in the state space.

We will next discuss the representation of a system model in a state variable form and its digital simulation.

6.1.1   NONLINEAR VECTOR DIFFERENTIAL EQUATION MODELS

The equations of motion derived in Chapter 5 for N-joint serial link manipulators usually represent a coupled nonlinear model. Such a general model can be written as a set of first-order differential equations:

$$\dot{x} = f[x(t), u(t)] \qquad (6.1.1)$$

where time $t \geq t_0$, a specified initial time, state x is a $2N$-dimensional vector, that is, $x(t) = [x_1(t), \ldots, x_{2N}(t)]'$, $f(x, u)$ is a differentiable vector-valued function of x and u, that is, $f(x, u) = [f_1(x, u), \ldots, f_{2N}(x, u)]'$, vector $u(t)$ is an input to the system, and the initial state is $x(t_0)$. Equation 6.1.1 is often called the plant equation for the system.

When the equations of motion for an N-joint manipulator are written in the form of Equation 6.1.1, it is often convenient to select the state vector $x(t)$ so that the first N components represent the positions and the last N components the velocity of the joints. In most robot manipulator systems, the positions of the joints are accessible for direct measurements. Moreover, the velocities of the joints may also be measured or calculated. These measurements are usually very accurate. Hence, they are directly available for control algorithms.

If the input to the system modeled by Equation 6.1.1 is zero, the system is assumed to have a steady-state position, which represents the equilibrium state. When the system state is suddenly disturbed, causing it to leave the equilibrium state to another state $x(t_0)$, then roughly stated, the system is called (asymptotically) stable if the ensuing state $x(t)$ returns to the equilibrium state as the time increases. When the input to a stable manipulator system is a step function of height $\bar{u}$ applied at $t = 0$, then the equilibrium state x_e is determined by

$$0 = f(x_e, \bar{u}) \qquad (6.1.2)$$

since the time derivative of the steady-state (the operating point) is zero. The real-valued solution of Equation 6.1.2 determines the equilibrium state x_e of the system.

The energy of the system at the equilibrium state is the minimum. For example, a planar two-joint manipulator moving on a vertical plane under the influence of the gravity and zero external input assumes the position where the gravity centers of both links are at the lowest positions. The equilibrium position of a manipulator is usually different from the parking position of the arm, which is usually maintained by means of the brakes on the joints.

The use of state variable models is illustrated by the following examples.

EXAMPLE 6.1.1

For the cylindrical robot manipulator shown schematically in Figures 6.1a and 6.1b, the origin of the base coordinate system $(x_0 y_0 z_0)$ is located at point O. The hub has a mass of m_h with the center of gravity on the z_0-axis, and its moment of inertia is I_θ about the z_0-axis. The length and mass of the prismatic link are ℓ and m_A, respectively, with the center of gravity at the midpoint of the moving part, as indicated in Figures 6.1a and 6.1b. The mass of the load is m_L. A spring with a constant k_s constrains the motion along the r-axis; it is at the resting position, that is, it produces a zero force when its radius is $2\ell/3$. Another spring acting on the z_0-axis produces a force proportional to placement $(z - \ell_0)$ with constant k_z.

The cylindrical manipulator described and shown in Figures 6.1a and 6.1b is different from the one modeled in Example 5.1.1 and displayed in Figures 5.1a and 5.1b. Namely, the part of the arm that can move linearly (the prismatic link) intersects the rotational axis z_0 in the latter case, but this is prevented in the former case. Consequently, the dynamical models of these manipulators will be different.

The equations of motion of the manipulator are obtained by writing first Lagrange's energy function for the system:

$$\mathscr{L}(r, \theta, z; \dot{r}, \dot{\theta}, \dot{z}) = K - P = \frac{1}{2} m \dot{r}^2 + \frac{1}{2} m_A \left[\left(r - \frac{\ell}{2} \right) \dot{\theta} \right]^2 + \frac{1}{2} m_L (r \dot{\theta})^2$$

$$+ \frac{1}{2} \overline{m}_A \dot{z}^2 + \frac{1}{2} I_\theta \dot{\theta}^2 - \overline{m}_A gz - \frac{1}{2} k_s \left(r - \frac{2}{3} \ell \right)^2 - \frac{1}{2} k_z (z - \ell_0)^2 \qquad (6.1.3)$$

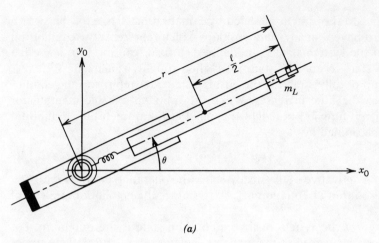

(a)

FIGURE 6.1*a*
Cylindrical robot manipulator (top view). (Compare with Figure
1.1*a*.)

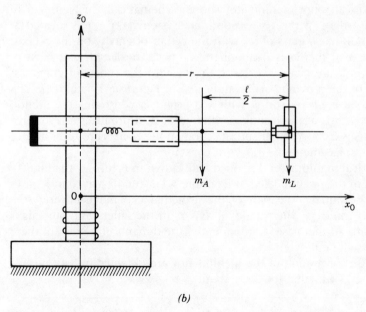

(b)

FIGURE 6.1*b*
Cylindrical robot manipulator (side view).

where $m = m_A + m_L$ and $\overline{m} = m_A + m_h + m_L$. Euler–Lagrange's equations can then be written (see Problem 5.11) in the following form by solving for the highest order derivatives:

$$
\begin{bmatrix} \ddot{r} \\ \ddot{\theta} \\ \ddot{z} \end{bmatrix} = \begin{bmatrix} \left(r - \dfrac{m_A \ell/2}{m}\right)\dot{\theta}^2 - \dfrac{k_s(r - 2\ell/3)}{m} + \dfrac{F_r - B_r(\dot{r})}{m} \\ \dfrac{1}{I_t}[m_A\ell - 2\,mr]\dot{r}\dot{\theta} + \dfrac{1}{I_t}[\tau_\theta - B_\theta(\dot{\theta})] \\ -g - \dfrac{k_z(z - \ell_0)}{\overline{m}} + \dfrac{F_z - B_z(\dot{z})}{\overline{m}} \end{bmatrix} \qquad (6.1.4)
$$

where the moment of inertia $I_t = I_t(r) = I_\theta + m_A\ell^2/4 + mr^2 - m_A r\ell$. Equation 6.1.4 reveals that the motion along the z-direction is not dependent on the motion along the other axes.

The problem is to describe the dynamical model of Equation 6.1.4 in the state variable form, that is, as a set of first-order differential equations in the form of Equation 6.1.1.

The state variable representation can be written by defining $r(t) = x_1(t)$, $\theta(t) = x_2(t)$, $z(t) = x_3(t)$, $\dot{x}_1(t) = x_4(t)$, $\dot{x}_2(t) = x_5(t)$, $\dot{x}_3 = x_6(t)$. It follows that $\dot{x}_4 = \ddot{r}$, $\dot{x}_5 = \ddot{\theta}$ and $\dot{x}_6 = \ddot{z}$. Equation 6.1.4 can then be converted to a first-order vector differential equation in which the input is written separately, and the Coriolis, the centripetal, and spring forces as well as the gravitational force are combined:

$$
\begin{bmatrix} \dot{x}_1 \\ \dot{x}_2 \\ \dot{x}_3 \\ \dot{x}_4 \\ \dot{x}_5 \\ \dot{x}_6 \end{bmatrix} = \begin{bmatrix} x_4 \\ x_5 \\ x_6 \\ [x_1 - m_A\ell/(2m)]x_5^2 - k_s(x_1 - 2\ell/3)/m \\ [m_A\ell - 2\,m\,x_1]x_4 x_5/I_t(x_1) \\ -g - k_z(x_3 - \ell_0)/\overline{m} \end{bmatrix}
$$

$$
+ \begin{bmatrix} 0 & 0 & 0 \\ 0 & 0 & 0 \\ 0 & 0 & 0 \\ \dfrac{1}{m} & 0 & 0 \\ 0 & \dfrac{1}{I_t(x_1)} & 0 \\ 0 & 0 & \dfrac{1}{\overline{m}} \end{bmatrix} \begin{bmatrix} F_r - B_r(x_4) \\ \tau_\theta - B_\theta(x_5) \\ F_z - B_z(x_6) \end{bmatrix} \qquad (6.1.5)
$$

where $I_t = I_t(x_1)$, $x = [x_1\, x_2\, x_3\, x_4\, x_5\, x_6]' = [r\,\theta\,z\,\dot{r}\,\dot{\theta}\,\dot{z}]'$ is the state vector, and $u = [F_r - B_r(x_4), \tau_\theta - B_\theta(x_5), F_z - B_z(x_6)]'$ the input vector. The left-hand side of Equation 6.1.5 contains only first-order derivatives. There are no explicit time derivatives on the right-hand side of Equation 6.1.5.

Equation 6.1.5 describes the manipulator model in Equation 6.1.4 in the state variable form. It may be rewritten in a concise form:

$$\begin{bmatrix} \dot{x}^1 \\ \dot{x}^2 \end{bmatrix} = \begin{bmatrix} x^2(t) \\ f^2[x(t)] \end{bmatrix} + \begin{bmatrix} 0 \\ B^2[x(t)] \end{bmatrix} u(t) \tag{6.1.6}$$

where $x^1 = [x_1 x_2 x_3]'$, $x^2 = [x_4 x_5 x_6]'$ and the right-hand side of Equation 6.1.6 contains the nonlinear vector-valued function $f^2[x(t)]$, that is, $f^2 = [f_1^2 f_2^2 f_3^2]'$ describing the centripetal, Coriolis, spring and gravity effects. The initial condition for Equation 6.1.5 is specified by the initial position and velocity of the gripper.

EXAMPLE 6.1.2

A compact model of a serial link manipulator is given by Equation 5.2.10 in the Lagrangian formulation. This general closed-form dynamical model for a planar two-joint manipulator moving on a vertical plane under the influence of the gravitational field assumes the form given in Equation 5.2.40, where the coefficients that are dependent on the joint positions are defined by Equations 5.2.32–5.2.39. The external forces and moments between the end-effector and the environment are assumed zero.

The problem is to determine the steady state (the operating point) of the system when step inputs of heights $\bar{\tau}$ and $\bar{\tau}_2$ are applied at $t = 0$ to the joints.

The state vector in the foregoing planar manipulator $x^1 = q = [\theta_1 \ \theta_2]'$ and $x^2 = \dot{q} = [\dot{\theta}_1 \ \dot{\theta}_2]'$. The equilibrium state $x_e^1 = \bar{q} = [\bar{\theta}_1 \ \bar{\theta}_2]'$ is determined by setting the time derivatives of the joint variables in the model Equation 5.2.40 equal to zero. Thus, Equation 5.2.40 reduces to

$$\begin{bmatrix} G_1(\bar{q}) \\ G_2(\bar{q}) \end{bmatrix} = \begin{bmatrix} \bar{\tau}_1 \\ \bar{\tau}_2 \end{bmatrix} \tag{6.1.7}$$

where $G_i(\bar{q})$, $i = 1, 2$ represents the gravity term evaluated at $\bar{q}$. Equation 6.1.7 can be rewritten on the basis of Equations 5.2.38 and 5.2.39:

$$(m_1 \ell_1/2 + m_2 \ell_1) \bar{c}_1 + (m_2 \ell_2/2) \bar{c}_{12} = \bar{\tau}_1/g_y \tag{6.1.8}$$

$$(m_2 \ell_2/2) \bar{c}_{12} = \bar{\tau}_2/g_y \tag{6.1.9}$$

where $\bar{c}_1 = \cos \bar{\theta}_1$, and $\bar{c}_{12} = \cos(\bar{\theta}_1 + \bar{\theta}_2)$.

Equations 6.1.8 and 6.1.9 are manipulated to obtain the equations for determining the steady-state values of the joint positions:

$$\cos \bar{\theta}_1 = \frac{\bar{\tau}_1 - \bar{\tau}_2}{(m_2 + m_1/2) g_y \ell_1} \tag{6.1.10}$$

$$\cos (\bar{\theta}_1 + \bar{\theta}_2) = \frac{\bar{\tau}_2}{m_2 g_y \ell_2/2} \tag{6.1.11}$$

The components $\bar{\theta}_1$ and $\bar{\theta}_2$ of the equilibrium state x_e are specified by Equations 6.1.10 and 6.1.11 providing that the absolute values of the right sides of these equations are less than or equal to one. If the magnitude of the right side of Equation 6.1.10 or 6.1.11 exceeds one, the joint angles of the manipulator will not assume a constant equilibrium state. This implies that the applied torques are such that one or two links of the planar manipulator continuously rotate without reaching a constant equilibrium state.

The dynamical behavior of an N-joint manipulator modeled by Equation 6.1.1 can be studied by solving this differential equation. However, closed-form solutions to nonlinear differential equations cannot in general be determined. In these cases, the system can be designed and tested using digital simulations that provide the designer with numerical solutions to the nonlinear differential equation models under specified conditions. Digital simulations of dynamical system models are usually based on discretized models represented by difference equations.

6.1.2 NONLINEAR VECTOR DIFFERENCE EQUATION MODELS

When a signal in a system occurs only at discrete instances of time, the system is called a discrete-time system. When the signal is also digitized (quantized), the system is referred to as a digital system. For example, a system containing a digital computer and/or encoders is termed a digital system. Discrete-time systems are modeled by difference equations. When digital simulations of a continuous-time system are performed, the differential equation model needs to be discretized with respect to time at some stage of the simulation. Digital computers used in controlling manipulators render the system discrete-time, although the signals in most components of the system are continuous-time (analog). Since robotic manipulator systems are mostly controlled by computers, discrete-time models play an important role in the analysis and design of these systems. We will next discuss discrete-time models and digital simulations of nonlinear systems.

A nonlinear vector differential equation cannot usually be solved explicitly in a closed form. A numerical solution, however, can be determined, for example, by properly approximating the derivatives using finite differences. The resulting equations are still nonlinear, but now they appear as difference equations, which are suitable for digital simulations.

A general form of a discrete-time nonlinear vector difference equation is

$$x[(k+1)T] = f[x(kT), u(kT)] \qquad (6.1.12)$$

where integer $k = k_0, k_0 + 1, \ldots$, T is the sampling period, vector $x(kT) = \mathrm{col}[x_1(kT) \ldots x_{2N}(kT)]$ is the state of the system at time kT, $x(k_0T)$ is given, and $f(x, u)$ is the vector-valued function of x and u, that is, $f[x, u] = \mathrm{col}[f_1(x, u), \ldots, f_{2N}(x, u)]$. The state vector x can be defined so that the first N components represent the positions, and the last N components the velocities, of the joints in an N-joint manipulator.

6.1.3 DIGITAL SIMULATIONS OF SYSTEM MODELS

The behavior of a system can conveniently be studied on the basis of a mathematical model, which describes the system dynamics. By solving the differential or difference equation model numerically, the behavior of the system can be determined under various circumstances. For example, changes in the manipulator motion due to varying loads carried by the end-effector can be determined by numerically integrating the dynamical equations. The process of the numerical integration of a system model on a digital computer is called digital simulation.

To determine the digital simulation of a continuous-time model, the continuous variable, time t, is replaced by a discrete variable that increases stepwise in finite increments Δt. When the differential equations of motion are discretized by this approximation, a sequence of snapshot pictures at the discrete intervals Δt instead of a continuous-time history of the variables is obtained. Several approximate expressions for obtaining digital simulations have been presented in the literature [2,3]. We will discuss in detail some common approximations that are often used in integrating differential equations numerically.

Euler's Forward Difference Approximations of Derivatives for Numerical Integration

Suppose that the equations of motion are described by Equation 6.1.1. Each component $\dot{x}_i$ of vector $\dot{x}$ can be replaced by a first-order forward approximation, which is based on the definition of a derivative:

$$\dot{x}_i(t) \approx \frac{x_i(t + \Delta t) - x_i(t)}{\Delta t} \qquad (6.1.13)$$

where $i = 1, \ldots, 2N$, and the subinterval Δt of time must be chosen sufficiently short for a good approximation. By substituting the approximate expression of Equation 6.1.13 for the derivatives, Equation 6.1.1 becomes

$$x(t + \Delta t) = x(t) + f[x(t), u(t)]\Delta t + e(t) \qquad (6.1.14)$$

where $e(t)$ represents a residual error. It occurs because of the substitutions of the finite difference approximations for the derivatives. Equation 6.1.14 represents Euler's forward difference approximation for integrating the differential Equation 6.1.1 numerically. It serves as the basis of an algorithm that gives a digital simulation of the dynamical model in Equation 6.1.1. According to Equation 6.1.14, the value $x(t + \Delta t)$ is calculated by adding to $x(t)$ the increment $\dot{x}(t)\Delta t$, where $\dot{x}(t)$ is the value of the slope (gradient) of $x(t)$ at time t. Thus, the first two terms on the right side of Equation 6.1.14 represent a straight-line extrapolation of $x(t)$ to $x(t + \Delta t)$ as shown in Figure 6.2.

By setting $\Delta t = T$ and $t = kT$, and assuming that the residual error $e(t)$ can be omitted, Equation 6.1.14 can be rewritten:

$$x[(k + 1)T] = x(kT) + f[x(kT), u(kT)]T \qquad (6.1.15)$$

Equation 6.1.15 is in the form of a general nonlinear difference equation that is suitable to the digital simulations of the system in Equation 6.1.1.

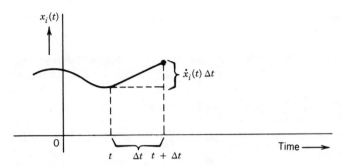

FIGURE 6.2

Integration using finite difference approximation of the
derivative of $x_i(t)$ at time t.

The dynamical model of a manipulator can be discretized and written in
the form of Equation 6.1.15. When the initial positions and velocities of the
manipulator and the inputs to the joint actuators, that is, $x(k_0T)$ and
$\{u(kT)\}, k = k_0, k_0 + 1, \ldots$ are given, the ensuing positions and velocities of
the motion can be calculated by the difference equation model.

Using Euler's finite difference approximation resulting in Equation 6.1.14,
the truncation error $e(t)$ can be shown to be of the order of $(\Delta t)^2$, indicated
by $o(\Delta t^2)$ when the first two derivatives of $x(t)$ are bounded and continuous
[3]. The selection of the sufficiently short subinterval Δt for the integration
is essential for the accuracy and numerical stability in digital simulations.
A short subinterval increases the accuracy of integration, but results in an
increase of total computation time. Usually, the designer must compromise
between these factors.

Trapezoidal Approximations of Integrals for Numerical Integration

If both sides of the dynamical model in Equation 6.1.1 are integrated over
the time interval $[kT, (k + 1)T)$, the result is

$$x[(k + 1)T] = x(kT) + \int_{kT}^{(k+1)T} f[x(\sigma), u(\sigma)] \, d\sigma \qquad (6.1.16)$$

where $k = 0, 1, \ldots$. To obtain digital simulations for the system, the problem
is then to approximate the integral in Equation 6.1.16.

A common approach to approximately evaluate the integral of a function
whose values are known or can be evaluated on the integration interval is to
pass an interpolating function through the adjacent points, and then to integrate the interpolating function. If a straight line is used as the interpolating
function between the values of the function evaluated at kT and at $(k + 1)T$,
the approximate value of the integral in Equation 6.1.16 is obtained as the
area of the trapezoid. Thus, the application of the trapezoidal approximation
to Equation 6.1.16 gives

$$x[(k+1)T] = x(kT) + \{f(x[(k+1)T], u[(k+1)T], +f[x(kT), u(kT)]\}T/2$$
$$(6.1.17)$$

where the second expression on the right side gives the area of the trapezoid. Equation 6.1.17 may be used to construct an algorithm for calculating an approximate solution to Equation 6.1.6. Particularly, the required control inputs, the generalized joint torques, can be computed by Equation 6.1.17, when the desired positions and velocities of the manipulator specify state $\{x(kT)\}$. Thus, the solution to the inverse dynamics determining the needed torques can be obtained.

Equation 6.1.17 gives an approximate solution to Equation 6.1.16. The approximation accuracy depends on the deviation of the interpolating function from the values of $f[x(t), u(t)]$ on the interval, and specifically on the magnitude of the curvature that is proportional to d^2f/dt^2. Indeed, it can be shown that the truncation error, that is, the difference between the right sides of Equation 6.1.16 and Equation 6.1.17 is proportional to the aforementioned curvature and the subinterval (T) squared [3]. An improved accuracy is achieved by selecting the interpolating function as a parabola between the points $(k-1)T$, kT, and $(k+1)T$. The resulting approximating expression of the integral is known as Simpson's rule [3]. For details, the reader is referred to a text on numerical analysis.

EXAMPLE 6.1.2

The dynamical model of the cylindrical robot manipulator described in Example 6.1.1 is given by Equation 6.1.5. By defining the position vector $x^1(t) = [r(t)\,\theta(t)\,z(t)]'$ and the velocity vector $x^2(t) = [\dot{r}(t)\,\dot{\theta}(t)\,\dot{z}(t)]'$, Equation 6.1.5 can be expressed as Equation 6.1.6:

$$\begin{bmatrix} \dot{x}^1 \\ \dot{x}^2 \end{bmatrix} = \begin{bmatrix} x^2(t) \\ f^2[x(t)] \end{bmatrix} + \begin{bmatrix} 0 \\ B^2[x(t)] \end{bmatrix} u(t) \qquad (6.1.18)$$

where the state vector $x = [(x^1)'(x^2)']'$, and the three-dimensional vector-valued function $f^2(x)$ contains the terms representing the centripetal, Coriolis, gravity, and spring forces in Equation 6.1.5. The control matrix $B^2(x)$ is a (3×3)-matrix in Equation 6.1.5, resulting from inverting the diagonal inertia matrix. The initial position and velocity of the manipulator at $t = 0$ is known, that is, $x^1(0)$ and $x^2(0)$ are given.

The task here is to determine the difference equations for digital simulations resulting from applying Euler's forward difference approximation to the equation specifying the acceleration vector $\dot{x}^2(t)$, and the trapezoidal rule to the velocity vector $\dot{x}^1(t)$.

The required difference equations are written by first applying Euler's formula given in Equation 6.1.13 to the acceleration vector $\dot{x}^2(t)$ to obtain

$$x^2[(k+1)T] = x^2(kT) + \{f^2[x(kT)] + B^2[x(kT)]u(kT)\}T \qquad (6.1.19)$$

The application of the trapezoidal rule described in Equation 6.1.17 to $\dot{x}^1(t)$ in Equation 6.1.18 gives

$$x^1[(k+1)T] = x^1(kT) + \{x^2[(k+1)T] + x^2(kT)\}T/2 \qquad (6.1.20)$$

Equations 6.1.19 and 6.1.20 describe the discretized dynamical model of Equation 6.1.18. They establish the basis of an algorithm to perform digital simulations.

For the digital simulations of the manipulator system, subinterval $T = \Delta t$, the initial condition $x(0)$ and the input $\{u(kT)\}$ are assumed to be given. The algorithm will generate the ensuing motion on the basis of Equations 6.1.19 and 6.1.20. Indeed, at time kT, $k = 0, 1, \ldots$, Equation 6.1.19 is first used to calculate $x^2[(k + 1)T]$, and then, Equation 6.1.20 to determine $x^1[(k + 1)T]$ since the values $x(kT)$ and $u(kT)$ are known. Thus, the velocity and position vectors can be generated recursively as integer k is increased, that is, in the forward direction of time when the initial conditions and the input are known.

The approximate discretized models that can be used to simulate digitally continuous-time systems contain stepsize $\Delta t = T$ explicitly. The residual (truncation) error due to the approximation depends on this subinterval, which is to be chosen sufficiently short. The stepsize of integration may be chosen as constant or variable. A fixed small stepsize in the integration increases the computation time considerably. A variable stepsize can be determined to adapt to changes in the dependent variable. It may be achieved, for example, by requiring that a change in the dependent variable from one time instant to the next exceed a specified percentage in order to be recorded. If this is not the case, the value is not recorded, and the algorithm moves to the next time instant and continues the calculations. Most standard integration algorithms such as the IMSL subroutines DVERK and DGEAR use variable stepsizes [4].

In addition to the foregoing discretization techniques, more accurate approximations are usually applied in standard subroutines to integrate differential equations. Indeed, many digital simulations are based on the recurrence formula of Runge–Kutta, or its modified versions. This approximation formula exhibits a residual error that is proportional to the fifth power of the integration subinterval [3]. Hence, the digital simulation will give very accurate numerical values.

In digital simulations, the accuracy and the computational efficiency are usually considered as the essential features of an algorithm. The convergence of the discretized model to the original equations of motion as the subinterval of time approaches zero and the numerical stability are additional important characteristics of digital simulations. They are, however, difficult to establish in general. In the simulation of manipulator systems, the designer may impose additional constraints on the simulations such as requiring that the energy and momentum or their functions remain constant from one sampling instant to the next one. This can be incorporated in the discretization scheme by using the integral-of-the-motion for a conservative system. If the motion of a manipulator is then observed only at the discrete times, the conservation of energy and momenta over each integration interval can be maintained. If the system is not lossless (conservative), a dissipating function can be introduced to accommodate energy dissipation caused, for example, by friction. Such an approach to the modeling of physical systems represents an application of discrete mechanics [5]. It can be applied to determine a discrete-time model for the motion of a manipulator [2].

6.2 METHODS FOR LINEARIZING NONLINEAR DYNAMICAL MODELS

The behavior of a manipulator described by a deterministic model with known parameters is completely predictable. In order to make a system response track a desired trajectory and satisfy certain design specifications, controllers are often constructed for the system. General procedures to design controllers such as the classical controllers (compensators) are based on linear time-invariant equations. Since the dynamical models of most manipulators are nonlinear and complex, a possible approach is to design one (primary) controller that makes the manipulator motion track the nominal trajectory and another (secondary) controller that compensates for small variations caused by disturbances acting on the system. The former controller can be specified by solving the dynamical equations for the torque (the inverse dynamics), and the latter controller may be determined on the basis of a model resulting from linearizing the nonlinear model. A linearized model is not only amenable to controller design but is also useful in studying locally other basic properties of the system such as the complete controllability. Indeed, it is tractable in the analysis, and it can also shed light on its dynamical behavior under various circumstances.

We will discuss here the linearization of the nonlinear equations of motion of robot manipulators that are derived in Chapter 5 and presented in the state variable form in Section 6.1. Specifically, we will present two methods to linearize nonlinear manipulator models. In the first one, the linearization is accomplished by determining the variational equations about a nominal trajectory. In the second method, nonlinear functions are generated by means of feedforward–feedback loops that will cancel the nonlinear terms in the model; thus, the overall system appears as linear.

6.2.1 VARIATIONAL (PERTURBATION) MODEL

The equations of motion for a manipulator are written in a state variable form:

$$\dot{x} = f\left[x(t), u(t)\right] \tag{6.2.1}$$

where time $t \geq t_0$, the initial time, state $x(t)$ of the system is a $2N$-dimensional vector, the initial state $x(t_0)$, input $u(t)$ is N-dimensional, and $f(x, u)$ is a vector valued function with continuous bounded derivatives.

The nominal trajectory $\bar{x}(t)$ is obtained by the application of the nominal input $\bar{u}(t)$ to the system. The variables of the nominal path satisfy the dynamical model equation:

$$\dot{\bar{x}} = f\left[\bar{x}(t), \bar{u}(t)\right] \qquad \bar{x}(t_0) = x(t_0) \tag{6.2.2}$$

The solution to Equation 6.2.2 determines the nominal trajectory. When input $\bar{u}(t)$ is solved from Equation 6.2.2 (the inverse dynamics) and applied to the manipulator system, the motion follows trajectory $\bar{x}(t)$, $t \geq t_0$ providing that the manipulator model in Equation 6.2.1 is sufficiently accurate. Because of imperfect modeling, and the effects of disturbances, the motion of the manipulator usually deviates from the nominal trajectory that is determined in the planning stage. It becomes then necessary to study deviations versus time from the nominal trajectory, and to attempt to compensate for the perturbations. These goals can be achieved by using a linearized model of the manipulator dynamics that describes small perturbations about the nominal trajectory. The determination of such a linear model is next presented.

Suppose that the motion of a robot manipulator evolving along the nominal trajectory $(\bar{x}(t), \bar{u}(t))$ is perturbed. It results in deviations $(\delta x(t), \delta u(t))$ from the nominal trajectory. When these deviations are sufficiently small and change smoothly, that is, they possess continuously turning tangents at all points, variables $(\delta x(t), \delta u(t))$ are called *weak variations* (see e.g., [6] p. 35). A typical variational trajectory for a component of vector $x(t) + \delta x(t)$ corresponding to input $u(t) = \bar{u}(t) + \delta u(t)$ is shown in Figure 6.3.

The perturbed variables also satisfy the equations of motion of the manipulator:

$$\frac{d}{dt}(\bar{x} + \delta x) = f[\bar{x}(t) + \delta x(t), \bar{u}(t) + \delta u(t)] \qquad (6.2.3)$$

The right side of Equation 6.2.3 may be expanded into Taylor's series about the nominal path $(\bar{x}(t), \bar{u}(t))$ to obtain

$$f[\bar{x}(t) + \delta x(t), \bar{u}(t) + \delta u(t)] = f[\bar{x}(t), \bar{u}(t)] + \left.\frac{\partial f}{\partial x}\right|_{(\bar{x},\bar{u})} \delta x + \left.\frac{\partial f}{\partial u}\right|_{(\bar{x},\bar{u})} \delta u(t) + \ldots$$

$$(6.2.4)$$

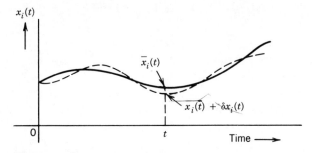

$x_i(t)$

$\bar{x}_i(t)$

$x_i(t) + \delta x_i(t)$

t

Time $\longrightarrow$

FIGURE 6.3
Weak variation $\delta x_i(t)$ (dotted line) about the nominal path $\bar{x}_i(t)$ vs. time.

where the derivatives $\partial f/\partial x$, $\partial f/\partial u$ are the Jacobian matrices evaluated at the nominal path $(\bar{x}, \bar{u})$. That is,

$$
\frac{\partial f}{\partial x} =
\begin{bmatrix}
\dfrac{\partial f_1}{\partial x_1} & \cdots & \dfrac{\partial f_1}{\partial x_{2N}} \\[2ex]
\dfrac{\partial f_2}{\partial x_1} & \cdots & \dfrac{\partial f_2}{\partial x_{2N}} \\
\vdots & & \vdots \\
\dfrac{\partial f_{2N}}{\partial x_1} & \cdots & \dfrac{\partial f_{2N}}{\partial x_{2N}}
\end{bmatrix}
\qquad
\frac{\partial f}{\partial u} =
\begin{bmatrix}
\dfrac{\partial f_1}{\partial u_1} & \cdots & \dfrac{\partial f_1}{\partial u_N} \\[2ex]
\dfrac{\partial f_2}{\partial u_1} & \cdots & \dfrac{\partial f_2}{\partial u_N} \\
\vdots & & \vdots \\
\dfrac{\partial f_N}{\partial u_1} & \cdots & \dfrac{\partial f_N}{\partial u_N}
\end{bmatrix}
\qquad (6.2.5)
$$

Equation 6.2.4 is substituted to the right side of Equation 6.2.3. By combining the resulting equation with Equation 6.2.2, the first-order variations $(\delta x, \delta u)$ will be governed by the following variational equation:

$$
\delta \dot{x} = \left.\frac{\partial f}{\partial x}\right|_{(\bar{x}, \bar{u})} \delta x(t) + \left.\frac{\partial f}{\partial u}\right|_{(\bar{x}, \bar{u})} \delta u(t) + e(t) \qquad (6.2.6)
$$

where $e(t)$ representing the high-order terms is assumed to be bounded, and $\delta x(t_0) = 0$. Equation 6.2.6 describes the evolution of the variations $(\delta x(t), \delta u(t))$ about the nominal trajectory $(\bar{x}(t), \bar{u}(t))$. It is a linear vector differential equation that often contains time-varying parameters, since $(\bar{x}, \bar{u})$ usually is time dependent. Although Equation 6.2.6 is difficult to determine for a general six-joint manipulator, it illustrates the effects of disturbances on the motion of a manipulator, and the task that a controller (a regulator) needs to perform to compensate for perturbations. Equation 6.2.6 can serve as a model in the design of such a controller. Moreover, certain local properties of the manipulator model can be studied by means of the linearized model in Equation 6.2.6.

EXAMPLE 6.2.1

The model for a cylindrical robot is presented in a state variable form in Equation 6.1.5. If the frictional terms are assumed to represent viscous friction, a comparison of Equation 6.1.5 with Equation 6.2.1 gives for the vector valued function

$$
f(x, u) =
\begin{bmatrix}
f_1(x, u) \\
f_2(x, u) \\
f_3(x, u) \\
f_4(x, u) \\
f_5(x, u) \\
f_6(x, u)
\end{bmatrix}
=
\begin{bmatrix}
x_4 \\
x_5 \\
x_6 \\
[x_1 - (m_A/m)\, \ell/2]\, x_5^2 - k_s(x_1 - 2\ell/3)/m + (F_r - B_r x_4)/m \\
[(m_A \ell - 2\, m\, x_1)\, x_4 x_5 + \tau_\theta - B_\theta x_5]/I_t(x_1) \\
-g - k_z(x_3 - \ell_0)/\bar{m} + (F_z - B_z x_6)/\bar{m}
\end{bmatrix}
\qquad (6.2.7)
$$

where the state vector $x = \mathrm{col}\,[r\,\theta\,z,\,\dot{r}\,\dot{\theta}\,\dot{z}]$, $I_t(x_1) = I_\theta + m_A \ell^2/4 + m x_1^2 - m_A \ell x_1$ and B_r, B_θ, B_z are coefficients for the viscous friction. Equation 6.2.7 describes a nonlinear (why?) vector differential equation model over the time interval $[t_0, t_f]$, where t_0 and t_f are the initial and final times of the motion, respectively. The problem is to determine a linearized model of the system about a given nominal trajectory $(\bar{x}, \bar{u})$.

The linearized model of Equation 6.2.7 is obtained on the basis of Equation 6.2.6. The Jacobian matrices in Equation 6.2.6 are first formed. Indeed, the evaluation of the matrices in Equation 6.2.5 along the nominal trajectory $(\bar{x}, \bar{u})$ gives:

$$\left.\frac{\partial f}{\partial x}\right|_{(\bar{x},\bar{u})} = \begin{bmatrix} 0 & 0 & 0 & 1 & 0 & 0 \\ 0 & 0 & 0 & 0 & 1 & 0 \\ 0 & 0 & 0 & 0 & 0 & 1 \\ \bar{x}_5^2 - k_s/m & 0 & 0 & -B_r/m & a_{45}(\bar{x}) & 0 \\ a_{51}(\bar{x}) & 0 & 0 & a_{54}(\bar{x}) & a_{55}(\bar{x}) & 0 \\ 0 & 0 & -k_z/\bar{m} & 0 & 0 & -B_z/\bar{m} \end{bmatrix} \quad (6.2.8)$$

where $a_{45}(\bar{x}) = [\bar{x}_1 - (m_A/m)\,\ell/2](2\bar{x}_5)$, $a_{51}(\bar{x}) = (m_A\ell - 2\,m\,\bar{x}_1)[(m_A\ell - 2\,m\,\bar{x}_1)\bar{x}_4\,\bar{x}_5 + \bar{\tau}_\theta - B_\theta\bar{x}_5]/I_t^2(\bar{x}_1) - 2\,m\,\bar{x}_4\,\bar{x}_5/I_t(\bar{x}_1)$, $I_t(\bar{x}_1) = I_\theta + m_A\ell^2/4 + m\,\bar{x}_1^2 - m_A\ell\bar{x}_1$, $a_{54}(\bar{x}) = (m_A\ell - 2\,m\,\bar{x}_1)\bar{x}_5/I_t(\bar{x}_1)$, and $a_{55}(\bar{x}) = [(m_A\ell - 2\,m\,\bar{x}_1)\bar{x}_4 - B_\theta]/I_t(\bar{x}_1)$; and by recognizing that $[F_r \quad \tau_\theta \quad F_z] = [u_1 \quad u_2 \quad u_3]$

$$\frac{\partial f}{\partial u} = \begin{bmatrix} 0 & 0 & 0 \\ 0 & 0 & 0 \\ 0 & 0 & 0 \\ 1/m & 0 & 0 \\ 0 & 1/I_t(\bar{x}_1) & 0 \\ 0 & 0 & 1/\bar{m} \end{bmatrix} \quad (6.2.9)$$

The linearized model for the motion of the cylindrical manipulator can be written concisely by defining $\partial f/\partial x = A(\bar{x})$ and $\partial f/\partial u = B(\bar{x})$. Equation 6.2.6 becomes

$$\delta \dot{x} = A(\bar{x})\,\delta x(t) + B(\bar{x})\,\delta u(t) + e(t) \quad (6.2.10)$$

where $t \in [t_0, t_f]$, the variational input $\delta u(t) = \mathrm{col}[\delta F_r(t)\ \delta\tau_\theta(t)\ \delta F_z(t)]$, the disturbance vector $e(t) = \mathrm{col}[e_1(t) \ldots e_6(t)]$, and the initial condition $\delta x(t_0) = 0$.

Equation 6.2.10 represents a linearized dynamical model. Since $(\bar{x}, \bar{u})$ usually varies with time along the trajectory, Equation 6.2.10 is a linear vector differential equation with time-varying parameters. The determination of the explicit general solutions to Equation 6.2.10 is usually cumbersome.

The control of the motion of a manipulator is sometimes discussed in the framework of transfer functions. They cannot be used in linear time-varying systems, except in some very special cases. If the model in Equation 6.2.10 is considered over only one sampling period at a time, then the trajectory $(\bar{x}, \bar{u})$ may not change appreciably over this time period. Under this *assumption*, the approximated model 6.2.10 can be viewed as a linear time-invariant differential equation. In this case, the eigenvalues of the plant matrix in Equation 6.2.10 over the particular sampling interval could be used to study the transient behavior of the system response. It should be emphasized that the properties of the system inferred on the basis of a model linearized about a nominal trajectory are only local properties.

6.2.2   LINEAR MODEL RESULTING FROM CANCELLATION OF NONLINEAR TERMS BY FEEDBACK–FEEDFORWARD LOOPS

The linearization procedure discussed in the previous section is based on weak variations about a nominal desired path. The variational equations are always linear. However, the expressions of the derivatives may be unwieldy in many applications, for example, in the models of six-link manipulators such as the Stanford/JPL arm, PUMA 560 and 600. An alternative method to linearize a nonlinear model is to introduce feedback and/or feedforward loops that generate from the available information nonlinear functions appearing in the model. These terms can be used to cause a total or partial cancellation of the nonlinear functions of the model by feeding them properly into the plant. Additional feedback and/or feedforward loops with adjustable gains can also be introduced as controllers into the same system.

We will discuss next a procedure to eliminate or reduce the effects of non-linearities by means of feedback–feedforward loops [7,8]. Before presenting the general method, we will describe the approach for a specific example.

EXAMPLE 6.2.2

A simple manipulator (a pendulum) with one joint and one link of length ℓ is shown in Figure 6.4. It can be modeled by the following scalar differential equation:

$$I\ddot{\theta} = \tau - B\dot{\theta} - mg(\ell/2)\cos\theta \qquad (6.2.11)$$

where the angular position of the joint is θ, the moment of inertia about the rotational axis I, the moment (torque) applied at the joint τ, the coefficient of the viscous friction B, and the mass of the link m. The gravitational term represents the nonlinearity in the system of Equation 6.2.11.

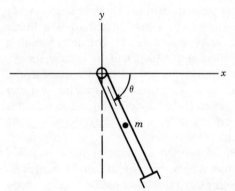

FIGURE 6.4
Simple one-joint arm (a pendulum).

The problem is to introduce a feedforward–feedback loop that generates a nonlinear function to cancel the nonlinear gravitational effect in Equation 6.2.11 [9].

The approach is to generate the input torque τ by means of a new input $\hat{u}$ and feedforward–feedback loop so as to cancel the nonlinear term. The designer may select for τ the following expression in which the gain is a (possibly nonlinear) function of the available measurement θ:

$$\tau = H(\theta)\hat{u} + h(\theta) \qquad (6.2.12)$$

where functions $H(\theta)$ and $h(\theta)$ are to be determined so that the nonlinear Equation 6.2.11 will become a linear model. This linear model is required to be of the following form:

$$\ddot{\theta}(t) + a_{22}\dot{\theta}(t) + a_{21}\theta(t) = \hat{u}(t) \qquad (6.2.13)$$

where a_{21} and a_{22} represent the position and velocity coefficients.

The substitution of Equation 6.2.12 into Equation 6.2.11 results in the following equation:

$$\ddot{\theta} + \frac{B}{I}\dot{\theta} = [H\hat{u} + h]/I - mg(\ell/2)\cos\theta/I \qquad (6.2.14)$$

Equation 6.2.14 can be made to assume the linear form of Equation 6.2.13 by satisfying the following constraints: $H/I = 1, h = mg(\ell/2)\cos\theta, a_{22} = B/I$, and a_{21} can assume an arbitrary positive value. The unknown functions $H(\theta)$ and $h(\theta)$ in Equation 6.2.12 are now specified:

$$H(\theta) = I$$

$$h(\theta) = mg(\ell/2)\cos\theta \qquad (6.2.15)$$

where the nonlinear function $h(\theta)$ represents a feedback loop, and constant H is the feedforward gain. By selecting the control input τ according to Equation 6.2.12, the nonlinear model given by Equation 6.2.11 assumes the linear form given by Equation 6.2.13. It results when the nonlinear function is accurately cancelled. This makes the total system appear linear.

The foregoing simple example demonstrates the basic approach to linearize a nonlinear system model by means of feedforward–feedback loops. We will next discuss the approach in the framework of multivariable system models.

The dynamical model of a serial-link manipulator may be written in the Lagrangian formulation as a second-order vector differential Equation 5.2.9. Such a model can be rewritten as a set of first-order differential equations:

$$\begin{bmatrix} \dot{x}^1 \\ \dot{x}^2 \end{bmatrix} = \begin{bmatrix} x^2(t) \\ f^2(x) \end{bmatrix} + \begin{bmatrix} 0 \\ B^2(x) \end{bmatrix} u(t) \qquad (6.2.16)$$

where $u(t)$ is an N-dimensional input. Vector $f^2(x)$ is an N-dimensional, bounded, continuous function of the arguments, such that $\lim[f^2(x)] \rightarrow 0$ as time t approaches infinity. Matrix $B^2(x)$ is of dimension $(N \times N)$ with entries that are bounded and continuous functions of the arguments. It represents the inverse of the inertial matrix in a manipulator dynamics. Hence, the inverse of $B^2(x)$ exists for all values of x and t in the operating domain. The state vector x has been decomposed into two N-dimensional vectors $x^1(t)$ and $x^2(t)$, that is, $x(t) = [x^1(t)', x^2(t)']'$. For example, the manipulator equations of motion in Equation 5.2.10 may be expressed as Equation 6.2.16 by defining $x^1(t)$ as the joint position vector $x^1(t) = [q_1(t), \ldots, q_N(t)]'$, and $x^2(t)$ as the corresponding velocity vector $x^2(t) = [\dot{q}_1(t), \ldots, \dot{q}_N(t)]'$.

The model Equation 6.2.16 will next be linearized by generating the input vector $u(t)$ by means of a new input vector $\hat{u}(t)$ and the nonlinear functions $H(x,t)$ and $h(x,t)$:

$$u(t) = H(x,t)\,\hat{u}(t) + h(x,t) \qquad (6.2.17)$$

where matrix $H(x,t)$ and vector $h(x,t)$ are to be determined. The terms on the right of Equation 6.2.17 can be implemented by means of feedforward and feedback loops when the state vector x is available from the measurements.

The substitution of Equation 6.2.17 into Equation 6.2.16 leads to

$$\begin{bmatrix} \dot{x}^1 \\ \dot{x}^2 \end{bmatrix} = \begin{bmatrix} x^2(t) \\ f^2(x) + B^2(x)h(x,t) \end{bmatrix} + \begin{bmatrix} 0 \\ B^2(x)H(x,t) \end{bmatrix} \hat{u}(t) \qquad (6.2.18)$$

The unspecified functions $H(x,t)$ and $h(x,t)$ in Equation 6.2.18 will be determined so that the nonlinear terms will be canceled, and the resulting model becomes linear.

The manipulator model may be required to assume the following linear time-invariant form:

$$\begin{bmatrix} \dot{x}^1 \\ \dot{x}^2 \end{bmatrix} = \begin{bmatrix} 0 & I \\ A_{21} & A_{22} \end{bmatrix} \begin{bmatrix} x^1(t) \\ x^2(t) \end{bmatrix} + \begin{bmatrix} 0 \\ B_2 \end{bmatrix} \hat{u}(t) \qquad (6.2.19)$$

where the constant submatrices A_{21}, A_{22}, and B_2 can be chosen to meet certain given design specifications. For example, matrices A_{21}, A_{22}, and B_2 can be chosen to be diagonal. Thus, the differential equations in Equation 6.2.19 can be uncoupled so that the dynamics of one joint do not depend on the variables of the other joints.

A straightforward comparison of Equations 6.2.18 and 6.2.19 gives

$$f^2(x) + B^2(x)h(x,t) = A_{21}x^1(t) + A_{22}x^2(t) \qquad (6.2.20)$$

$$B^2(x)H(x,t) = B_2 \qquad (6.2.21)$$

The unknown expressions $h(x,t)$ and $H(x,t)$ are now determined by Equations 6.2.20 and 6.2.21.

Equations 6.2.20 and 6.2.21 are solved explicitly for $h(x,t)$ and $H(x,t)$:

$$h(x,t) = [B^2(x)]^{-1}[A_{21}x^1(t) + A_{22}x^2(t) - f^2(x)] \qquad (6.2.22)$$

$$H(x,t) = [B^2(x)]^{-1}B_2 \qquad (6.2.23)$$

The original input vector $u(t)$ becomes specified after the substitution of Equations 6.2.22 and 6.2.23 into Equation 6.2.17:

$$u(t) = [B^2(x)]^{-1}[B_2\hat{u}(t) + A_{21}x^1(t) + A_{22}x^2(t) - f^2(x)] \qquad (6.2.24)$$

In Equations 6.2.24, and 6.2.19, submatrices A_{21} and A_{22} can be selected so that given design specifications will be satisfied; for example, the eigenvalues of the plant matrix in Equation 6.2.19 can be made to correspond to the acceptable step responses of the system. The selection of submatrix B_2 allows the adjustment of the gains in the control matrix. The expression of Equation 6.2.24 can be considered as consisting of two parts: the term that cancels the nonlinearity in Equation 6.2.16 (the primary controller) and the remaining terms that can be used to specify an additional controller. The block diagram in Figure 6.5 illustrates the implementation of the feedback–feedforward linearization procedure.

An example is next presented to illustrate the approach.

EXAMPLE 6.2.3

The dynamical model of the cylindrical manipulator in Example 6.1.1 is given by Equations 6.1.5 and 6.2.7; thus, $N = 3$. The equations of motion may be rewritten in the form of Equation 6.2.16 by defining $x^1 = [x_1\ x_2\ x_3]'$, and $x^2 = [x_4\ x_5\ x_6]'$. Moreover,

$$f^2(x) = \begin{bmatrix} [x_1 - (m_A/m)\,\ell/2]\,x_5^2 - k_s(x_1 - 2\ell/3)/m \\ (m_A\ell - 2\,m\,x_1)x_4\,x_5/I_t(x_1) \\ -g - k_z(x_3 - \ell_0)/\overline{m} \end{bmatrix} \quad B^2(x) = \begin{bmatrix} 1/m & 0 & 0 \\ 0 & 1/I_t(x_1) & 0 \\ 0 & 0 & 1/\overline{m} \end{bmatrix}$$

$$(6.2.25)$$

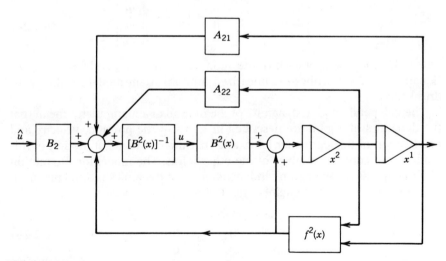

FIGURE 6.5
Realization of the feedforward–feedback linearization scheme.

The problem is to obtain a linear model in the form of Equation 6.2.19 for the manipulator dynamics by applying the procedure of the feedback–feedforward linearization to the model. Submatrix B_2 in Equation 6.2.19 is chosen as the (3×3) identity matrix, for convenience. Submatrices A_{21} and A_{22} are to be selected so that the dynamics of each joint be independent of the other joints.

The linear model is obtained by generating the input $u(t) = [F_r - B_r(x_4),$ $\tau_\theta - B_\theta(x_5), F_z - B_z(x_6)]'$ as determined by Equation 6.2.24. The inverse of $B^2(x)$ is calculated:

$$[B^2(x)]^{-1} = \begin{bmatrix} m & 0 & 0 \\ 0 & I_t(x_1) & 0 \\ 0 & 0 & \overline{m} \end{bmatrix} \qquad (6.2.26)$$

Equation 6.2.26 represents the diagonal inertial matrix that is typical for a cylindrical manipulator. The input in Equation 6.2.24 can now be rewritten:

$$u(t) = [B^2(x)]^{-1} \left\{ \begin{bmatrix} \hat{u}_1 \\ \hat{u}_2 \\ \hat{u}_3 \end{bmatrix} + A_{21}x^1(t) + A_{22}x^2(t) \right\}$$

$$- \begin{bmatrix} [mx_1 - m_A \, \ell/2]x_5^2 - k_s(x_1 - 2\ell/3) \\ [m_A \ell - 2 m x_1]x_4 x_5 \\ -\overline{m}g - k_z(x_3 - \ell_0) \end{bmatrix} \qquad (6.2.27)$$

In Equation 6.2.27, the terms on the right can be implemented by means of feedforward and feedback loops.

If Equation 6.2.27 is now substituted into Equation 6.1.5, the following equation results:

$$\begin{bmatrix} \dot{x}^1 \\ \dot{x}^2 \end{bmatrix} = \begin{bmatrix} 0 & I \\ A_{21} & A_{22} \end{bmatrix} \begin{bmatrix} x^1 \\ x^2 \end{bmatrix} + \begin{bmatrix} 0 \\ I \end{bmatrix} \begin{bmatrix} \hat{u}_1 \\ \hat{u}_2 \\ \hat{u}_3 \end{bmatrix} \qquad (6.2.28)$$

where the dimension of each of the submatrices A_{21}, A_{22} and I is (3×3). Equation 6.2.28 describes the linearized time-invariant model for the cylindrical robot.

The independent joint dynamics of the motion of a manipulator mean that the dynamics of a joint are described in terms of the position, velocity and acceleration of this particular joint, and that the equations of motion of this joint do not contain variables of the other joints. The model for the motion of a manipulator governed by independent joint dynamics is attractive from the viewpoint of control engineering. If the designer selects

$$A_{21} = \begin{bmatrix} -a_{41} & 0 & 0 \\ 0 & -a_{52} & 0 \\ 0 & 0 & -a_{63} \end{bmatrix} \qquad A_{22} = \begin{bmatrix} -a_{44} & 0 & 0 \\ 0 & -a_{55} & 0 \\ 0 & 0 & -a_{66} \end{bmatrix} \qquad (6.2.28)$$

then Equation 6.2.19 represents the independent joint dynamics. For example, the equations of motion for joint 1 can then be written from Equation 6.2.19 into the following form:

$$\ddot{x}_1 + a_{44}\dot{x}_1 + a_{41}x_1 = \hat{u}_1 \quad\quad\quad (6.2.29)$$

where x_1 represents the position of joint 1 and coefficients a_{44} and a_{41} are the velocity and position constants that the designer can select to achieve the desired characteristics in $x_1(t)$, the output response. Input $\hat{u}_1(t)$ may be chosen, for example, as the reference step input x_1^d for $t \geq 0$. Equation 6.2.29 reveals that the dynamics of joint 1 do not depend on the variables of the other joints. Expressions similar to Equation 6.2.29 govern also the dynamics of the other two joints.

The linearized model given in Equation 6.2.19 results after the exact cancellation of the nonlinear functions. Thus, the characteristics of the nonlinearities must be known accurately. Because of inaccuracies in the model, and particularly in the numerical values, the cancellation of terms in a model is seldom exact. To account for modeling errors, Equation 6.2.19 may be modified by adding a bounded three-dimensional error $e(t)$ into the model to obtain

$$\begin{bmatrix} \dot{x}^1 \\ \dot{x}^2 \end{bmatrix} = \begin{bmatrix} 0 & I \\ A_{21} & A_{22} \end{bmatrix} \begin{bmatrix} x^1 \\ x^2 \end{bmatrix} + \begin{bmatrix} 0 \\ B_2 \end{bmatrix} \hat{u}(t) + \begin{bmatrix} 0 \\ e(t) \end{bmatrix} \quad\quad (6.2.30)$$

Equation 6.2.30 can be used to design an appropriate controller for the manipulator system so as to compensate for the effects of disturbances exemplified by $e(t)$, the modeling error.

The implementation of the controller $u(t)$ in Equation 6.2.24 is accomplished by means of feedforward and feedback loops. A large number of mathematical operations may be needed in realizing the algorithm. Moreover, the designer may encounter difficulties in the implementation of Equation 6.2.24, mainly due to modeling errors. However, the cancellation of known nonlinearities by the controller is commonly achieved to a sufficient degree of accuracy in many practical applications. If some nonlinearities are not known exactly or their cancellation requires a considerable amount of computer time, the designer may attempt to cancel only some nonlinear terms in the model, for example, only the terms that represent the gravitational and/or frictional effects. Such an approach facilitates the design of a controller for the nonlinear system and often leads to an improved motion of a manipulator.

We have outlined the linearization procedure in which feedforward and feedback loops are used to generate certain nonlinear functions for attempting to cancel the nonlinear terms in the dynamical model. A more extensive exposition of the material is presented in [7,8,9].

We will next discuss linear systems and some of their fundamental characteristics.

6.3 LINEAR DYNAMICAL MODELS

We have discussed in Section 6.2 methods to determine linearized dynamical models for manipulator motions. Even though these linear models may not be valid in the entire state space, they are tractable and can be studied in detail. Particularly, the designer can determine the general transient behavior of the system and check whether or not the system possesses locally some basic desirable properties, such as the stability. The knowledge of these properties will help him to construct a controller for the linearized dynamical model using the general methods of controller design for linear systems.

We will next review some basic properties of linear continuous-time and discrete-time systems.

6.3.1   LINEAR VECTOR DIFFERENTIAL EQUATION MODELS

A linear time-invariant vector differential equation that can represent, under certain conditions, the linearized model of the motion of a manipulator is described in a general form:

$$\dot{x} = Ax(t) + Bu(t) \qquad x(t_0), \qquad t \geq t_0 \qquad (6.3.1)$$

$$y(t) = Cx(t) \qquad (6.3.2)$$

where vector $x(t) = [x_1(t), \ldots, x_{2N}(t)]'$ represents the state of the system at time t in the $2N$-dimensional real space, and vector $u(t) = [u_1(t), \ldots, u_N(t)]'$ is an N-dimensional input to the system. Matrix A is a $(2N \times 2N)$ plant matrix, and B, a $(2N \times N)$ control matrix. Matrices A, B, and C have constant elements. Vector $y(t) = [y_1(t), \ldots, y_m(t)]'$ is an m-dimensional output vector at time t. Equation 6.3.1 represents the plant or process model, and Equation 6.3.2 is the output or measurement equation.

The linear time-invariant system model in Equation 6.3.1 possesses the following properties:

Solution

The initial time t_0 can be chosen zero, that is, $t_0 = 0$, and then the solution to the constant parameter linear Equation 6.3.1 can be written as

$$x(t) = \Phi(t)x(0) + \int_0^t \Phi(t - \sigma) Bu(\sigma) \, d\sigma \qquad (6.3.3)$$

where each column of $\Phi(t)$ satisfies the homogeneous differential equation. The first term on the right side of Equation 6.3.3 is the zero-input response (the homogeneous solution), and the second part is the forced (particular) solution.

The fundamental matrix $\Phi(t)$ satisfies

$$\dot{\Phi}(t) = A\Phi(t) \qquad \Phi(0) = I \qquad (6.3.4)$$

The solution to Equation 6.3.4 specifies the state transition matrix $\Phi(t)$. It can be determined explicitly or numerically on a digital computer.

Alternatively, the fundamental matrix for the time-invariant system in Equation 6.3.1 can be computed by means of the series expansion of the exponential matrix:

$$\Phi(t) = \epsilon^{At}$$
$$= I + At + \frac{1}{2!}(At)^2 + \frac{1}{3!}(At)^3 + \dots \qquad (6.3.5)$$

The expression of the matrix exponential in Equation 6.3.5 is analogous to the power series of a scalar exponential. The speed of the convergence of the series in Equation 6.3.5 depends on the eigenvalues of the A-matrix.

Eigenvalues and Stability

The eigenvalue λ of the constant $(2N \times 2N)$ plant matrix A in Equation 6.3.1 satisfies the characteristic equation

$$|\lambda I - A| = 0 \qquad (6.3.6)$$

where $|\lambda I - A|$ signifies the determinant of matrix $(\lambda I - A)$. The solution of Equation 6.3.6 gives the eigenvalues λ_i, $i = 1, \dots, 2N$ of the plant matrix A appearing in Equation 6.3.1.

When the solution to the differential Equation 6.3.1 is written explicitly as shown in Equation 6.3.3, it will contain terms of the form $\exp(\lambda_i t)$, $t \geq 0$. Thus, the dynamic (transient) response of the system depends on the eigenvalues. The (asymptotic) stability of the system is guaranteed when the real part of each λ_i is negative, that is, $\text{Re}\{\lambda_i\} < 0$. The real part $\text{Re}\{\lambda_i\}$ specifies the rate of decay of the envelope of the step-response when the system is underdamped.

The linearized model of the motion of most manipulators can be considered as approximately time-invariant only over a sampling period. As a consequence, the eigenvalues of the plant matrix that is constant only over a sampling period cannot be used to determine the stability of the system over the entire time interval $[0, \infty)$. The eigenvalues of the foregoing plant matrix can, however, provide information about the local behavior of the system response on this interval.

Complete Controllability

When the dynamical state variable model of a manipulator system is given, it is often desirable to transfer the initial state of the system to a specified final state in a finite time by properly choosing input $u(t)$, $t \geq 0$. If such an input $u(t)$ transfers all components of the state vector $x(0)$ in the system of Equation 6.3.1 to a specified final state $x(t_f)$ in a finite time, then the system of Equation 6.3.1 is called completely controllable. A more detailed discussion of the complete controllability may be found in [1].

Necessary and sufficient conditions for the complete controllability of the time-invariant system in Equation 6.3.1 are

$$\text{Rank}\,[B, AB, \ldots, A^{2N-1}B] = 2N \qquad (6.3.7)$$

Thus, the controllability matrix in the bracket of Equation 6.3.7 must have full rank.

Necessary and sufficient conditions for the complete controllability of linear time-varying systems can also be determined, although the test is considerably more complicated to apply [1].

It is sometimes stated without proofs that a manipulator at a singular configuration is not (completely) controllable. It is naturally very difficult in general to establish whether or not a six-joint manipulator is completely controllable at the time when the singular condition occurs. Although a general study of the complete controllability of most manipulators is impractical, the designer can gain insight into the behavior of the manipulator motion by studying the complete controllability of some simplified manipulator models at the singular point.

If a linear time-invariant system is completely controllable, then it can be stabilized, if necessary, by means of a feedback controller that operates linearly on the state vector.

The following example illustrates the application of the complete controllability test to a manipulator system.

EXAMPLE 6.3.1

The cylindrical manipulator discussed in Example 6.1.1 is operated on the (rz)-plane by maintaining θ at a constant value. The dynamical model given by Equation 6.1.4 reduces to the following set of differential equations:

$$\ddot{r} = -k_s(r - 2\ell/3)/m - B_r\dot{r}/m + F_r/m \qquad (6.3.8)$$

$$\ddot{z} = -g - B_z\dot{z}/\overline{m} + F_z/\overline{m} \qquad (6.3.9)$$

where the friction is assumed to be viscous friction, and the spring constant $k_z = 0$. The coefficients m, k_s, B_r, and B_z are constants. The inputs to the system are represented by F_r and F_z describing the forces in the r- and z-directions, respectively.

The problem is to determine whether or not the system described by Equations 6.3.8 and 6.3.9 is completely controllable.

To solve the problem, the system in Equations 6.3.8 and 6.3.9 is first described in a state variable form by defining $x_1(t) = r(t)$, $x_2(t) = z(t)$, $x_3(t) = \dot{r}(t)$, and $x_4(t) = \dot{z}(t)$. Equations 6.3.8 and 6.3.9 can then be rewritten:

$$\begin{bmatrix} \dot{x}_1 \\ \dot{x}_2 \\ \dot{x}_3 \\ \dot{x}_4 \end{bmatrix} = \begin{bmatrix} 0 & 0 & 1 & 0 \\ 0 & 0 & 0 & 1 \\ -k_s/m & 0 & -B_r/m & 0 \\ 0 & 0 & 0 & -B_z/\overline{m} \end{bmatrix} \begin{bmatrix} x_1 \\ x_2 \\ x_3 \\ x_4 \end{bmatrix} + \begin{bmatrix} 0 & 0 \\ 0 & 0 \\ 1/m & 0 \\ 0 & 1/\overline{m} \end{bmatrix} \begin{bmatrix} \hat{F}_r \\ \hat{F}_z \end{bmatrix}$$

$$(6.3.10)$$

where $\hat{F}_r = F_r + 2\ell k_s/3$ and $\hat{F}_z = F_z - \overline{m}g$. Equation 6.3.10 is a linear time-invariant system in the form of Equation 6.3.1.

The complete controllability of system in Equation 6.3.10 is tested by forming the controllability matrix in Equation 6.3.7. In this case, it becomes

$$[B \quad AB \quad A^2B \quad A^3B]$$

$$= \begin{bmatrix} 0 & 0 & \dfrac{1}{m} & 0 & \dfrac{-B_r}{m^2} & 0 & \dfrac{(-k_s + B_r^2/m)}{m^2} & 0 \\[2ex] 0 & 0 & 0 & \dfrac{1}{m} & 0 & \dfrac{-B_z}{m^2} & 0 & \dfrac{B_z^2}{m^3} \\[2ex] \dfrac{1}{m} & 0 & \dfrac{-B_r}{m^2} & 0 & \dfrac{(-k_s + B_r^2/m)}{m^2} & 0 & \dfrac{B_r(2k_s + B_r^2/m)}{m^3} & 0 \\[2ex] 0 & \dfrac{1}{m} & 0 & \dfrac{-B_z}{m^2} & 0 & \dfrac{B_z^2}{m^3} & 0 & \dfrac{-B_z^3}{m^4} \end{bmatrix}$$

$$(6.3.11)$$

The maximum rank of the (4×8) controllability matrix in Equation 6.3.11 is four, that is, it has full rank. It is verified, for example, by choosing the first four columns and showing that they form a set of independent vectors. The maximum rank of the controllability matrix can also be determined on a computer. Hence, the planar manipulator in Equation 6.3.10 is completely controllable. This implies that the end-effector can be transformed in a finite time from a given initial position to a specified final state in the workspace by a properly chosen input.

6.3.2 LINEAR VECTOR DIFFERENCE EQUATION MODELS

The general discretization of continuous-time nonlinear models for digital simulations is discussed in Section 6.1 by means of various approximations. When the dynamical model of a system is linear such as Equation 6.3.1, a discrete-time model can be obtained exactly under the assumption that the input is piecewise constant which is indeed the case in most computer-controlled systems.

If the sampling period T is uniform, and the input $u(t) = u(kT)$ is constant over $kT \le t < (k + 1)T$, then the discretized form of Equation 6.3.1 is

$$x[(k + 1)T] = \overline{A}x(kT) + \overline{B}u(kT) \qquad (6.3.12)$$

where $k = 0, 1, 2, \ldots$, vector $x(kT) = [x_1(kT) \ldots x_{2N}(kT)]'$ is the system state at time kT, $x(0)$ is specified, and $u(kT)$ is the N-dimensional input vector at time kT. Moreover, the plant and control matrices are determined by the following expressions:

$$\overline{A} = \epsilon^{AT} \qquad \overline{B} = \int_0^T \epsilon^{A(T-\sigma)} B \, d\sigma \qquad (6.3.13)$$

The difference Equation 6.3.12 can be used to recursively compute the numerical solution.

The difference equation model in Equation 6.3.12 has some fundamental characteristics [1]:

Solution

The solution of a difference equation is described in terms of the input and the initial condition. The solution $x(kT)$ to Equation 6.3.12 at time kT can be written by induction as (the sampling period T is dropped in the sequel)

$$x(k) = \overline{A}^k x(0) + \sum_{i=0}^{k-1} \overline{A}^{k-1-i} \overline{B} u(i) \qquad (6.3.14)$$

The first term on the right side of Equation 6.2.14 represents the zero-input response (the homogeneous solution), and the second term, describing the summation convolution, is the forced response.

An alternative closed-form solution to Equation 6.3.12 can be written by means of the z-transformation ($z = \exp(sT)$, where s is the Laplacian variable). The z-transform $X(z)$ of $\{x(k)\}$ can indeed be determined

$$X(z) = (zI - \overline{A})^{-1} [\overline{B} U(z) + zx(0)] \qquad (6.3.15)$$

where $U(z)$ is the z-transform of $\{u(k)\}$. By determining the inverse transform of $X(z)$, the solution $x(k)$ on the time-domain can be obtained.

Eigenvalues and Stability

The eigenvalues of the plant matrix $\overline{A}$ in the linear time-invariant difference Equation 6.3.12 determine the stability of the system. The characteristic equation of the plant matrix in Equation 6.3.12 is

$$|zI - \overline{A}| = 0 \qquad (6.3.16)$$

where z signifies the eigenvalue of the system. The solution of Equation 6.3.16 gives $(2N)$ eigenvalues $(z_1, \ldots, z_{2N})$ for the $(2N)$-order system.

A discrete time-invariant system is (asymptotically) stable when the magnitude of each eigenvalue z_i is less than one, that is,

$$|z_i| < 1 \qquad (6.3.17)$$

for each $i = 1, \ldots, (2N)$. Thus, the stability region on the complex z-plane is specified by the inside of the unit circle. When the eigenvalues are close to the periphery of the unit circle, the output response of the system can be expected to be oscillatory.

Complete Controllability

The linear time-invariant system of Equation 6.3.12 is completely controllable if there exists a control sequence that transfers the initial state $x(0)$ to the specified final state $x(k)$ in finite time $(k < \infty)$.

Inspection of Equation 6.3.14 shows that a control sequence that transfers $x(0)$ to $x(k)$, $k = 2N$ can be found, when the following condition holds:

$$\text{Rank}[\,\overline{B}, \overline{A}\,\overline{B}, \ldots, \overline{A}^{\,2N-1}\,\overline{B}\,] = 2N \qquad (6.3.18)$$

Thus, the controllability matrix in the bracket of Equation 6.3.18 must have full rank. This is a necessary and sufficient condition for the complete controllability of the system in Equation 6.3.12.

To establish the foregoing properties for a general linear time-varying system is considerably more difficult [1] and will not be discussed here.

EXAMPLE 6.3.2

The system involving a cylindrical robot manipulator is designed in Example 6.2.3 so that the motion of the manipulator obeys the independent joint dynamics. Thus, the dynamical model of each joint is independent of the dynamics of the other joints, as exemplified by Equation 6.2.29 for the motion of joint 1. Suppose that the motion of joint 1 is described as

$$\ddot{x}_1 + 3\dot{x}_1 + 2x_1 = \hat{u}_1 \qquad (6.3.19)$$

where the joint position $x_1(t)$ can be measured. Thus, the output $y_1(t)$ of joint 1 is determined by

$$y_1(t) = x_1(t) \qquad (6.3.20)$$

The problem is to develop a discrete-time state variable model of the dynamics given in Equation 6.3.19 under the assumption that input $\hat{u}_1$ is (piecewise) constant over each sampling period $[kT, (k + 1)T)$. Then, the resulting model is to be converted to a single difference equation model.

A discretized form of Equation 6.3.19 can be obtained by expressing Equation 6.3.19 first in the state variable form, and then evaluating the expressions in Equation 6.3.13 for the difference Equation 6.3.12. By applying this procedure, the state variable equation representing Equation 6.3.19 is expressed in the following form after defining $x_{11}(t) = x_1(t)$ and $x_{12} = \dot{x}_1(t)$

$$\begin{bmatrix} \dot{x}_{11} \\ \dot{x}_{12} \end{bmatrix} = \begin{bmatrix} 0 & 1 \\ -2 & -3 \end{bmatrix} \begin{bmatrix} x_{11}(t) \\ x_{12}(t) \end{bmatrix} + \begin{bmatrix} 0 \\ 1 \end{bmatrix} \hat{u}_1(t) \qquad (6.3.21)$$

$$y_1(t) = x_{11}(t) \qquad (6.3.22)$$

Since $\hat{u}_1(t)$ is assumed to be piecewise constant, an exact discretized model can next be obtained from Equation 6.3.21. To determine Equation 6.3.12, the state transition matrix $\Phi(T) = \exp(AT)$ is first evaluated by Equation 6.3.4, and then the expressions in Equation 6.3.13 can be written. The resulting discrete-time model is as follows

$$\begin{bmatrix} x_{11}(k+1) \\ x_{12}(k+1) \end{bmatrix} \begin{bmatrix} 2\epsilon^{-T} - \epsilon^{-2T} & \epsilon^{-T} - \epsilon^{-2T} \\ -2\epsilon^{-T} + 2\epsilon^{-2t} & -\epsilon^{-T} + 2\epsilon^{-2T} \end{bmatrix} \begin{bmatrix} x_{11}(k) \\ x_{12}(k) \end{bmatrix}$$

$$+1 \begin{bmatrix} \frac{1}{2}(1-\epsilon^{-T})^2 \\ \epsilon^{-T}(1-\epsilon^{-T}) \end{bmatrix} \hat{u}_1(k) \quad (6.3.23)$$

$$y_1(k) = x_{11}(k) \quad (6.3.24)$$

where $y_1(k)$ is the measurement of the position of joint 1. The system in Equations 6.3.23 and 6.3.24 describes the discretized model of Equation 6.3.19. It can conveniently be used to study the stability as well as the complete controllability of the system.

If some of the parameters in the plant and/or control matrix are unknown and should be determined, a single difference equation model in terms of the measured variable is usually preferred in recursive on-line parameter estimation algorithms. That is, a discrete-time model expressed as a single difference equation is needed.

To convert the state variable Equations 6.3.23 and 6.3.24 to a single difference equation model, several approaches may be used [1]. For example, the z-transformation can be applied to both sides of Equations 6.3.23 and 6.3.24. The resulting equations can be solved for the z-transform $Y_1(z)$ of the output to obtain

$$Y_1(z^{-1}) = \frac{\frac{1}{2}(1-\epsilon^{-T})^2(z^{-1} + \epsilon^{-T}z^{-2})}{1 - (\epsilon^{-T} + \epsilon^{-2T})z^{-1} + \epsilon^{-3T}z^{-2}} \hat{U}(z^{-1}) \quad (6.3.25)$$

where the initial conditions $x_{11}(0)$ and $x_{12}(0)$ are assumed zero, and $\hat{U}(z)$ is the z-transform of the input $\{\hat{u}(k)\}$.

By considering variable z^{-1} as a delay operator, that is, $z^{-1}y_1(k) = y_1(k-1)$, Equation 6.3.25 can be expressed as a difference equation:

$$y_1(k) - (\epsilon^{-T} + \epsilon^{-2T})y_1(k-1) + \epsilon^{-3T}y_1(k-2) = K[\hat{u}_1(k-1) + \epsilon^{-T}\hat{u}_1(k-2)] \quad (6.3.26)$$

where $K = (1-\epsilon^{-T})^2/2$. The numerical solution to Equation 6.3.26 can be calculated recursively when the initial conditions $y_1(0) = x_1(0)$ and $y_1(1) = \dot{x}(0)$ are known by letting $k = 2, 3, \ldots$

Equation 6.3.26 is a single-input–single-output (SI-SO) discrete-time model describing the independent joint (one) dynamics. It may be written in a general form for joint 1:

$$A_1(z^{-1})y_1(k) = B_1(z^{-1})\hat{u}_1(k-1) \quad (6.3.27)$$

where the argument of the input has the delay of one sampling period, $A_1(z^{-1}) = 1 - (\epsilon^{-T} + \epsilon^{-2T})z^{-1} + \epsilon^{-3T}z^{-2}$ and $B_1(z^{-1}) = (1 - \epsilon^{-T})^2(1 + \epsilon^{-T}z^{-1})/2$. A time-series model in the form of Equation 6.3.27 is convenient in applying on-line estimation schemes such as the least squared-error algorithm to calculate the estimates of model parameters and in the adaptive self-tuning control algorithm to regulate the motion of a manipulator. These algorithms will be discussed in Chapter 9 and Appendix D.

6.4 SUMMARY

The dynamical models for the gross motion of robotic manipulators are usually described by nonlinear differential equations. They can be expressed conveniently as a set of first-order differential equations, usually referred to as a state variable representation. We have reviewed in this chapter some aspects of the state variable representations for differential and difference equation models.

Digital simulations play an important role in studying and designing non-linear systems such as manipulator dynamics. We have discussed common approximations that form the basis for many digital simulation algorithms.

In order to design a nonlinear system so that it meets design specifications, the nonlinear dynamical equations are often linearized. The system model can then be studied analytically. Moreover, the linearization of nonlinear system models is an important step in designing controllers for manipulator systems. Two methods for linearizing nonlinear state variable models have been presented. The one leads to the (weak) variational equations that describe the perturbed motion in a sufficiently small neighborhood about the nominal trajectory. The other relies on first generating the nonlinear functions that cause the cancellation of the nonlinear terms in the model. They are fed into the plant through feedforward and feedback loops. The resulting total system is linear if the cancellation of the nonlinear functions is exact. The system can then be made to behave in a desirable manner by using the basic design techniques known for linear systems.

We have also discussed some of the basic properties of linear time-invariant systems including the stability and the complete controllability. Although it is often difficult to test for these properties in many manipulator models, even in a small neighborhood of a trajectory, they can shed light on the behavior of the manipulator motion.

REFERENCES

[1]   G. M. SWISHER, *Introduction to Linear Systems Analysis*, Matrix Publishers, Champaign, IL, 1976.

[2] C. P. NEUMAN and V. D. TOURASSIS, "Discrete Dynamic Robot Models," *IEEE Transactions on Systems, Man and Cybernetics,* Vol. SMC-15, No. 2, pp. 193–204, March/April 1985.

[3] J. R. RICE, *Numerical Methods, Software and Analysis,* McGraw-Hill, New York, 1983.

[4] *IMSL Library Reference Manual,* International Mathematics Statistical Libraries, Inc., Houston, TX, June 1982.

[5] D. GREENSPAN, *Discrete Models,* Addison-Wesley, Reading, MA, 1973.

[6] S. E. DREYFUS, *Dynamic Programming and Calculus of Variations,* Academic Press, New York, 1965.

[7] C. REBOULET and C. CHAMPETIER, "A New Method for Linearizing Nonlinear Systems: the Pseudolinearization," *International Journal of Control,* Vol. 40, No. 4, pp. 631–638, 1984.

[8] H. HEMAMI and P. C. CAMANA, "Nonlinear Feedback in Simple Locomotion Systems," *IEEE Transactions on Automatic Control,* Vol. AC-21, No. 6 (short paper), pp. 855–859, December 1976.

[9] F. M. A. SALAM, "Feedback Stabilization of the Nonlinear Pendulum under Uncertainty: a Robustness Issue," *Systems & Control Letters,* Vol. 7, No. 3, pp. 199–206, June 1986.

PROBLEMS

6.1 *a.* Describe the model of a separately excited DC motor discussed in Chapter 1 in a state variable form.

    b. Determine the equilibrium state of the system determined in part *(a)* when the input is a step function of height $\theta_d = \bar{u}$ applied at $t = 0$.

6.2 *a.* The dynamical model of a general planar manipulator is given in Equation 5.2.41. Express the equations of motion in the state variable form.

    b. What is the dimension of the state vector? What is the order of the differential equation system given by Equation 5.2.40? Compare.

6.3 The model of the planar manipulator of Computer Problem 5.C.1 is given in Table 5.4 as second-order differential equations. The external forces and moments are assumed to be zero.

    a.   Determine the state variable representation of the system.

    b.   Determine the equilibrium state (steady state) of the system when step inputs of heights $\bar{\tau}_1$ and $\bar{\tau}_2$ are applied to the system, by assuming a stable operating point of the system. Discuss how the heights of the step inputs affect the equilibrium state.

6.4   The cylindrical robotic manipulator shown in Figures 6.1*a* and 6.1*b* is operated so that θ remains constant and $\tau_\theta = 0$ during the motion of the end-effector. The prismatic link moves inside the outer casing. The mass of the hub is m_h and it has center of gravity on the z_0-axis. The mass of the prismatic link is m_A with the center of gravity located as shown in the aforementioned figures. Mass m_L at the gripper describes the load. Assume that the frictional terms represent the sum of Coulomb and viscous friction. The springs with constants k_r and k_z act along the r- and z-axes, and they produce a zero force when $r = 2\ell_0/3$ and $z = \ell_1/3$.

    a.   Write the equations of motion for the case that θ is constant. Describe verbally the motion of the end-effector.

    b.   Write the dynamical model in the state variable form. Is the system linear?

    c.   Assume that the inputs are unit step functions of heights $\bar{F}_r$ and $\bar{F}_z$ applied at $t = 0$. Determine the equations for the equilibrium state $(\bar{r}, \bar{z})$.

6.5 *a.*   The difference Equations 6.1.19 and 6.1.20 result from discretizing the dynamical model given in Equation 6.1.6. Eliminate term x^2 and obtain an equation for x^1.

    b.   Give a physical interpretation to the resulting equation in terms of the translational motion.

6.6   The dynamical model for the cylindrical manipulator of Example 6.1.1 is given in Equation 6.1.4. The frictional effects are omitted here, and the spring constants are assumed to be zero.

    a.   Express the equations of motion as the total time derivatives in the following forms:

$$\frac{d}{dt}\{[I + i(r)]\,\omega\} = \tau_\theta(k)$$

$$\frac{d}{dt}(\overline{m}\dot{z}) = F_z(k) - \overline{m}g$$

$$\frac{1}{2}\frac{d}{dt}\{[I + i(r)]\omega + m\dot{r}^2\} = \frac{d}{dt}[\theta\tau_\theta(k) + rF_r(k)]$$

where the actuator inputs $\tau_\theta(k)$, $F_z(k)$, and $F_r(k)$ are constant over the kth sampling interval, $k = 0, 1, \ldots$ and $\omega = \dot{\theta}$. Determine $I, i(r), \overline{m}$ and m.

b. Obtain the discrete-time model for the cylindrical robot manipulator under consideration by applying the trapezoidal rule with Δt as the subinterval of integration to the following equations:

$$\dot{\theta} = \omega(t)$$

$$\dot{z} = v_z(t)$$

$$\dot{r} = v_r(t)$$

Then by applying Euler's forward difference approximation to the equations given in part *(a)*, complete the determination of the discretized model.

c. Discuss the conservation of energy and momenta from one sampling instant to the next one on the basis of the difference equations determined in part *(b)*.

6.7 The cylindrical manipulator discussed in Example 5.1.1 moves on a horizontal plane determined by $z = \bar{z}_1$ (constant). The friction is assumed to be viscous friction.

a. Determine the equations of motion for this (planar) manipulator.

b. Express the dynamical equations in the state variable form, that is, as a set of first-order differential equations. Are the equations linear? If not, why?

c. Determine the equations for the equilibrium state $(\bar{r}, \bar{\theta})$.

d. Determine the variational dynamical model about $(\bar{r}, \bar{\theta})$. Express the resulting equations in the form $\delta\dot{x} = A\,\delta x(t) + B\,\delta u(t)$. Are A and B constant matrices?

e. Linearize the nonlinear model by the feedback-feedforward linearization method. Select A_{21} and A_{22} in Equation 6.2.19 so that independent joint control can be applied. Draw a block diagram to show the feedback and/or feedforward loops needed in the implementation.

f. Comment on the imperfect cancellation of the nonlinear terms in part *(e)*.

6.8 The nonlinear model of the planar manipulator described in Computer Problem 5.C.1 is assumed to be asymptotically stable. It is to be linearized about a specific operating point $(\bar{F}, \bar{\tau}, \bar{r}, \bar{\theta})$.

a. Determine the equilibrium state $(\bar{r}, \bar{\theta})$ of the system when $\bar{F_r} = 5N$, $\bar{\tau_\theta} = 0$ Nm, $m_L = 1$ kg, $m_A = 2$ kg, $m_h = 0.5$ kg, $I_\theta = 1$ kgm^2, $\ell = 50$ cm, $k_s = 75$N/m, $B_r = 5.0$ Ns/m, and $B_\theta = 0.2$ Nms.

b. Suppose the value $\bar{F}$ is 12 N and the other values are as in part *(a)*. What is the equilibrium state? Give a physical explanation.

c. Determine the linearized equations of the system by obtaining the variational model about $(\bar{r}, \bar{\theta})$ calculated in *(a)*.

d. Introduce feedback–feedforward loops to linearize the system given. Let $A_{21} = 0$, $A_{22} = 0$, $B_2 = I$, and $\hat{u} = K_p x^1 + K_v x^2$. Determine the resulting dynamical model.

e. Determine the characteristic equation of the linear system of part *(d)*.

6.9 The transfer function matrix for a system with two inputs F and τ and two outputs r and θ is defined as follows:

$$G(s) = \begin{bmatrix} R(s)/F(s) & R(s)/\tau(s) \\ \theta(s)/F(s) & \theta(s)/\tau(s) \end{bmatrix}$$

where $R(s)$ is the Laplace transform of $r(t)$. The element $R(s)/F(s)$ is determined when $\tau = 0$; the other elements are defined similarly.

a. Can you determine the transfer function matrix $G(s)$ for the two-link manipulator system described in Equation 6.1.5 when z is constant? If yes, specify $G(s)$; if no, give reasons.

b. Can you obtain the transfer function matrix $G(s)$ for the system in Problem 6.4? If yes, specify $G(s)$; if no, give reasons.

c. Can you obtain the transfer function matrix $G(s)$ for the system in Problem 6.7 *(e)*? If yes, give $G(s)$; if no, explain.

d. What are the assumptions on the model of a system so that a transfer function can be used to represent

the system behavior? When do manipulator dynamics
satisfy these assumptions? Give examples.

6.10   The dynamical model for a two-link manipulator is given
in Computer Problem 5.C.1.

a.   Suppose that

$$m_1 = m_2 = 1, \quad L_1 = L_2 = 0.5, \quad \bar{\tau}_1 = 5\left(\frac{1}{\sqrt{2}} + \frac{1}{4}\right), \quad \bar{\tau}_2 = 1.25$$

and approximate $g_y = 10$ in consistent units. (i) What
are the consistent units of the variables? (ii) Deter-
mine the steady-state values $\bar{\theta}_1$, $\bar{\theta}_2$ of the joint variables
when the inputs are step functions of heights $\bar{\tau}_1$ and $\bar{\tau}_2$
applied at $t = 0$ to joints 1 and 2, respectively.

b.   If $\bar{\tau}_2 = 3$ Nm and the other values are the same as in
part (a), determine the equilibrium state $(\bar{\theta}_1, \bar{\theta}_2)$. Give
a physical interpretation for this state.

6.11   The model of a separately excited DC motor is obtained in
the state variable form in Problem 6.1.

a.   Is the model completely controllable?

b.   Determine the eigenvalues of the plant matrix.
Compare them with the poles of the transfer function.

6.12   Check the complete controllability of system described
in Equations 6.2.8–6.2.10 under the assumption that the
plant and control matrices are time invariant. Notice that
you need to find only a set of column vectors that form a
square matrix with full rank (i.e., the determinant of the
square matrix must be nonzero).

6.13   To design a controller for the two-link manipulator, the
model described in Computer Problem 5.C.1 is linearized.

a.   Introduce appropriate feedback and/or feedforward
loops so as to generate the nonlinear terms needed
to cancel the nonlinear terms in the model. Make
the resulting system model time invariant and linear.
Comment on the practicality of the approach.

b.   Modify the feedback loop proposed in part (a) by
adding a PD-controller (position, velocity feedback)
with adjustable gains K_p and K_v into the system.
Determine the characteristic equation for the closed-
loop system. What is the order of the characteristic
equation? How many eigenvalues (natural frequencies) does the system possess?

    c.   Determine a desired characteristic equation of the system so that the step responses of the joint variables in the linearized model exhibit an overshoot of less than 5% and a settling time of 10 time units.

    d.   What is the main drawback in the procedure of the feedforward-feedback linearization?

6.14 A digital simulation of a linear second-order model in Equation 6.3.19 is to be performed.

    a.   Write the difference equation model that results from applying Euler's finite difference approximation for the derivatives. Express the model first as a set of first-order difference equations, that is, in the state variable form.

    b.   The difference equation model obtained in *(a)* contains the subinterval Δt of the integration in the parameters. Determine the range of Δt so that the eigenvalues of this difference equation model are inside the unit circle. The values of Δt in the range determined will then guarantee that the difference equation model results in a numerically stable response.

COMPUTER PROBLEMS

6.C.1   Given:

$$\text{input } u(t) \rightarrow \boxed{\frac{K(\alpha s + 1)}{s^2 + 2\xi\omega_n s + \omega_n^2}} \rightarrow \text{output } y(t)$$

where the natural frequency $\omega_n = 5$ rad/sec, the damping factor is ξ, α is a real number, and $K = 25$.

    a.   Determine the single differential equation that describes the input-output relation.

    b.   Express the input–output relation as a set of first-order differential equations. (No derivatives of the input must appear in the model.)

    c.   Write a digital simulation program for the system. Determine the unit step response for the system for $t \in [0, \infty]$, when $\xi = 30, 2, 1, 0.707$, and 0.1, and $\alpha = 0$. Graph the responses on the same coordinate system.
        Note: Use (at least) three different values for the subinterval Δt of integration when $\xi = 0.1$. Make

sure that the value of Δt used in obtaining the graphs is sufficiently small. (Discuss the criterion used.)

d.   Assume $\alpha = 0$. (i) What is the value of the damping factor ξ for the critical damping ξ in case of step inputs? (ii) Obtain the unit step response for the case of critical damping. (iii) What is the steady-state output in this case? How much does it deviate from the input value? (iv) What is the value of the output at the time $T_s = 4/\xi_{cr}\omega_n$? How much does it deviate from the input?

e.   Let $\alpha = 10$. Obtain the graphs for the unit step responses for the values of ξ 1 and 0.1. Compare the responses thus obtained with the corresponding graphs obtained in (c). What is the effect of α on the step response?

6.C.2   A mathematical model for a planar manipulator is given in Table 5.4, and the numerical values for the parameters are specified in Computer Problem 5.C.1.

a.   Express the mathematical model as a set of first-order differential equations.

b.   For the integration, you need to specify the total time interval $N\Delta t$, where Δt is the subinterval of integration and N is the number of subintervals. Since you cannot select $N\Delta t = \infty$, discuss possible choices for specifying $N\Delta t$, as well as the choice of N. What are the consequences computationally when Δt is a very small number?

c.   Perform digital simulation by integrating the resulting equations. Choose the subinterval Δt for the integration so that the step response outputs are numerically stable. The value of Δt should be sufficiently small so that it would not affect the system response. Graph the response $\theta_1(t)$ versus time and $\theta_2(t)$ versus time for two different values of Δt. Compare simulation results.

CHAPTER SEVEN

TRAJECTORY PLANNING
FOR MANIPULATOR MOTION

For many tasks, the end-effector of a manipulator is made to move from a given initial position and orientation to a specified terminal position and orientation. The homogeneous transformation matrix can be used to describe these points in the Cartesian base (world) coordinate system. The curve that the center of the end-effector traces in moving from the initial point to the terminal state determines a translational path. The curve that describes the motion of the hand orientation forms a rotational path. The set of the points that specifies the translational and rotational paths of the manipulator gripper as a function of time is referred to as the *trajectory*. The positions, velocities, and even accelerations of the generalized coordinates as functions of time are often included in the description of a trajectory.

When the desired trajectory for a manipulator motion in the workspace is being planned, the maximum velocity and acceleration of the manipulator will impose constraints on the planned motion. They may be specified either in the Cartesian base coordinate system or for each generalized coordinate in the joint space. These restrictions need to be taken into account in the desired velocity profile when the desired trajectory for the motion is determined.

A path of an end-effector can easily be visualized in the Cartesian space. However, it is usually difficult to think of the shape of the corresponding path in the joint space. The path descriptions in the two different spaces are related by the kinematic equations. For a specific task, the desired trajectory to be tracked by the manipulator is usually determined in advance by off-line computations.

If the desired trajectory is specified by discrete points in the Cartesian base coordinate system and the control is performed in the joint space, the designer may convert the discrete trajectory points to the points of the joint space using the kinematic transformation equations. When the line segments joining the discrete trajectory points in the Cartesian space are converted into the joint space, the resulting graphs will not, in general, consist of line segments. Since the conversion of the line segments from the Cartesian space into the joint space usually involves a large number of points, this approach becomes unattractive for real-time control because of the memory requirements and computation time. An alternative is to introduce a function such as a polynomial to describe the path between two adjacent trajectory points in the joint space. The concatenated functions can be splined together by satisfying certain continuity conditions that guarantee a smooth motion for the manipulator. Indeed, a smooth motion of the manipulator is achieved when the trajectory has the property that the positions, the velocities, and often even the accelerations are continuous on the desired trajectory. The derivative of the acceleration may also be required to be continuous along the trajectory so that jerks (i.e., jumps in accelerations) are avoided. These requirements can be taken into account in splining the functions together by off-line calculations. The spline functions can then be used to determine necessary intermediate points on the desired trajectory of the generalized coordinate.

When points are chosen to describe the desired motion of a manipulator in the Cartesian space, the location of obstacles in the workspace must be considered. Since the object moved by the manipulator occupies a certain volume, it is important for avoiding collisions that this volume does not intersect with the volumes occupied by the obstacles. Because it is usually difficult to determine the intersections of these volumes, a possible approach for the analysis and design is to shrink the object being transferred to a point while enlarging the volumes of the obstacles so as to account for the volume of the object. The desired trajectory can then be determined for the point object around the fictitious enlarged obstacles so that the manipulator will avoid collisions with the obstacles while moving along the specified path.

We will discuss first the trajectory planning both in the Cartesian base coordinate system and the joint space for a manipulator that is to track a straight line between given initial and terminal points when the velocity profile is specified. For given task specifications, we will introduce polynomial functions that are splined together to describe the desired trajectory for a smooth motion. We will then discuss the generation of desired trajectories for the manipulator motion in an environment that contains stationary obstacles.

7.1 FORMULATION OF PLANNING PROBLEM

The purpose in the planning of the motion for a manipulator is to convert the specifications of a given task to a *desired* trajectory that helps the manipulator perform the task. A controller produces the inputs to the manipulator so that the manipulator will track the desired path. The block diagram in Figure 7.1 illustrates the functional organization of the system.

The desired trajectory may be specified as a sequence of discrete points or as a continuous function of time. If the desired trajectory is specified by discrete points in the Cartesian base (world) coordinate system and if the controller is designed in the joint space, it is a common practice to convert the desired trajectory points into the joint space using the kinematic equations. If this conversion is needed at every sampling instant, the computations impose a considerable burden because the sampling rate in most digitally controlled manipulators is in the range of 20 Hz to 200 Hz. When these trajectory points are precomputed, the storage of the trajectory points may be troublesome, particularly in small computers. To avoid the problems caused by the large number of computations and/or storage requirements, a possible approach is to convert only certain points from the Cartesian space into the joint space, and then use a properly chosen function between these points to approximate the desired trajectory. Thus, only the discrete points and the parameters describing the approximating function need to be stored. This function can then be used to compute intermediate points on the specified trajectory.

Suppose that the position and orientation of the gripper of an N-joint manipulator are given by a sequence of n homogeneous matrices specified at the discrete times t_k, $k = 1, \ldots, n$

$$T(t_k) = \begin{bmatrix} n(t_k) & s(t_k) & a(t_k) & p(t_k) \\ 0 & 0 & 0 & 1 \end{bmatrix} \qquad (7.1.1)$$

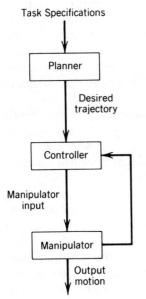

Task Specifications

Planner

Desired trajectory

Controller

Manipulator input

Manipulator

Output motion

FIGURE 7.1
Architectural organization of a manipulator control.

where $p(t_k)$ determines the position, and $n(t_k)$, $s(t_k)$, $a(t_k)$ give the orientation of the end-effector in the Cartesian base coordinate system. The corresponding values of the joint variables $q_i(t_k)$, $i = 1, \ldots, N$ can be calculated using the kinematic equations in the backward direction, as discussed in Chapter 3. The aforementioned variables define the desired trajectory at discrete points in their respective spaces.

The values of the desired trajectory are usually needed also between the specified discrete points because of the high sampling rates in the computer controlled manipulator systems. For this purpose, functional approximations for the desired trajectory between the discrete points can be introduced [1, 2, 3]. To describe the entire desired trajectory between the initial and terminal points, the approximating functions are splined together so that certain continuity conditions are satisfied to guarantee that the motion of the manipulator will be smooth. The spline functions are then used for interpolating additional trajectory points, for example, for a control algorithm. The aforementioned problems are thus reduced considerably, because only the parameters of the spline functions are stored, and many of the necessary calculations can be performed in the planning stages off-line and in advance. The effective determination of the spline functions and their parameters is essential for successful planning, particularly in applications.

7.2 GENERATION OF DESIRED TRAJECTORY FOR STRAIGHT LINE MOTION

For a specific task of a manipulator, only the initial and terminal positions and orientations of the gripper in the Cartesian base coordinate system are commonly specified. These points may be connected by a straight line in many applications. This line segment represents the desired trajectory of the motion in the base coordinate system for a control algorithm. Such a straight line is an attractive choice for the path because it is the shortest distance between the given two points, although it may not represent the minimum time path. It is also suitable to many applications.

The desired motion of an end-effector is in some cases conveniently specified by a sequence of discrete points. A typical task of a manipulator transferring an object is illustrated in Figure 7.2. The manipulator is to move to the object, grasp it, pick it up, and move it to a new specified location. The desired path for the end-effector is then described by a sequence of discrete points $P_0 - P_1 - P_2 - P_3 - P_4 - P_5$. The straight line segments joining two adjacent points describe the desired trajectory as a polygonal path. The desired trajectory can similarly be described in the joint space; Figure 7.3 illustrates such a desired trajectory as a polygonal path for a joint variable.

In order for a manipulator to move smoothly during the transition from one line segment to the next one, the ends of the line segments may be bent around the corner points. The functions describing the transitions are to be

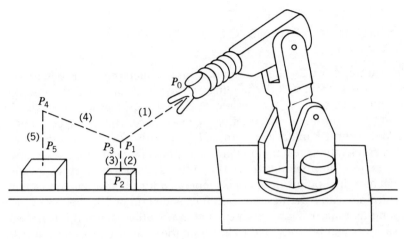

FIGURE 7.2
Manipulator end-effector is to move $P_0-P_1-P_2-P_3-P_4-P_5$ to transfer an object.

splined together with the appropriate polygonal line segments by satisfying certain well-defined continuity conditions. The entire desired trajectory can thus be described by a sequence of functions splined together so that a smooth motion for the end-effector is achieved.

We will first illustrate the generation of the desired straight line trajectory between two given points by assuming that the velocity profile has a special shape. We will then describe an approach to determine the bent trajectory around the corner points to achieve a smooth transition from one line segment to the next one.

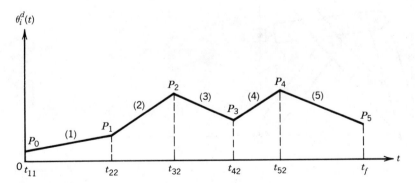

FIGURE 7.3
Desired trajectory $\Theta_i^d(t)$ for joint variable i may consist of polygonal line segments.

7.2.1 DESIRED TRAJECTORY BASED ON SPECIFICATIONS OF POSITION AND VELOCITY

The desired trajectory for the motion of a robotic manipulator is determined on the basis of a given task. Depending on the specifications of the task and control, the generation of the desired trajectory is performed either in the Cartesian base coordinate system or in the joint space. The task specifications may define only the initial and terminal states. Also the maximum speeds and even accelerations of a manipulator in certain directions are often known and/or given by the manufacturer. The entire desired trajectory connecting the initial and final points needs to be generated for the task while satisfying the constraints. After the desired trajectory is determined in one space, the corresponding trajectory can be calculated in the other space, if it is needed. Two examples will be presented to illustrate the planning procedure first for the Cartesian base coordinates, and then for the joint variables.

EXAMPLE 7.2.1

The planar revolute manipulator shown in Figure 7.4a has the end-effector initially at the location 1 given as $[p_{x_0}^d(0) \quad p_{y_0}^d(0) \quad p_{z_0}^d(0)] = [30 \text{ cm} \quad 0 \text{ cm} \quad 0 \text{ cm}]$ expressed in the base coordinates $(x_0 \ y_0 \ z_0)$. It is to move in $t_f = 2$ s along a vertical straight line to the position 3 specified as $[p_{x_0}^d(t_f) \quad p_{y_0}^d(t_f) \quad p_{z_0}^d(t_f)] = [30 \text{ cm} \quad 30\sqrt{3} \text{ cm} \quad 0 \text{ cm}]$. The lengths of the

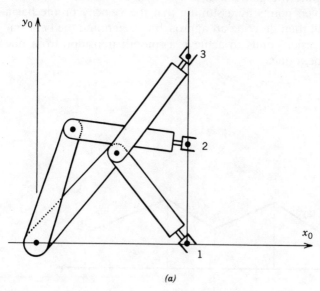

(a)

FIGURE 7.4a

Revolute planar manipulator in Example 7.2.1; motion 1–2–3.

links are $\ell_1 = \ell_2 = 30$ cm. The problem is to generate the desired trajectory for a controller both in the Cartesian base coordinate system and in the joint space ($q_1 = \theta_1$, $q_2 = \theta_2$) when the sampling period T is 20 ms (milliseconds).

The profile of the (total) velocity $v_0^d = v_0^d(t)$ of the end-effector in the Cartesian base coordinate system is chosen to have the shape of an equilateral trapezoid as shown in Figure 7.4b. The time interval for the nonzero acceleration lasts $t_2 - t_1 = 0.2$ s, that is, 10 sampling periods. The maximum velocity of the end-effector is denoted by v_0^m.

Since the distance traveled by the end-effector is equal to the integral of the velocity over time, the area of the trapezoid in Figure 7.4b equals the distance D between the initial and terminal points, that is,

$$v_0^m \frac{1.6 + 2.0}{2} s = D \tag{7.2.1}$$

where $D = \{[p_{x_0}^d(t_f) - p_{x_0}^d(0)]^2 + [p_{y_0}^d(t_f) - p_{y_0}^d(0)]^2 + [p_{z_0}^d(t_f) - p_{z_0}^d(0)]^2\}^{1/2}$. Equation 7.2.1 can be solved for v_0^m, giving $v_0^m = 50\sqrt{3}/3$ (cm/s).

The v_0^d-vector is next decomposed into the components $v_{x_0}^d$, $v_{y_0}^d$, and $v_{z_0}^d$ along the x_0-, y_0-, and z_0-axes, respectively. Since the path is a straight line in the base coordinates, the decomposition is accomplished by using the angles that the velocity vector v_0^d forms with the coordinate axes. Thus

$$v_{x_0}^d = v_0^d \frac{p_{x_0}^d(t_f) - p_{x_0}^d(0)}{D} \tag{7.2.2}$$

$$v_{y_0}^d = v_0^d \frac{p_{y_0}^d(t_f) - p_{y_0}^d(0)}{D} \tag{7.2.3}$$

$$v_{z_0}^d = v_0^d \frac{p_{z_0}^d(t_f) - p_{z_0}^d(0)}{D} \tag{7.2.4}$$

Each velocity component specified by Equations 7.2.2–7.2.4 has trapezoidal shape and is continuous in time, since v_0^d has these characteristics. Moreover, the initial and final velocities of all components are zero. In the particular problem under consideration, $v_{x_0}^d = 0$, $v_{y_0}^d(t) = v_0^d(t)$, and $v_{z_0}^d = 0$ for $t \in [0,2]$.

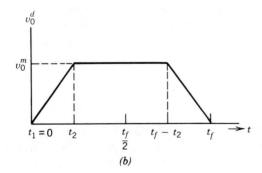

FIGURE 7.4b
Velocity profile of manipulator
in Example 7.2.1.

The acceleration a_0^d is determined by the slope of the velocity graph in Figure 7.4b:

$$a_0^d = \begin{cases} \dfrac{250\sqrt{3}}{3}\,\text{cm/s}^2 & \text{for}\quad 0\ \le t < 0.2\ \text{s} \\[2mm] 0 & \text{for}\ 0.2\ \text{s} \le t < 1.8\ \text{s} \qquad (7.2.5) \\[2mm] \dfrac{-250\sqrt{3}}{3}\,\text{cm/s}^2 & \text{for}\ 1.8\ \text{s} \le t < 2.0\ \text{s} \end{cases}$$

Equation 7.2.5 and Figure 7.4c reveal that the acceleration a_0^d versus time is discontinuous, exhibiting jumps at times 0.2 s and 1.8 s. The acceleration vector a_0^d is decomposed into components $a_{x_0}^d$, $a_{y_0}^d$, and $a_{z_0}^d$ along the x_0-, y_0-, and z_0-axes, respectively, in the same manner as velocity v_0^d. Thus,

$$a_{x_0}^d = a_0^d\,\frac{p_{x_0}^d(t_f) - p_{x_0}^d(0)}{D} \qquad (7.2.6)$$

$$a_{y_0}^d = a_0^d\,\frac{p_{y_0}^d(t_f) - p_{y_0}^d(0)}{D} \qquad (7.2.7)$$

$$a_{z_0}^d = a_0^d\,\frac{p_{z_0}^d(t_f) - p_{z_0}^d(0)}{D} \qquad (7.2.8)$$

The components of the acceleration given in Equations 7.2.6–7.2.8 have the same characteristics as the total acceleration vector a_0^d.

The desired velocity and acceleration components are now known by Equations 7.2.2–7.2.8. To complete the description of the desired trajectory, the desired discrete positions $p_0^d(kT)$ are next determined. Vector $p_0^d(k) = [p_{x_0}^d(k)\ \ p_{y_0}^d(k)\ \ p_{z_0}^d(k)]'$ is computed at each sampling instant kT, where $k =$

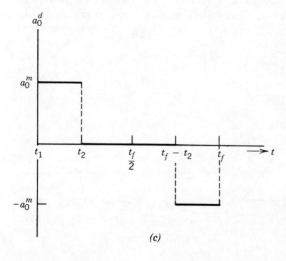

FIGURE 7.4c
Acceleration profile of manipulator in Example 7.2.1.

0, 1, ... , 100, and the sampling period T is dropped in the arguments. The components of the position vector $p_0^d(k)$ are determined by the following relations:

$$p_{x_0}^d(k + 1) = p_{x_0}^d(k) + Tv_{x_0}^d(k) + a_{x_0}^d(k)T^2/2 \qquad (7.2.9)$$

$$p_{y_0}^d(k + 1) = p_{y_0}^d(k) + Tv_{y_0}^d(k) + a_{y_0}^d(k)T^2/2 \qquad (7.2.10)$$

$$p_{z_0}^d(k + 1) = p_{z_0}^d(k) + Tv_{z_0}^d(k) + a_{z_0}^d(k)T^2/2 \qquad (7.2.11)$$

The desired trajectory at discrete points in the Cartesian base coordinate system is now known at the sampling instances: Equations 7.2.9–7.2.11 specify the desired position vector $p_0^d(k)$, Equations 7.2.2–7.2.4 the desired velocity vector $v_0^d(k) = [v_{x_0}^d(k) \; v_{y_0}^d(k) \; v_{z_0}^d(k)]'$, Equations 7.2.6–7.2.8 the acceleration vector $a_0^d(k) = [a_{x_0}^d(k) \; a_{y_0}^d(k) \; a_{z_0}^d(k)]'$.

The desired discrete-time trajectory is next determined in the joint space by converting the desired trajectory points in the Cartesian space to the corresponding points in the joint space. The independent joint variables $q_1^d(k) = \theta_1^d(k)$ and $q_2^d(k) = \theta_2^d(k)$, $0 \le k \le 100$ are computed by means of the kinematic equations and the foregoing desired trajectory known in the Cartesian space. For the planar manipulator in question, the inverse solution to the kinematic equations determine the expressions for θ_1^d and θ_2^d as given by Equations 3.1.23 and 3.1.24.

$$\theta_1^d = \tan^{-1}\left\{ \sqrt{\frac{4\ell_1^2[(p_{x_0}^d)^2 + (p_{y_0}^d)^2] - [(p_{x_0}^d)^2 + (p_{y_0}^d)^2 + \ell_1^2 - \ell_2^2]^2}{\pm[(p_{x_0}^d)^2 + (p_{y_0}^d)^2 + \ell_1^2 - \ell_2^2]}} \right.$$

$$\left. + \tan^{-1}\left[\frac{p_{y_0}^d}{p_{x_0}^d}\right] \right\} \pm \pi \qquad (7.2.12)$$

$$\theta_2^d = \tan^{-1}\left[\frac{-p_{x_0}s_1^d + p_{y_0}c_1^d}{p_{x_0}c_1^d + p_{y_0}s_1^d - \ell_1}\right] \pm \pi \qquad (7.2.13)$$

where $s_1^d = \sin\theta_1^d$, $c_1^d = \cos\theta_1^d$, and the angle $\pm\pi$ rad has been added to emphasize the degeneracy of the solution. Equations 7.2.12 and 7.2.13 are valid at all times, and specifically at the sampling instances. Thus, sequence $\{p_0^d(k)\}$ can be converted to the desired sequence $\{\theta^d(k)\}$ of points $\theta^d(k) = [\theta_1^d(k) \; \theta_2^d(k)]'$ in the joint space.

The desired angular velocity vector $\dot{\theta}^d(k) = [\dot{\theta}_1^d(k) \; \dot{\theta}_2^d(k)]'$ can be computed either from the desired position vectors using a sufficiently accurate approximation for the derivatives (see Section 6.1.3), or from Equations 7.2.2–7.2.4 by means of the Jacobian matrix as discussed in Section 4.2. If the desired acceleration vector is needed, necessary equations are obtained from Equation 7.2.5 and the Jacobian matrix (see Problem 4.10b). Thus, the desired discrete-time trajectory that is first determined in the Cartesian base coordinate system can be converted to the desired discrete-time trajectory in the joint space.

In the problem discussed, the duration of the motion is specified in the problem, and the maximum velocity is computed. The designer should check that this velocity is attainable by the manipulator. The manufacturer of a manipulator often specifies the maximum velocities of the end-effector and the joints. An alternative approach to the trajectory planning problem is obtained, if the maximum velocities and accelerations are assumed when the manipulator moves from an initial position to a given terminal location. If the velocity profile has again a trapezoidal shape, the duration t_f of the motion is first calculated. Then, the desired trajectory can be determined as discussed.

EXAMPLE 7.2.2

In the six-joint revolute manipulator PUMA 600, the initial and the final positions and orientations of the gripper correspond to the following readings of the joints [expressed in encoder units (eu)]: $[\theta_1(0) \quad \theta_2(0)$ $\theta_3(0) \quad \theta_4(0) \quad \theta_5(0) \quad \theta_6(0)]$ = [10000 12000 12000 12000 10000 15000]; and $[\theta_1(t_f) \quad \theta_2(t_f) \quad \theta_3(t_f) \quad \theta_4(t_f) \quad \theta_5(t_f) \quad \theta_6(t_f)]$ = [15000 12000 14500 14500 10000 17500]. In the encoders used here, 65536 eu correspond to 2π rad. Each joint variable is to follow a straight line from the initial value to the final value [4]. Moreover, *all* joints start moving at time $t = 0$ and stop moving at time $t_f = 60$ sampling periods, where the sampling period T equals 28 ms. For each joint variable, the velocity profile $\dot{\theta}_i = \dot{\theta}_i(t), i = 1, \ldots, 6$ is chosen to have the shape of an equilateral trapezoid, as shown in Figure 7.5a, where the duration of the acceleration (deceleration) is $10T$. The maximum attainable angular velocity in each joint is assumed to be 100 eu/sampling period.

Since the angular position θ_1 of joint 1 will change the largest amount (the change in θ_1 is $\Delta\theta_1 = 5000$ eu), the maximum attainable angular velocity is applied to this joint. Thus, $\dot{\theta}_1^m = 100$ eu/T. Therefore, the maximum angular acceleration is $\ddot{\theta}_1^m = (100\text{eu}/T)/(10T) = 10\text{eu}/T^2$.

The discrete-time values of the angular acceleration, velocity, and position for joint i are determined by the following equations, in which the sampling period T is dropped in the arguments:

$$\ddot{\theta}_i(k) = \begin{cases} \ddot{\theta}_i^m & \text{for } k = 0, \ldots, 9 \\ 0 & \text{for } k = 10, \ldots, 49 \\ -\ddot{\theta}_i^m & \text{for } k = 50, \ldots, 59 \end{cases} \quad (7.2.14)$$

$$\dot{\theta}_i(k+1) = \dot{\theta}_i(k) + \ddot{\theta}_i(k)T \quad k = 0, \ldots, 59 \quad (7.2.15)$$

$$\theta_i(k+1) = \theta_i(k) + \dot{\theta}_i(k)T + \frac{1}{2}\ddot{\theta}_i(k)T^2 \quad k = 0, \ldots, 59 \quad (7.2.16)$$

Equations 7.2.14–7.2.16 specify the acceleration, velocity and position sequences for joint 1, when $i = 1$. The initial velocity and position are $\dot{\theta}_1(0) = 0$ and $\theta_1(0) = 10000$.

The changes in the angular positions of joints 3, 4, and 6 are 2500 eu. Since the duration of the motion of each joint must be the same, it follows

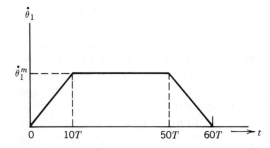

FIGURE 7.5a
Velocity profile of joint variable 1; Example 7.2.2.

that the maximum angular velocity $\dot{\theta}_i^m$, $i = 3$, 4, and 6 equals 50 eu/T. The maximum acceleration is $\ddot{\theta}_i^m = (50\text{eu}/T)(10T) = 5$ eu/T^2, which is half of the value of $\ddot{\theta}_1^m$. For $i = 3$, 4, and 6, the angular acceleration is specified by

$$\ddot{\theta}_i(k) = \begin{cases} 5eu/T^2 & \text{for } k = 0, \ldots, 9 \\ 0 & \text{for } k = 10, \ldots, 49 \\ -5eu/T^2 & \text{for } k = 50, \ldots, 59 \end{cases} \quad (7.2.17)$$

The angular velocities are given by Equation 7.2.15 for $i = 3$, 4, and 6 with $\dot{\theta}_i(0) = 0$. Similarly, the angular position $\theta_i(k)$ for $i = 3$, 4, and 6 is calculated using Equation 7.2.16, but now the initial conditions are $\theta_3(0) = \theta_4(0) = 12000$ eu for $i = 3$ and 4, and $\theta_6(0) = 15000$ eu for $i = 6$. The velocity profile of $\dot{\theta}_3$ is shown in Figure 7.5b. It is also an equilateral trapezoid. The computed maximum velocity and acceleration make the manipulator complete the motion in the specified time. The velocities of joints 4 and 6 follow the same profile as that shown in Figure 7.5b. The angular positions of joints 2 and 5 remain unchanged.

In Example 7.2.2, the desired trajectory is directly expressed in encoder units. The desired trajectory in Example 7.2.1 is described in terms of the joint variables given in radians. In a control program, these values must be expressed in consistent units so that they will be compatible with the readings of the positional indicators and those used in the control algorithm.
The readings for the initial and final values of the joint variables in Example 7.2.2 are obtained by placing the end-effector of the manipulator to the

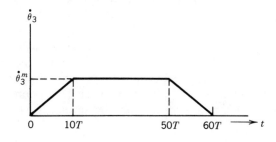

FIGURE 7.5b
Velocity profile of joint variable 3; Example 7.2.2.

desired locations and orientations. Thus, the manipulator is taught off-line to assume the specified positions and orientations (by doing so). Intermediate points can also be read by this "learning by doing" procedure from the encoders along the trajectory that the manipulator is made to follow. The values of the desired velocities are difficult to achieve by this approach.

The desired trajectories generated in Examples 7.2.1 and 7.2.2 in the work (world) space and in the joint space represent straight line segments that join the initial and terminal points. In addition to specifying the initial and final points for the desired trajectory, the task may require that the designer (or the operator) determine other points via which the end-effector should move; these points may be called the via or knot points. By connecting any two adjacent points, a polygonal path is formed that can be used to describe the entire desired trajectory. Because the velocity at the corner (knot) points that join two adjacent line segments together is usually discontinuous, the corresponding motion may not be sufficiently smooth. The desired trajectory can specifically be designed around the corner points for a smooth transition. Such a planning procedure is described next.

7.2.2   DESIRED TRAJECTORY FOR SMOOTH TRANSITION FROM ONE LINE SEGMENT TO THE NEXT

If a planning algorithm is constructed by applying the approach of Examples 7.2.1 and 7.2.2 to each segment of the desired trajectory using the assumed velocity profile, then the speed of the end-effector at the via points would be equal to zero. The resulting motion as a whole would be inefficient and could lead to troublesome vibrations in the neighborhood of the via points. Therefore, it is advisable to design the desired path separately for the regions around the specified points. Several approaches have been proposed to obtain a smooth motion during the transition from one line segment to the next one on a polygonal path.

Suppose that the desired trajectory $\theta_i^d(t)$ for joint variable i includes straight line segments $(j-1)$, (j), and $(j+1)$, for $j = 2, \ldots, n-2$, as shown in Figure 7.6a. The velocity of the joint variable moving along a line segment is constant determined by the slope of the line. The velocity profile corresponding to the desired trajectory $\theta_i^d(t)$ is displayed in Figure 7.6b; thus, the velocity function $\dot{\theta}_i^d(t)$ is discontinuous at time t_{j2} (point P_j in Figure 7.6a). The problem is to change the desired straight line path about point P_j in Figure 7.6a so as to achieve a smooth transition from one line segment to the next.

During the transition interval $[t_{j1}, t_{j3}]$, the speed of the motion is assumed to change linearly between the two constant velocity values associated with the straight line motion along segments $(j-1)$ and (j). Thus, the jump in the velocity occurring at t_{j2} is replaced by a gradually changing velocity profile that corresponds to a constant acceleration. The mathematical description for the velocity and position of the joint variable can then be developed [1, 2, 3].

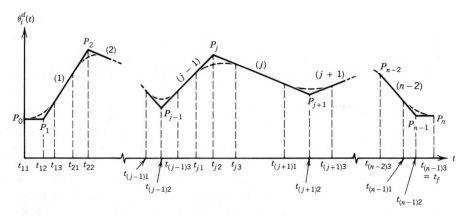

FIGURE 7.6a

Straight line segments joining discrete points determine desired trajectory for joint i variable.

The velocity of joint i on the line segment (j) is assumed to be constant denoted by $\dot{\theta}_i{}^j$. Moreover, the maximum (minimum) achievable value $\ddot{\theta}_i^m$ of the acceleration (deceleration) can be used on the joint during the transition from one line segment to the next one. The problem is then to determine the time instants t_{j1} and t_{j3} for the beginning and end of the trajectory bending and the equation for the position of the joint variable during the transition period.

The equation for the velocity of joint variable i can be written on the basis of Figures 7.6a and 7.6b:

$$\dot{\theta}_1^{j-1} = \left[\theta_1(t_{j2}) - \theta(t_{(j-1)2})\right]/(t_{j2} - t_{(j-1)2}) \qquad t_{(j-1)3} \leq t < t_{j1} \qquad (7.2.18)$$

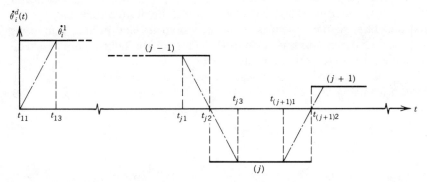

FIGURE 7.6b

Velocity profile of joint i variable before and after smoothing (bending).

$$\dot{\theta}_i(t) = \dot{\theta}_i^{j-1} + \ddot{\theta}_i(t - t_{j1}) \qquad t_{j1} \leq t \leq t_{j3} \tag{7.2.19}$$

where $j = 2, \ldots, (n-2)$ and $\ddot{\theta}_i = |\ddot{\theta}_i^m|$ if $\dot{\theta}_i^{j-1} < \dot{\theta}_i^j$, and $-|\ddot{\theta}_i^m|$ if $\dot{\theta}_i^{j-1} > \dot{\theta}_i^j$, that is, $\ddot{\theta}_i = -|\ddot{\theta}_i^m| \, \mathrm{sgn} \, (\dot{\theta}_i^{j-1} - \dot{\theta}_i^j)$. By evaluating Equation 7.2.19 at t_{j3}, the resulting equation may be solved for the duration $(t_{j3} - t_{j1})$ of the transition to obtain

$$t_{j3} - t_{j1} = [\dot{\theta}_i^j - \dot{\theta}_i^{j-1}]/\ddot{\theta}_i^m \tag{7.2.20}$$

where variable $\dot{\theta}_i^{j-1}$ represents the velocity of joint i on the line segment $(j-1)$. Equation 7.2.20 determines the duration of the transition (bending) of the trajectory. It follows that $t_{j1} = t_{j2} - (t_{j3} - t_{j1})/2$ and $t_{j3} = t_{j2} + (t_{j3} - t_{j1})/2$ if velocity at t_{j2} is assumed to be the average of velocities $\dot{\theta}_i^{j-1}$ and $\dot{\theta}_i^j$.

The duration of the transition depends on the value of the acceleration (deceleration). It is assumed that time $(t_{j2} - t_{j1}) + (t_{(j-1)3} - t_{(j-1)2})$ is shorter than the duration $(t_{j2} - t_{(j-1)2})$ of the movement along the line segment $(j-1)$. The position of joint i is then governed by the equations that result from Equations 7.2.18 and 7.2.19:

$$\theta_i(t) = \theta_i(t_{(j-1)3}) + \dot{\theta}_i^{j-1}(t - t_{(j-1)3}) + \ddot{\theta}(t)(t - t_{j1})^2/2 \qquad t_{(j-1)3} \leq t \leq t_{j3} \tag{7.2.21}$$

where the joint acceleration $\ddot{\theta}_i$ is

$$\ddot{\theta}_i(t) = \begin{cases} 0 & t_{(j-1)3} \leq t < t_{j1} \\ \ddot{\theta}_i^m & t_{j1} \leq t < t_{j3} \end{cases} \tag{7.2.22}$$

Equation 7.2.21 describes a parabola that is splined at point t_{j1} together with the line segment $(j-1)$ and at t_{j3} with the line segment (j) so that $\theta_i(t)$ and $\dot{\theta}_i(t)$ are continuous. This continuous curve will enhance the smoothness of the motion on the transition interval as compared with the original trajectory.

When the entire trajectory consists of straight line segments as shown in Figure 7.6a, the bending of the trajectory around the via points for a smooth motion can be designed on the basis of Equations 7.2.18–7.2.22. Since the desired trajectories for the initial and final segments are straight lines, they also give rise in general to discontinuous velocity profiles at the initial and final times. To circumvent these discontinuities, the desired trajectory for the initial and final segments can be bent as shown in Figure 7.6a. Equations 7.2.18–7.2.22 describing the desired trajectory for the bent regions need to be modified for these particular intervals.

The motion of the manipulator is assumed to start at $t_{11} = 0$ with the velocity increasing at a constant acceleration from zero up to time t_{13} after which the speed stays at a constant level of $\dot{\theta}_i^1$, as displayed in Figure 7.5b. At $t = t_{13}$, the velocity of the end-effector is

$$\dot{\theta}_i^1 = \ddot{\theta}_i^m(t_{13} - t_{11})$$

$$= [\theta_i(t_{22}) - \theta_i(t_{12})]/(t_{22} - t_{12}) \tag{7.2.23}$$

where $\dot{\theta}_i^1$ is the slope of the line segment (1) through the point $[t_{22}, \theta_i(t_{22})]$, and t_{12} is such that the fictitious knot point $[t_{12}, \theta_i(0)]$ lies on this line. By assuming that $t_{12} = t_{13}/2$, Equation 7.2.23 can be solved for t_{13}:

$$t_{13} = t_{22} - \sqrt{t_{22}^2 - 2\left[\theta_i(t_{22}) - \theta_i(t_{12})\right]/\ddot{\theta}_i^m} \qquad (7.2.24)$$

where $\theta_i(t_{12}) = \theta_i(0)$. Time interval $(t_{13} - t_{12}) + (t_{22} - t_{21})$ is assumed to be shorter than the duration $(t_{22} - t_{12})$ of the motion.

The desired trajectory during the initial bending is determined by

$$\theta_i(t) = \theta_i(0) + \ddot{\theta}_i t^2/2 \qquad 0 \le t \le t_{13} \qquad (7.2.25)$$

where $\ddot{\theta}_i$ is specified in conjunction with Equation 7.2.19. The trajectory in Equation 7.2.25 is splined together with the straight line (1) at time t_{13} so that $\theta_i(t)$ and $\dot{\theta}_i(t)$ are continuous at this point.

The motion during the last segment can also be described by equations similar to Equations 7.2.23–7.2.25. Velocity $\dot{\theta}_i(t)$ is assumed to decrease linearly to zero over the interval $[t_{(n-1)1}, t_{(n-1)3}]$, where $t_{(n-1)3} = t_f$, the final time. On the basis of Figure 7.6a, the velocity relation at $t_{(n-1)1}$ is

$$\dot{\theta}_i^{n-2} = \ddot{\theta}_i^m(t_{(n-1)1} - t_f)$$

$$= \left[\theta_i(t_{(n-2)2}) - \theta_i(t_{(n-1)2})\right]/(t_{(n-2)2} - t_{(n-1)2}) \qquad (7.2.26)$$

since $t_{(n-1)2} = (t_f + t_{(n-1)1})/2$. Equation 7.2.26 can be solved for the duration of the bending, that is, $(t_f - t_{(n-1)1})$ to obtain

$$(t_f - t_{(n-1)1}) = (t_f - t_{(n-2)2}) - \sqrt{(t_f - t_{(n-2)2})^2 + 2\left[\theta_i(t_f) - \theta_i(t_{(n-2)2})\right]/\ddot{\theta}_i^m} \qquad (7.2.27)$$

where the $(n-1)$st knot point is $[t_{(n-2)2}, \theta_i(t_{(n-2)2})]$ and the last knot point is $[t_f, \theta_i(t_f)]$. Time interval $(t_f - t_{(n-1)1}) + (t_{(n-2)3} - t_{(n-2)2})$ is assumed to be shorter than time $(t_f - t_{(n-2)2})$ between the endpoint and the $(n-1)$st knot point.

The desired trajectory corresponding to the constant acceleration that determines the bending is

$$\theta_i(t) = \theta_i(t_f) + \ddot{\theta}_i(t - t_f)^2/2 \qquad t_{(n-1)1} \le t \le t_f \qquad (7.2.28)$$

Equation 7.2.28 describes a parabola that is splined with the straight line at time $t_{(n-1)1}$. It connects the last straight line segment to the final point.

In order to achieve a smooth motion, the desired trajectory can be planned so that the straight lines joining the via points are bent (rounded) around the corner points that join two adjacent line segments. Equations 7.2.23–7.2.25 determine the bent trajectory for the initial segment when the joint variable starts at a given initial value. The bending of the desired trajectory is followed by a straight line segment on which the velocity is constant. Equations 7.2.18–7.2.22 determine the variables for the desired bent trajectory around the intermediate knot points. Equations 7.2.26–7.2.28 specify the bending of the desired trajectory on the last segment. As indicated in Figure 7.6a, the desired trajectory does not usually pass through the knot point, unless the procedure described is modified, for example, by using fictitious knot points.

The foregoing procedure for the smoothing of the trajectory around the via points can be applied equally well in the joint space as in the Cartesian (world) space.

The procedure described for the bending of the desired trajectory is based on the following information:

1. The knot (via) points are given to the designer, and two adjacent points are connected by a line segment.

2. The velocities on the line segments joining two adjacent knot points are constant.

3. The acceleration (deceleration) used on the bent trajectory is maximum (minimum) and constant.

4. The positions and velocities of the joint variables are required to be continuous throughout the desired trajectory, particularly, at the points where the line segments and the bent trajectories are splined together.

Based on 1–4, the desired trajectory for each joint variable can be planned as a sequence of functions splined together. The positional functions are composed of quadratic and linear time functions in the procedure described. The desired positions and velocities are continuous over the duration of the motion, but the accelerations of the joint variables will, in general, be discontinuous.

The generation of the desired trajectory by the procedure of bending the positional path around the knot points may be considered as a generalization of the approach presented in Examples 7.2.1 and 7.2.2. In these examples, the initial and final points and in addition, the trapezoidal velocity profile over the duration of the motion are known. Both methods described generate a desired trajectory with acceleration discontinuities, which may lead to undesirable vibrations in the motion of a manipulator in some applications.

The bending of the positional trajectory is introduced here so that discontinuities in the velocities are avoided. This is necessary for a manipulator to usually achieve a smooth motion. The same procedure could also be applied to the velocity profile, such as shown in Figure 7.6b. The transition from one linear line segment to the next on a velocity profile can be described by means of quadratic time functions; thus, the velocity profile is bent around the corner points. The graph for the acceleration of a joint variable versus time is then of the form shown in Figure 7.6c that is similar to the velocity graph in Figure 7.6b. As the consequence, the positional joint variable would be described by a sequence of third- and second-order polynomials that are splined together by satisfying appropriate continuity conditions.

We will next describe the use of third-order polynomial spline functions that are chosen to directly describe the desired trajectory for a positional joint variable.

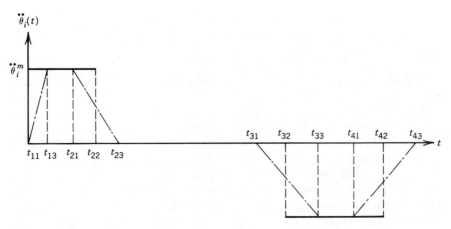

FIGURE 7.6c
Acceleration profile of joint i variable before and after smoothing (bending).

7.3 SPLINE FUNCTIONS DESCRIBE DESIRED TRAJECTORY FOR MANIPULATOR MOTION

When the velocity profile for a manipulator is constrained to be of trapezoidal shape, the position and the velocity of the end-effector are continuous in the Cartesian base coordinate system and in the joint space. The acceleration, however, is discontinuous in time exhibiting jumps at the vertices of the velocity trapezoid. This results in motion that may not be sufficiently smooth in some applications. The discontinuities in the acceleration can be avoided by a different design. For example, the velocity profile may be selected so that it possesses a continuous derivative at all points as discussed at the end of Section 7.2.2. Alternatively, a function describing the desired positional trajectory can be required to have continuous second derivatives.

In the following, we will discuss approximate expressions for a polygonal path connecting given via points. These functions can be appropriately splined together to describe the entire trajectory. Specifically, we will discuss the use of polynomials to approximate the desired trajectory between the given via (knot) points. These functions are concatenated by requiring that the positions, velocities, and accelerations are continuous at the points where the splining occurs. The sequence of these functions splined to each other represents the desired trajectory that will represent a smooth motion of the manipulator.

7.3.1 THIRD- AND FIFTH-ORDER POLYNOMIAL SPLINE FUNCTIONS FOR DESCRIBING DESIRED TRAJECTORY

When the joint variables of a manipulator with N joints are controlled, the desired trajectory should be specified for the joint variables. The initial and terminal positions and orientations of the end-effector in the manipulator are assumed to be given in the Cartesian base coordinate system by means of the homogeneous transformation matrix $T_0^N(t)$ that is evaluated at times t_1 and t_n. These endpoints $T_0^N(t_1)$ and $T_0^N(t_n)$ are converted into the joint space onto points $\{q_i(t_1)\}$ and $\{q_i(t_n)\}$, $i = 1, \ldots, N$. When the straight line segment joining the initial and terminal positions of the end-effector in a certain orientation is transformed into the joint space, the graph connecting $q_i(t_1)$ to $q_i(t_n)$ of joint i is usually not a straight line, but a curve. A function can be introduced to approximate this curve sufficiently closely. Polynomials in time t are common choices for such a function.

The order of the polynomial chosen to represent the desired trajectory in the joint space depends on the specifications of the manipulator task. When the position and velocity of a joint variable are given at the initial and terminal times, these four conditions can be satisfied by a third-order polynomial with four unknown parameters. If the acceleration is also given at those times, the six conditions to be satisfied suggest a fifth-order polynomial with six unknown parameters.

Suppose that the initial values of the position and velocity for joint i are $q_i(t_1)$ and $\dot{q}_i(t_1)$, respectively, $i = 1, \ldots, N$ and the corresponding values at the terminal time $q_i(t_n)$ and $\dot{q}_i(t_n)$. A polynomial of order three is needed to satisfy the four aforementioned constraints.

Suppose that the desired trajectory joining the initial and final points of the joint i variable is represented as the third-order polynomial:

$$Q_1^1(t) = a_0 + a_1t + a_2t^2 + a_3t^3 \tag{7.3.1}$$

where $t_1 \leq t \leq t_n$ and parameter $a_j, j = 0, 1, 2, 3$ is to be determined so as to satisfy the initial and terminal conditions.

When the expression in Equation 7.3.1 and its derivative are evaluated at t_1 and t_n, the following equations result:

$$Q_i^1(t_1) = a_0 + a_1t_1 + a_2t_1^2 + a_3t_1^3 = q_i(t_1) \tag{7.3.2}$$

$$\dot{Q}_i^1(t_1) = a_1 + 2a_2t_1 + 3a_3t_1^2 = \dot{q}_i(t_1) \tag{7.3.3}$$

$$Q_i^1(t_n) = a_0 + a_1t_n + a_2t_n^2 + a_3t_n^3 = q_i(t_n) \tag{7.3.4}$$

$$\dot{Q}_i^1(t_n) = a_1 + 2a_2t_n + 3a_3t_n^2 = \dot{q}_i(t_n) \tag{7.3.5}$$

Equations 7.3.2–7.3.5 represent four equations in four unknown parameters. A unique solution for the unknown parameters can be obtained since determinant Δ consisting of the coefficients of the unknown variables is equal to $\Delta = (t_n - t_1)^4$. The solution to Equations 7.3.2–7.3.5 is

$$a_0 = \{ (t_n - t_1) [q_i(t_1) t_n^2(t_n - 3t_1) - q_i(t_n) t_1^2(t_1 - 3t_n)]$$
$$+ (t_n - t_1)^2 t_1 t_n [-\dot{q}_i(t_1) t_n - \dot{q}_i(t_n) t_1]\} /\Delta \quad (7.3.6)$$

$$a_1 = \{ (t_n - t_1) t_1 t_n [6q_i(t_1) - 6q_i(t_n)] + (t_n - t_1)^2$$
$$\times [\dot{q}_i(t_1) t_n(t_n + 2t_1) + \dot{q}_i(t_n) t_1(t_1 + 2t_n)]\} /\Delta \quad (7.3.7)$$

$$a_2 = \{ (t_n - t_1)(t_n + t_1)[-3q_i(t_1) + 3q_i(t_n)] + (t_n - t_1)^2$$
$$\times [-\dot{q}_i(t_1)(t_1 + 2t_n) - \dot{q}_i(t_n)(t_n + 2t_1)]\} /\Delta \quad (7.3.8)$$

$$a_3 = \{ (t_n - t_1)[2q_i(t_1) - 2q_i(t_n)] + (t_n - t_1)^2[\dot{q}_i(t_1) + \dot{q}_i(t_n)]\} /\Delta \quad (7.3.9)$$

Thus, polynomial $Q_i^1(t)$ in Equation 7.3.1 is now specified.

Coefficients a_i, $i = 0, \dots, 3$ given by Equations 7.3.6–7.3.9 can be substituted into Equation 7.3.1. The resulting expression can be simplified considerably by introducing a normalized time $t' = (t-t_1)/(t_n-t_1)$. By eliminating t in Equation 7.3.1, the resulting expression may be combined with Equations 7.3.6–7.3.9 to obtain

$$Q_i^1(t') = (1 - t')^2\{q_i(t_1) + [2q_i(t_1) + \dot{q}_i^s(t_1)]t'\} + (t')^2$$
$$\times \{q_i(t_n) + [2q_i(t_n) - \dot{q}_i^s(t_n)](1 - t')\} \quad (7.3.10)$$

where $q_i^s(t_j) = (t_n - t_1)\dot{q}_i(t_j), j = 1, n$ and the normalized time is such that $0 \le t' \le 1$. The third-order polynomial in Equation 7.3.10 approximates the desired trajectory between the two points $[t_1, q(t_1), \dot{q}(t_1)]$ and $[t_n, q(t_n), \dot{q}(t_n)]$. The values of intermediate points can be calculated by Equation 7.3.10.

If the designer specifies several via points for the desired trajectory, Equation 7.3.10 can be applied to generate trajectory points between any two adjacent via points. For example, suppose that two successive via points $[t_k, q_i(t_k), \dot{q}_k(t_k)]$ and $[t_{k+1}, q_i(t_{k+1}), \dot{q}_i(t_{k+1})]$ for joint i are specified. The approximate expression $Q_i^k(t')$ for the desired trajectory can then be written from Equation 7.3.10 by replacing 1 by k:

$$Q_i^k(t') = (1 - t')^2\{q_i(t_k) + [2q_i(t_k) + \dot{q}_i^s(t_k)]t'\}$$
$$+ (t')^2\{q_i(t_{k+1}) + [2q_i(t_{k+1}) - \dot{q}_i^s(t_{k+1})](1 - t')\} \quad (7.3.11)$$

where $\dot{q}_i^s(t_j) = (t_{k+1} - t_k)\dot{q}_i(t_j), j = k, k + 1, t' = (t - t_k)/(t_{k+1} - t_k)$, integer k satisfies $1 \le k \le (n - 1)$, and n is the number of the via points specified for the desired trajectory.

Equation 7.3.11 determines the desired trajectory between the two adjacent via points. It is a third-order polynomial that satisfies at times $t' = 0$ and 1 the given position and velocity specifications. Moreover, when polynomial $Q_i^k(t')$ is splined together with another adjacent function, for example, $Q_i^{k+1}(t')$, then the position and velocity specifications at the via point are met by both functions. That is, the function obtained by joining the two polynomials together at time t_{k+1} represents a continuous transition in the position and velocity through the via point at time t_{k+1} over time interval $[t_k, t_{k+2}]$. The polynomial sequence $\{Q_i^k(t')\}_k$ of the spline functions describes the desired trajectory for the ith joint.

In addition to position $q_i(t_k)$ and velocity $\dot{q}_i(t_k)$, the designer may also be given acceleration $\ddot{q}_i(t_k)$ at the via points. Then a function with six adjustable parameters will be used to describe the desired trajectory between the via points. Such a function can be chosen to be a fifth-order polynomial with six unknown coefficients. By following a procedure similar to the one described in determining the coefficients of the third-order polynomial in Equation 7.3.1, the coefficients of the fifth-order polynomial become specified. After determining the parameters, the polynomial may be expressed in the following form [2]:

$$Q_i^1 = (1 - t')^3 \left\{ q_i(t_1) + [3q_i(t_1) + \dot{q}_i^s(t_1)] \, t' + [\ddot{q}_i^s(t_1) + 6\dot{q}_i^s(t_1) \right.$$

$$\left. + 12q_i(t_1)] \frac{(t')^2}{2} \right\} + (t')^3 \left\{ q_i(t_n) + [3q_i(t_n) - \dot{q}_i^s(t_n)](1 - t') \right.$$

$$\left. + [\ddot{q}_i^s(t_n) - 6\dot{q}_i^s(t_n) + 12q(t_n)] \frac{(1 - t')^2}{2} \right\} \quad (7.3.12)$$

where $\dot{q}_i^s(t_j) = (t_n - t_1)\dot{q}_i(t_j)$, $\ddot{q}_i^s(t_j) = (t_n - t_1)^2 \ddot{q}_i(t_j)$, $j = 1, n, 0 \le t' \le 1$ and $t' = (t - t_1)/(t_n - t_1)$. Equation 7.3.12 corresponds to Equation 7.3.10. It represents the trajectory between two via points of the ith joint variable. It is readily checked that polynomial $Q_i^1(t')$ at $t' = 0$ and $t' = 1$ has the specified values of the position, and that the first and second time derivatives of $Q_i^1(t')$ satisfy the velocity and acceleration constraints, respectively, at the initial and terminal times. [Note that $dQ_i^1(t')/dt = dQ_i^1(t')/dt'(dt'/dt)$]. Intermediate values of the joint variable q_i can be computed using $Q_i^1(t')$; for example, at time $t' = 1/2$, the value of the joint variable $q_i[(t_n - t_1)/2] = Q_i^1(1/2)$.

If the task specifications give position $q_i(t_k)$, $k = 1, \ldots, n$, at n via points including the initial and terminal points and also velocities $\dot{q}_i(t_k)$ and accelerations $\ddot{q}_i(t_k)$, then a fifth-order polynomial similar to the one in Equation 7.3.12 can be used to describe the desired trajectory between any two adjacent knot points. Indeed, the polynomial expression for the kth time interval may be written as follows:

$$Q_i^k = (1 - t')^3 \left\{ q_i(t_k) + [3q_i(t_k) + \dot{q}_i^s(t_k)] \, t' + [\ddot{q}_i^s(t_k) + 6\dot{q}_i^s(t_k) \right.$$

$$\left. + 12q_i(t_k)] \frac{(t')^2}{2} \right\} + (t')^3 \left\{ q_i(t_{k+1}) + [3q_i(t_{k+1}) - \dot{q}_i^s(t_{k+1})](1 - t') \right.$$

$$\left. + [\ddot{q}_i^s(t_{k+1}) - 6\dot{q}_i^s(t_{k+1}) + 12q_i(t_{k+1})] \frac{(1 - t')^2}{2} \right\} \quad (7.3.13)$$

where integer k satisfies $1 \le k \le (n-1)$, $\dot{q}_i^s(t_j) = (t_{k+1} - t_k)\dot{q}_i(t_j)$, $\ddot{q}_i^s(t_j) = (t_{k+1} - t_k)^2 \ddot{q}_i(t_j)$, $j = k, k+1$, $t' \in [0,1]$, and $t' = (t - t_k)/(t_{k+1} - t_k)$.

Polynomial $Q_i^k(t')$ in Equation 7.3.13 is so constructed that at time t_k, or $t' = 0$, its value and the values of the first and second time derivatives equal $q_i(t_k)$, $\dot{q}_i(t_k)$, and $\ddot{q}_i(t_k)$, respectively. Moreover, the polynomial spline functions $Q_i^k(t')$ and $Q_k^{k+1}(t')$ satisfy at time t_{k+1} the continuity conditions $Q_i^k(1) = Q_i^{k+1}(0)$, $\dot{Q}_i^k(1) = \dot{Q}_i^{k+1}(0)$, and $\ddot{Q}_i^k(1) = \ddot{Q}_i^{k+1}(0)$ that guarantee a smooth transition of the motion through the via point. Thus, the desired trajectory for the variable of joint i is described by the sequence $\{Q_i^k(t')\}$ of the fifth-order polynomials that are splined together at the via points so that the position, velocity, and acceleration of the joint variable are continuous. Moreover, the spline functions also satisfy the specifications at the initial and terminal times. The number of the polynomials in the sequence is determined by the number of the via points. It is specified by the designer, usually so that Equation 7.3.11 or 7.3.13 when it is converted into the world coordinate system provides a sufficiently close approximation of a straight line.

The described method is applied next to the generation of the desired trajectory for a cylindrical robotic manipulator.

EXAMPLE 7.3.1

For a cylindrical robot, the workspace is defined in terms of r, θ, and z by specifying the ranges of these variables: 9 cm $\le r \le$ 30 cm, $0 \le \theta \le \pi/2$ rad, 10 cm $\le z \le$ 44 cm. A Cartesian coordinate system has been chosen so that the origin is located on the shoulder of the manipulator. The initial position of the gripper is represented by $[p_x(0) \quad p_y(0) \quad p_z(0)]' = [10\ \text{cm} \quad 0\ \text{cm} \quad 10\ \text{cm}]'$ in the Cartesian base coordinate system. The manipulator is to move from the initial point along the straight line to the terminal point $p(t_f) = [p_x(t_f) \quad p_y(t_f) \quad p_z(t_f)]' = [0\ \text{cm} \quad 30\ \text{cm} \quad 40\ \text{cm}]'$ in time $t_f = 4$ s. The Cartesian terminal points correspond to the joint space points $[r(0) \quad \theta(0) \quad z(0)]' = [10\ \text{cm} \quad 0° \quad 10\ \text{cm}]'$ and $[r(t_f) \quad \theta(t_f) \quad z(t_f)]' = [30\ \text{cm} \quad 90° \quad 40\ \text{cm}]'$.

The problem is to determine the sequences $\{Q_i^k(t')\}$ of the fifth-order polynomials ($i = 1, 2, 3$) that describe the desired trajectories for the joint variables r, θ, and z. The desired trajectory in the Cartesian space is a straight line drawn between the initial and terminal points. Five via points are introduced. The end-effector assumes the coordinates of these points at the times 0, 1, 2, 3, and 4 s. The velocities of the gripper at these times are $v_0^d(0) = 0$, $v_0^d(1) = v_0^d(2) = v_0^d(3) = 13.0$ cm/s, $v_0^d(4) = 0$, and the accelerations $a_0^d(0) = 0.1233$ cm/s^2, $a_0^d(1) = a_0^d(2) = a_0^d(3) = 0.0$, and $a_0^d(4) = 0.7107$ cm/s^2. The desired trajectories of the joints are to be generated by means of the fifth-order polynomial spline functions that have the form of Equation 7.3.13. The desired trajectory is generated only for the gripper position, and the orientation is not considered.

The generalized coordinate vector is $q(t) = [q_1(t) \quad q_2(t) \quad q_3(t)]' = [r(t) \quad \theta(t) \quad z(t)]'$. The relationships between the Cartesian base coordinates and the generalized coordinates (the joint variables), their velocities and accelerations are

$$p_x(t) = q_1(t) \cos [q_2(t)] \tag{7.3.14}$$

$$p_y(t) = q_1(t) \sin [q_2(t)] \tag{7.3.15}$$

$$p_z(t) = q_3(t) \tag{7.3.16}$$

$$v(t) = J[q(t)]\dot{q}(t) \tag{7.3.17}$$

$$\dot{v}(t) = \dot{J}[q(t)]\dot{q}(t) + J[q(t)]\ddot{q}(t) \tag{7.3.18}$$

The velocity and acceleration vectors as well as the Jacobian matrix are defined, respectively, as

$$v(t) = [\dot{p}_x(t) \quad \dot{p}_y(t) \quad \dot{p}_z(t)]' \qquad \dot{v}(t) = [\ddot{p}_x(t) \quad \ddot{p}_y(t) \quad \ddot{p}_z(t)]' \tag{7.3.19}$$

$$J[q(t)] = \begin{bmatrix} \cos [q_2(t)] & -q_1(t) \sin [q_2(t)] & 0 \\ \sin [q_2(t)] & q_1(t) \cos [q_2(t)] & 0 \\ 0 & 0 & 1 \end{bmatrix} \tag{7.3.20}$$

Thus, it follows that the position, velocity, and acceleration vectors of the joints are

$$[q_1(t) \quad q_2(t) \quad q_3(t)]' = \left\{ \sqrt{p_x^2(t) + p_y^2(t)} \quad \tan^{-1}[p_y(t)/p_x(t)| \quad p_z(t) \right\}'$$
$$\tag{7.3.21}$$

$$\dot{q}(t) = J^{-1}[q(t)]v(t) \tag{7.3.22}$$

$$\ddot{q}(t) = \{J[q(t)]\}^{-1}\{\dot{v}(t) - \dot{J}[q(t)]\dot{q}(t)\} \tag{7.3.23}$$

Equations 7.3.21–7.3.23 are used to calculate the positions, velocities and accelerations of the joint variables at the via points.

The five via points in the base coordinates are given (in cm): $p(0) = [10 \ 0 \ 10]'$, $p(1) = [7.5 \ 7.5 \ 17.5]'$, $p(2) = [5.0 \ 15.0 \ 25.0]'$, $p(3) = [2.5 \ 22.5 \ 32.5]'$, and $p(4) = [0 \ 30 \ 40]'$. These points are mapped into the joint space, and the following five points are obtained: $[10 \ cm \ 0° \ 10 \ cm]'$, $[10.61 \ cm \ 45° \ 17.50 \ cm]'$, $[15.81 \ cm \ 71.57° \ 25.0 \ cm]'$, $[22.64 \ cm \ 83.66° \ 32.50 \ cm]'$ and $[30.0 \ cm \ 90.0° \ 40.0 \ cm]'$. The velocity and acceleration vectors are then determined at these via points using Equations 7.3.22 and 7.3.23. For example, the velocities and accelerations at the initial and terminal times are

$$\dot{q}(0) = 0 \qquad \dot{q}(t_f) = 0 \tag{7.3.24}$$

$$\ddot{q}(0) = \begin{bmatrix} \ddot{p}_x(0) & \ddot{p}_y(0)/q_1(0) & \ddot{p}_z(0) \end{bmatrix}' \qquad \ddot{q}(t_f) = \begin{bmatrix} \ddot{p}_y(t_f) & -\ddot{p}_x(t_f)/q_1(t_f) & \ddot{p}_z(t_f) \end{bmatrix}' \tag{7.3.25}$$

The spline function sequence can now be established. It is comprised of the following polynomials specified by Equation 7.3.13:

(i) $Q_1^k(t')$ between the via points $q_1(t_k) = r(t_k)$ and $q_1(t_{k+1}) = r(t_{k+1})$, $k = 1, 2, 3, 4$, for the joint variable r;

(ii) $Q_2^k(t')$ between the via points $q_2(t_k) = \theta(t_k)$ and $q_2(t_{k+1}) = \theta(t_{k+1})$, $k = 1, 2, 3, 4$, for the joint variable θ;

(iii) $Q_3^k(t')$ between the via points $q_3(t_k) = z(t_k)$ and $q_3(t_{k+1}) = z(t_{k+1})$, $k = 1, 2, 3, 4$, for the joint variable z;

where $t' = (t - t_k)/(t_{k+1} - t_k)$. Thus, four polynomials form the spline function sequence for each joint variable. The velocities and accelerations at the knot points $q_i(t_k)$ are evaluated by using Equations 7.3.22 and 7.3.23.

The desired trajectories $r^d(t)$ and $\theta^d(t)$ for the joint variables r and θ are graphed in Figures 7.7a and 7.7b using the approximating spline functions $\{Q_1^k(t')\}$ and $\{Q_2^k(t')\}$, $k = 1, \ldots, 4$, respectively. The foregoing trajectories are then mapped (pointwise) into the Cartesian base coordinate system to obtain $p_x^d(t)$, $p_y^d(t)$, and $p_z^d(t)$, which represent the desired trajectory in the Cartesian base coordinate system. They are displayed in Figure 7.7c. The graphs illustrate that the spline functions provide reasonable approximations to the desired straight line trajectory. They also demonstrate that if the via points specifying the desired trajectory in the Cartesian base coordinate system are located at equidistant intervals, the corresponding points in the joint space, in general, are not equally spaced, and vice versa.

The spline functions can now be used to calculate intermediate points in the joint space for a controller. Only the polynomial expressions with their coefficients need to be stored for these calculations.

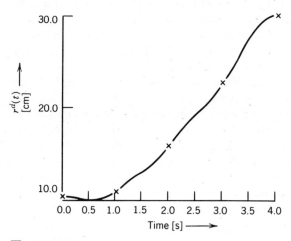

FIGURE 7.7a
Joint variable $r^d(t')$ versus time using spline function $Q_1^k(t')$, $k = 1, 2, 3, 4$; Example 7.3.1.

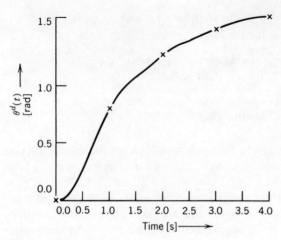

FIGURE 7.7*b*

Joint variable $\Theta^d(t')$ versus time using spline function $Q_2^k(t'), k = 1, 2, 3, 4$; Example 7.3.1.

The spline functions used in describing the desired trajectory in Example 7.3.1 consist of fifth-order polynomials that satisfy six boundary conditions on the position, velocity, and acceleration. The third-order polynomial in Equation 7.3.11 is based on satisfying the boundary conditions on the position and velocity only. It is possible to determine cubic polynomials to describe the desired trajectory that satisfies the specified position, velocity, and acceleration conditions at the knot points [5]. These spline functions are derived by assuming that the acceleration changes linearly with time. By integrating the expression of the acceleration, and determining the constants of integration, a cubic polynomial is obtained that meets the boundary conditions. This approach requires a considerable amount of computations, but they can be performed off-line and in advance.

When the via points describing the desired trajectory in the world space are placed at sufficiently short intervals apart, the mapping of the line segments of the polygonal path into the joint space produces curves that can be approximated closely by polynomial spine functions. If an error between the spline functions and the desired trajectory is not acceptable, additional via points can be added along the desired trajectory, for example, at the midpoints of the line segments. Digital simulations indicate that this approach leads to acceptable solutions in most cases.

The spline functions can thus be determined to describe the desired trajectories for all joint variables. They provide continuous representations for the desired trajectories over the entire time interval. This information can then be used by a controller so that a smooth motion of the manipulator is obtained.

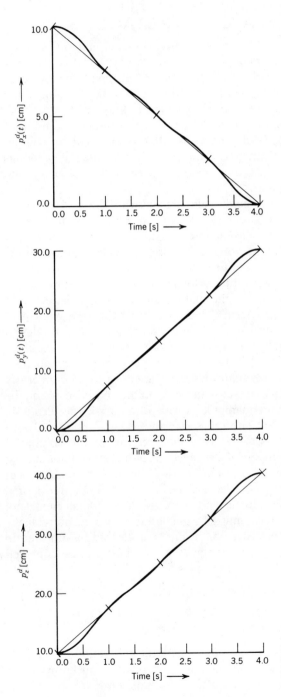

FIGURE 7.7c
Base coordinate components
of desired trajectory versus
time; Example 7.3.1.

7.4 GENERATION OF DESIRED TRAJECTORY TO AVOID OBSTACLES

The generation of a desired trajectory is discussed in the previous sections for the smooth motion of a manipulator in the workspace that does not contain obstacles. The desired trajectory is described for the translational and rotational motion of the *point* of the end-effector. The skeleton of the desired trajectory is specified by a graph joining the knot (via) points that are chosen for a specific task. If the workspace contains stationary obstacles, the desired trajectory should be determined so that the manipulator will not collide with the obstacles. It can be accomplished by selecting the via points for the desired trajectory so that interferences between the manipulator and the environment will not occur.

When the end-effector of a manipulator transfers an object, it sweeps a certain volume. The volume swept should not intersect with the volume occupied by the obstacles if collisions are to be avoided. The determination of the intersections of these volumes is difficult. When changes in the desired path are necessary to avoid interferences between the obstacles and the manipulator, it is often difficult to systematically change the desired trajectory to another one that makes the manipulator avoid the obstacles. In view of such difficulties, the system that moves an object in a workspace containing stationary obstacles can be modified for the analysis and design by shrinking the object held by the gripper to a point while *growing the obstacles* by a corresponding amount [6]. The desired trajectory is then planned for the point so that the enlarged fictitious obstacles are avoided. The trajectory generated in this way is such that collisions between the end-effector and the obstacles are avoided while the manipulator is performing the specified task.

When an object in the end-effector is transformed to a point, called a reference point, the obstacles are enlarged to account for the volume of the object. If the reference point of the object does not enter the enlarged forbidden regions (volumes) at any time during the motion, interferences between the object and the obstacles will be avoided. Since the desired trajectory that prevents collisions depends strongly on the detail models of the obstacles, it is common to circumscribe the obstacles and the object being moved by convex polyhedra. Any object and obstacle can be modeled to a desired degree of accuracy in this way. We will illustrate the approach outlined in the framework of three examples.

7.4.1 DESIRED TRAJECTORY AROUND ENLARGED FORBIDDEN REGIONS

It is convenient to study collision avoidance and the generation of a desired trajectory first on a plane that contains stationary obstacles at known locations. The approach is then discussed in a three-dimensional space. The task of a manipulator is to transfer an object from a given initial location to a specified terminal position while avoiding collisions with known obstacles [6].

EXAMPLE 7.4.1

A *point* object is to be transferred without collisions along the shortest path from the initial point P_0 to the terminal point P_f. The operation takes place on a plane shown in Figure 7.8a. The three obstacles O1, O2, and O3 are indicated by the shaded regions. The straight line from P_0 to P_f intersects with the area occupied by an obstacle. The task is to find the shortest path that consists of straight line segments joining P_0 to P_f around the obstacles.

As shown in Figure 7.8a, the shortest path from P_0 to P_f is formed by connecting P_0 to vertex 1 of obstacle O2, then by following a side of O2 to vertex 2, and to vertex 3 of obstacle O3, and finally, a straight line drawn from vertex 3 of obstacle O3 to terminal point P_f. Other feasible (without collisions) paths can also be found similarly. The shortest path can then be selected among the feasible trajectories.

EXAMPLE 7.4.2

The problem of Example 7.4.1 is changed so that the object to be moved occupies a *circular* area. A circle *(C)* of known radius and with center at point P_0 is to be moved to a new position, where its center will be at point P_f. The problem is to determine the shortest path for the transfer of the circle from P_0 to P_f while avoiding interferences with the obstacles.

When the motion takes place, the circle will cover a roadlike area in which the midline is traced by the center of the circle. The shortest path from P_0 to P_f will not be obtained in this case by connecting the center point P_0 to the vertex of an obstacle, and then to another vertex, and so on. Therefore, the problem of avoiding the obstacles is modified as follows: the original obstacles are first enlarged to describe the forbidden areas while shrinking the circle to a point (center point). The operation of determining the new forbidden areas on the basis of the original obstacles and the size of the moving object *(C)* is called the *growing of the obstacle* by object *(C)*. The enlarged forbidden areas represent fictitious obstacles to the reference point of the object. The desired trajectory around the enlarged forbidden regions is then determined only for the reference point of the circle.

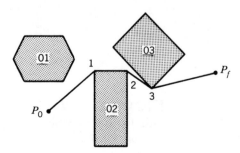

FIGURE 7.8a
Path for transferring point object from P_0 to P_f on plane.

The area of object *(C)* is taken into account by enlarging the areas of the original obstacles. The growing of obstacle *O*1 by object *(C)* is accomplished by moving the circle on the edges of the obstacle while graphing the trace of the reference point, the center of the circle. This trace determines the boundary of the enlarged forbidden region. These boundaries are determined by moving the sides of the original obstacles outward by the amount of the radius of the circle. When the circle moves from one edge to another around the vertex, the center of the circle makes an arc. By including these arcs in the boundary, the enlarged forbidden regions can be made somewhat smaller than the areas of the figures that are similar to the original regions. Obstacles *O*2 and *O*3 are similarly grown by object *(C)*. The enlarged regions represent the fictitious obstacles. They are shown in Figure 7.8*b*.

If the center point of the circle follows a path that does not enter the enlarged forbidden regions, the collisions between object *(C)* and the original polyhydral obstacles are avoided. The desired path representing the shortest path starting from the initial point and ending in the terminal point is obtained by following a procedure similar to the one described in Example 7.4.1. The desired path from point P_0 to P_f is shown in Figure 7.8*b*.

The example discussed demonstrates the conversion of the object to a reference point for the planning by performing the growing operations on the obstacles. The desired path is then generated for the reference point so that the interferences of the object with the enlarged (fictitious) obstacles are avoided.

The desired path can be expressed as a polygonal path connecting the via points, some of which are on the edges of the enlarged polyhedra. The via points are selected so as to avoid collisions between the object and the obstacles. The desired trajectory can then be expressed by means of the spline functions in either the joint or the Cartesian workspace.

The growing operation enlarges the regions occupied by the obstacles. The sizes of the resulting grown obstacles are determined by the positions of the reference points on the objects. The position of the reference point depends on the orientation of the object that can change or be fixed throughout

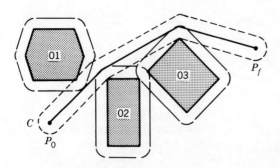

FIGURE 7.8*b*
Path for transferring circular object from P_0 to P_f on plane.

the motion. As the consequence, the enlarged forbidden regions are also dependent on the orientation of the object being moved. This aspect is demonstrated in Example 7.4.3.

EXAMPLE 7.4.3

The workspace of a manipulator contains three obstacles $O1$, $O2$, and $O3$, as shown in Figure 7.8c. The object to be moved has a rectangular shape. A reference point P has been selected on the object. The orientation can change so that the long side of the object can assume an angle between $0°$ and $45°$ with the horizontal line; the extreme orientations at the initial position of the object are shown in Figure 7.8c. The orientation of the object is such that at the initial time the long edges make an angle $45°$ with the horizontal line and they are horizontal at the final time. The reference point is shown in Figure 7.8c at the initial time as P and at the final time as P_f. The task is to generate a shortest path to move the object from the initial position to the final position while avoiding collisions with the obstacles.

The system is modified so that the desired trajectory around the enlarged obstacles is determined only for the reference point. A new fictitious object

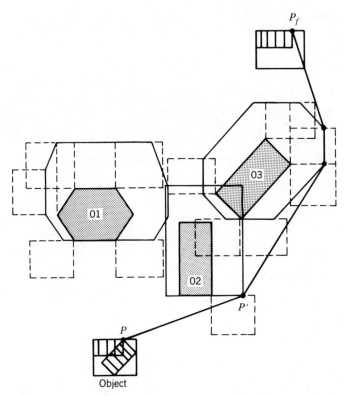

FIGURE 7.8c
Path for transferring object at changing orientation on plane.

is first determined so that it encloses the original object regardless of its orientation. It is a rectangle that is called the envelope of the object. This envelope representing the boundaries of the fictitious object is used to grow the obstacles. The enlarged forbidden regions are drawn by moving the boundaries of the original obstacles by the amount determined by tracking the reference point when the fictitious object moves on the sides of the obstacles. Thus, the enlarged forbidden regions shown in Figure 7.8c are obtained.

The shortest path for the reference point of the object from P to P_f is then determined by using the vertices of the enlarged forbidden regions. The desired trajectory connects P to P_f along a polygonal path, as shown in Figure 7.8c. The path of the reference point stays all the time outside the enlarged forbidden regions. The path determined is such that there will be no collisions between the obstacles and the object during the transfer of the object.

The growing of the convex polyhedral obstacles by a given object is accomplished in the two-dimensional case by sliding the object along the edges of the obstacles while tracing the path of a reference point chosen on the object. This determines the boundaries of the enlarged forbidden regions. The same basic approach is applicable also in the three-dimensional space, although the boundaries of the enlarged forbidden regions are usually difficult to describe graphically and/or analytically.

In the three-dimensional space, the growing of a convex polyhedron by an object can be accomplished by moving each face of the polyhedron independently outward by an amount determined by the reference point on the object. Thus, a fictitious new polyhedron with new faces is obtained. The faces of the grown polyhedron can then be connected in a way which is similar to the two-dimensional case. Figure 7.9a shows a polyhedral obstacle, and the faces of the fictitious obstacle obtained by growing the original obstacle by a cuboidal object. The grown obstacle results when the object is made to slide on the surfaces of the original obstacle.

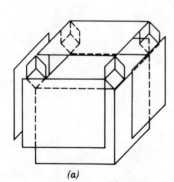

(a)

FIGURE 7.9a
A three-dimensional obstacle.

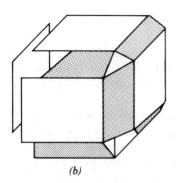

FIGURE 7.9*b*
Enlarged forbidden volume resulting from growing the obstacle by a cube.

The faces of the grown obstacle intersect and they form the edges of the fictitious obstacle. Thus, the forbidden volume around the polyhydral obstacle is determined.   An alternative way to connect the faces of the grown fictitious obstacle is to determine the surfaces the reference point traces when the object moves on and around the edges of the obstacles. Figure 7.9*b* illustrates the fictitious enlarged obstacle in which the faces of the enlarged obstacle form a new polyhedron.   Figure 7.9*c* results when appropriate patches have been added around the edges and corners of the obstacle to keep the volume of the forbidden region small. When the specified object is moved by the manipulator, the reference point of the object must not enter the enlarged three-dimensional forbidden region. As a consequence, the interferences of the object with the obstacles are avoided.

In the examples presented for generating desired trajectories for the motion of manipulators, the locations of the stationary obstacles are assumed to be available. Thus, the information about the environment is completely known, which is often the case in a structured environment such as a manufacturing plant.   In many applications, however, the description of the workspace is incomplete [7]. Moreover, the environment may contain moving obstacles such as another independent robot manipulator or a human.

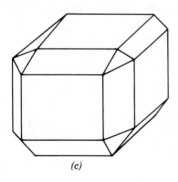

FIGURE 7.9*c*
Enlarged forbidden volume obtained by patching the edges.

It is also possible, depending on the available information about the environment, that the generation of the desired trajectory may have to be performed on-line while the manipulator is moving. These situations require on-line testing for collisions and generation of collision-free trajectory points while the manipulator is moving. They represent challenging problems for researchers at the present.

The approaches described to solve the planning of the trajectories for manipulator motions are based on kinematic equations. They do not take into account the dynamics of the manipulators that play an important role, particularly when the environment contains obstacles. The problem of the trajectory planning for manipulators can be formulated as a dynamic optimization problem [8]. The determination of the solution is difficult, particularly from the computational viewpoint. Progress on this research area should be forthcoming.

7.5 SUMMARY

The problem of generating a desired trajectory (path) for the position and orientation of the end-effector is posed in this chapter. The generation of the desired straight line trajectories for the Cartesian base coordinates and for the joint variables on the basis of the velocity profiles is presented. In order to avoid jumps in the derivatives of the generalized coordinates, the desired trajectory can be bent around the knot points. The desired positional trajectory during the transition intervals can then be described by quadratic polynomials that are splined together with the linear line segments. Thus, the desired trajectory is described as a sequence of quadratic and linear polynomials. Alternatively, the desired trajectories for smooth motions can be generated by means of polynomial spline functions. They must satisfy certain continuity conditions at the points where the functions are splined together. We discussed the use of polynomial spline functions for computing desired trajectory values between the via points. This procedure reduces the computing time and memory requirement. The spline functions are precomputed during the planning phase, and only the polynomials with their coefficients need to be stored for future calculations. The approaches discussed are also applicable to the generation of the desired trajectories for the orientation of the end-effector.

Since the workspace in many applications contains stationary obstacles, the trajectory generation in such an environment needs special attention. A possible approach is to shrink the object (and the end-effector) to a point while enlarging the obstacles by an amount determined by the object and the reference point chosen on the object. The desired trajectory is then generated for the reference point in such a way that it avoids entering the enlarged forbidden regions. As a result, the possible interferences of the manipulator with the obstacles are avoided.

REFERENCES

[1] M. BRADY, "Trajectory Planning," in *Robot Motion: Planning and Control,* edited by M. Brady et al., MIT Press, Cambridge, MA, 1982.

[2] R. H. TAYLOR, "Planning and Execution of Straight-Line Manipulator Trajectories," *Robot Motion: Planning and Control,* edited by M. Brady et al., MIT Press, Cambridge, MA, 1982, pp. 265–286.

[3] J. J. CRAIG, *Introduction to Robotics, Mechanics and Control,* Addison-Wesley Publishing Company, Reading, MA, 1986.

[4] T. H. CHIU, A. J. KOIVO, and R. LEWCZYK, "Experiments on Manipulator Gross Motion Using Self-Tuning Controller and Visual Information," *Journal of Robotic Systems,* Vol. 3, No. 1, pp. 59–70, Spring 1986.

[5] C.-S. LIN, P.-R. CHANG, and J. Y. S. LUH, "Formulation and Optimization of Cubic Polynomial Joint Trajectories for Industrial Robots," *IEEE Trans. on Automatic Control,* Vol. AC-28, No. 12, pp. 1066–1074, December 1983.

[6] T. LOZANO-PEREZ and M. A. WESLEY, "An Algorithm for Planning Collision-Free Paths Among Polyhedral Obstacles," *Communications of the ACM,* Vol. 22, No. 10, October 1979, pp. 560–570.

[7] V. J. LUMELSKY, "Effect of Kinematics on Motion Planning for Planar Robot Arms Moving Amidst Unknown Obstacles," *IEEE Journal on Robotics and Automation,* pp. 207–223, June 1987.

[8] E. G. GILBERT and D. W. JOHNSON, "Distance Functions and Their Application to Robot Path Planning in the Presence of Obstacles," *IEEE Journal on Robotics and Automation,* Vol. RA-1, No. 1, pp. 21–30, March 1985.

PROBLEMS

7.1 The gripper of a planar manipulator used in Problem 5.C.1 is to move along a straight line from point $P_0[50(1 + \sqrt{3}), 50(1 + \sqrt{3})]$ to point $P_f(0, 100)$, where the units are in cm. The velocity profile has the shape of an equilateral trapezoid. The maximum Cartesian velocity of the

gripper is 20 cm/s and the maximum acceleration 40 cm/s^2.
The initial and final velocities are zero.

a.   Determine the time t_f needed for the transfer of the
gripper from P_0 to P_f when the maximum velocity and
acceleration are used.

b.   Choose a via (knot) point in addition to the end points
along the desired trajectory in the base coordinate system. Graph the via points on the $x_0 y_0$-plane.

c.   Determine the via points in the joint space that correspond to the three via points specified in part (b).

d.   Determine the joint velocities at the via points. Sketch
the joint velocities versus time.

e.   Determine the joint accelerations at the via points.
Sketch the joint accelerations versus time.

7.2   Assume a first-order polynomial representation for a spline
function to describe the positional motion of a joint variable

$$\theta(t) = a_0 + a_1 t$$

a.   Determine coefficients a_0 and a_1 so as to satisfy $\theta(t_1) = q(t_1)$ and $\theta(t_2) = q(t_2)$ at the initial and final times.

b.   Assume that the velocity is constant along the line segment joining the initial and final points. Specify this
velocity. Sketch the velocity profile. Is it continuous
on $[t_1, t_2]$?

c.   If also the initial and final velocity values are given,
suggest a polynomial description for generating the
desired trajectory path.

7.3   For the joint variable trajectory $\theta_i(t)$, the first two via points
$[0, \theta_i(0)]$ and $[t_{22}, \theta_i(t_{22})]$ are joined by a straight line
segment.

a.   Determine the equation for this straight line.

b.   The end point $[t_{13}, \theta_i(t_{13})]$ of the bent trajectory is
determined by Equation 7.2.25. Verify that this point
is on the line determined in (a).

7.4   Assume a third-order polynomial for a spline function:

$$\theta(t) = a_0 + a_1 t + a_2 t^2 + a_3 t^3$$

a.   Determine parameters a_i, $i = 0, \ldots, 3$ so that $\theta(t)$
satisfies: $\theta(t_1) = q(t_1)$, and $\dot{\theta}(t_1) = \dot{q}(t_1)$ where the ini-

tial time $t_1 = 0$. Moreover, $\theta(t_n) = q(t_n)$ and $\dot{\theta}(t_n) = \dot{q}(t_n)$, where t_n is the final time.

b. Determine the polynomial $\theta(t)$ obtained in (a) in the form similar to Equation 7.3.10.

7.5 The velocity profile for a joint variable θ_i is given in Figure 7.5b. The maximum acceleration of the joint is $\ddot{\theta}^m$.

a. Determine the equation that describes a smooth transition from one line segment to the next one in the velocity profile by assuming that the acceleration used during the transition is linear with respect to time. Use the notations of Figures 7.6a and 7.6b.

b. Specify the conditions for a smooth motion at the points where the straight line segment is splined together with the bent part.

7.6 In Equation 7.3.13, verify that

a. at $t' = 0$, or $t = t_1$, $Q_i^k(0) = q_i(t_k)$; $\dot{Q}_i^k(0) = \dot{q}_i(t_k)$ and $\ddot{Q}_i^k(0) = \ddot{q}_i(t_k)$;

b. at $t' = 1$, $Q_i^k(1) = q_i(t_{k+1})$, $\dot{Q}_i^k(1) = \dot{q}_i(t_{k+1})$, $\ddot{Q}_i^k(1) = \ddot{q}_i(t_{k+1})$.

7.7 The transition from one line segment to the next one is started at time t_{k-1}. The times when the knot points occur are $t_{k-1}, t_k,$ and t_{k+1} and Δt is a time increment. Suppose that the bending of the trajectory is performed around the point $[t_k, q(t_k)]$. The position q_i and the velocity $\dot{q}_i$ of joint i are governed by $q_i(t_k - \Delta t) = q_i(t_k) - \Delta t \Delta q_i(t_k)/T_1$; $\dot{q}_i(t_k - \Delta t) = \Delta q_i(t_k)/T_1$, where $T_1 = t_k - t_{k-1}$ and $\Delta q_i(t_k) = q_i(t_k) - q_i(t_{k-1})$. Moreover, the corresponding variables at time $t = t_k + \Delta t$ are $q_i(t_k + \Delta t) = q_i(t_k) + \Delta t \Delta q_i(t_{k+1})/T_2$; $\dot{q}_i(t_k + \Delta t) = \Delta q_i(t_{k+1})/T_2$, where $T_2 = t_{k+1} - t_k$. During the transition, a constant but unknown acceleration a_i is applied to the joint, that is, $\ddot{q}_i = a_i$ [2].

a. Integrate the acceleration equation twice, and use the boundary conditions at time $t_k - \Delta t$ to determine the constants of integration. Then, eliminate a_i and show that the transition trajectory during the bending period can be described by

$$q_i(t' + t_k) = q_i(t_k) - (\Delta t - t')^2 \Delta q_i(t_k)/(4\Delta t T_1)$$
$$+ (\Delta t + t')^2 \Delta q_i(t_{k+1})/(4\Delta t T_2)$$

where $t' = t - t_k$.

b.   Sketch $q_i(t')$ versus t'. Mark the knot points $q_i(t_k)$ and $q_i(t_{k+1})$ on the graph.

c.   Evaluate the position, velocity, and acceleration of joint i at times $t = t_k - \Delta t$, t_k, and $t_k + \Delta t$.

d.   Does $q_i(t' + t_k)$ pass through the knot points $[t_k, q_i(t_k)]$ and $[t_{k+1}, q_i(t_{k+1})]$? Explain.

7.8   Show that the Expression 7.3.1 with the parameters given by Equations 7.3.6–7.3.9 can be expressed in the form of Equation 7.3.10.

7.9   The planar manipulator specified in Computer Problem 5.C.1 must move its gripper along a straight line from point $P_0(-10, 50)$ to point $P_f(50, 70)$, where the units are in cm. The velocity profile has the shape of an equilateral trapezoid. The initial and terminal velocities are zero. The time for the acceleration is 2 s, and the total time for the motion is 6 s.

a.   Determine the maximum (Cartesian) velocity and acceleration of the gripper in the base coordinate system.

b.   Assume that there is an obstacle in the environment. Its boundary is specified by the equation $(x - 30)^2 + (y - 60)^2 = 400$. Design a path from point P_0 to point P_f so that no collision occurs during the motion. Introduce via (knot) points so that interferences with the obstacles are avoided.

c.   Graph the trajectory generated in part (b) in the base coordinate system.

d.   Graph the desired trajectories of the joint variables $\theta_1^d(t)$ and $\theta_2^d(t)$ that correspond to the desired trajectory determined in (b) in the base coordinate system. What is the shape of the obstacle in the joint space?

7.10   In the xy-plane, three polyhedral obstacles have vertices at the following points:

(i) Obstacle $O1$: rectangle (15,2), (19,2), (15,10), (19,10) ; (ii) Obstacle $O2$: hexagon (6,10), (6,11), (12,13), (11,16), (6,16), (5,13) ; (iii) Obstacle $O3$: rectangle (25,6), (29,10), (24,15), (20,11). Object (C) to be moved has a rectangular shape. Its initial location is specified by the vertices (3,2), (4,1), (6,3), (5,4). The final positions of the vertices are (34,16), (35,15), (37,17), (36,18).
   The task is to generate the desired trajectory for trans-

ferring object (C) from the initial position to the terminal location. Notice that the orientation of object (C) cannot be changed.

To solve the problem, use point (5,4) on object (C) as the reference point.

a. Determine the enlarged forbidden regions by growing the three obstacles by the object.

b. Graph the desired trajectory for the transfer of the object from the initial position to the final position along the shortest path.

c. Draw the position of the object when it is in contact with the obstacles $O1, O2$, and $O3$ for the first time, and for the last time during the motion specified by the shortest path.

7.11 Repeat Problem 7.10 with a different object. Assume that the vertices of a rectangular object are initially located at (0,0), (3,0), (3,2), (0,2). It is to be moved to the final position (35,10), (38,10), (38,12), (35,12).

7.12 A polyhedral obstacle in the workspace is defined as a triangle with vertices at points (0,0), (10,0), (10,−6). The object to be transferred is a circle with radius of 2 units.

a. Determine the forbidden region of the fictitious obstacle when the center of the circle is used as a reference point. Assume that the enlarged forbidden region is similar to the original obstacle.

b. Determine the new forbidden region of minimal area obtained by sliding the circle along the edges of the original obstacle. Indicate the areas that demonstrate the difference between the minimal forbidden area and the forbidden area determined in (a). Shade the space wasted in part (a).

COMPUTER PROBLEMS

7.C.1 Assume the base coordinate system for the vertical planar manipulator in Problem 5.C.1 is specified. The end-effector is to follow a straight line defined by $y = 30$ cm in the positive direction of the x_0-axis starting from $x_0(0) = 0$, $y_0(0) = 30$ cm to the point $x(t_f) = 95$ cm, $y(t_f) = 30$ cm. The gripper velocity is to be designed using an equilateral trapezoidal velocity profile. The gripper is to accelerate

to the maximum velocity, v_{max}, in 10 sampling periods
(i.e., $10T$), and decelerate to zero velocity in 10 sampling
periods where $v_{max} = 1$ m/s and the sampling period $T = 20$ ms.

a. Compute the desired trajectory for the joint variables
$\theta_i^d(t)$, $\dot{\theta}_i^d(t)$, and $\ddot{\theta}_i^d(t)$, $i = 1, 2$. Graph the desired trajectories, that is, (i) $\theta_i^d(t)$ versus time, (ii) $\dot{\theta}_i^d(t)$ versus
time, and (iii) $\ddot{\theta}_i^d(t)$ versus time for $i = 1, 2$.

b. Compute the joint torques needed to make the end-
effector track the desired trajectory. To solve
the problem, substitute the desired values of $\theta_i^d(t)$,
and $\dot{\theta}_i^d(t)$, $\ddot{\theta}_i^d(t)$ into the model equations and obtain
the torque values. This control method is often re-
ferred to as the computed torque method, that is, the
calculation of the inverse dynamics. Graph the computed torques $\tau_1(t)$ and $\tau_2(t)$ versus time.

7.C.2 For the cylindrical robot manipulator specified in Example
7.3.1, three via (knot) points are chosen at times $t_1 = 0$,
$t_3 = 2$ s, and $t_5 = 4$ s. Use the specifications of Example
7.3.1.

a. Generate third-order polynomial spline functions for
each joint variable. Specify the conditions that will be
met at the via points. Graph the desired trajectory as
a function of time using the spline functions.

b. Convert the spline functions pointwise into the Cartesian base coordinate system and graph the results.

7.C.3 The end-effector of the cylindrical robot manipulator
described in Example 7.3.1 should be transferred from P_0
(0.0, 0.8m, 0.5m) to P_f (0.7m, -0.6m, 0.0) expressed in
the Cartesian base coordinate system.

a. The desired trajectories for the joint variables are to
be generated by assuming that the velocity
profiles of the joint velocities $\dot{r}(t)$, $\dot{\theta}(t)$, and $\dot{z}(t)$ are
equilateral trapezoids with the maximum (minimum)
accelerations (decelerations) of 0.4m/s², 0.6 rad/s²,
and 0.4 m/s², and the maximum velocities of 0.8 m/s,
1.2 rad/s and 0.8 m/s in the r-, θ- and z-directions,
respectively.

b. Design the bent trajectories for $r(t)$ around the corner
points of the velocity profile. Calculate the resulting
desired trajectory for the entire duration of motion.

Verify that the desired trajectory designed is continuous in the position, velocity, and acceleration.

c. Generate the desired position trajectory $[r^d(t) \; \theta^d(t) \; z^d(t)]'$
    versus time in the joint space, and in the Cartesian
    space.

d. Generate the corresponding desired velocity trajectory
    $[\dot{r}^d(t) \; \dot{\theta}^d(t) \; \dot{z}^d(t)]'$ versus time.

e. Graph the desired position trajectory in the joint space
    for each component versus time.

f. Graph the desired velocity trajectory in the joint space
    for each component versus time.

7.C.4 For a cylindrical manipulator, the desired trajectory
$p^d(t) = [p_x^d(t), \; p_y^d(t)]'$ expressed in the Cartesian base
coordinate system is given in Table 7.1 (page 266).
It is such that the end-effector is transferred from
point $P_0 = (0.0, 0.7\text{m}, 0.5\text{m})$ to point $P_f = (0.7\text{m},$
$-0.6\text{m}, 0.5\text{m})$. The desired trajectory was generated
by assuming that the profiles of the joint velocities
$\dot{r}(t)$ and $\dot{\theta}(t)$ are equilateral trapezoids. The maximum
acceleration (deceleration) is 0.8 m/s^2 and the maximum
velocity 0.2 m/s in the r-direction.

a. Compute the desired positional trajectory $[r^d(t) \; \theta^d(t) \; z^d(t)]'$
    versus time in the joint space. Note: $p_x^d(t) = r^d(t) \cos[\theta^d(t)]$
    and $p_y^d(t) = r^d(t) \sin[\theta^d(t)]$.

b. Determine the maximum velocity and acceleration
    in the θ-direction. Generate the desired velocity
    trajectory $[\dot{r}^d(t) \; \dot{\theta}^d(t) \; \dot{z}^d(t)]'$.

c. Graph (i) the desired trajectory in the $p_y \, p_x$-plane, (ii)
    $r^d(t)$ versus t, and (iii) $\dot{r}^d(t)$ versus t.

TABLE 7.1
Desired Trajectory $p^d(iT) = [p_x^d(iT), p_y^d(iT)]'$ for Computer Problem 7.C.4

iT				
1	2	3	4	5
$p_x^d(iT)$ 0	5.09804E-04	2.04786E-03	4.64039E-03	8.33091E-03
$p_y^d(iT)$ 0.700000	0.701000	0.703997	0.708985	0.715952
6	7	8	9	10
$p_x^d(iT)$ 1.31802E-02	1.92401E-02	2.65416E-02	3.51267E-02	4.50362E-02
$p_y^d(iT)$ 0.724880	0.734748	0.744527	0.754182	0.763673
11	12	13	14	15
$p_x^d(iT)$ 5.63100E-02	6.89860E-02	8.31002E-02	9.86859E-02	0.115773
$p_y^d(iT)$ 0.772952	0.781963	0.790645	0.798928	0.806735
16	17	18	19	20
$p_x^d(iT)$ 0.134389	0.154554	0.176286	0.199596	0.224486
$p_y^d(iT)$ 0.813981	0.820572	0.826407	0.831376	0.835363
21	22	23	24	25
$p_x^d(iT)$ 0.250953	0.278372	0.306108	0.333841	0.361081
$p_y^d(iT)$ 0.838241	0.840080	0.841025	0.840327	0.837079
26	27	28	29	30
$p_x^d(iT)$ 0.387567	0.413113	0.437539	0.460977	0.484005
$p_y^d(iT)$ 0.831139	0.822575	0.811466	0.798436	0.784691
31	32	33	34	35
$p_x^d(iT)$ 0.506623	0.528812	0.550553	0.571829	0.592621
$p_y^d(iT)$ 0.770282	0.755221	0.739521	0.723195	0.706258

iT	$p_x^d(iT)$	$p_y^d(iT)$
36	0.612912	0.688723
37	0.632684	0.670605
38	0.651920	0.651920
39	0.670605	0.632683
40	0.688723	0.612911
41	0.706258	0.592621
42	0.723196	0.571829
43	0.739521	0.550553
44	0.755221	0.528811
45	0.770282	0.506622
46	0.784691	0.484005
47	0.798436	0.460977
48	0.811506	0.437560
49	0.823889	0.413772
50	0.835574	0.389635
51	0.846553	0.365168
52	0.856816	0.340391
53	0.866354	0.315327
54	0.875158	0.289996
55	0.883223	0.264420
56	0.890540	0.238619
57	0.897103	0.212617
58	0.902907	0.186435
59	0.907948	0.160096
60	0.912220	0.133621
61	0.915721	0.107032
62	0.918446	8.03536E-02
63	0.920395	5.36069E-02
64	0.921564	2.68148E-02
65	0.921954	3.09109E-08
66	0.921564	-2.68148E-02
67	0.920395	-5.36068E-02
68	0.918446	-8.03536E-02
69	0.915721	-0.107032
70	0.912220	-0.133621

(continued)

TABLE 7.1 *(Continued)*

Desired Trajectory $p^d(iT) = [p_x^d(iT), p_y^d(iT)]'$ for Computer Problem 7.C.4

iT	71	72	73	74	75
$p_x^d(iT)$	0.907948	0.902907	0.897103	0.890540	0.883223
$p_y^d(iT)$	-0.160096	-0.186435	-0.212617	-0.238619	-0.264419

iT	76	77	78	79	80
$p_x^d(iT)$	0.875158	0.866354	0.856816	0.846553	0.835754
$p_y^d(iT)$	-0.289996	-0.315327	-0.340391	-0.365168	-0.389250

iT	81	82	83	84	85
$p_x^d(iT)$	0.824759	0.813775	0.802901	0.792231	0.781850
$p_y^d(iT)$	-0.412034	-0.433324	-0.453155	-0.471561	-0.488580

iT	86	87	88	89	90
$p_x^d(iT)$	0.771838	0.762267	0.753203	0.744707	0.736831
$p_y^d(iT)$	-0.504248	-0.518603	-0.531681	-0.543518	-0.554149

iT	91	92	93	94	95
$p_x^d(iT)$	0.729623	0.723127	0.717379	0.712412	0.708252
$p_y^d(iT)$	-0.563605	-0.571915	-0.579109	-0.585209	-0.590237

iT	96	97	98	99	100
$p_x^d(iT)$	0.704922	0.702438	0.700814	0.700059	0.700000
$p_y^d(iT)$	-0.594210	-0.597144	-0.599049	-0.599932	-0.600000

CHAPTER EIGHT

PRIMARY AND SECONDARY CONTROLLER DESIGN FOR GROSS MOTION OF MANIPULATORS

The desired trajectory for the motion of a manipulator is generated by path planner. The end-effector of the manipulator can alternatively be made to move, for example, manually after the brakes of the joints have been released along the desired path, and the positions of the joints are recorded. Thus, the desired trajectory is "learned (taught) by doing." Having the desired trajectory with the initial and final positions for the centerpoint of the end-effector, controllers need to be constructed for the actuators that make the end-effector follow the specified trajectory as closely as possible. This design task is usually called the *gross motion* control problem. If the motion of the end-effector is constrained, for example, to move along the surface of a table, then the motion is referred to as the *constrained motion*. The motion of the manipulator must comply with the environmental constraints. Often the compliant motion is associated with a task such as an assembly requiring *fine motion* control.

The design of a controller for a given manipulator is based on the design objectives and a dynamical model. The objective in designing a controller for the free or gross motion is to make the centerpoint and the orientation of the end-effector follow the desired trajectory. It is achieved by determining the inputs to the joint shafts or to the joint actuators so that the system follows the desired trajectory as closely as possible. The inputs are thus applied in the joint space. This makes it convenient to have also the desired trajectory expressed in terms of the joint variables. The dynamical model determines the system response. The ideal transient behavior can be described by means of a reference model that represents an ideal model of the manipulator system in regard to the dynamical behavior, as defined by the designer.

The equations of motion for a general manipulator represent a multiple-input–multiple-output (MIMO) system, in which the equations are coupled. The total dynamical model of an N-joint serial link manipulator written in Lagrange's formulation shows the effects of the interactions of the joints. For example, the Coriolis terms in the model of a six-joint manipulator describe the interactions between the joints. These terms represent coupling between the differential equation models of the joints. Under certain conditions, the coupling terms may be partially compensated or even ignored in the model equations. This results in a set of equations that govern the motion of a joint without interacting effects of the other joints. Thus, the equation of motion for joint i contains only the joint variable q_i, its derivatives, and the input u_i to this joint. If the output is the joint position, the equation of motion governing the joint variable q_i, $i = 1, \ldots, N$ is then a single-input–single-output (SISO) model. The control designed for such a model is commonly termed the *independent joint variable control*.

We will first discuss possible specifications for the controller design based on the basic characteristics of the step response of a second-order linear underdamped reference model. Then, we will present several methods to design controllers for the gross motion of robotic manipulators. Specifically, we will construct primary controllers for manipulators to track the desired trajectory under ideal conditions and secondary controllers to compensate for small deviations from the desired trajectory caused by disturbances.

8.1 DESIGN SPECIFICATIONS BASED ON SECOND-ORDER LINEAR UNDERDAMPED SISO REFERENCE MODEL

In many applications of the so-called classical control, a common practice is to evaluate the performance of a SISO system on the basis of its step response. Step inputs are, in fact, common inputs to many physical processes. The response of a system to a step input may be sluggish (i.e., overdamped), which is often unacceptable. Or it may rise quickly to the height of the step input, but exhibit *overshoot* (i.e., underdamped). The overshoot is the (absolute) maximum deviation in the step response of a stable system relative to its steady-state value (obtained as t or $kT \rightarrow \infty$). Large oscillations in the step response are undesirable. They should be made to stay within acceptable limits and die out sufficiently fast.

The decay of the transients can be measured by means of the *settling time* in the step response. The settling time in the step response is understood as the time the step response needs to come and stay within 2–5% of its steady-state value.

Although the aforementioned characteristics can be observed in the step responses of any system, it is convenient to use a second-order linear underdamped SISO model as a reference model. This reference system can be described by the transfer function $G(s)$ that has a complex conjugate pair of poles:

$$G(s) = \frac{Y(s)}{U(s)}$$

$$= \frac{\omega_n^2}{s^2 + 2\xi\omega_n s + \omega_n^2} \qquad (8.1.1)$$

where $U(s)$ and $Y(s)$ are the Laplace transforms of the input and output, respectively, the damping factor ξ is in the range $(0,1]$ for the underdamped system, and ω_n is the (undamped) natural frequency. Equation 8.1.1 specifies the second-order linear time invariant reference model.

The unit step response $y(t)$ of the reference model in Equation 8.1.1 may be determined as the function of time:

$$y(t) = 1 - \frac{1}{\sqrt{1 - \xi^2}} \epsilon^{-\xi\omega_n t} \sin\left(\omega_n \sqrt{1 - \xi^2}\, t + \phi\right) \qquad (8.1.2)$$

where angle $\phi = \cos^{-1}(\xi)$ and $\omega_n(1 - \xi^2)^{1/2}$ is called the damped natural

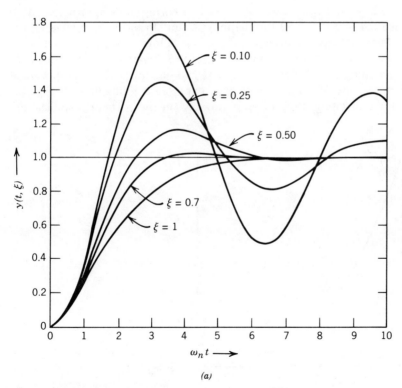

(a)

FIGURE 8.1a
Family of step responses of the second-order reference model.

frequency. A family of unit step responses $y(t, \xi)$ versus time for different values of the damping factor ξ is shown in Figure 8.1a. The overshoot (os) of a step response is dependent on the damping factor ξ through the following relation:

$$os = y_{peak} - y_{ss} = \epsilon^{-\xi\pi/\sqrt{1-\xi^2}} \qquad (8.1.3)$$

where y_{ss} indicates the steady-state value of the output $y(t)$, and y_{peak} is the (absolute) maximum value of $y(t)$. The graph of os versus ξ is displayed in Figure 8.1b. Equations 8.1.3 and 8.1.1 reveal that there is no overshoot in the step response when the damping factor equals one; this value is called the critical damping. If the settling time T_s in the step response is the time after which the step response deviates by less than 2% from its steady-state value, then T_s is determined by

$$T_s = \frac{4}{\xi\omega_n} \qquad (8.1.4)$$

Indeed, time T_s in Equation 8.1.4 makes the envelope function $\exp(-\xi\omega_n t)$ < 0.02 when $t \geq T_s$. If the two successive peak values are measured relative to the steady-state value, the ratio of these values can also be used as a measure of the settling time.

For the design of a control system, the overshoot and the settling time in the step response of the system can be specified. The overshoot in many applications should not exceed 25%, which corresponds approximately to the range $0.4 < \xi < 1.0$ of the damping factor. In most manipulators, the overshoot of the step responses should be kept below 15%. As Equation 8.1.3 shows, the overshoot in the step response of a linear system depends only on the damping factor. The settling time in Equation 8.1.4 is proportional to the time constant of the envelope $(\exp(-\xi\omega_n t))$ function. Acceptable values for the settling time depend strongly on the particular application.

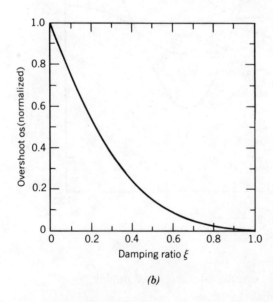

(b)

FIGURE 8.1b
Overshoot os versus damping factor ξ in step responses of the second-order model.

In many manipulator systems, the settling time should be chosen so that the oscillations have settled down before the next sampled value becomes available, that is, before the sampling period ends.

The step response of a second-order reference model represents a reference (ideal) output relative to which the step responses of an *actual* system can be evaluated. In fact, a manipulator system could be designed so that its step response is as close as possible to the step response of the reference model specified by the aforementioned parameters. It can be accomplished by forming the difference (error) of the two outputs, as shown in Figure 8.2, and feeding it into a controller that operates on it so as to force it to zero. The preferred values for the overshoot and the settling time as the design parameters determine the desired values for the damping factor and the natural frequency; they will be denoted by ξ^d and ω_n^d, respectively. Thus, the transfer function of the reference model in Equation 8.1.1 becomes specified. The poles of this reference model are

$$s_{1,2}^d = -\xi^d \omega_n^d \pm j\omega_n^d \sqrt{1 - (\xi^d)^2} \qquad (8.1.5)$$

A controller can then be designed so that the resulting system has the (dominant) poles at the locations specified by s_1^d and s_2^d. These values can also be considered as the desired eigenvalues of the plant matrix of a linearized time invariant model in most manipulator applications.

The reference input to a manipulator system is usually not a single step function applied at $t = 0$; rather, it consists of a sequence of time functions defined over a certain time period. For example, spline functions or singularity functions such as step and ramp inputs may serve as such functions. If a sequence of step functions (a staircase function) is used to approximate the desired trajectory of a joint variable, the response should exhibit an acceptable overshoot and settling time over each time interval of the constant input. It can be accomplished by a proper design of the system.

An example is next presented to illustrate an acceptable output response of the second-order reference model to a piecewise constant input.

EXAMPLE 8.1.1

A piecewise constant input of the staircase shape is applied to a second-order system described by the transfer function $G(s)$ in Equation 8.1.1, where

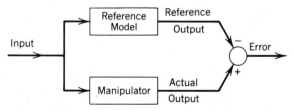

FIGURE 8.2
Comparison of manipulator output with reference model output.

ξ and ω_n are unknown parameters to be determined. The heights of the staircase function are $u(t) = 1.0$ for $0 \le t < 1.0$, $u(t) = 2.0$ for $1.0 \le t < 2.0$, $u(t) = 1.0$ for $2.0 \le t < 3.0$, and $u(t) = 0.05$ for $3.0 \le t < 4.0$ time units. The problem is to determine the desired values for the damping factor ξ^d and the natural frequency ω_n^d, respectively, so that the overshoot of the response at all intervals is less than 5%, and that the settling time T_s does not exceed $0.4T$, where T is the time interval over which $u(t)$ remains constant. Moreover, digital simulations should be performed to verify that the design specifications are met.

The requirement on the settling time implies that the steady-state value on each time interval has been reached before the occurrence of a new step in the input. Moreover, since the system is linear and time invariant, the desired values ξ^d and ω_n^d determined for one time interval on the basis of the design specifications will result in acceptable step responses also on the other time intervals.

The family of the step responses of the second-order reference model in Figure 8.1a shows that the value of the damping factor $\xi^d = 0.707$ corresponds to a step response with overshoot not exceeding 5%. The settling time T_s is constrained so that $T_s \le 0.40T$, where the sampling period T equals one time unit. This constraint and Equation 8.1.4 give $\omega_n^d \ge 4/(0.40\xi^d)$. The value of $\omega_n^d = 15$ rad/(time units) is chosen. The (desired) transfer function of the reference model is now specified:

$$G^d(s) = \frac{225}{s^2 + 21.2s + 225} \tag{8.1.6}$$

The design specifications for the step responses are satisfied by the model specified by $G^d(s)$. The desired poles (i.e., the eigenvalues) of the reference model are at the locations: $s_{1,2}^d = -10.60 \pm j10.613$ on the complex s-plane.

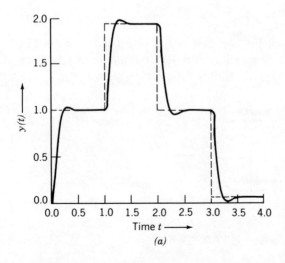

FIGURE 8.3a

Second-order linear system response when $s^d = 0.707$, $\omega_n^d = 15$ rad/(time unit); Example 8.1.1.

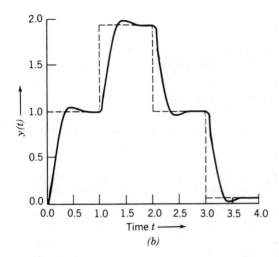

FIGURE 8.3b
Second-order linear system
response when $s^d = 0.707$,
$\omega_n^d = 10$ rad/(time unit); Example 8.1.1.

To simulate the response of the reference model, the SISO-model in Equation
8.1.6 is presented in the form of two first-order differential equations:

$$\begin{bmatrix} \dot{x}_1 \\ \dot{x}_2 \end{bmatrix} = \begin{bmatrix} 0 & 1 \\ -(\omega_n^d)^2 & -2\xi^d\omega_n^d \end{bmatrix} \begin{bmatrix} x_1(t) \\ x_2(t) \end{bmatrix} + \begin{bmatrix} 0 \\ (\omega_n^d)^2 \end{bmatrix} u(t) \qquad (8.1.7)$$

where $x(0) = 0$, $\xi^d = 0.707$, $\omega_n^d = 15$ rad/(time unit), and $x_1(t)$ is the
output.

The vector Equation 8.1.7 is simulated on a digital computer. When the
specified values of ξ^d and ω_n^d are used, the response shown in Figure 8.3a
is obtained. If the damping factor is kept at the value of ξ^d, but the natural
frequency ω_n is chosen to assume values 10 and 20 rad/(time units), then
the responses of the system shown in Figures 8.3b and 8.3c, respectively,

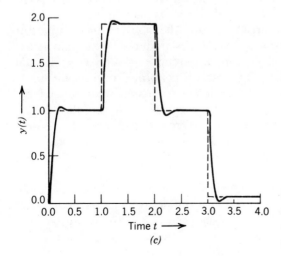

FIGURE 8.3c
Second-order linear system
response when $s^d = 0.707$,
$\omega_n^d = 20$ rad/(time unit); Example 8.1.1.

are recorded. The settling time in Figure 8.3*b* does not satisfy the given specifications. On the other hand, the system response in Figure 8.3*c* does meet the given condition on the settling time.

If the piecewise constant reference input in Example 8.1.1 is considered as an approximation for the desired positional motion of a joint variable, then the velocity toward the end of each sampling interval approaches zero. This fact may render such an approximation unattractive in some applications. An alternative representation of the desired positional path is a sequence of concatenated ramp functions. The controller is then designed in view of such a reference input.

In order to apply transfer function approaches to the dynamical models of manipulators, the nonlinear coupled differential equation models described in Chapter 5 need to be converted to linear time invariant differential equations by a linearization method. If the coupling terms in the dynamical model of the manipulator can be at least partially compensated or eliminated, for example, by a proper design [1], then the dynamics of each joint variable are governed by a differential equation that is independent of the dynamics of the other joints. That is, the model is a SISO system, and the independent joint control is applicable to each joint. In this case, a joint variable can be made to behave like the foregoing SISO reference model with the desired poles by designing an appropriate controller. A controller can thus be constructed to meet the design specifications in regard to the step response. Although the inputs may be different from the step functions, the system designed in regard to step responses usually behaves in an acceptable manner for other inputs.

8.2 CONTROLLER DESIGN FOR MANIPULATOR GROSS MOTION

The manipulator control problem for the gross (free) motion is a *servo* problem in which the centerpoint of the end-effector is made to track the reference values representing the desired trajectory. The inputs to the manipulator system can be chosen as the generalized torques produced by the joint actuators, and the outputs to be controlled are the positions, possibly velocities and accelerations of the joints when the control is performed in the joint space. The actual positions and velocities of the joints in a manipulator can be measured by encoders and tachometers, respectively. The values of the accelerations can be calculated by hardware or software programming. These variables can then be compared with their desired values using feedback loops to form the position, velocity, and acceleration errors for driving the plant.

To make the end-effector of a manipulator track the desired nominal trajectory, the generalized torques applied to the system should have the appropriate (nominal) values that result in the desired motion under ideal conditions. The controller generating these values can be referred to as the *primary controller*. Thus, it compensates for the nonlinear effects, and attempts to cancel the nonlinear terms in the model. Since the mathematical model used is usually not exact, and since the system is subject to disturbances, undesirable deviations (errors) of the actual motion from the nominal trajectory can be corrected by means of an additional controller called a *secondary controller*. The block diagram showing the total control system is displayed in Figure 8.4*a*.

Many general methods are known to design controllers for linear time invariant models. For example, the poles of the transfer function or the eigenvalues of the plant matrix in a linear time invariant system can be assigned to specified locations so that the desired characteristics in the step responses are obtained. Even though the parameters in the plant matrix cannot usually be chosen, the introduction of a feedback controller with adjustable gains will allow the designer to solve the eigenvalue assignment problem. This design procedure can be applied to the linear time invariant models.

Another common controller in feedback systems is a proportional integral and derivative (PID) controller. This controller, shown in Figure 8.4*b*, corrects for the errors in the trajectory tracking. A PID-controller performs the operations of amplification (P = proportional), integration (I = integral), and differentiation (D = derivative) on the error signal fed into the controller as the input. In general, the D-part speeds up the response by performing a predictive-type function, the I-part influences the steady-state error, and the P-part influences the open-loop steady-state gain. Each part of the controller has gains to be adjusted or tuned experimentally so that the desirable responses of the system are obtained. The gains of a PID-controller can also be determined on the basis of the eigenvalue assignment.

We will next discuss the determination of a primary controller for the gross motion of the manipulators in the joint space. We will then present the design of a secondary PD-controller to compliment the primary controller. We will also discuss the construction of a secondary state feedback controller.

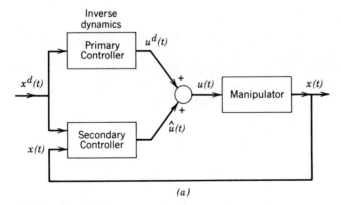

(a)

FIGURE 8.4*a*
Manipulator system with primary and secondary controllers.

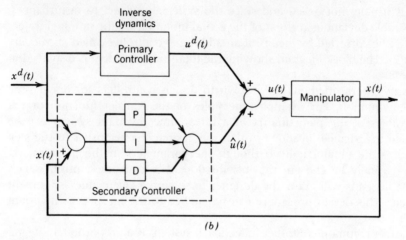

(b)

FIGURE 8.4*b*

Manipulator system driven by primary controller and secondary PID-controller.

8.2.1 PRIMARY FEEDBACK—FEEDFORWARD LOOP CONTROLLER

The dynamical model of most manipulators is nonlinear and computationally complex. Most general methods for controller design are based on linear time invariant dynamical models. Therefore, the design of controllers for the gross motion of the manipulators may be considered in two stages: (1) to design a primary controller that under ideal conditions makes the end-effector track the desired trajectory; (2) based on a linearized model, to design a secondary controller that compensates for undesirable deviations of the motion from the desired trajectory caused by external and/or internal disturbances.

We will first apply the feedback–feedforward linearization method to the nonlinear model to determine a primary controller. Then, we introduce a PD-controller as the secondary controller and determine the gains on the basis of an eigenvalue assignment.

The dynamical model for a manipulator is assumed to be in the form similar to Equation 6.2.16 that is used in conjunction with the feedback-feedforward linearization method in Section 6.2.2:

$$\begin{bmatrix} \dot{x}^1 \\ \dot{x}^2 \end{bmatrix} = \begin{bmatrix} x^2(t) \\ f^2(x) \end{bmatrix} + \begin{bmatrix} 0 \\ B^2(x) \end{bmatrix} u(t) \qquad (8.2.1)$$

where the 2N-dimensional state vector $x(t)$ is decomposed into two N-dimensional vectors $x^1(t)$ and $x^2(t)$, and the input vector $u(t)$ is N-dimensional. The vector-valued functions $f^2(x)$ and $B^2(x)$ have been defined in conjunction with Equation 6.2.16. By eliminating $x^2(t)$ in Equation 8.2.1, the resulting expression may be written as follows:

$$\ddot{x}^1 = f^2(x) + B^2(x)u(t) \tag{8.2.2}$$

The problem is to determine the input vector $u(t)$ so that $x(t)$ will track the desired trajectory $x^d(t)$. The input will be generated as $u(t) = u^d(t) + \hat{u}(t)$ by constructing first a primary controller specified by $u^d(t)$ for tracking $x^d(t)$ under ideal conditions and then a secondary controller specified by $\hat{u}(t)$ to compensate for the effects of disturbances.

The primary controller is specified so that system in Equation 8.2.2 will be linearized by the feedback–feedforward linearization method. The designer may choose $u^d(t)$ such that $(x^d(t), u^d(t))$ satisfy Equation 8.2.2:

$$\ddot{x}^{1d} = f^2(x^d) + B^2(x^d)u^d(t) \tag{8.2.3}$$

Since the inverse of $B^2(x)$ is the inertia matrix, it exists for all time, and Equation 8.2.3 can be solved for $u^d(t)$. The resulting equation specifies the primary controller. It also represents the (ideal) inverse dynamics $(x^d(t) \rightarrow u^d(t))$ of the manipulator. This approach is often referred to as the computed torque method [2] or the calculation of the inverse dynamics. It can be realized as illustrated in Figures 8.4a and 8.4b when a sufficiently accurate model and the numerical values of the parameters are known.

An example illustrates the design of a primary controller.

EXAMPLE 8.2.1

The dynamical model of a cylindrical robot manipulator may be expressed in the form of Equation 8.2.2 obtained from Equation 6.1.4:

$$\begin{bmatrix} \ddot{x}_1 \\ \ddot{x}_2 \\ \ddot{x}_3 \end{bmatrix} = \begin{bmatrix} [x_1 - m_A(\ell/2)/m](\dot{x}_2)^2 - k_s(x_1 - 2\ell/3)/m \\ (m_A\ell - 2mx_1)\dot{x}_1 \dot{x}_2/I_t(x_1) \\ -g - k_z(x_3 - \ell_0)/\overline{m} \end{bmatrix}$$

$$+ \begin{bmatrix} 1/m & 0 & 0 \\ 0 & 1/I_t(x_1) & 0 \\ 0 & 0 & 1/\overline{m} \end{bmatrix} \begin{bmatrix} u_1(t) \\ u_2(t) \\ u_3(t) \end{bmatrix} \tag{8.2.4}$$

where $x^1 = [x_1 \ x_2 \ x_3]' = [r \ \theta \ z]'$, $u = [u_1 \ u_2 \ u_3]' = [F_r - B_r(\dot{x}_1) \quad \tau_\theta - B_\theta(\dot{x}_2) \quad F_z - B_z(\dot{x}_3)]'$, $m = m_A + m_L$, $\overline{m} = m_A + m_L + m_h$, $I_t(x_1) = mx_1^2 - m_A\ell x_1 + m_A\ell^2/4 + I_\theta$ with I_θ a constant, and g specifies the gravitational acceleration. By comparing Equations 8.2.4 and 8.2.2, the vector functions $f^2(x)$ and $B^2(x)$ become specified.

The desired trajectory specified by $x^{1d}(t) = [r^d(t) \quad \theta^d(t) \quad z^d(t)]'$ and $\ddot{x}^{1d}(t)$ for the end-effector is given. The problem is to design a primary controller so as to make the end-effector track the desired trajectory.

The primary controller for this manipulator is determined by solving Equation 8.2.3 for $u^d(t)$:

$$u^d(t) = [B^2(x^d)]^{-1}[\ddot{x}^{1d} - f^2(x^d)] \tag{8.2.5}$$

Specifically, Equation 8.2.5 gives in this case

$$
\begin{bmatrix} u_1^d(t) \\ u_2^d(t) \\ u_3^d(t) \end{bmatrix} = \begin{bmatrix} m & 0 & 0 \\ 0 & I_t^d & 0 \\ 0 & 0 & \overline{m} \end{bmatrix}
$$

$$
\times \left\{ \begin{bmatrix} \ddot{x}_1^d \\ \ddot{x}_2^d \\ \ddot{x}_3^d \end{bmatrix} - \begin{bmatrix} [x_1^d - m_A(\ell/2)/m](\dot{x}_2^d)^2 - k_s(x_1^d - 2\ell/3)/m \\ (m_A\ell - 2mx_1^d)\dot{x}_1^d \dot{x}_2^d/I_t^d \\ -g - k_z(x_3 - \ell_0)/\overline{m} \end{bmatrix} \right\}
\tag{8.2.6}
$$

where the inertia term $I_t^d = I_t(x_1^d)$ is evaluated along the desired trajectory. Under ideal conditions, controller 8.2.6 causes the exact cancellation of the nonlinear expressions in Equation 8.2.4, that is, if $B^2(x^d) = B^2(x), f^2(x^d) = f^2(x)$ and $x(0) = x^d(0)$, then $x(t)$ equals $x^d(t)$ for all $t \geq 0$.

The implementation and the proper operation of the primary controller in Equation 8.2.6 requires that the dynamical model and the numerical values of the parameters are accurately known. If this is not the case, the cancellation of the nonlinear terms may not be perfect. Therefore, a secondary controller is needed to compensate for the errors and the effects of other disturbances. Its design is next presented.

8.2.2 SECONDARY PD-CONTROLLER DESIGN BY EIGENVALUE ASSIGNMENT METHOD

To compensate for the effects of internal modeling errors and disturbances, a secondary controller is inserted into the system. It will *regulate* small deviations of the system response about a nominal trajectory. Hence, it can be designed on the basis of a model obtained by linearizing Equation 8.2.2 about the prespecified trajectory.

The manipulator model in Equation 8.2.2 is linearized by means of the primary controller $u^d(t)$, as described in the previous section. It is implemented by introducing appropriate feedback–feedforward loops in the system. If $f^2(x) = f^2(x^d)$ and $B^2(x) = B^2(x^d)$, the primary controller causes the exact cancellation of nonlinear terms. The resulting model can be made to assume the specific linear form of Equation 6.2.19 for the design of a secondary controller $\hat{u}(t)$. For example, a PD-controller can be realized by the choice of $A_{21} = A_{22} = 0$ and $B_2\hat{u}(t) = -K_p(x^1 - x^{1d}) - K_v(\dot{x}^1 - \dot{x}^{1d})$ in Equation 6.2.24, where matrices K_p and K_v are the controller gains. Thus, the secondary controller $\hat{u}(t)$ is proportional to the position and velocity errors:

$$
\hat{u}(t) = [B^2(x)]^{-1}[-K_v(\dot{x}^1 - \dot{x}^{1d}) - K_p(x^1 - x^{1d})]
\tag{8.2.7}
$$

where the elements of matrices K_v and K_p are constants to be determined so that the specified system responses of all joint variables are obtained. When the derivative of the correction in Equation 8.2.7 refers to the velocity of revolute joints, this term represents the resolved motion rate control [3].

The total input $u(t)$ to the manipulator system is determined by the combination of the primary and secondary controllers, that is, $u(t) = u^d(t) + \hat{u}(t)$

as shown in Figures 8.4*a* and 8.4*b*. Under the assumption that $B^2(x) = B^2(x^d)$ and $f^2(x) = f^2(x^d)$, this input can be expressed as follows:

$$u(t) = [B^2(x)]^{-1}[\ddot{x}^{1d} - f^2(x^d) - K_v(\dot{x}^1 - \dot{x}^{1d}) - K_p(x^1 - x^{1d})] \qquad (8.2.8)$$

The desired acceleration $\ddot{x}^{1d}$ in Equation 8.2.8 helps to produce a faster system response (and makes the error analysis tractable).

When the composite controller 8.2.8 is substituted into the manipulator Equation 8.2.2, the dynamical model for the error $\Delta x^1(t) = x^1(t) - x^{1d}(t)$ becomes:

$$\Delta \ddot{x}^1 + K_v \Delta \dot{x}^1 + K_p \Delta x^1 = 0 \qquad (8.2.9)$$

where the N-dimensional vector $\Delta x^1(t)$ represents the positional error vector.

In a linear time invariant second-order reference model, the characteristics of the step responses are determined by the location of the eigenvalues of the plant matrix. Indeed, desirable and/or acceptable step responses are obtained when the dominant eigenvalues of the plant matrix are placed to the proper locations. It can be realized by determining the gains of the secondary feedback controller appropriately. The values of gains K_v and K_p can be adjusted (tuned) so that the eigenvalues (the natural frequencies) of system in Equation 8.2.9 assume the values that result in the acceptable responses of the system states.

Two examples are next presented to illustrate the design of a secondary controller.

EXAMPLE 8.2.2

For the cylindrical manipulator discussed in Example 8.2.1, a controller is to be designed so that the state of the system tracks the desired prespecified trajectory as closely as possible. This controller consists of two parts: (1) the primary controller that is designed in Example 8.2.1 and given by Equation 8.2.6 for the tracking under ideal conditions, and (2) a secondary controller that compensates for the modeling errors and disturbance effects. The gains of the secondary controller are to be determined so that the overshoots in the step responses of the perturbed joint variables do not exceed 15%, and the settling times should be less than 40% of the sampling period.

The composite controller for this manipulator is written on the basis of Equation 8.2.8 as the combination of the primary and secondary controllers and assuming $f^2(x) = f^2(x^d)$ and $B^2(x) = B^2(x^d)$.

$$\begin{bmatrix} u_1(t) \\ u_2(t) \\ u_3(t) \end{bmatrix} = \begin{bmatrix} m & 0 & 0 \\ 0 & I_t(x_1) & 0 \\ 0 & 0 & \overline{m} \end{bmatrix}$$

$$\times \left\{ \begin{bmatrix} \ddot{x}_1^d \\ \ddot{x}_2^d \\ \ddot{x}_3^d \end{bmatrix} - \begin{bmatrix} f_1^2(x^d) \\ f_2^2(x^d) \\ f_3^2(x^d) \end{bmatrix} - K_v \begin{bmatrix} \Delta \dot{x}_1(t) \\ \Delta \dot{x}_2(t) \\ \Delta \dot{x}_3(t) \end{bmatrix} - K_p \begin{bmatrix} \Delta x_1(t) \\ \Delta x_2(t) \\ \Delta x_3(t) \end{bmatrix} \right\} \qquad (8.2.10)$$

where $f_1^2(x) = [x_1 - m_A(\ell/2)/m](\dot{x}_2)^2 - k_s(x_1 - 2\ell/3)/m$, $f_2^2(x) = (m_A\ell - 2mx_1)\dot{x}_1\dot{x}_2/I_t(x_1)$, $f_3^2(x) = -g - k_z(x_3 - \ell_0)/\overline{m}$, with the desired value $x^d = [x_1^d \ x_2^d \ x_3^d]'$ substituted for $x = [x_1 \ x_2 \ x_3]'$, the error vector $\Delta x^1(t) = [\Delta x_1(t) \ \Delta x_2(t) \ \Delta x_3(t)]' = [x_1 - x_1^d \ x_2 - x_2^d \ x_3 - x_3^d]'$. The gain matrices K_p and K_v are to be determined.

When the controller in Equation 8.2.10 is used in the system modeled by Equation 8.2.4, it is assumed that the nonlinear terms are cancelled exactly. The dynamical behavior of the error vector governed by Equation 8.2.9 can be written in terms of the joint-variable errors in the following form:

$$\begin{bmatrix} \ddot{r} - \ddot{r}^d \\ \ddot{\theta} - \ddot{\theta}^d \\ \ddot{z} - \ddot{z}^d \end{bmatrix} + K_v \begin{bmatrix} \dot{r} - \dot{r}^d \\ \dot{\theta} - \dot{\theta}^d \\ \dot{z} - \dot{z}^d \end{bmatrix} + K_p \begin{bmatrix} r - r^d \\ \theta - \theta^d \\ z - z^d \end{bmatrix} = 0 \qquad (8.2.11)$$

The (3×3) matrices K_p and K_v may be selected so that independent joint control results. It is achieved by choosing K_p and K_v as the diagonal matrices: $K_p = \text{diag}\,[K_p^1 \ K_p^2 \ K_p^3]$ and $K_v = \text{diag}\,[K_v^1 \ K_v^2 \ K_v^3]$. With these gain matrices, the error of each joint in Equation 8.2.11 will evolve in time independently of the errors of the other joints.

The values K_p^i and K_v^i, $i = 1, 2, 3$ are chosen so as to guarantee the eigenvalue assignment corresponding to the design specifications. The characteristic equation of the dynamical model in Equation 8.2.11 for the ith joint is

$$s^2 + K_v^i s + K_p^i = 0 \qquad (8.2.12)$$

Equation 8.2.12 is compared with the desired characteristic equation that corresponds to the desired eigenvalues in Equation 8.1.5 for the second-order linear underdamped reference model. It follows that

$$K_p^i = (\omega_{ni}^d)^2 \qquad K_v^i = 2\xi_i^d \omega_{ni}^d \qquad (8.2.13)$$

where ξ_i^d and ω_{ni}^d are the desired damping and the desired natural frequency, respectively, describing the desired dynamics of the ith joint.

For the simulation, the desired (nominal) trajectory $[p_x^d(t) \ p_y^d(t) \ p_z^d(t)]$ is given; it specifies the line segment between the end points in the Cartesian base coordinate system. Using the kinematic Equation 7.3.21, the corresponding nominal values of the variables $[r^d(t) \ \theta^d(t) \ z^d(t)]$ in the joint space can be determined pointwise. As an example, the graph of the desired trajectory $r^d(t)$ is shown in Figure 8.5a. The nominal trajectory may be approximated by a staircase function that is piecewise constant. The heights of the staircase function are obtained here by evaluating the continuous trajectory variables at the time instances $kT + T/2$, where $k = 0, 1, \ldots, 49$, and the sampling period T equals 0.5 time units.

To make the end-effector follow the specified trajectory, the foregoing primary and the secondary controllers are inserted into the system. The gain matrices of the secondary controller are determined so that the independent joint control is applicable. The design specifications in this example are as follows: an overshoot less than 15% requires that $\xi_i^d \geq 0.5$ and the settling time requirement is satisfied by $\omega_{ni}^d \geq 4/(\xi_i^d 0.4T)$. By Equation 8.2.13, these specifications are met when $K_p^i = 1600$ and $K_v^i = 40$ for $i = 1, 2, 3$. The

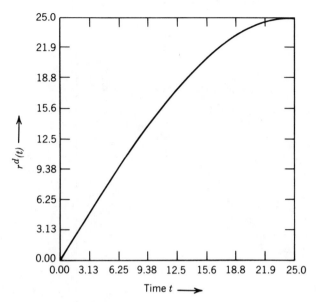

FIGURE 8.5*a*
Desired trajectory $r^d(t)$ versus time t; Example 8.2.2.

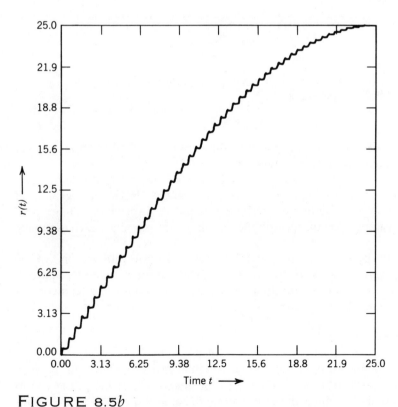

FIGURE 8.5*b*
Response $r(t)$ versus t of manipulator driven by primary and secondary (PD) controllers; Example 8.2.2.

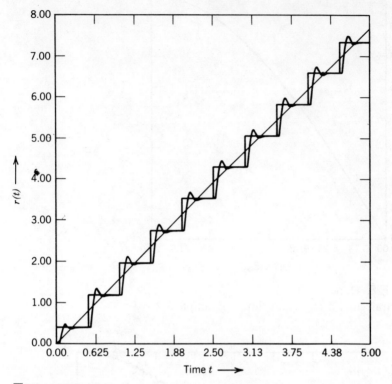

FIGURE 8.5c

Response curve during the first ten sampling periods enlarged from
Figure 8.5b; Example 8.2.2.

same values of the gains are used in the controllers of all three joints over
the entire time interval of the motion.

The system with the controller included is simulated. The simulation
results for $r(t)$ versus time are graphed in Figure 8.5b. The enlargement of
the response over the first ten sampling periods is displayed in Figure 8.5c.
It demonstrates that the design specifications are met over each sampling
period.

The velocity of each joint variable becomes zero toward the end of the
sampling periods, as seen in Figure 8.5c. This can be avoided by approximating the desired trajectory by concatenated continuous line segments, or
by the spline functions. The desired controller will still be applicable, and
acceptable system responses will be obtained.

In the design of the secondary controller for the gross motion of the
manipulator, the system model is here linearized by means of the feedback–
feedforward linearization method. It is achieved by selecting the primary
controller specified in Equation 8.2.5. The realization of the primary controller described requires that an accurate dynamical model of the system is
available. Under ideal conditions, the primary controller makes the manip-

ulator track the desired trajectory. A secondary controller compensates for modeling errors and disturbance effects. The composite controller given by Equation 8.2.8 consists of the primary and secondary controllers as shown in Figure 8.4a.

EXAMPLE 8.2.3

The dynamical model of a serial link N-joint manipulator given in Equation 5.2.9 is expressed in Lagrange's formulation as the following set of second-order vector differential equations:

$$D(q)\ddot{q} + C(q,\dot{q}) + G(q) = F \qquad (8.2.14)$$

where the dependent variable q is an N-dimensional vector, and the inertia matrix D is of dimension $(N \times N)$ having entries D_{nk} with $n, k = 1, \ldots, N$. The vector-valued function $C(q, \dot{q})$ represents the Coriolis and centrifugal terms and $G(q)$ the gravitational terms. The vector input F to the system is an N-dimensional generalized torque.

After the desired trajectory $\{ q^d(t) \; \dot{q}^d(t) \; \ddot{q}^d(t) \}$ for $t_0 \le t \le t_f$ has been determined in the planning stage, the dynamical model 8.2.14 can be used to determine the desired generalized torque vector F^d that is needed to make the manipulator track the desired trajectory. The following relation holds under ideal circumstances between the desired trajectory and the corresponding generalized torque:

$$D(q^d)\ddot{q}^d + C(q^d, \dot{q}^d) + G(q^d) = F^d \qquad (8.2.15)$$

Equation 8.2.15 specifies the primary controller. If the computed generalized torque F^d is applied to the manipulator modeled by Equation 8.2.14, the resulting motion tracks the desired trajectory under ideal circumstances, that is, when the nonlinear terms are precisely equal.

Because of disturbances and inaccuracies in the system model particularly in the parameter values, the motion of the manipulator usually deviates from the desired trajectory. Therefore, a secondary controller of PD-type is designed to compensate for these deviations.

To design a secondary controller for system 8.2.14, the dynamical model could be represented in the form of Equation 8.2.2, and then Equation 8.2.8 would specify the controller. In this example, a more direct approach will be described. It is systematic and also analytically tractable.

By comparing Equation 8.2.14 with Equation 8.2.2, the controller for this case may be written in the form similar to Equation 8.2.8:

$$F(t) = F^d(t) - D(q)[K_v(\dot{q} - \dot{q}^d) + K_p(q - q^d)] \qquad (8.2.16)$$

where the $(N \times N)$ matrices K_v and K_p are the controller gains to be determined. In Equation 8.2.16, the first term on the right side is the computed desired generalized torque. It represents the primary controller F^d that is given by Equation 8.2.15. The last two terms on the right side of Equation 8.2.16 describe the secondary controller. The controller in Equation 8.2.16 is of the PD-type with compensation for nonlinearities. The block diagram for the implementation of the controller in Equation 8.2.16 is shown in Figure 8.6a.

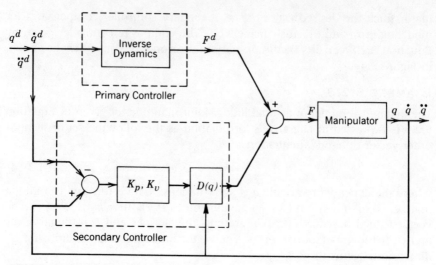

FIGURE 8.6*a*

General manipulator system with primary and secondary (PD) controllers; Example 8.2.3.

For the analysis of the control system specified by Equations 8.2.14–8.2.16, the following equalities are assumed:

$$G(q) = G(q^d) \qquad C(q, \dot{q}) = C(q^d, \dot{q}^d) \qquad D(q) = D(q^d) \qquad (8.2.17)$$

After substituting Equation 8.2.16 into Equation 8.2.14 and recognizing relations 8.2.17 in the resulting equation, the dynamical model for the error $(q - q^d)$ is obtained:

$$D(q)[(\ddot{q} - \ddot{q}^d) + K_v(\dot{q} - \dot{q}^d) + K_p(q - q^d)] = 0 \qquad (8.2.18)$$

Equation 8.2.18 can be premultiplied by the inverse of the inertia matrix that exists for all time. The substitution of $\Delta q(t) = q(t) - q^d(t)$ leads to

$$\Delta \ddot{q} + K_v \Delta \dot{q} + K_p \Delta q = 0 \qquad (8.2.19)$$

where $\Delta q(t) = \text{col}[\Delta q_1(t) \cdots \Delta q_N(t)]$. The initial conditions $\Delta q(0)$ and $\Delta \dot{q}(0)$ are assumed to be known for Equation 8.2.19. For this specific system, the block diagram is redrawn in Figure 8.6*b*.

The gain matrices K_v and K_p are adjusted so as to force the error $\Delta q(t)$ to decay to zero in a desirable manner. For example, matrices K_v and K_p can be chosen so that the perturbation variables Δq_i, $i = 1, \ldots, N$ of the joints in Equation 8.2.19 become uncoupled. Thus, the independent joint control can be applied, as in Example 8.2.1. The characteristic equation of the ith joint can be expressed in the form of Equation 8.2.12. The controller gains can then be designed in the same manner as in Example 8.2.2.

The gross motion of the end-effector in a manipulator is here controlled by designing a primary controller mainly for tracking and a secondary controller

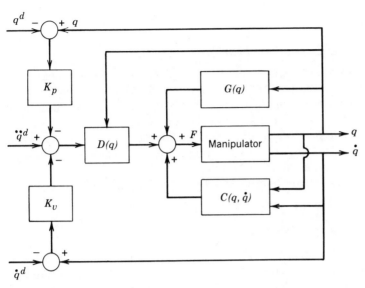

FIGURE 8.6*b*

Implementation of (secondary) PD-controller with nonlinear compensation; Example 8.2.3.

for regulation. The primary controller makes the motion of the end-effector follow the desired trajectory under ideal circumstances. If the cancellation of the nonlinear terms is exact, the primary controller inserted makes the total system linear. This model can serve as the basis for designing a secondary controller that is here selected to be a PD-controller. The gains of this controller are adjusted by solving an eigenvalue assignment problem or simply by experimentation [4]. The values of the gains are determined for the entire time interval of the motion. The designer could alternatively choose a PID-controller with adjustable gains. It appears, however, that the addition of the I-component in Equation 8.2.7 may not result in noticeable improvements in the performance. On the other hand, a multivariable PD-controller that takes into account interactions between the joint variables may significantly improve in certain applications the dynamic behavior of the manipulator, particularly at high speeds [5].

When the primary controller in Equation 8.2.5 causes the exact cancellation of the nonlinear terms in the Equations 8.2.2 of the manipulator dynamics, then the dynamical behavior of the tracking error is governed by Equation 8.2.9. However, if the cancellation of the nonlinear terms by the primary controller is not perfect, Taylor's series expansions can be used to determine first-order approximations for the imperfectly cancelled terms. By including these terms in the model, a linearized variational model results. It can also serve as the basis to design a secondary controller. We will next discuss the construction of a secondary state feedback controller using such a variational model.

8.2.3 SECONDARY VARIATIONAL FEEDBACK CONTROLLER BY EIGENVALUE ASSIGNMENT

In the previous section, the design of the secondary controller is based on the model obtained by the feedback–feedforward linearization method. When the expression for the primary control from Equation 8.2.5 is substituted into Equation 8.2.2 of the manipulator motion, a linear model is obtained if the nonlinear terms are precisely equal. However, if the cancellation is not exact, then first-order approximations for the nonlinear terms can be determined to obtain a variational model.

Equations 8.2.2 for the motion of a serial link N-joint manipulator is in the form

$$\ddot{x}^1 = f^2(x) + B^2(x)u(t) \qquad (8.2.20)$$

where $f^2(x)$ and $B^2(x)$ are vector-valued continuous functions of the state variable and $x = [(x^1)'(x^2)']'$. The dynamical model in Equation 8.2.4 is in the form of Equation 8.2.20. If control $u(t) = u^d(t) + \delta u(t)$ is substituted into Equation 8.2.20, and Equation 8.2.3 is subtracted from the resulting equation, the following model for the error is obtained:

$$\ddot{x}^1 - \ddot{x}^{1d} = [f^2(x) - f^2(x^d)] + [B^2(x) - B^2(x^d)]u^d(t) + B^2(x)\delta u \qquad (8.2.21)$$

where $\delta u(t)$ represents a variational control. If the nonlinear terms satisfy $f(x) = f(x^d)$ and $B(x) = B(x^d)$, the choice for $\delta u(t)$ in Equation 8.2.21 equal to $\hat{u}(t)$ in Equation 8.2.7 leads to Equation 8.2.9. However, the cancellation of the nonlinear terms in Equation 8.2.21 is usually not exact.

Because the dynamical model given in Equation 8.2.1 is not precise, and due to disturbances acting on the system, a secondary controller is usually necessary to compliment the primary controller and to compensate for the errors. Since the disturbed motion can be expected to stay in the small neighborhood of the nominal desired trajectory, a linearized model should suffice to describe this motion for the controller design. If $f^2(x)$ and $B^2(x)$ in Equation 8.2.21 are expanded into Taylor's series about (x^d, u^d), and only the first-order terms in the expansions are retained, the resulting equation can be written in a state variable form:

$$\delta\dot{x} = \overline{A}\delta x(t) + \overline{B}\delta u(t) + e(t) \qquad (8.2.22)$$

where the variables $\delta x(t)$ and $\delta u(t)$ represent first-order (weak) variations of $x(t)$ and $u(t)$ about $x^d(t)$ and $u^d(t)$, respectively. The plant matrix $\overline{A} = A(x^d)$ contains the Jacobian matrix of $[f^2(x) + B^2(x)u^d]$ with respect to the state vector x evaluated at the desired trajectory. The control matrix $\overline{B} = B^2(x^d)$, and $e(t)$ signifies bounded modeling errors that may often be neglected in the design. Equation 8.2.22 reveals that the imperfect cancellation of the first two terms on the right side of Equation 8.2.21 gives rise to the $\overline{A}$-matrix in Equation 8.2.22 by Taylor's series expansion. The $(2N \times 2N)$ matrix $\overline{A}$ and the $(2N \times N)$ matrix $\overline{B}$ in the model of an N-joint manipulator usually vary with time. However, they may be approximated as constant matrices over a sufficiently short sampling period. The variational input $\delta u(t)$ can then be determined so as to control the behavior of $\delta x(t)$, the variational variable.

The dynamical model in Equation 8.2.22 is linear, and it is assumed to be completely controllable. A secondary controller can be designed on the basis of the linear system theory. Particularly, a secondary controller can be constructed by regarding the model in Equation 8.2.22 as a linear time invariant system over a sampling period.

The problem considered here is to design a state feedback controller so that the initial state $\delta x(0)$ will asymptotically (as $t \to \infty$) be transferred to the origin. This is commonly known as a regulator problem. The controller gains are determined so that the eigenvalues of the constant plant matrix of the system with the controller will assume the values $s_1^d, \ldots, s_{2N}^d$ that correspond to acceptable characteristics in the step response.

The state feedback controller is chosen under the assumption that all components of the state vector are accessible for measurements:

$$\delta u = K \delta x \qquad (8.2.23)$$

where the constant gain matrix K is of dimension $(N \times 2N)$. The controller defined by Equation 8.2.23 exhibits a static behavior. Thus, the dimension of the state vector of the overall system does not change due to the controller. In a manipulator, vector δx usually contains the position and velocity variables of the joints making the control δu proportional to the position and velocity of the joints. The controller 8.2.23 can be considered as a PD-controller. The block diagram of the system is shown in Figure 8.7.

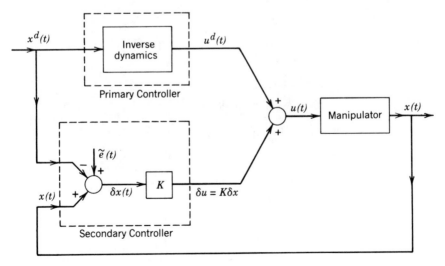

FIGURE 8.7

Manipulator system driven by primary controller and secondary controller designed by eigenvalue assignment.

The substitution of Equation 8.2.23 into Equation 8.2.22 gives the closed-loop variational model

$$\delta \dot{x} = (\overline{A} + \overline{B}K)\, \delta x + e(t) \qquad (8.2.24)$$

where matrix $(\overline{A} + \overline{B}K)$ is assumed to be time invariant and the initial value $\delta x(0)$ is given.

Since matrices $\overline{A}$ and $\overline{B}$ are assumed to form a completely controllable pair, a constant matrix K can be determined so that the eigenvalues of $(\overline{A} + \overline{B}K)$ will be $s_1^d, \ldots, s_{2N}^d$. Indeed, the unknown parameter matrix K of the controller is determined so that the characteristic equation of the system will be the same as the desired characteristic equation:

$$|sI - \overline{A} - \overline{B}K| = (s - s_1^d) \cdots (s - s_{2N}^d) \qquad (8.2.25)$$

By equating the coefficients of equal powers of s, the equations to solve for the unknown elements of matrix K are obtained. Thus, the feedback controller in Equation 8.2.23 becomes specified. When the controller 8.2.23 is used in system 8.2.22, the initial state $\delta x(0)$ is driven to zero while the transients exhibit an acceptable behavior, as determined by the desired eigenvalues.

The approach described is next illustrated by an example.

EXAMPLE 8.2.4

The dynamical model of a three-joint cylindrical robot manipulator discussed in Example 8.2.1 is given by Equations 8.2.4 and 6.1.5 in the framework of Lagrange's formulation, where $N = 3$. Viscous friction is assumed; it opposes the motion. The model is in the general form of Equations 8.2.1:

$$
\begin{bmatrix} \dot{x}_1 \\ \dot{x}_2 \\ \dot{x}_3 \\ \dot{x}_4 \\ \dot{x}_5 \\ \dot{x}_6 \end{bmatrix} =
\begin{bmatrix}
x_4 \\
x_5 \\
x_6 \\
[x_1 - m_A \ell/(2m)]\, x_5^2 - k_s(x_1 - 2\ell/3)/m \\
[m_A \ell - 2mx_1]\, x_4 x_5/I_t(x_1) \\
-g - k_z(x_3 - \ell_0)/\overline{m}
\end{bmatrix}
$$

$$
+ \begin{bmatrix}
0 & 0 & 0 \\
0 & 0 & 0 \\
0 & 0 & 0 \\
1/m & 0 & 0 \\
0 & 1/I_t(x_1) & 0 \\
0 & 0 & 1/\overline{m}
\end{bmatrix}
\begin{bmatrix}
F_r - B_r(x_4) \\
\tau_\theta - B_\theta(x_5) \\
F_z - B_z(x_6)
\end{bmatrix}
\qquad (8.2.26)
$$

where the state vector $x = \mathrm{col}[x_1\, x_2\, x_3\, x_4\, x_5\, x_6] = \mathrm{col}[r\, \theta\, z, \dot{r}\, \dot{\theta}\, \dot{z}]$, $m = m_A + m_L$, $\overline{m} = m_A + m_h + m_L$, and $I_t(x_1) = mx_1^2 - m_A \ell x_1 + I_\theta + m_A \ell^2/4$

with I_θ a constant. The desired path $x^d = \text{col } [r^d\ \theta^d\ z^d,\ \dot{r}^d\ \dot{\theta}^d\ \dot{z}^d]$ is given as a function of time for $0 \le t \le t_f$.

The problem is to design a primary controller and a secondary controller for the gross motion of the given manipulator. The primary controller should make the end-effector track the desired nominal trajectory $x^d(t)$ under ideal conditions. The secondary controller should compensate for the effects of disturbances, and it should keep the variational state vector close to zero.

The given values of the desired trajectory $x^d(t)$ are substituted into Equation 8.2.26. The resulting equations can then be solved for the generalized torque vector $\text{col}[F_r^d(t)\ \ \tau_\theta^d(t)\ \ F_z^d(t)]$ needed to satisfy the equations of the desired motion. Thus, the primary controller is determined by the (ideal) inverse dynamics. It is given by Equation 8.2.6.

The design of a secondary state feedback controller is based on a linearized model determined under the assumption that the cancellation of the nonlinear terms in the model is not exact. A linear model in the form of Equation 8.2.22 is obtained by writing the variational equations of motion about the desired trajectory. Such a variational model for the cylindrical robot manipulator under consideration is derived in Example 6.2.1 and given by Equation 6.2.10. The variational model is rewritten here:

$$
\begin{bmatrix} \delta\dot{x}_1 \\ \delta\dot{x}_2 \\ \delta\dot{x}_3 \\ \delta\dot{x}_4 \\ \delta\dot{x}_5 \\ \delta\dot{x}_6 \end{bmatrix}
=
\begin{bmatrix}
0 & 0 & 0 & 1 & 0 & 0 \\
0 & 0 & 0 & 0 & 1 & 0 \\
0 & 0 & 0 & 0 & 0 & 1 \\
a_{41}(x^d) & 0 & 0 & -B_r/m & a_{45}(x^d) & 0 \\
a_{51}(x^d) & 0 & 0 & a_{54}(x^d) & a_{55}(x^d) & 1 \\
0 & 0 & -k_z/\overline{m} & 0 & 0 & -B_z/\overline{m}
\end{bmatrix}
\begin{bmatrix} \delta x_1 \\ \delta x_2 \\ \delta x_3 \\ \delta x_4 \\ \delta x_5 \\ \delta x_6 \end{bmatrix}
$$

$$
+
\begin{bmatrix}
0 & 0 & 0 \\
0 & 0 & 0 \\
0 & 0 & 0 \\
1/m & 0 & 0 \\
0 & 1/I_t(x_1^d) & 0 \\
0 & 0 & 1/\overline{m}
\end{bmatrix}
\begin{bmatrix} \delta F_r \\ \delta \tau_\theta \\ \delta F_z \end{bmatrix}
\quad (8.2.27)
$$

where δx_i represents a (weak) variation in the ith component x_i of the x-vector, $i = 1, \ldots, 6$, $a_{41}(x^d) = (x_5^d)^2 - k_s/m$, $a_{45}(x^d) = [x_1^d - (m_A/m)\ell/2](2x_5^d)$, $a_{51}(x^d) = (m_A\ell - 2mx_1^d)\ [(m_A\ell - 2mx_1^d)x_4^d x_5^d + \tau_\theta^d - B_\theta x_5^d]/I_t^2(x_1^d) - 2mx_4^d x_5^d/I_t(x_1^d)$, $I_t(x_1^d) = I_\theta + m_A\ell^2/4 + m(x_1^d)^2 - m_A\ell x_1^d$, $a_{54}(x^d) = (m_A\ell - 2mx_1^d)x_5^d/I_t(x_1^d)$, $a_{55}(x^d) = [(m_A\ell - 2mx_1^d)x_4^d - B_\theta]/I_t(x_1^d)$, and input $[\delta F_r\ \ \delta\tau_\theta\ \ \delta F_z]' = \delta u$ describes the variation of $[F_r\ \tau_\theta\ F_z]$ about the desired (nominal) values. Equation 8.2.27 is valid in a sufficiently small neighborhood of the desired trajectory.

A comparison of Equation 8.2.27 with Equation 8.2.21 reveals that the

plant matrix in Equation 8.2.27 arises due to the imperfect cancellation of
the nonlinear terms in Equation 8.2.21. Moreover, if the variational control in Equation 8.2.27 is chosen in the form of $\hat{u}(t)$ in Equation 8.2.7,
then Equation 8.2.27 for the first-order variations represents an error model
corresponding to Equation 8.2.9.

The dependence of the parameters on the desired trajectory is indicated in
Equation 8.2.27. Since the nominal path usually changes with time, Equation 8.2.27 represents a linear time varying system. As a first-order approximation, the coefficients in Equation 8.2.27 may be considered as constants
over a sufficiently short sampling period. Thus, the dynamical model 8.2.27
would appear as a linear time invariant system. Its behavior over a sampling
period may then be designed using the methods applicable to linear time
invariant systems, such as the eigenvalue assignment, root locus, Bode's diagram, Nyquist plot, or Nichols chart, which are based on the transfer function
of a system.

The problem here is to design a secondary state feedback controller for the
system in Equation 8.2.27 so that the joints are independently controlled,
that is, the coupling terms are zero, and the eigenvalues are $s_{i,1}^d$, $s_{i,2}^d =
-\xi_i^d \omega_{ni}^d \pm j\omega_{ni}^d [1 - (\xi_i^d)^2]^{1/2}$ where ω_{ni}^d and ξ_i^d are known design parameters
and $i = 1,2,3$.

The problem is solved by introducing the state feedback controller for the
variational model in the form

$$
\begin{bmatrix} \delta F_r \\ \delta \tau_\theta \\ \delta F_z \end{bmatrix} =
\begin{bmatrix}
K_{11} & K_{12} & K_{13} & K_{14} & K_{15} & K_{16} \\
K_{21} & K_{22} & K_{23} & K_{24} & K_{25} & K_{26} \\
K_{31} & K_{32} & K_{33} & K_{34} & K_{35} & K_{36}
\end{bmatrix}
\begin{bmatrix} \delta x_1 \\ \delta x_2 \\ \delta x_3 \\ \delta x_4 \\ \delta x_5 \\ \delta x_6 \end{bmatrix}
\tag{8.2.28}
$$

where the gains K_{ij}, $i = 1, 2, 3, j = 1, \ldots, 6$ are so determined that the
eigenvalue assignments will be fulfilled.

The substitution of Equation 8.2.28 into Equation 8.2.27 leads to the
homogeneous vector differential equation

$$
\begin{bmatrix} \delta \dot{x}_1 \\ \delta \dot{x}_2 \\ \delta \dot{x}_3 \\ \delta \dot{x}_4 \\ \delta \dot{x}_5 \\ \delta \dot{x}_6 \end{bmatrix}
= (\bar{A} + \bar{B}K)
\begin{bmatrix} \delta x_1 \\ \delta x_2 \\ \delta x_3 \\ \delta x_4 \\ \delta x_5 \\ \delta x_6 \end{bmatrix}
\tag{8.2.29}
$$

$$\bar{A} - \bar{B}K = \begin{bmatrix} 0 & 0 & 0 & 1 & 0 & 0 \\ 0 & 0 & 0 & 0 & 1 & 0 \\ 0 & 0 & 0 & 0 & 0 & 1 \\ \bar{a}_{41} + \dfrac{K_{11}}{m} & \dfrac{K_{12}}{m} & \dfrac{K_{13}}{m} & \dfrac{(K_{14} - B_r)}{m} & \bar{a}_{45} + \dfrac{K_{15}}{m} & \dfrac{K_{16}}{m} \\ \bar{a}_{51} + \dfrac{K_{21}}{I_t^d} & \dfrac{K_{22}}{I_t^d} & \dfrac{K_{23}}{I_t^d} & \bar{a}_{54} + \dfrac{K_{24}}{I_t^d} & \bar{a}_{55} + \dfrac{K_{25}}{I_t^d} & \dfrac{K_{26}}{I_t^d} \\ \dfrac{K_{31}}{\bar{m}} & \dfrac{K_{32}}{m} & \dfrac{(K_{33} - k_z)}{\bar{m}} & \dfrac{K_{34}}{\bar{m}} & \dfrac{K_{35}}{\bar{m}} & \dfrac{(K_{36} - B_z)}{\bar{m}} \end{bmatrix}$$

where $I_t^d = I_t(x^d)$, $\bar{a}_{41} = a_{41}(x^d)$, $\bar{a}_{45} = a_{45}(x^d)$, $\bar{a}_{51} = a_{51}(x^d)$, $\bar{a}_{54} = a_{54}(x^d)$, and $a_{55} = a_{55}(x^d)$. The differential equations for $\delta \dot{x}_i$ and $\delta \dot{x}_{i+3}$ with $i = 1, 2, 3$ describe the dynamics of the ith joint. The equations of a joint must not contain variables of the other joints, since the dynamics of the joint variables should be decoupled.

The controller gains K_{ij} are now chosen so that each joint will be controlled independently of the motions of the other two joints. Since $\delta \dot{x}_i$ and $\delta \dot{x}_{i+3}$ describe the equations of motion for joint i, these two equations can be coupled with each other, but not with the other joint equations. It implies that the lower left (3 × 3) submatrix and the lower right (3 × 3) submatrix in the plant matrix of Equation 8.2.29 should be diagonal. It is achieved by selecting $K_{21} = -I_t^d \bar{a}_{51}$, $K_{31} = 0$, $K_{12} = 0$, $K_{32} = 0$, $K_{13} = 0$, $K_{23} = 0$, and $K_{24} = -I_t^d \bar{a}_{54}$, $K_{34} = 0$, $K_{15} = -m\bar{a}_{45}$, $K_{35} = 0$, $K_{16} = 0$, and $K_{26} = 0$. With these values for the gains, Equation 8.2.29 is rewritten

$$\bar{A} - \bar{B}K = \begin{bmatrix} 0 & 0 & 0 & 1 & 0 & 0 \\ 0 & 0 & 0 & 0 & 1 & 0 \\ 0 & 0 & 0 & 0 & 0 & 1 \\ \bar{a}_{41} + \dfrac{K_{11}}{m} & 0 & 0 & \dfrac{(K_{14} - B_r)}{m} & 0 & 0 \\ 0 & \dfrac{K_{22}}{I_t^d} & 0 & 0 & \bar{a}_{55} + \dfrac{K_{25}}{I_t^d} & 0 \\ 0 & 0 & \dfrac{(K_{33} - k_z)}{\bar{m}} & 0 & 0 & \dfrac{(K_{36} - B_z)}{\bar{m}} \end{bmatrix}$$

$$(8.2.30)$$

Equation 8.2.30 represents a variational system model with a secondary controller. The independent joint dynamics is evident in Equation 8.2.30. Indeed, the interactions between the three joints have been decoupled.

The transient behavior of each joint can next be adjusted by determining the remaining gain constants so that the eigenvalues are placed to the desired locations. The characteristic equation of the plant matrix in Equation 8.2.30 can be determined to obtain

$$\left(s^2 + \frac{B_r - K_{14}}{m}s - \bar{a}_{41} - \frac{K_{11}}{m}\right)\left(s^2 - a_1 s - \frac{K_{22}}{I_t^d}\right)$$

$$(s^2 + a_2 s + a_3) = 0 \quad (8.2.31)$$

where $a_1 = \bar{a}_{55} + K_{25}/I_t^d$, $a_2 = (B_z - K_{36})/\bar{m}$ and $a_3 = (k_z - K_{33}/\bar{m}$. In Equation 8.2.31, each term in the product represents the characteristic polynomial of a joint. For example, the characteristic equation for joint 1 is

$$s^2 + \frac{B_r - K_{14}}{m}s - \bar{a}_{41} - \frac{K_{11}}{m} = 0 \quad (8.2.32)$$

Using the assigned eigenvalues $s_{1,1}^d$ and $s_{1,2}^d$ for the dynamics of joint one, the desired characteristic equation can be written

$$(s - s_{1,1}^d)(s - s_{1,2}^d) = 0 \quad (8.2.33)$$

Equating the coefficients of the equal powers of s in Equations 8.2.32 and 8.2.33 leads to

$$\frac{B_r - K_{14}}{m} = -(s_{1,1}^d + s_{1,2}^d) \quad (8.2.34)$$

$$-\bar{a}_{41} - \frac{K_{11}}{m} = s_{1,1}^d s_{1,2}^d \quad (8.2.35)$$

where $s_{1,1}^d + s_{1,2}^d = -2\xi_1^d \omega_{n1}^d$, and $s_{1,1}^d s_{1,2}^d = (\omega_{n1}^d)^2$. Equations 8.2.34 and 8.2.35 can now be solved for K_{11} and K_{14}:

$$K_{11} = m[-\bar{a}_{41} - (\omega_{n1}^d)^2] \qquad K_{14} = B_r - 2\xi_1^d \omega_{n1}^d m \quad (8.2.36)$$

By a similar reasoning, the remaining gain constants are determined:

$$K_{22} = -(\omega_{n2}^d)^2 I_t(x^d) \qquad K_{25} = -[\bar{a}_{55} + 2\xi_2^d \omega_{n2}^d] I_t(x^d) \quad (8.2.37)$$

$$K_{33} = k_z - \bar{m}(\omega_{n3}^d)^2 \qquad K_{36} = B_z - 2\xi_3^d \omega_{n3}^d \bar{m} \quad (8.2.38)$$

The gains in the state feedback controller in Equation 8.2.28 are now determined. The values of the gains depend on the (operating) point (x^d, u^d). If the desired trajectory changes from one sampling period to the next, the gain values must be recalculated for each sampling interval. Indeed, the plant matrix $\bar{A}$, the control matrix $\bar{B}$, and the gain matrix K should be updated at each sampling instant if it is computationally possible. For the implementation, the primary and the secondary controllers are combined to generate $u(t)$ as shown in Figure 8.7.

The approach described can be applied to the design of controllers for linear time invariant systems. Usually, the position and even velocity components of the joint variables can accurately be measured. If all components of the state vector, for some reason, are not directly available for the controller in Equation 8.2.23, only the accessible components of the state vector can be used in the controller. However, it is possible to reconstruct the components of the state vector on the basis of the output measurements under certain conditions. These reconstructed components can then be used for forming the state feedback controller.

We have now discussed the design of primary and secondary controllers for the gross motion of the end-effector in a manipulator. The realization of the primary controller requires that the dynamical model and the numerical values of the parameters are known at a sufficient degree of accuracy. The calculations of the values of the nonlinear expressions often require a considerable amount of computer time and memory. The realization of the secondary controller is straightforward, although the determination of a linearized model could be a tedious task.

When the methods discussed previously are used to construct the controllers, the design specifications are not usually very specific. Rather, they represent "ball park figures" based on sound engineering judgments; this is typical in the conventional (classical) controller design. No optimality criterion on the performance of the system is used explicitly in the design. If the model is linear and time invariant, and if a linear controller such as a PD-controller is used, it is possible to find a quadratic criterion that the controller minimizes. It can be determined by solving the inverse optimum control problem [6]. If the optimum control theory is applied to the design of controllers for manipulators, the design specifications must be expressed mathematically in a precise manner.

8.3 DISCUSSION ON OPTIMUM CONTROLLER DESIGN

The optimum control theory in the framework of Pontryagin's minimum (maximum) principle or of the dynamic programming is a powerful tool for designing controllers in industrial applications. Because of the complexity of the dynamical constraint equations, the direct use of the optimum control theory to design controllers for the gross motion of six-joint robot manipulators is not very appealing. However, some approaches in the framework of optimum control theory have been reported [7–11].

Dynamic programming is applied in [7] to the design of an approximately optimal controller for the gross motion of a manipulator; however, its applicability is limited. The solution to the minimum energy problem with a fixed terminal time is reported in [8] with a description of implementation.

When the plant equations are linear and the criterion to be minimized is quadratic in the state variables and the control inputs, the resulting optimal controller is linear in the state variables. Since the variational (perturbation) model is linear, the designer can determine an optimal secondary controller by minimizing a criterion functional that is quadratic in the control input (i.e., the total energy of the input) and in the error describing the difference between the actual and desired trajectory. The resulting controller is in the form of a state feedback controller. Thus, in applying the optimum control theory, the mathematical solution specifies the linear form for the secondary controller. In Section 8.2.3, the linear form is selected by the designer, and the gains are determined by the eigenvalue assignment method. It should be noticed, however, that the determination of a variational model for a six-joint manipulator is tedious, and in most cases impractical. On the other hand, when the feedforward–feedback linearization method is used, an optimal controller could be designed on the basis of the optimum control theory. If the imperfect cancellation is modeled as a deterministic disturbance, its form should be known in advance for designing the optimum controller. This requirement hinders the use of the optimum control theory for designing an optimal secondary controller for the gross motion of a manipulator.

The optimum control theory is applied in [9] to determine a near-minimum time solution for a dynamical model that results from linearizing the model about the final target state. The minimum time solution is determined on the phase plane for the decoupled model representing independent joint dynamics. The secondary controller in this case is an open loop control, which is a drawback. (Why?) The difficulty of the approach is in the implementation of the switching times. Moreover, the linear model is valid only in the neighborhood of the terminal point.

A variable structure controller proposed in [10] also makes use of switching curves for bounded inputs. When the state variables change causing the switching function to change sign, the input is switched from one boundary to the other one. As the result of this bang-bang control, the state variables stay close to the switching curve and "slide" toward the desired final point, that is, the system operates in a sliding mode. The practicality of this approach is still to be tested.

An alternative to the minimum-time problem is presented in [11] by combining the duration of the motion and the measure of the fuel in a performance functional. The near-optimal control is determined using a simplified decoupled dynamics resulting from an interesting averaging procedure applied to the nonlinear terms. The consequence is that the optimal switching curves are approximated based on an averaging procedure, and the approximating terms are recalculated at each sampling instant. Simulation experiments are presented in [11] to illustrate the applicability of the approach.

The aforementioned techniques illustrate attempts to use the deterministic (continuous-time) optimum control theory to design controllers for the gross motion of robot manipulators. At the present, the practicality of the optimal solutions proposed is uncertain. It is difficult to critically evaluate advantages and disadvantages of optimal controllers relative to the conventional controllers. The future research hopefully will provide guidelines to select controllers for the gross motion of robot manipulators.

8.4 SUMMARY

We have discussed the specifications for designing controllers for the gross motion of robot manipulators. A primary controller is designed so that the manipulator tracks the desired trajectory under ideal circumstances. It is determined so that the nonlinear terms in the dynamical model cancel and so that the resulting system model is linear. A secondary controller is then designed as a regulator on the basis of the linearized model. Its purpose is to control the transients by compensating for the effects caused by inaccuracies in the model and disturbances acting on the system. The secondary controller compliments the primary controller.

If the cancellation of the nonlinear terms is not exact, a variational model that is always linear can be determined as a first-order approximation. A secondary controller can then be designed on the basis of this model as, for example, a PD (or PID) controller that operates on the position and/or velocity error, or as a variational state feedback controller that is a linear function of the variational state vector. The gains of the secondary controller can be adjusted by the eigenvalue assignment method or experimentally.

The implementation of the control schemes presented is in some aspects different from that of many computer-controlled systems. The primary controller requires that the model with the numerical parameters is accurately known. Moreover, a considerable amount of calculation time is often consumed on a digital computer during every sampling period. It determines a lower bound on the sampling period. The implementation of the controller algorithms discussed is straightforward, although the communication channels for interfacing the manipulator and the digital computer may require special design considerations. Experienced computer engineers will make the system operate smoothly and successfully.

REFERENCES:

[1] J. Y. LUH, "Conventional Controller Design for Industrial Robots — A Tutorial," *IEEE Transactions on Systems, Man and Cybernetics*, Vol. SMC-13, No. 3, May/June 1983.

[2] R. C. PAUL, "Modeling Trajectory, Calculation and Servoing of a Computer Controlled Arm," Artificial Intelligence Lab, Stanford University, Stanford, CA, A.I. Memo 177, September 1972.

[3] D. E. WHITNEY, "Resolved Motion Rate Control of Manip-
     ulators and Human Prostheses," *IEEE Transactions Man-
     Machine Systems*, Vol. MMS-10, pp. 47–53, June 1969.

[4] J. PENTTINEN and H. N. KOIVO, "Multivariable Tuning
     Regulators for Unknown Systems," *Automatica*, Vol. 16, pp.
     393–398, 1980.

[5] A. PELTOMAA and A. J. KOIVO, "Compensation of Non-
     linearities and Interactions in Industrial Robots," *Symposium
     on Robot Control '85*, Barcelona, Spain, November 1985.

[6] R. E. KALMAN, "When is a Linear Control System Opti-
     mal?" *Transactions of the ASME, Journal of Basic Engineering*,
     pp. 51–60, March 1964.

[7] G. S. SARIDIS and C. S. LEE, "An Approximation Theory
     of Optimal Control for Trainable Manipulators," *IEEE Tran-
     sactions on Systems, Man and Cybernetics*, Vol. SMC-9, pp.
     152–159, March 1979.

[8] W. E. SNYDER and W. A. GRUVER, "Microprocessor
     Implementation of Optimal Control for Robotic Manipu-
     lator Systems," *Proc. of the 18th IEEE Conference on Deci-
     sion and Control*, Vol. 2, Fort Lauderdale, FL, pp. 839–841,
     December 1979.

[9] M. E. KAHN and B. ROTH, "The Near-Minimum Time Con-
     trol of Open-Loop Articulated Kinematic Chains," *Journal
     of Dynamic Systems, Measurements and Control (ASME)*,
     pp. 164–172, September 1971.

[10] K. K. D. YOUNG, "Controller Design for a Manipulator
     Using Theory of Variable Structure Systems," *IEEE Transa-
     ctions on Systems, Man and Cybernetics*, Vol. SMC-8, pp.
     101–109, February 1978.

[11] B. K. KIM and K. G. SHIN, "Suboptimal Control of Indus-
     trial Manipulators with a Weighted Time-Fuel Criterion,"
     IEEE Transactions on Automatic Control, Vol. AC-30, No. 1,
     pp. 1–10, January 1985.

PROBLEMS

8.1 The step response of a joint variable in a manipulator
     should be critically damped, and the settling time should
     not exceed 30% of the sampling period T, which is 28 ms.

     a. Determine the desired second-order transfer function
           of the reference model.

b.   Specify the location of the desired poles (eigenvalues) for the reference model.

c.   State two consequences resulting from choosing a short settling time.

8.2   The state feedback controller is given in Equation 8.2.23 in a general form.

a.   Rewrite the state feedback controller equation by separating the part of the controller that operates on the joint position and the part that operates on the velocity.

b.   Show that the result of part (a) can be expressed in the form of a PID-controller. Specify each part explicitly.

8.3   The variational model of a cylindrical manipulator with the controller is given as the homogeneous differential Equation 8.2.30. Determine the eigenvalues of the plant matrix.

8.4   If the Coriolis, centripetal, and gravity terms are cancelled, the dynamical model in Equation 8.2.14 can be written:

$$\ddot{q} = D^{-1}(q)F$$

Assume a two-joint manipulator (see Equation 5.2.41), that is, $N = 2$.

a.   Express this simplified model as a state variable equation.

b.   Can a transfer function matrix be determined for the system? Which assumption is needed to solve this problem?

c.   Suggest primary and secondary controllers for the given manipulator system.

8.5   The variational model of the cylindrical manipulator in Example 8.2.4 given in Equation 8.2.27 is assumed to be time invariant over a sampling period. The variational output $\delta y(t)$ is

$$\delta y(t) = \begin{bmatrix} 1 & 0 & 0 & 0 & 0 & 0 \\ 0 & 0 & 1 & 0 & 0 & 0 \\ 0 & 0 & 0 & 0 & 1 & 0 \end{bmatrix} \delta x(t)$$

a.   Determine the transfer function matrix that relates the variational input vector $\delta U(s)$ and the variational output vector $\delta Y(s)$.

b.   Is it possible to determine the transfer function matrix for the dynamical model given in Equation 8.2.26? Explain.

8.6   A cylindrical manipulator is modeled by Equation 6.1.4.
Define the state vector as $x = \text{col}[r \; \dot{r} \; \theta \; \dot{\theta} \; z \; \dot{z}]$.

   a.   Determine the state variable representation for the
equations of motion.

   b.   For the manipulator a desired trajectory

$$x^d = [r^d \; \dot{r}^d \; \theta^d \; \dot{\theta}^d \; z^d \; \dot{z}^d]'$$

is given. Determine a primary controller.

   c.   Obtain the variational model about the desired trajectory in the form of Equation 8.2.27.

   d.   Introduce the state feedback controller given in Equation 8.2.28 for the system. Determine the gains so
that the independent joint control is achieved. Obtain
the state variable model for the system. Compare this
model with Equation 8.2.30.

   e.   Determine the remaining gains so that the eigenvalues
are placed to the locations $s_{i,1}^d$ and $s_{i,2}^d$, $i = 1, 2, 3$ corresponding to the poles of a second-order underdamped
transfer function, as specified in Example 8.2.4.

8.7   For a two-joint planar manipulator, the dynamical model
is given in Table 5.4, where τ_i is the input torque on the
motor shaft of joint i, and $i = 1$ and 2. Assume that the
actuators are separately excited DC motors. The input to
the motor of joint i is V_i and the output is τ_i.

   a.   Write the manipulator model given in Table 5.4 in the
form of Equation 8.2.1. Identify x^1, x^2, f^2, and B^2.

   b.   Determine the dynamical model in the state variable
form for the entire system, where the input is the motor
voltage vector, and the output is the joint position
vector.

   c.   Design a primary controller for the system so that the
output tracks the desired trajectory $\{ \theta_i^d(t), \dot{\theta}_i^d(t), \ddot{\theta}_i^d(t) \}$
under ideal conditions. Moreover, the model should be
linearized by means of the primary controller.

   d.   Suggest a secondary controller for the linearized system. Use a PD-controller. Determine the gains so that
the independent joint control applies.

   e.   Determine the remaining gains of the joint controllers
so that the eigenvalues of the system with the chosen
controller are at the locations that correspond to 5%
overshoot and a settling time of $0.6T$ ($T =$ sampling
period) in the step response.

      f. If only two eigenvalues are given to specify the behav-
ior of the step response, how would you solve the
problem when the actual system has more eigenvalues? Discuss.

8.8 Repeat parts *(a)–(d)* of Problem 8.7 when the model in
Equation 5.2.41 is used.

8.9 For a two-joint manipulator described by Equation 5.2.41,
a PD-controller is proposed in the following form:

$$F(t) = D(q^d)\ddot{q}^d + C(q^d, \dot{q}^d) + G(q^d) - K_v(\dot{q} - \dot{q}^d) - K_p(q - q^d)$$

where q^d represents the desired trajectory for the joint variables.

      a. Derive the first-order perturbation model about
(q^d, F^d) for the system. State your assumptions.

      b. Obtain the characteristic equation for the perturbation
model.

      c. Draw a block diagram for the implementation of the
controller.

      d. Specify the equation for the primary controller and that
for a secondary controller.

8.10 The optimum control problem is to minimize

$$J[u] = \frac{1}{2}\|x(t_f) - x^d(t_f)\|^2_{W_f} + \frac{1}{2}\int_{t_o}^{t_f}[\|x(t) - x^d(t)\|^2_{Q(t)} + \|u(t)\|^2_{R(t)}]dt$$

while satisfying the equations of motion:

$$\dot{x} = Ax(t) + Bu(t) + e(t)$$

where $R(t) = R'(t) > 0, Q(t) = Q'(t) \geq 0$, for all $t \geq t_o$,
and $W_f = W_f' \geq 0$; $x(t_o)$ is given, the input is unrestricted
and $u \in \Omega$, and $e(t)$ is a deterministic known disturbance
function.

    The problem can be solved by means of Hamilton-
Jacobi-Bellman (HJB) equation

$$-\frac{\partial V[x, t]}{\partial t} = \underset{u \in \Omega}{\text{Min}}\left\{\frac{1}{2}\|x - x^d\|^2_Q + \frac{1}{2}\|u\|^2_R + \frac{\partial V(x, t)}{\partial x}(Ax + Bu + e)\right\}$$

where the return function $V[x, t]$ defined as

$$V[x, t] = \underset{u}{\text{Min}}\frac{1}{2}\int_t^{t_f}[\|x - x^d\|^2_Q + \|u\|^2_R]dt$$

is the minimum value of the cost functional for the system
starting from state $x(t)$ at time t. The HJB-equation is a

partial differential equation. It may be solved by assuming
the form of the solution $V[x, t]$, which is substituted into
the HJB-equation.

    a.   Determine the optimal solution by assuming that the
return function is $V[x, t] = \frac{1}{2}\|x(t)\|_{P_0}^2 + P_1[x(t)] + P_2$
where $P_0 = P_0(t)$ is symmetric, $P_1 = P_1(t)$ is a row vector,
and $P_2 = P_2(t)$ is a scalar. Determine the differential
equations for P_0, P_1, and P_2 and their boundary values.

    b.   Determine the value of the performance index $J[u]$ in
terms of $V[x, t]$.

    c.   Determine the optimal controller by specifying the
defining equation for $u = u(t, x)$.

    d.   Discuss the use of this approach to design a secondary
controller for a manipulator motion. List advantages
and disadvantages of this method relative to the eigenvalue assignment method.

8.12   Repeat Problem 8.11 when A, B, Q, R, and W_f are time
invariant and $t_f = \infty$.

8.13   After the linearization and the uncoupling of the state variables in a manipulator model, the dynamics of a single joint
are described as follows:

$$\delta\ddot{x} = \delta u$$

where $\delta x(t)$ describes the variations in the joint position
from the desired path, and δu represents the variational
input variable about the desired (nominal) input. A feedback controller has been determined by the eigenvalue
assignment method:

$$\delta u = -1600\delta x_1 - 40\delta x_2$$

where $\delta x_1 = \delta x$, and $\delta x_2 = \delta\dot{x}$.

    a.   Determine the state variable representation for the
foregoing variational model.

    b.   Determine the damping factor, the (undamped) natural
frequency, and the eigenvalues of the closed loop system.

    c.   Determine a possible quadratic performance index with
$t_f = \infty$ that the given controller minimizes.

COMPUTER PROBLEM

8.C.1   A cylindrical robot manipulator shown in Figure 5.1*a* and
5.1*b* is discussed in Example 5.1.1. Its dynamical model

given in Equation 5.1.13 can be expressed in the state variable representation:

$$
\begin{bmatrix} \dot{x}_1 \\ \dot{x}_2 \\ \dot{x}_3 \\ \dot{x}_4 \\ \dot{x}_5 \\ \dot{x}_6 \end{bmatrix} = \begin{bmatrix} x_4 \\ x_5 \\ x_6 \\ x_1 \overline{m}_e(x_1) x_5^2/(2m) - k_s(x_1 - 2\ell/3)/m \\ x_1 \overline{m}_e(x_1) x_4 x_5 / I_t(x_1) \\ -g - k_z(z - \ell_0)/\overline{m} \end{bmatrix}
$$

$$
+ \begin{bmatrix} 0 & 0 & 0 \\ 0 & 0 & 0 \\ 0 & 0 & 0 \\ 1/m & 0 & 0 \\ 0 & 1/I_t(x_1) & 0 \\ 0 & 0 & 1/\overline{m} \end{bmatrix} \begin{bmatrix} F_r - B_r x_4 \\ \tau_\theta - B_\theta x_5 \\ F_z - B_z x_6 \end{bmatrix}
$$

where $\mathrm{col}[x_1, x_2, x_3, x_4, x_5, x_6] = \mathrm{col}[r, \theta, z, \dot{r}, \dot{\theta}, \dot{z}]$, $m = m_A + m_L$, $\overline{m} = m_A + m_L + m_h$, $\overline{m}_e(x_1) = 3m_A(2 - \ell/x_1)/4 + 2m_L$, $I_t(x_1) = [x_1^3 - (\ell - x_1)^3]m_A/(4\ell) + m_L x_1^2 + I_\theta$, and B_r, B_θ, B_z are the coefficients for the viscous friction.

Assume the following hypothetical numerical values: $m_A = 10.0$ kg, $m_L = 5.0$ kg, $m_h = 5.0$ kg, $I_\theta = 1.0$ kgm^2, $\ell = 1.0$ m, $\ell_o = \ell/3$, $k_z = k_s = 100$N/m, $B_r = B_z = 5.0$ Ns/m, and $B_\theta = 0.001$ Nms. Program the digital simulation of the manipulator using a subinterval of 0.05 s or smaller for integration. The duration of motion is 10 s.

a. Construct a primary controller specified by $u^d(t) = [F_r^d(t) \quad \tau_\theta^d(t) \quad F_z^d(t)]'$ on the basis of the ideal inverse dynamics (the computed torque method). The desired trajectory is given in Computer Problem 7.C.4.

b. Apply the step inputs: $u^1 = [2.0N \quad 0 \quad 0]'$ and $u^2 = [0 \quad 0.015 \text{ Nm} \quad 0]'$ to the system when $r(0) = 0.7$ m, $\theta(0) = \pi/2$ rad, and $\dot{r}(0) = \dot{\theta}(0) = 0$. Record the response vector $[r(t) \quad \theta(t) \quad z(t)]'$ for each case. Graph $r(t)$ versus time t for the aforementioned two cases.

c. The perturbed trajectory $\{r(t)\,\theta(t)z(t)\}$ of the manipulator is obtained in the simulation by integrating the system equation when the components of the inputs are $F_r(t) = F_r^d(t) + \delta F_r(t)$, and $\tau_\theta(t) = \tau_\theta^d(t) + \delta \tau_\theta(t)$. The specific values for the (desired) input $[F_r^d(t), \tau_\theta^d(t), F_z^d(t)]'$ are determined in part (a). The

perturbation about the nominal (desired) trajectory is
generated by the disturbance inputs $\delta F_r(t) = 1.5$ N for
$0 \le kT < 1.0$ s, $\delta F_r(t) = -1.5$ N for 1.0 s $\le kT < 2.0$ s
and zero elsewhere. Moreover, $\delta \tau_\theta(t) \equiv 0$ and $T = 20$
ms is the sampling period. The perturbed variables are
then obtained by generating $\Delta r(t) = r(t) - r^d(t)$ and
$\Delta \theta(t) = \theta(t) - \theta^d(t)$. Graph $\Delta r(t)$, $\Delta \theta(t)$ versus t.

d. Repeat part (c) when $\delta F_r \equiv 0$ and $\delta \tau_\theta = 0.075$ Nm for
$0 \le kT < 1.0$ s, $\delta \tau_\theta = -0.075$ Nm for 1.0 s $\le kT <$
2.0 s, and zero elsewhere.

e. Design a secondary controller to achieve independent
joint control and to regulate the perturbed trajectory.
Specify the controller gains for 10% overshoot and
$0.60T$ settling time. Realize it in the simulation
program. Test the system with the controller by graph-
ing $\Delta r_{\text{reg}}(t)$, $\Delta \theta_{\text{reg}}(t)$, and $\Delta r_{\text{reg}}(t)$ versus time. Comment on the controller performance.

CHAPTER NINE

ADAPTIVE CONTROL OF MANIPULATOR GROSS MOTION

The design of primary and secondary controllers for the gross motion of robot manipulators is presented in Chapter 8 using the joint-space variables. The primary controller is constructed so as to cause the cancellation of nonlinear terms, which renders the overall system linear. If the dynamical model is not known accurately, for example, due to the unknown or varying masses of the payloads, the foregoing cancellation is not exact. As a consequence, the linearized model for designing a secondary controller is inaccurate. It will then be difficult for a control engineer to tune the controller so that even the frequent readjustments of the gains may not lead to improvements in the system performance.

Under such varying operating conditions, the designer may opt to apply an adaptive controller. It functions so as to recognize changes via measurements, and after evaluating the system performance (relative to the desired goals), it will generate corrective actions in the system. Thus, the parameters in the system, mainly in the controller, and/or the system performance are adjusted so as to comply with changing operating conditions. Indeed, the entire system attempts to adapt to changes whereby the uncertainties that pertain to the system are reduced [1].

To construct an adaptive controller, a time-series difference equation model may be chosen to describe the input–output measurements in the manipulator. The parameters in such a model can be computed recursively and on-line at each sampling instant. The controller gains are adjusted according to the updated parameter estimates and measurements so as to make the system achieve the desired goals that are given as the design specifications. Thus, it inherently adapts to unknown operating conditions that are conveyed by the measurements to the controller. Such an adaptive controller can be designed as a self-tuner [2], in which the parameters of a difference equation model are estimated and the controller gains calculated at each sampling instant using available measurements.

A self-tuning controller has been applied to the gross motion control of a manipulator. It can be designed to operate on the variables in the joint space [3–5] or in the Cartesian world space [6]. For a manipulator operating at slow or moderate speeds, the independent joint dynamics represented by a single-input–single-output (SISO) model serve as the basis of the controller design. For a manipulator operating at high speeds, a model should account for the interacting generalized forces between the joints. Therefore, a multiple-input–multiple-output (MIMO) model must be used in the design of a self-tuning controller [3,6].

9.1   ADAPTIVE SELF-TUNING CONTROL OF JOINT SPACE VARIABLES

The end effector of a manipulator can be made to track the desired trajectory of the gross motion by properly designing a controller for the joint variables. The design of an adaptive controller will be presented here in a framework similar to that of Chapter 8. If the dynamics of the manipulator are known at a sufficient degree of accuracy, a primary controller is first designed as described in Chapter 8. A primary controller will make the joint variables track the desired trajectory under ideal circumstances. A secondary controller is then constructed to compensate for the imperfect cancellations of non-linear terms and the effects of modeling errors and disturbances. It will be designed here by the explicit method of the self-tuning controller design: the parameters of a discrete time-series model are first estimated and then used to determine the controller gains.

The design of a self-tuning controller is discussed first for independently controlled joint variables using discrete-time SISO-models. If the designer does not have accurate information about the behavior of the manipulator dynamics, a self-tuning controller can still be designed for the manipulator; and it then functions as the sole controller in the system. The self-tuning controller can be determined by minimizing a quadratic criterion, or alternatively, by the pole-zero placement technique. Both approaches will be discussed. Examples are presented to demonstrate the applicability of the design methods.

9.1.1   PRIMARY CONTROLLER CAN BE DETERMINED BY ACCURATELY KNOWN TERMS IN DYNAMICAL MODEL

The dynamical model for the gross motion of a serial link N-joint manipulator is assumed to be given in Lagrange's formulation by Equation 8.2.14.

It may be described in the state variable form by defining the position vector $x^1 = [q_1 \ldots q_N]'$, the velocity vector $x^2 = [\dot{q}_1 \ldots \dot{q}_N]'$ and the input vector $u(t) = F(t)$:

$$\begin{bmatrix} \dot{x}^1 \\ \dot{x}^2 \end{bmatrix} = \begin{bmatrix} x^2(t) \\ -D^{-1}(x^1)[C(x^1,x^2) + G(x^1) + B(x^2)] \end{bmatrix} + \begin{bmatrix} 0 \\ D^{-1}(x^1) \end{bmatrix} u(t) \quad (9.1.1)$$

where $D(x^1)$ is the inertia matrix, $C(x^1,x^2)$ signifies Coriolis' and centripetal generalized forces, $G(x^1)$ describes the gravitational effects on the motion, $B(x^2)$ indicates the friction that may depend on the generalized velocity, and the N-dimensional vector F is the input to the system.

For a prespecified desired trajectory, a primary controller $u^d(t)$ is determined by Equation 8.2.5. After comparing Equations 9.1.1 and 8.2.1, the primary controller is obtained:

$$\begin{aligned} u^d(t) &= [B^2(x^d)]^{-1}[\ddot{x}^{1d} - f^2(x^d)] \\ &= D(x^{1d})\ddot{x}^{1d} + C(x^{1d},\dot{x}^{1d}) + G(x^{1d}) + B(\dot{x}^{1d}) \end{aligned} \quad (9.1.2)$$

The realization of the primary controller $u^d(t)$ in Equation 9.1.2 presumes that all expressions and the numerical values of the parameters are accurately known. Unfortunately, this is not usually the case. Moreover, the calculations of the expressions in Equation 9.1.2 require a considerable amount of computer time and memory.

If only some terms in Equation 9.1.2 are available at a sufficient degree of accuracy, a primary controller can be constructed using these terms only. For example, suppose that only the expressions in the gravitational term $G(x^1)$ are accurately known. A primary controller $u_r^d(t)$ can then be chosen as

$$u_r^d(t) = G(x^{1d}) \quad (9.1.3)$$

where $G(x^{1d})$ signifies the gravitational terms evaluated along the desired trajectory.

The input $u(t)$ to the system is now chosen as follows:

$$u(t) = u_r^d(t) + \hat{u}(t) \quad (9.1.4)$$

where $u_r^d(t)$ is obtained from the primary controller and $\hat{u}(t)$ is determined by the secondary controller. As a consequence of Equations 9.1.3 and 9.1.4, the system model is simplified. After the substitution of Equations 9.1.4 and 9.1.3 into Equation 9.1.1, and assuming that $G(x^{1d}) = G(x^1)$, the model for the manipulator becomes

$$\begin{bmatrix} \dot{x}^1 \\ \dot{x}^2 \end{bmatrix} = \begin{bmatrix} x^2(t) \\ -D^{-1}(x^1)[C(x^1,x^2) + B(x^2)] \end{bmatrix} + \begin{bmatrix} 0 \\ D^{-1}(x^1) \end{bmatrix} \hat{u}(t) \quad (9.1.5)$$

Equation 9.1.5, the design model for the secondary controller, is nonlinear and thus more complicated than, say, the linearized model in Equation 8.2.22. On the other hand, the primary controller based on *a priori* accurate information of only the gravitational effects is simple to implement. The composite control input to the manipulator system is given by Equation 9.1.4.

If only some terms contained in vector $G(x^1)$ are accurately known, then only they are used in specifying the primary controller. If the Coriolis terms $C(x^1, x^2)$ with their numerical parameters are accurate and available, they can also be included in the expression of the primary controller. In fact, any reliable information of the terms in the dynamical model of Equation 9.1.1 can be extracted and used in determining the primary controller. This is a well known procedure in any estimation problem. The use of *a priori* information will simplify necessary calculations and increase confidence levels of the model estimates. This is true particularly in applying self-tuning controllers to a manipulator system.

An example is next presented to illustrate the design of a primary controller in the case that only certain terms in the dynamical model of a manipulator are known accurately.

EXAMPLE 9.1.1

A revolute three-joint planar manipulator shown in Figure 9.1 can move on the vertical plane. In view of the gross motion, the manipulator is redundant. The third joint is used to orient the gripper for grasping on the plane. The gripper cannot rotate, but the opening of the grasping plates can be changed. The length of link i is ℓ_i, its mass m_i, and moment of inertia I_{0i} about the center of gravity, $i = 1, 2, 3$.

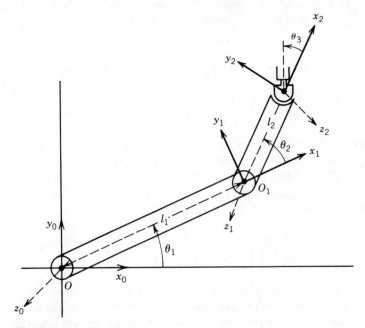

FIGURE 9.1
A planar (vertical) manipulator. Example 9.1.1.

The motions of the joint variables θ_1, θ_2, and θ_3 in this planar manipulator are governed by three nonlinear second-order differential equations (compare with Equation 5.3.28)

$$
D(\theta)\begin{bmatrix} \ddot{\theta}_1 \\ \ddot{\theta}_1 + \ddot{\theta}_2 \\ \ddot{\theta}_1 + \ddot{\theta}_2 + \ddot{\theta}_3 \end{bmatrix} + \begin{bmatrix} d_4 s_2 + d_5 s_{23} & -d_4 s_2 + d_6 s_3 & -d_5 s_{23} - d_6 s_3 \\ d_4 s_2 + d_5 s_{23} & d_6 s_3 & -d_6 s_3 \\ d_5 s_{23} & d_6 s_3 & 0 \end{bmatrix}
$$

$$
\begin{bmatrix} \dot{\theta}_1^2 \\ (\dot{\theta}_1 + \dot{\theta}_2)^2 \\ (\dot{\theta}_1 + \dot{\theta}_2 + \dot{\theta}_3)^2 \end{bmatrix} + \begin{bmatrix} B_1 \dot{\theta}_1 \\ B_2 \dot{\theta}_2 \\ B_3 \dot{\theta}_3 \end{bmatrix} +
$$

$$
\begin{bmatrix} -\frac{1}{2}\ell_1 c_1 & -\ell_1 c_1 - \frac{1}{2}\ell_2 c_{12} & -\ell_1 c_1 - \ell_2 c_{12} - \frac{1}{2}\ell_3 c_{123} \\ 0 & -\frac{1}{2}\ell_2 c_{12} & -\ell_2 c_{12} - \frac{1}{2}\ell_3 c_{123} \\ 0 & 0 & -\frac{1}{2}\ell_3 c_{123} \end{bmatrix} \begin{bmatrix} m_1 g \\ m_2 g \\ m_3 g \end{bmatrix} = \begin{bmatrix} \tau_1 \\ \tau_2 \\ \tau_3 \end{bmatrix} \quad (9.1.6)
$$

$$
D(\theta) = \begin{bmatrix} 2d_1 + d_4 c_2 + d_5 c_{23} & 2d_2 + d_4 c_2 + d_6 c_3 & 2d_3 + d_5 c_{23} + d_6 c_3 \\ d_4 c_2 + d_5 c_{23} & 2d_2 + d_6 c_3 & 2d_3 + d_6 c_3 \\ d_5 c_{23} & d_6 c_3 & 2d_3 \end{bmatrix}
$$

$$(9.1.7)$$

where $d_1 = m_1(\ell_1/2)^2/2 + m_2\ell_1^2/2 + m_3\ell_1^2/2 + I_{01}/2$, $d_2 = I_{02}/2 + (m_3 + m_2/4)\ell_2^2/2$, $d_3 = I_{03}/2 + m_3(\ell_3/2)^2/2$, $d_4 = (m_3 + m_2/2)\ell_1\ell_2$, $d_5 = m_3\ell_1\ell_3/2$, $d_6 = m_3\ell_2\ell_3/2$, B_i, $i = 1, 2, 3$ represents the coefficient of viscous friction, $\theta = [\theta_1 \quad \theta_1 + \theta_2 \quad \theta_1 + \theta_2 + \theta_3]'$ and $c_{123} = \cos(\theta_1 + \theta_2 + \theta_3)$. The inputs τ_1, τ_2, and τ_3 to the system are generated by the actuators.

Equation 9.1.6 may be expressed succinctly:

$$D(\theta)\ddot{\theta} + C(\theta, \dot{\theta}) + B(\dot{\theta}) + G(\theta) = F(t) \quad (9.1.8)$$

where $F(t) = [\tau_1(t) \quad \tau_2(t) \quad \tau_3(t)]'$, $C(\theta, \dot{\theta})$ and $G(\theta)$ represent the Coriolis and gravity terms, respectively, and the constant friction vector $B(\dot{\theta}) = [B_1\dot{\theta}_1 \quad B_2\dot{\theta}_2 \quad B_3\dot{\theta}_3]'$.

The problem here is to derive an expression for the primary controller under the assumption that the gravitational and Coriolis terms $G(\theta)$ and $C(\theta, \dot{\theta})$ are accurately known. The desired trajectory $[\theta^d(t), \dot{\theta}^d(t)]$ for the gross motion of the manipulator is assumed to be given for $0 \le t \le t_f$ (final time). The given orientation angle specifies the sum $(\theta_1 + \theta_2 + \theta_3)$; this condition removes the redundancy in this manipulator.

The primary controller $u_r^d(t)$ is now chosen as follows:

$$u_r^d(t) = C(\theta^d, \dot{\theta}^d) + G(\theta^d) \quad (9.1.9)$$

where $C(\theta^d, \dot{\theta}^d)$ and $G(\theta^d)$, the Coriolis and gravity terms, respectively, are evaluated along the desired trajectory. The output of the primary controller attempts to cancel the Coriolis and gravity effects in the model.

The input to the system is composed of the outputs of the primary and secondary controllers:

$$F(t) = u_r^d(t) + \hat{u}(t) \tag{9.1.10}$$

where $\hat{u}(t)$ is generated by the secondary controller.

Under the assumption that

$$C(\theta, \dot{\theta}) = C(\theta^d, \dot{\theta}^d) \quad G(\theta) = G(\theta^d) \tag{9.1.11}$$

the dynamical model in Equation 9.1.8 becomes

$$D(\theta)\ddot{\theta} + B(\dot{\theta}) = \hat{u}(t) \tag{9.1.12}$$

Matrices $D(\theta)$ and $B(\dot{\theta})$ are not used in the primary controller, since by the assumption their numerical values are not known at a sufficient degree of accuracy.

The observations are performed according to

$$y(t) = \mu\theta(t) \tag{9.1.13}$$

where μ is a constant (3×3) matrix, $y(t) = [y_1(t) \ y_2(t) \ y_3(t)]'$ refers to the output position vector when the manipulator described in Equation 9.1.8 is driven by input $u_r^d(t) + \hat{u}(t)$, and the equality relations 9.1.11 hold.

The primary controller given in Equation 9.1.9 is reduced in complexity as compared with Equation 9.1.2. Thus, the needed computer time is reduced in the realization of this controller.

A secondary controller $\hat{u}(t)$ is then to be designed for the nonlinear model given in Equations 9.1.12 and 9.1.13. In this case, however, applicable design methods are limited. A possible approach is to design a self-tuning controller for $\hat{u}(t)$ so that the desired tracking in the gross motion is achieved.

9.1.2 SELF-TUNING LQG-CONTROLLER FOR INDEPENDENT JOINT DYNAMICS

A secondary controller is designed in Sections 8.2.2 and 8.2.3 on the basis of the linear models that result from the perfect cancellation of nonlinear terms by means of a primary controller. When this cancellation is not exact, matrices $\overline{A}$ and $\overline{B}$ in Equation 8.2.22 must be determined. For many manipulators, this is difficult and often hampered by the lack of sufficiently accurate information. A possible alternative approach [3–5] is to discretize the linearized differential Equation 8.2.22 with respect to time and express it in terms of the input and the measured output variables. Since the resulting linear difference equation will contain several parameters that are often inaccurate, and their evaluation is usually difficult and tedious, the designer may omit the explicit linearization, and *assume* already at the beginning of the design a linear discrete time series model [3]. The parameters of this model are then determined so that the best fit of the model-generated values to the measured input–output values is obtained in the sense of the sum of the least-squared errors. Naturally, the designer should keep in mind that if reliable information on the parameters in the differential equation model 9.1.1 is available, it should be used in the primary controller, and the subsequent calculations will be reduced. The primary controller specified by Equation 9.1.2 incorporates the dynamical interactions between the joint variables. A secondary controller may, however, be designed on the basis of independent joint dynamics, particularly for manipulators operating at moderate speeds.

To design a controller under the assumption of independent joint dynamics, an SISO-model is chosen to represent the input–output measurements of each joint i, $i = 1, \ldots, N$. Indeed, a linear discrete time-series model of autoregressive type with external excitation (an ARX-model) can be assumed for the variables $\{y_i(k), \hat{u}_i(k)\}$:

$$A_i(z^{-1})y_i(k) = a_0^i + B_i(z^{-1})\hat{u}_i(k-1) + e_i(k) \qquad (9.1.14)$$

where $y_i(k)$ and $\hat{u}_i(k)$ represent the measured output and input of joint i at time kT, respectively, and T is the sampling period. The real polynomials $A_i(z^{-1})$ and $B_i(z^{-1})$ are expressions in the delay operator z^{-1}, that is, $A_i(z^{-1}) = 1 - a_1^i z^{-1} - \ldots - a_n^i z^{-n}$ and $B_i(z^{-1}) = b_0^i + b_1^i z^{-1} + \ldots + b_n^i z^{-n}$. Term a_0^i signifies a possible bias, for example, due to gravity effects. The equation error $e_i(k)$ represents a sample from a white Gaussian zero-mean noise process. The measurements are used to estimate the unknown parameters a_j^i and b_j^i, $j = 0, \ldots, n$ in Equation 9.1.14. The estimates of the parameters are substituted into Equation 9.1.14 for the unknown parameters, and the resulting model is used in the design of the controller. This approach to the design of an adaptive self-tuning controller is called the indirect (explicit) method [2,7,8].

We next present the equations for estimating the unknown parameters in the model, Equation 9.1.14, and then the design of a self-tuning controller. The details of the method are reviewed in Appendix D.

Estimation Algorithm for Parameters in ARX-Model

The unknown parameters in the model given by Equation 9.1.14 are estimated by minimizing the sum of the squared errors, that is,

$$E_i(\alpha_i) = \frac{1}{N+1} \sum_{k=n}^{N+n} \gamma_i^{N+n-k} e_i^2(k) \qquad (9.1.15)$$

where vector $\alpha_i = [a_1^i \cdots a_n^i; b_0^i \cdots b_n^i; a_0^i]'$ contains the unknown parameters, integer n specifies the order of the model in Equation 9.1.14, and $(n + N)T$ indicates the time of the last measurement. The forgetting factor γ_i often chosen between 0.95 and 1.0 assigns different weights on the errors depending on their relative importance, for example, past errors are usually less important than the present ones.

The best estimate $\hat{\alpha}_i(k)$ of the unknown parameter α_i at time kT is obtained by minimizing the error criterion $E_i(\alpha_i)$. The details of the minimization are presented in Appendices D.1.2. and D.1.3. The resulting equations for the on-line calculations are

$$\hat{\alpha}_i(k) = \hat{\alpha}_i(k-1) + P_i(k)\phi_i(k-1)[y_i(k) - \phi_i'(k-1)\hat{\alpha}_i(k-1)] \qquad (9.1.16)$$

$$P_i(k) = \frac{1}{\gamma_i}\left[P_i(k-1) - \frac{P_i(k-1)\phi_i(k-1)\phi_i'(k-1)P_i(k-1)}{\gamma_i + \phi_i'(k-1)P_i(k-1)\phi_i(k-1)}\right] \quad (9.1.17)$$

where $\phi_i'(k-1) = [y_i(k-1)\ldots y_i(k-n); \hat{u}_i(k-1)\ldots \hat{u}_i(k-n-1); 1]$, and P is a symmetric matrix. The second term on the right side of Equation 9.1.16 describes the correction term. It contains a gain and in the bracket the one-step-ahead prediction error. The initial values $\hat{\alpha}_i(0)$ and $P_i(0)$ may be approximated. Then, the parameters can be estimated recursively by on-line calculations using Equations 9.1.16 and 9.1.17.

The estimates from Equation 9.1.16 are substituted for the unknown parameters into Equation 9.1.14 to obtain a model for the controller design:

$$\hat{A}_i(z^{-1})y_i(k) = \hat{a}_0^i + \hat{B}_i(z^{-1})\hat{u}_i(k-1) + e_i(k) \quad (9.1.18)$$

where $\hat{A}_i(z^{-1}) = 1 - \hat{a}_1^i z^{-1} - \ldots - \hat{a}_n^i z^{-n}$ and $\hat{B}_i(z^{-1}) = \hat{b}_0^i + \hat{b}_1^i z^{-1} + \ldots + \hat{b}_n^i z^{-n}$. In Equation 9.1.18, the error in the parameter estimates has been ignored. The difference equation model 9.1.18 with known numerical values for the parameters is next used to design an adaptive self-tuning controller so that the system tracks the desired trajectory $\{y_i^d(k)\}$.

Self-Tuning Control Algorithm for Independent Joint Dynamics

The goal of the controller is to make the output $\{y_i(k)\}$ of the ith joint $i = 1, \ldots, N$ track the sequence of the desired values $\{y_i^d(k)\}$.

The tracking of the desired output can be achieved when the controller $\hat{u}_i(k)$ is so determined that a chosen well-defined performance criterion is minimized. A convenient functional that can be used to measure the system performance is quadratic in the error $(y_i - y_i^d)$ and control input $\hat{u}_i$:

$$I_k[\hat{u}_i] = E\{[y_i(k+1) - y_i^d(k+1)]^2 + \epsilon_i \hat{u}_i^2(k)|\Sigma_i(k)\} \quad (9.1.19)$$

where the weighting factor ϵ_i is a nonnegative number, and $E\{\cdot|\Sigma_i(k)\}$ signifies a conditional expectation operation given the past values of the output and the input, that is, $\Sigma_i(k) = \{y_i(j), \hat{u}_i(j-1), j \le k\}$. If the equation error $e_i(k) \equiv 0$, then the expectation operation is not needed. The problem posed involves a linear plant, a quadratic criterion and a Gaussian additive noise disturbance; hence, it is usually called an LQG-problem. The solution to an LQG-problem is discussed in detail in Appendix D.2.

The performance index in Equation 9.1.19 is minimized over the admissible inputs $\{\hat{u}_i(k)\}$ while satisfying the model Equation 9.1.18. The control input $\hat{u}_i(k)$ at time kT is admissible when it is a function of available measurements up to and including time kT. The minimizing value of $\hat{u}_i(k)$ determines the structure of the controller.

To minimize $I_k[\hat{u}_i]$ with respect to admissible control inputs, Equation 9.1.18 is solved for $y_i(k+1)$, and the resulting expression is substituted into Equation 9.1.19 to obtain

$$I_k[\hat{u}_i] = E\{[z\hat{A}_{1i}(z^{-1})y_i(k) + \hat{a}_0^i + \hat{b}_0^i \hat{u}_i(k) + \hat{B}_{1i}(z^{-1})\hat{u}_i(k) - y_i^d(k+1)]^2$$

$$+ \epsilon_i \hat{u}_i^2(k)|\Sigma_i(k)\} \quad (9.1.20)$$

where $\hat{A}_{1i}(z^{-1}) = \hat{a}_1^i z^{-1} + \ldots + \hat{a}_n^i z^{-n}$ and $\hat{B}_{1i}(z^{-1}) = \hat{b}_1^i z^{-1} + \ldots + \hat{b}_n^i z^{-n}$.
The minimization of $I_k[\hat{u}_i]$ in Equation 9.1.20 with respect to $\hat{u}_i(k)$ gives (see
Appendix D.2.1):

$$\hat{u}_i(k) = \frac{\hat{b}_0^i}{(\hat{b}_0^i)^2 + \epsilon_i}[y_i^d(k+1) - z\hat{A}_{1i}(z^{-1})y_i(k) - \hat{a}_0^i - \hat{B}_{1i}(z^{-1})\hat{u}_i(k)|$$

(9.1.21)

Equation 9.1.21 specifies the controller that makes variable $y_i(k)$ in Equation
9.1.18 track the desired trajectory $y_i^d(k)$.

In the case that $\epsilon_i = 0$, the controller of Equation 9.1.21 is called a
minimum variance controller. If this minimum variance control is substituted
into Equation 9.1.18, the resulting equation assumes the following form:

$$y_i(k+1) = y_i^d(k+1) + e_i(k+1)$$

(9.1.22)

Thus, the mean value of y_i is equal to y_i^d, since the mean value of residual
e_i is zero. The minimum variance controller corresponding to $\epsilon_i = 0$ in
Equation 9.1.21 is often varying fast in time and assumes large values. It
causes difficulties in the implementation, for example, due to saturation. By
appropriately choosing the weighting factor $\epsilon_i > 0$, the magnitude of input
$\hat{u}_i(k)$ can advantageously be reduced. The system with $\epsilon_i \neq 0$, however, will
exhibit a steady-state tracking error (see Problem 9.2).

The self-tuning controller based on the model in Equation 9.1.18 is
specified by Equations 9.1.16, 9.1.17, and 9.1.21. The block diagram shown
in Figure 9.2 illustrates the implementation of the self-tuning controller.

The control algorithm may now be summarized:

Realization of self-tuning LQG-controller

1. Choose the initial approximations $\hat{\alpha}_i(0)$, $P_i(0)$, and $\hat{u}_i(0) \ldots$,
    $\hat{u}_i(k - n - 1)$ as well as the values of ϵ_i and γ_i for each $i = 1, \ldots, N$. Set $k = n$.

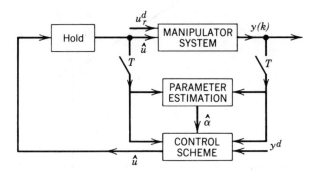

FIGURE 9.2
Self-tuning controller implementation by the explicit
method.

2. Compute the parameter estimate $\hat{\alpha}_i(k)$ using Equations 9.1.16
    and 9.1.17 and the control input $u_i(k)$ using Equations 9.1.4
    and 9.1.21 for each i.

3. Apply the computed values $u_i(k)$ to the manipulator system.
    Obtain the current measurements. Replace k by $k + 1$.

4. Repeat the procedure from step 2 on, until the final time of
    the motion is reached.

Two examples are presented to demonstrate the applicability of the self-
tuning controller.

EXAMPLE 9.1.2

The planar three joint manipulator discussed in Example 9.1.1 and shown
in Figure 9.1 is to move from the resting position $P_a = (0, 1.0$ m) to position
$P_b = (0.6403$ m, -0.1 m) and to position P_c to pick up an object. Point
P_b is 0.5 m above position P_c of the object. The end-effector grasps the
object at P_c and transfers it through position P_b to the final position $P_e =$

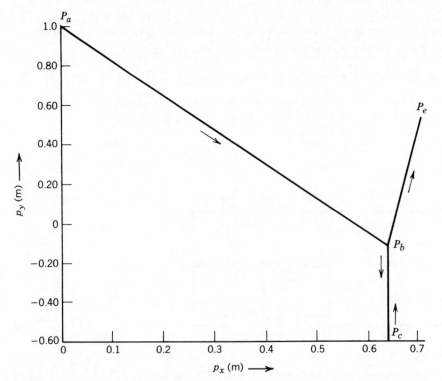

FIGURE 9.3
The desired trajectory of the end-effector. Example 9.1.2.

(0.7 m, 0.5 m). The trajectory for the gross motion of the end-effector is shown in Figure 9.3. The orientation of the gripper is initially $\pi/2$ rad at P_a. It changes during the motion to $-\pi/2$ rad by point P_b and stays at this value during the motion from P_b to P_c and back to P_b. When the end-effector moves the object from P_b to P_e, the orientation is changed to 0 rad.

The trajectory $P_a - P_b - P_c - P_b - P_e$ consists of straight line segments. Along each segment, the velocity profile in the Cartesian base coordinate system is described as an equilateral trapezoid. The maximum velocity $v_0^m = 0.7$ m/s for each segment, the time for the acceleration is $10T$ and the sampling period $T = 30$ ms. When the lengths ℓ_1, ℓ_2, and ℓ_3 of the links are given, the desired trajectory $\theta^d(t) = [\theta_1^d(t)\ \theta_2^d(t)\ \theta_3^d(t)]'$ for the joint variable vector can be generated as described in Chapter 7.

The equations of motion for this manipulator are given in Equations 9.1.6 and 9.1.7. The parameters in the expressions of the Coriolis and gravity terms are assumed to be known accurately. The problem is to design a controller for the system so that the motion of the end-effector tracks the desired trajectory as closely as possible.

The controller $u(t)$ for the gross motion of the manipulator is constructed so that it is comprised of the primary controller $u_r^d(t)$ and the secondary controller $\hat{u}(t)$, that is, $u(t)$ is specified by Equation 9.1.4. The primary controller is determined by the Coriolis and gravity terms of Equation 9.1.6; it is given in the general form in Equation 9.1.9.

When the control Equation 9.1.10 is substituted into the manipulator model in Equation 9.1.6, and a perfect cancellation of the Coriolis and gravity terms is assumed, the following model is obtained after observing Equations 9.1.9 and 9.1.11 and rearranging:

$$D(\theta)\begin{bmatrix} 1 & 0 & 0 \\ 1 & 1 & 0 \\ 1 & 1 & 1 \end{bmatrix}\begin{bmatrix} \ddot{\theta}_1 \\ \ddot{\theta}_2 \\ \ddot{\theta}_3 \end{bmatrix} + \begin{bmatrix} B_1\dot{\theta}_1 \\ B_2\dot{\theta}_2 \\ B_3\dot{\theta}_3 \end{bmatrix} = \begin{bmatrix} \hat{u}_1 \\ \hat{u}_2 \\ \hat{u}_3 \end{bmatrix} \tag{9.1.23}$$

$$y(t) = \mu\theta(t) \tag{9.1.24}$$

where $D(\theta)$ is given in Equation 9.1.7 and $\hat{u}(t) = [\hat{u}_1(t)\ \hat{u}_2(t)\ \hat{u}_3(t)]'$. The output vector y is three-dimensional and μ is now an identity matrix. A secondary controller with self-tuning will next be designed for system 9.1.23 and 9.1.24.

To design a controller with self-tuning for system 9.1.23 and 9.1.24, an autoregressive exogenous difference equation model (an ARX-model) needs to be chosen. It is selected for each joint i under the assumption of independent joint dynamics:

$$y_i(k) = a_0^i + a_1^i y_i(k-1) + a_2^i y_i(k-2) + b_0^i \hat{u}_i(k-1) + e_i(k) \tag{9.1.25}$$

where integer i refers to the ith joint, $i = 1, 2, 3$, $y_i(k)$ is the output $\theta_i(k)$ of the ith joint when input $u_i = u_{ri}^d + \hat{u}_i$ is applied to this joint at time kT, and $e_i(k)$ represents a white Gaussian zero-mean noise process. Parameters a_j^i, $j = 0, 1, 2$ and b_0^i are the unknowns to be determined by the parameter estimation algorithm. Equation 9.1.25 contains only the variables of the ith joint, since the independent joint dynamics are assumed.

The problem is to design a secondary controller $\hat{u}(t)$ on the basis of Equation 9.1.25 for system in Equations 9.1.23 and 9.1.24 so as to make the output of system in Equations 9.1.6, 9.1.7 moving under the influence of $u_r^d(t) + \hat{u}(t)$ track the desired trajectory $\{\theta^d(k)\}$.

The design objective can be achieved by minimizing the squared tracking errors and the energy of the input:

$$I_k[\hat{u}_i] = E\{[y_i(k+1) - y_i^d(k+1)]^2 + \epsilon_i\hat{u}_i^2(k)|\Sigma_i(k)\} \qquad (9.1.26)$$

where $E\{\cdot\,|\Sigma_i(k)\}$ represents the expectation operation conditioned on the information available up to and including time kT, and ϵ_i is a nonnegative weighting factor.

To solve the minimization problem subject to the constraints, the unknown parameters in Equation 9.1.25 are first estimated. For this purpose, Equation 9.1.25 is rewritten as

$$y_i(k) = \phi_i'(k-1)\alpha_i + e_i(k) \qquad (9.1.27)$$

where $\phi_i'(k-1) = [y_i(k-1)\; y_i(k-2); \hat{u}_i(k-1); 1]$ and $\alpha_i = [a_1^i\; a_2^i; b_0^i; a_0^i]'$. The parameter vector α_i in the ARX-model of joint i can now be calculated recursively according to Equations 9.1.16 and 9.1.17 to obtain the least-squares estimate $\hat{\alpha}_i(k)$ at time kT.

The design of the self-tuning controller will be based on the model obtained from Equation 9.1.25 by replacing the unknown parameters by the numerical values of the estimates and by ignoring the errors due to the estimated values:

$$y_i(k+1) = \hat{a}_1^i y_i(k) + \hat{a}_2^i y_i(k-1) + \hat{a}_0^i + \hat{b}_0^i \hat{u}_i(k) + e_i(k+1) \qquad (9.1.28)$$

where the parameter estimates indicated with the carets are known numerical values. When Equation 9.1.28 is substituted into Equation 9.1.26, the expectation operation and the minimization can be performed. It results in the expression that specifies the controller

$$\hat{u}_i(k) = \frac{-\hat{b}_0^i}{(\hat{b}_0^i)^2 + \epsilon_i}[\hat{a}_1^i y_i(k) + \hat{a}_2^i y_i(k-1) + \hat{a}_0^i - y_i^d(k+1)] \qquad (9.1.29)$$

Equation 9.1.29 describes an adaptive controller in which the parameters of the model and the gains of the controller are updated at each sampling instant. The controller in Equation 9.1.29 exhibits the self-tuning property, if the parameter estimates converge to the true values. That is, the controller gains converge to the optimal values that could have been determined if the true parameter values of the model equation were known [2].

If $\epsilon_i = 0$, then Equation 9.1.29 describes a minimum variance controller. In this case, the steady-state value of the output is determined by Equation 9.1.22. When $\epsilon_i \neq 0$, the mean value of the output will deviate from the desired value in the steady state.

As a summary of the adaptive solution, the control input to system 9.1.6 is $u = u_r^d + \hat{u}$. The components of input $u_r^d(t)$ are produced by the primary controller specified by Equation 9.1.9. The components of input $\hat{u}$ are generated by the secondary controller. They are piecewise constant and determined by Equation 9.1.29.

A digital simulation is performed on system 9.1.6 using the following numerical values: $m_1 = 1.0$ kg, $m_2 = 0.7$ kg, $m_3 = 1.4$ kg, $\ell_1 = 0.5$ m, $\ell_2 = 0.4$ m, $\ell_3 = 0.1$ m, $I_{01} = 43.33 \ 10^{-3}$ kgm^2, $I_{02} = 25.08 \ 10^{-3}$ kgm^2, and $I_{03} = 32.67 \ 10^{-3}$ kgm^2. Initially, the end-effector of the manipulator is at P_a. The task is to transfer a part specified by the location of point P_c to point P_e. The object has a shape similar to that of the third link. It is grasped on one end at point P_c for assembly so that it appears as an extension of the third link. It has a mass of 0.5 kg, inertia of $11.67 \ 10^{-3}$ kgm^2, and its length is 0.1 m. The viscous friction is described by coefficient $B_i = 20$ Nms for $i = 1, 2, 3$. The forgetting factor is $\gamma_i = 0.95$ and $\epsilon_i = 10^{-6}$ for all i.

The system Equation 9.1.6 was simulated by using the subroutine DVERK in the integration. In the parameter estimation, the initial approximations were $\hat{\alpha}_i(0) = [0 \ 0; 0.5; 0]'$, $P_i(0) = 100 \ I$ for $i = 1, 2, 3$.

Some typical graphs determined by this simulation study are displayed in Figures 9.4a–9.4g. Figures 9.4a, 9.4b, and 9.4c show the tracking of the joint variables $\theta_1(k)$, $\theta_2(k)$ and $\theta_3(k)$, respectively, when input $(u_r^d + \hat{u})$ is applied. The system with a self-tuning controller exhibits large tracking errors initially.

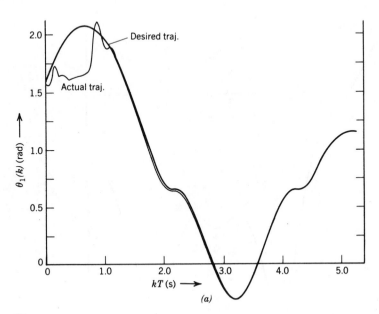

FIGURE 9.4 *a*

Responses of manipulator system with primary controller and secondary self-tuning controller. Initial parameter estimates arbitrary in *a–f*, determined by averaging in *g–i*. Example 9.1.2. *a*. Actual (θ_1) and desired (θ_1^d) joint one positions versus time.

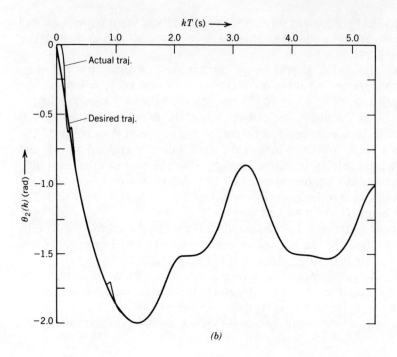

(b)

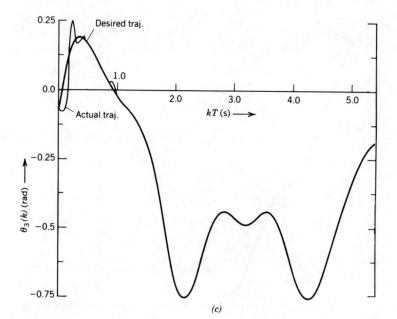

(c)

FIGURE 9.4 b AND c

b. Actual (θ_2) and desired (θ_2^d) joint one positions versus time.

c. Actual (θ_3) and desired (θ_3^d) joint one positions versus time.

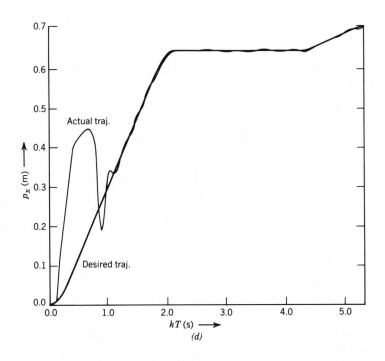

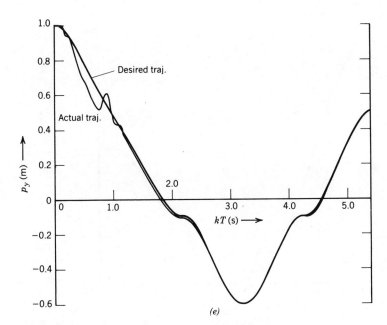

FIGURE 9.4 *d* AND *e*

d. Actual (p_x) and desired (p_x^d) joint one positions versus time.

e. Actual (p_y) and desired (p_y^d) joint one positions versus time.

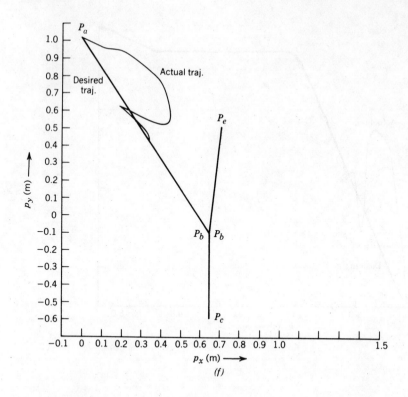

(f)

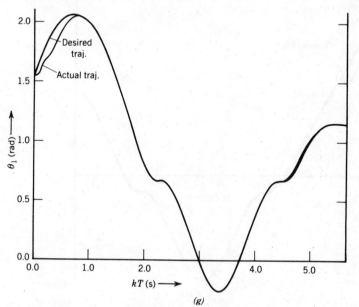

(g)

FIGURE 9.4 *f* AND *g*

f. Actual and desired end-effector trajectories on phase-plane $p_x p_y$.

g. Actual (Θ_1) and desired (Θ_1^d) joint one positions versus time.

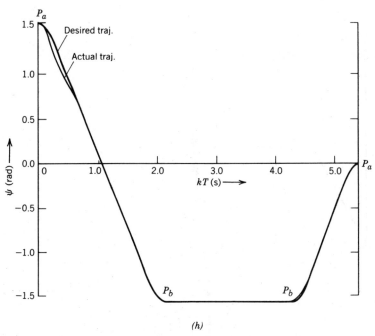

(h)

FIGURE 9.4*h*

Actual (ψ) and desired (ψ^d) orientation of end-effector versus
time. $\psi = \theta_1 + \theta_2 + \theta_3$.

During this time, the adaptive algorithm is searching proper estimates and
controller gains. After this learning period, the outputs of the joints start
tracking closely the desired values. Similar behavior is observed in Figures
9.4*d* and 9.4*e*, which show the *x*- and *y*-components of the trajectory in
the Cartesian coordinate system versus time. Figure 9.4*f* displays the phase
plane ($p_x p_y$) trajectory for the motion: the actual (simulated) and desired
trajectories corresponding to Figure 9.3 are shown. The simulation study
demonstrates that the controller designed performs well in the described task.

The determination of the initial approximations $\hat{\alpha}_i(0)$ and $P_i(0)$ is often
difficult, particularly because of the finite duration of the motion. (Explain!)
The self-tuning controller for the gross motion of a manipulator was first
proposed for repetitive motions [3, 4] to allow a sufficient time for the
convergence of the estimates and controller gains. The initial oscillations and
thus the convergence can be improved when the designer selects the initial
approximations on the basis of the learned knowledge from the previous
runs. It is demonstrated by the following simulation.

An approximation for $\hat{\alpha}_i(0)$, $i = 1, 2, 3$ can be obtained, for example,
by averaging the parameter estimates over the duration of the motion
in a previous run. On the basis of a previous simulation run, the
following initial values for the estimates are determined: $\hat{\alpha}_1(0) \quad =$

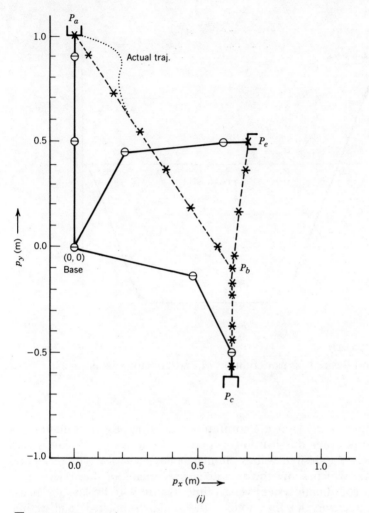

FIGURE 9.4*i*
Actual and desired end-effector trajectories on phase-plane
$(p_x p_y)$.

$[1.124, -0.114; 0.50; -0.0239]'$, $\hat{\alpha}_2(0) = [1.360, -0.361; 0.50; -0.806 \cdot 10^{-3}]'$,
$\hat{\alpha}_3(0) = [1.087, -0.0905; -0.50; -0.797 \cdot 10^{-3}]'$, and $P_i(0) = 100\,I$, $i =$
$1, 2, 3$. The controller for the motion is otherwise determined by the same
expressions as in the previous simulations. The graphs resulting from this
simulation show typically an improved behavior. For example, Figure 9.4*g*
displays θ_1 versus time, Figure 9.4*h* shows the orientation $\psi(k) = \theta_1(k) +$
$\theta_2(k) + \theta_3(k)$ versus time and Figure 9.4*i* the phase-plane graph. Clear
improvements in the tracking are observed, which is due to the better initial approximations of the parameters. It is feasible that a task-dependent

knowledge base could be constructed to facilitate the selection of the initial approximations of the parameters.

EXAMPLE 9.1.3

The end-effector of the planar manipulator discussed in Example 9.1.2 is to travel along the trajectory described and shown in Figure 9.3. The inputs and the output positions of the joints in the manipulator system are assumed to be available. However, the model and the numerical values of the parameters are not known at a sufficient degree of accuracy so that a primary controller based on the nonlinear terms in the dynamical model could be constructed. The problem is to design a self-tuning LQG-controller that will make the system track the prespecified desired trajectory $\{\theta^d(k)\}$. Thus, this controller will be the sole controller in the system.

Because of the absence of the input component from the primary controller, that is, $u_r^d(t) \equiv 0$ in Equation 9.1.10, the input to the manipulator is:

$$F(t) = [\hat{u}_1(t) \ \hat{u}_2(t) \ \hat{u}_3(t)]' \tag{9.1.30}$$

A controller specifying $\hat{u}_i(t)$, $i = 1, 2, 3$ for system 9.1.6 is again designed by following the steps described in Example 9.1.2. Assuming that the independent joint dynamics are applicable, the ARX-model in Equation 9.1.25 with $u(k) = \hat{u}(k)$ serves as the model. To design a controller with self-tuning property, the criterion in Equation 9.1.26 is minimized. Thus, the controller given in Equation 9.1.29 is obtained. When the controller is implemented, no information about the dynamics of the manipulator is assumed to be available.

To test the system with the designed adaptive controller, digital simulations are performed using the same numerical values as in Example 9.1.2 for the parameters in Equation 9.1.6. The initial approximations for the parameters to be estimated are $\hat{\alpha}_i(0) = [0, 0; 0.5; 0]'$, and $P_i(0) = 100 \, I$ for $i = 1, 2, 3$. Typical graphs recorded in a simulation study are shown in Figures 9.5a–9.5g. Figures 9.5a–9.5c are the graphs for the joint variables $\theta_1(t)$, $\theta_2(t)$ and $\theta_3(t)$ versus time. Figures 9.5d–9.5f demonstrate the end-effector tracking in the Cartesian base coordinate system. Figure 9.5g shows the graph for orientation $\psi(k) = \theta_1(k) + \theta_2(k) + \theta_3(k)$ versus time. Similar behavior is observed in all variables: initially large tracking errors are reduced quickly to small error values. The self-tuning controller constructed without using *a priori* knowledge of the manipulator dynamics or the parameters can indeed make the system perform the servo control satisfactorily.

In order to improve the tracking during the initial time interval in repetitive motions, the last estimates of the parameters obtained in one run can be used as the initial approximations of these parameters in the following run. The last estimates in the simulation corresponding to Figures 9.5a–9.5g are the initial approximations for the next run: $\hat{\alpha}_1(0) = [1.391, -0.389; 0.5; -0.013]'$, $\hat{\alpha}_2(0) = [1.485, -0.489; 0.5; -0.00938]'$ and $\hat{\alpha}_3(0) = [1.112, -0.115; 0.5; -0.00166]'$. The value of $P_i(0)$ is the same as in the previous run. The resulting graphs in Figures 9.6a–9.6c illustrate that the tracking is improved when the initial parameter estimates are chosen as the last estimates of the previous run.

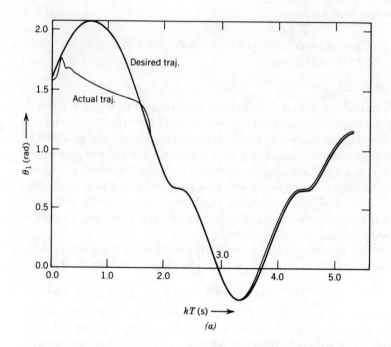

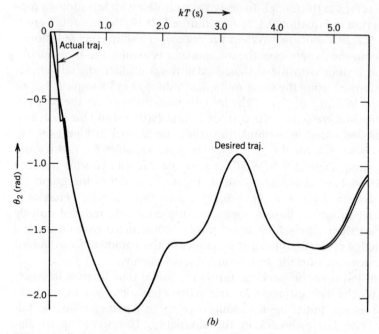

FIGURE 9.5 *a* AND *b*

Responses of joint positions when only self-tuning controller is used. Initial parameter estimates arbitrary. Example 9.1.3.

a. Actual (θ_1) and desired (θ_1^d) joint one positions versus time.

b. Actual (θ_2) and desired (θ_2^d) joint two positions versus time.

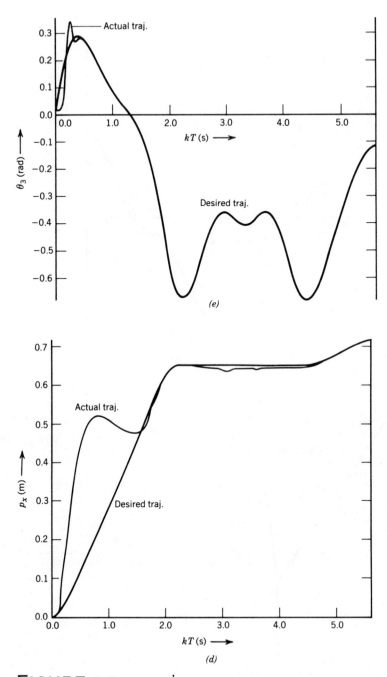

FIGURE 9.5 c AND d

c. Actual (θ_3) and desired (θ_3^d) joint three positions versus time.
d. Actual (p_x) and desired (p_x^d) end-effector x-coordinates versus time.

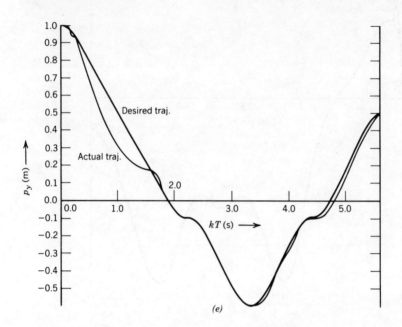

(e)

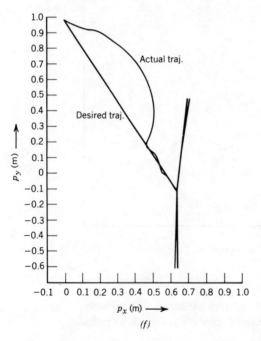

(f)

FIGURE 9.5 e AND f
e. Actual (p_y) and desired (p_y^d)
end-effector y-coordinates versus time.
f. Actual and desired end-
effector trajectories on phase-
plane $(p_x p_y)$.

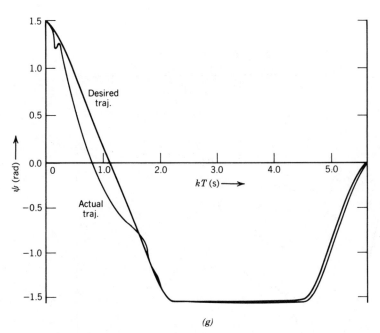

(g)

FIGURE 9.5g

Actual (ψ) and desired (ψ^d) orientation of end-effector versus time. $\psi = \theta_1 + \theta_2 + \theta_3$.

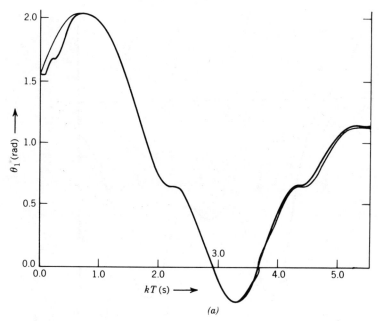

(a)

FIGURE 9.6a

Responses of joint positions when only self-tuning controller is used. Initial parameter estimates equal to last estimates of previous run. Example 9.1.3.

a. Actual (θ_1) and desired (θ_1^d) joint one positions versus time.

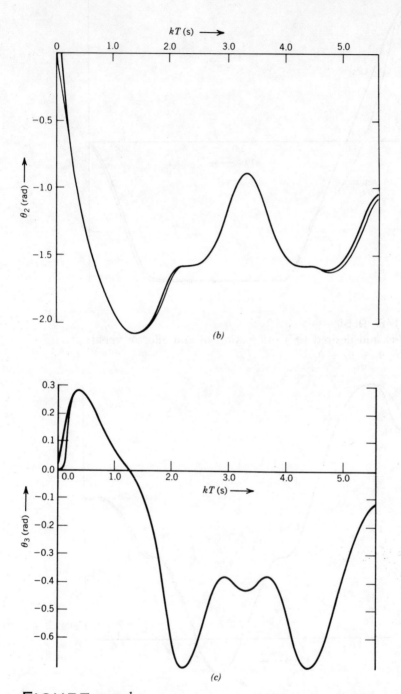

FIGURE 9.6 *b* AND *c*

b. Actual (θ_2) and desired (θ_2^d) joint two positions versus time.

c. Actual (θ_3) and desired (θ_3^d) joint three positions versus time.

Example 9.1.3 demonstrates that the self-tuning LQG-controller can successfully be applied to control the joint variable motions when the independent joint dynamics may be assumed. No *a priori* information about the model or the parameter values of the system is used in the approach discussed. If, however, some parameters in the model are accurately known, they should be used to construct a primary controller. In this way, uncertainties about the motion and also the computational burden will be reduced.

9.1.3 SELF-TUNING CONTROLLER BY POLE-ZERO PLACEMENT METHOD FOR INDEPENDENT JOINT DYNAMICS

An adaptive self-tuning controller is designed in Section 9.1.2 by minimizing a performance criterion involving a one-step-ahead prediction error. The control thus determined is often rapidly varying and large in magnitude, particularly for a small value of the weighting factor of the energy term. Moreover, this (optimal) controller may experience difficulties when the system exhibits nonminimum phase type behavior, and/or has constraints on the output response. Some of these difficulties can be circumvented by designing the self-tuning controller by the pole placement (eigenvalue assignment) method. Thus, the closed-loop poles will be placed to prespecified locations [7–9]. This approach has an inherent appeal to control engineers because it bears resemblance to the classical controller design techniques. Moreover, the poles at the desired locations can conveniently be related to the desired characteristics of the system responses as discussed in Sections 8.1, 8.2.2 and 8.2.3. We will next present the design of a self-tuning controller for a manipulator using the pole-zero placement method [5].

When the motion of a manipulator is assumed to be described by independent joint dynamics, an autoregressive exogenous (ARX) discrete-time SISO-model given in Equation 9.1.14 can be chosen for designing a self-tuning controller. Since the parameters in Equation 9.1.14 are not known, they are estimated, for example, by the recursive least-squares method specified by Equations 9.1.16 and 9.1.17. Using the estimated parameter values, the ARX-model for output $y_i(k)$ of the ith joint may then be described as follows:

$$y_i(k) = \hat{a}_0^i + \hat{a}_1^i y_i(k-1) + \cdots + \hat{a}_n^i y_i(k-n) + \hat{b}_0^i u_i(k-1)$$
$$+ \cdots + \hat{b}_n^i u_i(k-n-1) + e_i(k) \quad (9.1.31)$$

Using the delay operator z^{-1}, Equation 9.1.31 is expressed concisely as

$$\hat{A}_i(z^{-1}) y_i(k) = z^{-1} \hat{B}_i(z^{-1}) u_i(k) + \hat{a}_0^i + e_i(k) \quad (9.1.32)$$

where $k = n, n+1, \ldots, i = 1, \ldots, N$, $\hat{a}_0^i$ is a bias, $\hat{A}_i(z^{-1}) = 1 - \hat{a}_1^i z^{-1} - \cdots - \hat{a}_n^i z^{-n}$ and $\hat{B}_i(z^{-1}) = \hat{b}_0^i + \hat{b}_1^i z^{-1} + \cdots + \hat{b}_n^i z^{-n}$ are polynomials that have real coefficients and are relatively prime (that is, they have no common factors). The equation error $e_i(k)$ represents a white Gaussian zero-mean noise process. System 9.1.32 is assumed to represent an asymptotically stable system, that is, all zeros of $\hat{A}(z^{-1}) = 0$ are inside the unit circle. We should notice that parameters $\hat{a}_j^i$, $\hat{b}_j^i$, $j = 0, \ldots, n$ are usually time-varying, merely describing the values obtained from Equations 9.1.16 and 9.1.17.

Each joint variable is to track the prespecified reference trajectory that will make the end-effector of the manipulator follow the desired trajectory in the base coordinate system. The task is to design control $u_i(k)$ such that the resulting closed-loop system for each joint will be described by the following (desired) pulse transfer function:

$$G_i^d(z^{-1}) = \frac{z^{-1}B_i^d(z^{-1})}{A_i^d(z^{-1})} \tag{9.1.33}$$

where $A_i^d(z^{-1}) = 1 + a_{1i}^d z^{-1} + \cdots + a_{ni}^d z^{-n}$ and $B_i^d(z^{-1}) = \overline{K}_i(b_{0i}^d + b_{1i}^d z^{-1} + \cdots + b_{ni}^d z^{-n})$ are real polynomials that are relatively prime. They are determined so that the poles and zeros of $G_i^d(z^{-1})$ are at the desired locations. Moreover, A_i^d is monic, and the roots of $A_i^d(z^{-1}) = 0$ are assumed to be inside the unit circle. The low-frequency gain of $G_i^d(z^{-1})$ can be chosen to be one, that is, $B_i^d(1)/A_i^d(1) = 1$.

The desired poles of $G_i^d(z^{-1})$ can be specified on the basis of the variables that characterize the system step responses, as discussed in Section 8.1. For example, the specified overshoot and settling time of an underdamped step-response of the ith joint determine the desired damping coefficient ξ_i^d and the desired natural frequency ω_{ni}^d. Thus, the closed-loop transfer function $G_i^d(s)$ of the linear time invariant continuous-time reference model can be expressed as

$$G_i^d(s) = \frac{\overline{K}_i(\omega_{ni}^d)^2 \epsilon^{-dT}}{s^2 + 2\xi_i^d\omega_{ni}^d s + (\omega_{ni}^d)^2} \tag{9.1.34}$$

where $\overline{K}_i$ is a constant gain, and the time delay dT is an integer multiple of the sampling period. The z-transfer function corresponding to $G_i^d(s)$ for the sampling period T is

$$G_i^d(z^{-1}) = \frac{\overline{K}_i z^{-d}(b_{1i}^d z^{-1})}{1 + a_{1i}^d z^{-1} + a_{2i}^d z^{-2}} \tag{9.1.35}$$

where $\overline{K}_i = (1 + a_{1i}^d + a_{2i}^d)/b_{1i}^d$, $a_{1i}^d = -2\exp(-\xi_i^d\omega_{ni}^d T)\cos(\omega_{ni}^d T\{1 - (\xi_i^d)^2\}^{1/2})$, $a_{2i}^d = \exp(-2\xi_i^d\omega_{ni}^d T)$ and $b_{1i}^d = \omega_{ni}^d\exp(-\xi_i^d\omega_{ni}^d T)\sin(\omega_{ni}^d T\{1 - (\xi_i^d)^2\}^{1/2})/\{1 - (\xi_i^d)^2\}^{1/2}$. The specific problem would be to design a self-tuning controller for Model 9.1.31 such that the resulting feedback system will behave like the reference model given by Equation 9.1.35.

To solve the general problem posed, the compensation of bias $\hat{a}_0^i$ in Equation 9.1.32 is first performed by generating $u_i(k)$ according to the following equation:

$$u_i(k) = \overline{u}_i(k) - \frac{\hat{a}_0^i}{z^{-1}\hat{B}_i(z^{-1})} \tag{9.1.36}$$

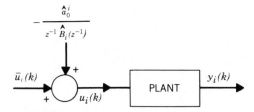

FIGURE 9.7
Compensation of bias-term in designing self-tuning controller; Equation 9.1.36.

where $\bar{u}(k)$ represents a new input. The block diagram of this system is shown in Figure 9.7.

The input–output model of the system in Equation 9.1.32 can now be expressed as

$$y_i(k) = \frac{z^{-1}\hat{B}_i(z^{-1})}{\hat{A}_i(z^{-1})}\bar{u}_i(k) + \frac{1}{\hat{A}_i(z^{-1})}e_i(k) \qquad (9.1.37)$$

The servo controller is sought in the following form:

$$\bar{u}_i(k) = \frac{K_i(z^{-1})}{R_i(z^{-1})}y_i^r(k) - \frac{S_i(z^{-1})}{R_i(z^{-1})}y_i(k) \qquad (9.1.38)$$

where the reference input $y_i^r(k) = y_i^d(k)$ specifies the desired trajectory. The real polynomials $K_i(z^{-1})$, $R_i(z^{-1})$, and $S_i(z^{-1})$ are to be determined so that the poles and zeros of the open-loop system will be moved to the locations specified by Equation 9.1.33. The block diagram of system 9.1.36–9.1.38 is shown in Figure 9.8.

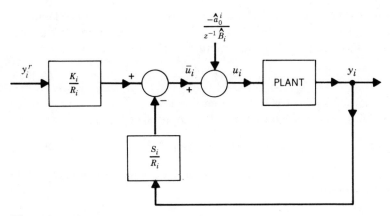

FIGURE 9.8
Realization of self-tuning controller designed by pole-placement method with bias compensation.

The input–output relation of the closed-loop system for the ith joint is governed by Equations 9.1.37 and 9.1.38. They are combined to obtain

$$y_i = \frac{z^{-1}\hat{B}_i K_i}{\hat{A}_i R_i + z^{-1}\hat{B}_i S_i} y_i^r + \frac{R_i}{\hat{A}_i R_i + z^{-1}\hat{B}_i S_i} e_i \qquad (9.1.39)$$

where the arguments have been dropped for convenience.

The comparison of Equations 9.1.39 and 9.1.33 gives the following equation for determining the unknown polynomials:

$$\frac{z^{-1}\hat{B}_i K_i}{\hat{A}_i R_i + z^{-1}\hat{B}_i S_i} = \frac{z^{-1}B_i^d}{A_i^d} \qquad (9.1.40)$$

where polynomials B_i^d and A_i^d are specified. Polynomials R_i, S_i, and K_i are determined by satisfying Equation 9.1.40, that specifies the desired pole-zero placement.

In order to avoid a possible attempt to cancel unstable modes, the factors of $\hat{B}_i$ corresponding to the finite zeros inside the unit circle, or well-damped modes, are grouped into polynomial $\hat{B}_i^+$ and those representing the finite zeros outside the unit circle, or poorly damped modes, into $\hat{B}_i^-$; thus,

$$\hat{B}_i(z^{-1}) = \hat{B}_i^+(z^{-1})\hat{B}_i^-(z^{-1}) \qquad (9.1.41)$$

To determine a stable servo controller, the poorly damped modes should not be canceled, and hence they appear in the numerator of the closed-loop pulse transfer function. Therefore, this (desired) numerator polynomial is:

$$\overline{B}_i^d(z^{-1}) = B_i^d(z^{-1})\hat{B}_i^-(z^{-1}) \qquad (9.1.42)$$

Consequently, the zeros of the closed-loop pulse transfer function consist of those of $\hat{B}_i^-(z^{-1})$ and the desired zeros in $B_i^d(z^{-1})$. In this case, $\overline{B}_i^d$ replaces B_i^d in Equation 9.1.40.

Equation 9.1.40 reveals that the factors of $\hat{B}_i(z^{-1})$ that are not factors of $B_i^d(z^{-1})$ are canceled in the closed-loop pulse transfer function, that is, they must also be factors of R_i. It is convenient to write

$$R_i(z^{-1}) = R_1^i(z^{-1})\hat{B}_i^+(z^{-1}) \qquad (9.1.43)$$

where polynomial $R_1^i(z^{-1})$ is to be determined.

Polynomial $K_i(z^{-1})$ in Equation 9.1.40 may be specified as

$$K_i(z^{-1}) = K_i^0(z^{-1})B_i^d(z^{-1}) \qquad (9.1.44)$$

where $K_i^0(z^{-1})$ can be chosen [8]. By employing Equations 9.1.41–9.1.44, Equation 9.1.40 gives:

$$\hat{A}_i(z^{-1})R_1^i(z^{-1}) + z^{-1}\hat{B}_i^-(z^{-1})S_i(z^{-1}) = A_i^d(z^{-1})K_i^0(z^{-1}) \qquad (9.1.45)$$

Equation 9.1.45 can be solved for $R_1^i(z^{-1})$, and $S_i(z^{-1})$, when $A_i^d(z^{-1})$, $K_i^0(z^{-1})$, $\hat{A}_i(z^{-1})$ and $\hat{B}_i^-(z^{-1})$ are known.

The solutions for R_1^i and S_i are usually not unique. It is practical to assume that K_i/R_i and S_i/R_i in Equation 9.1.38 are causal. Moreover, the polynomials for R_1^i, S_i, and K_i can be constrained; for example, the degrees (deg) of the polynomials in Equation 9.1.45 can satisfy [7, 8]: deg S_i = deg $\hat{A}_i$ − 1 and deg R_1^i = deg A_i^d + deg K_i^0 − deg $\hat{A}_i$.

The self-tuning control algorithm for joint i, $i = 1, \ldots, N$ based on the specific pole-zero placement may now be summarized. Assume that polynomials A_i^d, B_i^d, $\hat{A}_i$, $\hat{B}_i$, and K_i^0 are given. Then, the algorithm is described by the following steps:

Realization of self-tuning (pole-zero placement) controller

1. The parameters in the model Equation 9.1.32 are estimated by means of Equations 9.1.16 and 9.1.17, that is, by the recursive least-squared error method.

2. Polynomial $\hat{B}_i(z^{-1})$ is factored to $\hat{B}_i^+(z^{-1})$ and $\hat{B}_i^-(z^{-1})$ according to Equation 9.1.41, where the zeros of $\hat{B}_i^+(z^{-1})$ represent well-damped modes, and the zeros of $\hat{B}_i^-(z^{-1})$ correspond to poorly damped or unstable modes. Polynomial $\overline{B}_i^d$ is determined by Equation 9.1.42.

3. Equation 9.1.45 is solved for $R_1^i(z^{-1})$ and $S_i(z^{-1})$. Polynomial $R_i(z^{-1})$ is calculated by Equation 9.1.43 and $K_i(z^{-1})$ by Equation 9.1.44.

4. The control input to joint i is calculated by Equations 9.1.38 and 9.1.36.

5. Steps 1–4 are repeated for each $i = 1, \ldots, N$ in every sampling period during the motion of the manipulator.

In the special case that the model in Equation 9.1.32 does not contain (process) zeros besides the known delay, step 2 is not necessary. Similarly, step 2 is not needed, if all zeros of Equation 9.1.32 are included in $\hat{B}_i^-(z^{-1})$, that is, $\hat{B}_i(z^{-1}) = \hat{B}_i^-(z^{-1})$.

The general pulse transfer function of the closed-loop system of joint i will be $z^{-1}B_i^d(z^{-1})\hat{B}_i^-(z^{-1})/A_i^d(z^{-1})$, since the zeros in $\hat{B}_i^-(z^{-1})$ will not be canceled.

An example is next presented to illustrate the design procedure.

EXAMPLE 9.1.4

The motion of the planar manipulator discussed in Example 9.1.3 is assumed to obey independent joint dynamics. The input–output model for the design of the controller of the ith joint variable $\theta_i = y_i$, $i = 1, 2, 3$ is described by Equation 9.1.28, where the parameters with the carets represent the numerical values calculated using the parameter estimator Equations 9.1.16 and 9.1.17. The joint variables should track the desired trajectories $\{\theta_i^d(k)\}$

that will give rise to the motion P_a–P_b–P_c–P_b–P_e of the end-effector in the Cartesian base coordinate system as shown in Figure 9.3.

The problem is to design a self-tuning controller for the ith joint using the pole-zero placement method. The desired pulse transfer function $G_i^d(z^{-1})$ for the closed-loop system representing the input–output relation of joint i can be determined from the transfer function of the second-order reference model given in Equation 9.1.35. Suppose that the desired pulse transfer function is

$$G_i^d(z^{-1}) = \frac{z^{-1}(\overline{K}_i b_{0i}^d)}{1 + a_{1i}^d z^{-1} + a_{2i}^d z^{-2}} \tag{9.1.46}$$

where $\overline{K}_i = (1 + a_{1i}^d + a_{2i}^d)/b_{0i}^d$, $b_{0i}^d = 0.25$, $a_{1i}^d = -1.0$ and $a_{2i}^d = 0.25$ in this case. Thus, the comparison of Equations 9.1.33 and 9.1.46 reveals that $B_i^d(z^{-1}) = 0.25\overline{K}_i$, $\overline{K}_i = 1.0$, and $A_i^d(z^{-1}) = 1 - z^{-1} + 0.25z^{-2}$.

The bias $\hat{a}_0^i$ in the ARX-model given by Equation 9.1.28 is first compensated by applying Equation 9.1.36. The new input to the ith actuator is $\overline{u}_i(k)$. The input–output relation in Equation 9.1.28 for the ith joint becomes

$$y_i(k) = \frac{z^{-1}(\hat{b}_0^i)}{1 - \hat{a}_1^i z^{-1} - \hat{a}_2^i z^{-2}}\overline{u}_i(k) + \frac{1}{1 - \hat{a}_1^i z^{-1} - \hat{a}_2^i z^{-2}}e_i(k) \tag{9.1.47}$$

Thus, the polynomials in the open-loop transfer function are:

$$\hat{A}_i(z^{-1}) = 1 - \hat{a}_1^i z^{-1} - \hat{a}_2^i z^{-2} \qquad \hat{B}_i(z^{-1}) = \hat{b}_0^i \tag{9.1.48}$$

Since the desired discrete values for the joint variables are given, that is, $\theta_i^d(k) = y_i^r(k)$, the controller for self-tuning servoing is specified by Equation 9.1.38. Polynomials $R_i(z^{-1})$ and $S_i(z^{-1})$ are next to be determined. By choosing $K_i^0(z^{-1}) = 1$ in Equation 9.1.44, $R_1^i(z^{-1}) = r_0^i$ in Equation 9.1.43, and $S_i(z^{-1})$ as a first-order polynomial in z^{-1}, the unknown coefficients r_0^i, s_0^i, and s_1^i in these polynomials can be calculated using Equation 9.1.40 or 9.1.45:

$$(1 - \hat{a}_1^i z^{-1} - \hat{a}_2^i z^{-2})(r_0^i) + z^{-1}(1.0)(s_0^i + s_1^i z^{-1}) = 1 + a_{1i}^d z^{-1} + a_{2i}^d z^{-2} \tag{9.1.49}$$

After expanding the left side of Equation 9.1.49 into powers of z^{-1}, the coefficients of the equal powers of z^{-1} on both sides of this equation are set equal. It gives

$$r_0^i = 1.0 \tag{9.1.50}$$

$$-r_0^i \hat{a}_1^i + s_0^i = a_{1i}^d \tag{9.1.51}$$

$$s_1^i - \hat{a}_2^i r_0^i = a_{2i}^d \tag{9.1.52}$$

The algebraic Equations 9.1.50–9.1.52 can be solved for the unknown coefficients:

$$r_0^i = 1.0 \tag{9.1.53}$$

$$s_0^i = (a_{1i}^d + \hat{a}_1^i) \tag{9.1.54}$$

$$s_1^i = (a_{2i}^d + \hat{a}_2^i) \tag{9.1.55}$$

Polynomials $R_i(z^{-1})$, $S_i(z^{-1})$, and $K_i(z^{-1})$ in Equation 9.1.38 are now known. The control input $u_i(k)$ to the ith joint can be computed from Equations 9.1.38 and 9.1.36; thus

$$u_i(k) = \frac{\overline{K}_i b_{0i}^d}{r_0^i \hat{b}_0^i} y_i^r(k) - \frac{s_0^i + s_1^i z^{-1}}{r_0^i \hat{b}_0^i} y_i(k) - \frac{\hat{a}_0^i}{\hat{b}_0^i}$$

$$= \frac{0.25}{\hat{b}_0^i} y_i^r(k) - \frac{(a_{1i}^d + \hat{a}_1^i) + (a_{2i}^d + \hat{a}_2^i)z^{-1}}{\hat{b}_0^i} y_i(k) - \frac{\hat{a}_0^i}{\hat{b}_0^i} \tag{9.1.56}$$

The output of joint i under the influence of input $u_i(k)$ evolves in time k according to Equation 9.1.28, which may be simplified to the following form:

$$(1 + a_{1i}^d z^{-1} + a_{2i}^d z^{-2})y_i(k) = 0.25y^r(k-1) + e(k+1) \tag{9.1.57}$$

Equation 9.1.57 is the time-domain representation of the pulse transfer function $G_i^d(z^{-1})$ in Equation 9.1.46 that specifies the desired pole-zero locations when $e(k+1) \equiv 0$. Hence, the input–output relation of joint i is governed by the given pulse transfer function.

Each joint of the manipulator has a controller in the same form; naturally, the parameter values are different. The realization of the control algorithm is accomplished by following the previously described steps 1–5 for all three joints.

The control algorithm designed for this example is tested by digital simulations using the same numerical values on the gross motion of the planar manipulator as in Example 9.1.3. Figures 9.9a–9.9c display the simulation results for the following initial values: $\hat{a}_0^i(0) = 0$, $\hat{a}_1^i(0) = 1.0$, $\hat{a}_2^i(0) = 0$, $\overline{u}_i(0) = 0.1$, and $\hat{b}_0^i(0) = 0.5$. The values of the three joint variables $\theta_1(t)$, $\theta_2(t)$, and $\theta_3(t)$ fluctuate considerably over the initial part of the motion, after which the transients settle down. The initial values of the parameters were adjusted so as to reduce the oscillations at the beginning of the motion. The following initial values were finally used: $\hat{a}_0^1(0) = -0.00034, \hat{a}_1^1(0) = 1.580, \hat{a}_2^1(0) = -0.580, \hat{b}_0^1(0) = 0.0055$ (joint one); $\hat{a}_0^2(0) = -0.0081, \hat{a}_1^2(0) = 1.360, \hat{a}_2^2(0) = -0.360, \hat{b}_0^2(0) = 0.013$ (joint two); $\hat{a}_0^3(0) = -0.00042, \hat{a}_1^3(0) = 1.260, \hat{a}_2^3(0) = -0.260, \hat{b}_0^3(0) = 0.0116$ (joint three); and $\overline{u}_1(0) = 2.210, \overline{u}_2(0) = -2.430, \overline{u}_3(0) = 0.440$. Figures 9.9d–9.9f show the responses of the joint variables when the foregoing initial approximations for the parameters and inputs are chosen. The corresponding tracking error is graphed in Figure 9.9g. The transients in the outputs are considerably reduced by selecting the initial values on the basis of the behavior of the parameter estimates in the previous runs. The graphs in Figures 9.9a–9.9g demonstrate that a good tracking is achieved for each joint variable.

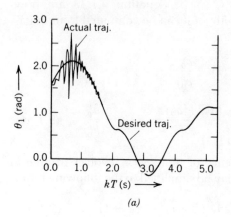

(a)

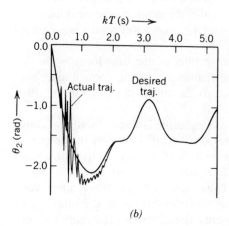

(b)

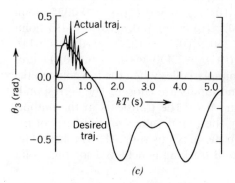

(c)

FIGURE 9.9 *a–c*

Responses of joint positions versus time when only self-tuning pole-zero placement controller is used. Initial parameter estimates arbitrary in *a–c*, adjusted using values of previous runs in *d–g*. Example 9.1.4.
a. Actual (θ_1) and desired (θ_1^d) joint one positions versus time.
b. Actual (θ_2) and desired (θ_2^d) joint two positions versus time.
c. Actual (θ_3) and desired (θ_3^d) joint three positions versus time.

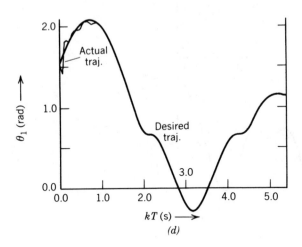

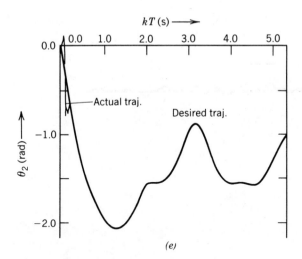

FIGURE 9.9 *d* AND *e*

d. Actual (θ_1) and desired (θ_1^d) joint one positions versus time.

e. Actual (θ_2) and desired (θ_2^d) joint two positions versus time.

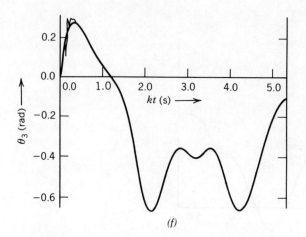

(f)

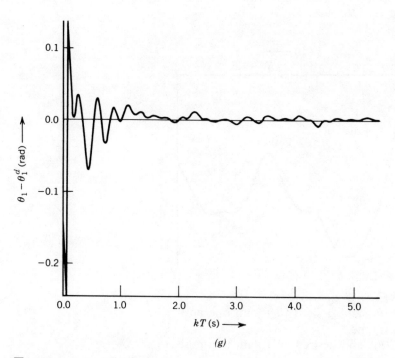

(g)

FIGURE 9.9 f AND g

f. Actual (θ_3) and desired (θ_3^d) joint three positions versus time.

g. Tracking error in joint one position versus time.

9.2 STOCHASTIC TIME-SERIES MODEL FOR JOINT VELOCITY TO DESIGN SELF-TUNING CONTROLLER

The control for the gross motion of a manipulator is discussed in Section 9.1 by designing a controller with self-tuning property for each joint position on the basis of an ARX-model. In this stochastic SISO-model, the input can be a voltage applied to the actuator or a generalized torque acting on the joint. The measured output in the model is usually the joint position. When a self-tuning controller is designed using such a model, difficulties in tracking may be experienced [3] when the ARX-model for the input and the positional output is of low order, that is, when the model for the joint position apparently is not sufficiently rich in the (natural) frequency content. Similar problems in tracking could be encountered if the noise process deviates considerably from the Gaussian assumption that is needed in the controller design.

The aforementioned difficulties may often be circumvented by constructing a time-series model for the joint velocities. These values can directly be measured or generated from two adjacent positional readings by means of software or hardware. The values of the parameters in this model can be estimated by the least-squared error method. The model with the estimated numerical values as the coefficients can then serve as the constraint equation in the design of a controller with self-tuning. We will next discuss this design approach.

By assuming independent joint dynamics, the velocity of joint i in a manipulator may be modeled for a controller design as a time-series SISO-model:

$$\hat{A}_i(z^{-1})\omega_i(k) = z^{-1}\hat{B}_i(z^{-1})u_i(k) + \hat{a}_0^i + e_i(k) \qquad (9.2.1)$$

where $\omega_i(k) \doteq \dot{\theta}_i(k)$ is the generalized velocity of the ith joint variable at time $kT, k = 0, 1, \ldots$, polynomials $\hat{A}_i(z^{-1}) = 1 - \hat{a}_1^i z^{-1} - \ldots - \hat{a}_n^i z^{-n}$, $\hat{B}_i(z^{-1}) = \hat{b}_0^i + \ldots + \hat{b}_n^i z^{-n}$ have real coefficients and $\hat{a}_0^i$ accounts for a possible bias. The parameters with the carets refer to the numerical values calculated according to the parameter estimation algorithm described in Equations 9.1.16 and 9.1.17. Equation 9.2.1 for each $i = 1, \ldots, N$ contains only variables of the ith joint; thus, the interactions between the joints are neglected.

The problem is to design a controller with self-tuning property for each joint i so that the position joint variable $\theta_i(k)$ will track the specified discrete points $\{\theta_i^d(k)\}$ and the velocity joint variable $\omega_i(k)$ the desired velocity $\{\omega_i^d(k)\}$ as closely as possible while satisfying the velocity model in Equation 9.2.1.

The controller is constructed as an LQG-controller by minimizing a well-defined single-stage criterion. This performance index can be chosen so that the position and also velocity errors are penalized. A possible form for the performance criterion is [3]

$$I_k[u_i] = E\{([\omega_i(k+1) - \omega_i^d(k+1)] + \rho_i[\theta_i(k+1) - \theta_i^d(k+1)])^2 + \epsilon_i u_i^2(k)|\Sigma_i(k)\}$$
$$(9.2.2)$$

where ϵ_i is a nonnegative weighting factor, and the nonnegative number ρ_i expresses the relative importance of the position error with respect to the velocity error. The expectation operation $E\{\cdot \,|\Sigma_i(k)\}$ is conditioned on the available information up to and including time kT designated by $\Sigma_i(k)$. The problem is then to minimize the performance criterion in Equation 9.2.2 over the admissible controls $u_i(k)$ while satisfying Equation 9.2.1. A control input $u_i(k)$ is called admissible when it is a function of the current and/or past measurements $\omega_i(j)$, $\theta_i(j)$ and possibly of the past controls $u_i(j-1)$, $j = k, k-1, \ldots$.

The controller can be determined after substituting Equation 9.2.1 with k replaced by $(k+1)$ for $\omega_i(k+1)$ in Equation 9.2.2 and approximating the joint position $\theta_i(k+1)$ by the following expression:

$$\theta_i(k+1) = \theta_i(k) + \omega_i(k)T \qquad (9.2.3)$$

where the sampling period T should be sufficiently short for an acceptable approximation. By performing the minimization of $I_k[u_i]$ in Equation 9.2.2 with respect to $u_i(k)$, the controller is obtained:

$$u_i(k) = \hat{K}_i\{[\omega_i^d(k+1) - z\hat{A}_{1i}(z^{-1})\omega_i(k) - \hat{a}_0^i - \hat{B}_{1i}(z^{-1})u_i(k)]$$
$$+ \rho_i[\theta_i^d(k+1) - \theta_i(k+1)]\} \qquad (9.2.4)$$

where $\overline{K}_i = \hat{b}_0^i/[(\hat{b}_0^i)^2 + \epsilon_i]$, $\hat{A}_{1i}(z^{-1}) = \hat{a}_1^i z^{-1} + \ldots + \hat{a}_n^i z^{-n}$, and $\hat{B}_{1i}(z^{-1}) = \hat{b}_1^i z^{-1} + \ldots + \hat{b}_n^i z^{-n}$. Position $\theta_i(k+1)$ is determined by Equation 9.2.3. Equation 9.2.4 specifies the control input $u_i(k)$ that consists of the feedback loops from ω_i and θ_i as well as of the feedforward loops for ω_i^d and θ_i^d.

The minimum variance controller is obtained from Equation 9.2.4 by setting $\epsilon_i = 0$. If the expression of the minimum variance controller is substituted into Equation 9.2.1, the resulting equation can be expressed as follows:

$$\omega_i(k+1) = -\rho_i[\theta_i(k) + T\omega_i(k) - \theta_i^d(k+1)] + \omega_i^d(k+1) + e_i(k+1) \quad (9.2.5)$$

Equation 9.2.5 corresponds to Equation 9.1.22 that is obtained by using the position ARX-model. If the mean velocity $\omega_i(k+1)$ becomes equal to $\omega_i^d(k+1)$, then the mean of the joint position assumes also the desired value, and vice versa. Moreover, it can be shown that the system consisting of Equations 9.2.3 and 9.2.5 is (asymptotically) stable when $0 < \rho_i T < 2$ (Problem 9.3). This condition should be observed in selecting the weighting factor ρ_i and the sampling period T for the manipulator system.

The gross motion control of a Stanford/JPL arm using the described controller with self-tuning to be presented next will illustrate the design procedure.

EXAMPLE 9.2.1

The dynamical model for the six-joint Stanford/JPL arm can be expressed in the form of Equation 8.2.14. It is assumed that no accurate *a priori* information is available to design a primary controller. An adaptive controller with self-tuning is to be designed so that the end-effector tracks a desired trajectory. It is to be determined so that the average of the squared errors in the tracking and the control energy appropriately weighted is minimized, where the errors refer to the deviations of the actual values of the motion from the desired values. Then, the resulting controller is to be tested by digital simulations.

For the system, the input of joint i is voltage u_i applied to the joint motor, and the outputs are position θ_i and velocity ω_i, $i = 1, \ldots, 6$. Independent joint dynamics are assumed to apply during the manipulator motion. On the basis of the experimental simulations, a second-order autoregressive model for the joint velocity is chosen with $a_1^i = 0$; thus, Equation 9.2.1 gives [3]:

$$(1 - \hat{a}_2^i z^{-2}) \omega_i(k) = z^{-1}(\hat{b}_0^i + \hat{b}_1^i z^{-1}) u_i(k) + \hat{a}_0^i + e_i(k) \qquad (9.2.6)$$

where $1 - \hat{a}_2^i z^{-2} = \hat{A}_i(z^{-1})$ and $\hat{b}_0^i + \hat{b}_1^i z^{-1} = \hat{B}_i(z^{-1})$. The coefficients in these polynomials are the numerical values determined by the parameter estimation Equations 9.1.16 and 9.1.17 with $\hat{\alpha} = [\hat{a}_2^i; \hat{b}_0^i \hat{b}_1^i; \hat{a}_0^i]'$.

The design goals can be achieved by minimizing the following performance criterion (Equation 9.2.2):

$$I_k[u_i] = E\{([\omega_i(k+1) - \omega_i^d(k+1)] + \rho_i[\theta_i(k+1) - \theta_i^d(k+1)])^2 + \epsilon_i u_i^2(k) | \Sigma_i(k)\} \qquad (9.2.7)$$

where $\{\theta_i^d(k), \omega_i^d(k)\}$ specify the desired discrete-time trajectory for the ith joint. The real numbers ρ_i and ϵ_i are chosen nonnegative weighting factors. Position $\theta_i(k + 1)$ in the performance criterion $I_k[u_i]$ is replaced by the expression given in Equation 9.2.3. The expression in Equation 9.2.7 is so chosen that the tracking of the position is achieved even though the ARX-model is in terms of the joint velocity.

The performance criterion in Equation 9.2.7 is minimized relative to the admissible controls $u_i(k)$ while satisfying the difference Equation 9.2.6 constraint. It results in the controller

$$u_i(k) = \overline{K}_i\{[\omega_i^d(k + 1) - \hat{a}_2^i \omega_i(k - 1) - \hat{a}_0^i - \hat{b}_1^i u_i(k - 1)]$$

$$+ \rho_i[\theta_i^d(k + 1) - \theta_i(k) - T\omega_i(k)]\} \qquad (9.2.8)$$

where $\overline{K}_i = \hat{b}_0^i / [(\hat{b}_0^i)^2 + \epsilon_i]$. The control algorithm is then realized on the basis of Equations 9.2.6, 9.1.16, 9.1.17, and 9.2.8.

The performance of the control system is tested by digital simulation studies [3]. The dynamical model in Equation 8.2.14 is simulated using the numerical values of [10]. The measurements are generated by digital simulations using 0.0025 s for the subinterval of integration. These values are superimposed by sample values from a white Gaussian noise process with zero mean and variance σ^2. In this example, σ was chosen as 0.005 rad for the rotational movements and 0.05 in for the translational motion.

Two cases of the desired trajectory are studied: (1) a straight line and (2) a circle specified by discrete points in the Cartesian coordinate system. The desired trajectory points in the joint space are calculated using the inverse kinematic equations presented in Section 3.3. The simulation is programmed on CDC 6500. Due to the D/A conversion, a constraint $|u_i(k)| \leq 40\text{V}$ is imposed. Constant ρ_i of the performance criterion in Equation 9.2.7 is chosen as $0.5/T$. The weighting factor ϵ_i is set equal to $(\hat{b}_0^i)^2$ that varies with time. It provides a relative weighting between control $u_i(k)$ and the position and velocity errors. The forgetting factor γ_i in Equation 9.1.17 is set equal to 0.95 for every joint. The initial value of the parameter vector is $\hat{\alpha}_i(0) = [1; 0\ 1; 0]'$ and $P_i(0) = 10^4 I$.

Case 1: The desired trajectory is a straight line. The task is to move the end-effector from point P_1 to point P_2 in 0.6 s. The numerical values are $P_1 = (-42.5\ 2.4\ 23.2)$ and $P_2 = (-12.3\ -10.5\ 20.0)$ expressed in inches in the Cartesian base coordinate system. The desired path of the gripper is a straight line in the Cartesian coordinate system. Moreover, the speed of the end-effector is to follow an isosceles triangle at a constant acceleration (deceleration) starting from zero initial speed until it reaches a maximum speed in 0.3 s and then decelerates to zero speed at the terminal state P_2. The total desired rotational motion is 84°.

Typical simulation results for the system with the adaptive self-tuning type controller are shown for the third and fourth joint variables in Figures 9.10a and 9.10b in the joint space and for the trajectory of the gripper on the yz-plane of the Cartesian base coordinate system in Figure 9.10c. The estimates of parameter $\hat{b}_0^1$ versus time k are graphed in Figure 9.10d.

Case 2: The desired trajectory is a circle. The task of the gripper is to follow a circle with a diameter of 25 in. centered at point $(-30\ 8.5\ 20)$ in. relative to the Cartesian base coordinate system. The plane of the circle is specified by $x = -30$ in. The hand starts at the point $(-30\ 8.5\ 32.5)$ in. with zero speed. The orientation rotates 90° during the motion, while the angular speed of the gripper is kept constant.

The graphs describing the behavior of the system with the adaptive controller are computed. Typical simulation results are shown in Figures 9.11a and 9.11b for the third and fourth joint variables and for the gripper position on the yz-plane of the Cartesian coordinate system in Figure 9.11c. The estimates of parameter $\hat{b}_0^1$ are graphed in Figure 9.11d. The responses of the other joints are similar.

In both cases, the adaptation capability of the controller can be observed. Initially, the system response oscillates while trying to follow the desired values and then settles down. Moreover, the graphs demonstrate how the system attempts to learn the values of the parameters along the trajectory by on-line adjustments. No *a priori* information about manipulation dynamics is utilized in the controller. When the manipulator moves payloads of different weights (0–4 lbs) along the same desired path, similar responses for the manipulator system with the controller are recorded. The simulations indicate that the system can adapt to changing operating conditions. Indeed, the self-tuning controller performs well in the case studied when the independent joint control is applied. The adaptive self-tuning controller can be implemented using microprocessors or a small computer.

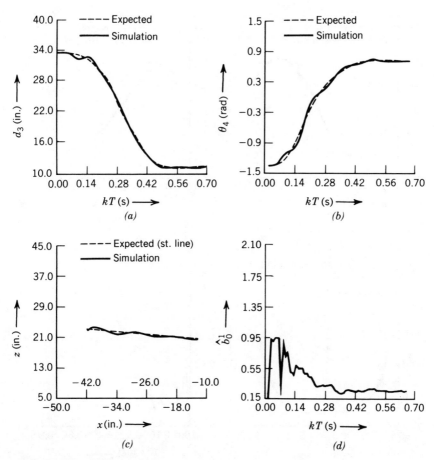

FIGURE 9.10 *a–d*
Actual and desired trajectories of Stanford/JPL arm with self-tuning controller Example 9.2.1.
a. Joint three positions $\theta_3 = d_3$ and $\theta_3^d = d_3^d$ versus time.
b. Joint four positions θ_4 and θ_4^d versus time.
c. Gripper position on (xz)-plane.
d. Parameter estimate $\hat{b}_0^1$ versus time in joint one model.

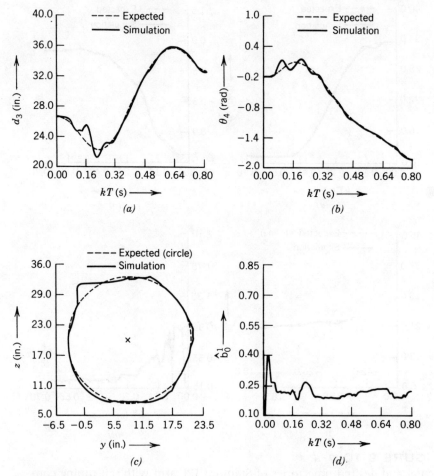

FIGURE 9.11 *a–d*
Actual and desired trajectories of Stanford/JPL arm with self-tuning controller.
a. Joint three positions θ_3 and θ_3^d versus time.
b. Joint four positions θ_4 and θ_4^d versus time.
c. Gripper trajectory on (yz)-plane.
d. Estimated parameter $\hat{b}_0^1$ versus time in joint one model.

9.3 MULTIVARIABLE SELF-TUNING CONTROLLER FOR MANIPULATOR GROSS MOTION

The gross motion of a manipulator is made to track a desired trajectory by means of an appropriately designed controller. Such a controller is constructed in the previous sections as a self-tuner under the assumption of the independent joint dynamics. The controller with self-tuning is thus designed on the basis of a stochastic SISO-model. Even for the case that there is no *a priori* knowledge of the manipulator model or dynamical parameters, the designed system functions well at the moderate operating speeds, when the effects of Coriolis and centripetal forces on the motion of the manipulator are insignificant. The controller designed for the independent joint dynamics can be expected to exhibit difficulties in the tracking when this assumption does not hold. This can be demonstrated, for example, by simulating the manipulator at high operating speeds. Interactions between the joint variables can be included in the ARX-model describing the input–output relation of a manipulator by using an appropriately chosen multivariate time-series model [3, 11]. A multivariate time-series model is also needed when a self-tuning controller for a manipulator is designed to operate on the variables describing the end-effector position in the Cartesian base coordinate system, whereby the calculations for the solution to the inverse kinematic equations are avoided.

We will present next the design of a multivariate self-tuning controller for the gross motion of manipulators. We will thus generalize the method of the self-tuning controller design presented in Section 9.1.2 for SISO-systems to MIMO-systems [12]. Interacting effects between the manipulator joints can then be incorporated in the ARX-model, and thus in the self-tuning controller design. Specifically, we will present first the recursive least-squares equations for the parameter estimation in multivariable discrete-time systems. Then, we will construct a multivariable LQG-controller with self-tuning for the gross motion of a manipulator.

The multivariable autoregressive difference equations may be written in the general form as

$$A(z^{-1})y(k) = z^{-1}B(z^{-1})u(k) + a_0 + e(k) \tag{9.3.1}$$

where the sampling period is omitted in the arguments, and integer k refers to the sampling instant, $k = 0, 1, 2 \dots$. The m-dimensional vector a_0 is a forcing term that accounts for the effects of the generalized gravitational forces. The m-dimensional output vector y and input vector u have as the ith component output y_i and input u_i of joint i, respectively, and $i = 1, \dots, m$. The equation error vector $e(k)$ represents a random white Gaussian zero-mean noise with a finite covariance; it is independent of current and past outputs and inputs. The $(m \times m)$ matrices A and B are polynomials defined by

$$A(z^{-1}) = I - (A_1 z^{-1} + \dots + A_n z^{-n}) \tag{9.3.2}$$

$$B(z^{-1}) = B_0 + B_1 z^{-1} + \ldots + B_n z^{-n} \qquad (9.3.3)$$

where n is a positive integer specifying the order of the model. B_0 is nonsingular, and $\det(B)$ has all zeros strictly inside the unit circle.

To estimate the parameters in Equation 9.3.1, matrix α and vector $\phi(k-1)$ are defined:

$$\alpha = [A_1 \ldots A_n; B_0 \ldots B_n; a_0]' = [\alpha_1 \ldots \alpha_{m'}] \qquad (9.3.4)$$

$$\phi(k-1) = [y'(k-1) \ldots y'(k-n); u'(k-1) \ldots u'(k-n-1); 1]' \quad (9.3.5)$$

Then for $i = 1, \ldots, m'$,

$$\alpha_i = [a_1^{i1} \ldots a_1^{im} a_2^{i1} \ldots a_2^{im} \ldots; b_0^{i1} \ldots b_0^{im} b_1^{i1} \ldots b_n^{im} \ldots; a_0^i]' \qquad (9.3.6)$$

The numerical values of the components of parameter α are calculated by the least-squared error method by estimating one vector α_i at a time. The algorithm for computing the estimate $\hat{\alpha}$ of parameter α is based on the recursive equations described in detail in Appendix D.1.3:

$$\hat{\alpha}_i(k) = \hat{\alpha}_i(k-1) + P(k)\phi(k-1)[y_i(k) - \phi'(k-1)\hat{\alpha}_i(k-1)] \qquad (9.3.7)$$

$$P(k) =$$
$$\frac{1}{\gamma}[P(k-1) - P(k-1)\phi(k-1)[\gamma + \phi'(k-1)P(k-1)\phi(k-1)]^{-1}\phi'(k-1)P(k-1)] \qquad (9.3.8)$$

The self-tuning LQG-controller is constructed by minimizing the following quadratic criterion:

$$I_k(u) = E\{\|y(k+1) - y^d(k+1)\|_Q^2 + \|u(k)\|_R^2 |\Sigma(k)\} \qquad (9.3.9)$$

where $\|\cdot\|_R$ indicates the generalized norm with weight R, for example, $\|u\|_R^2 = u'Ru$. Matrices R and Q are symmetric and positive semidefinite, and $\{y^d(k)\}$ describes the desired trajectory vector as a sequence of discrete points. The expectation operation is conditioned on the available measurements up to and including time kT indicated by $\Sigma(k)$. The admissible controls considered in the minimization are the ones that are functions of the measurements $y(k-j)$, $j = 0, 1, \ldots$ and the past controls.

The problem is to minimize the performance criterion 9.3.9 relative to the admissible controls while satisfying the following constraint equation:

$$\hat{A}(z^{-1})y(k) = z^{-1}\hat{B}(z^{-1})u(k) + \hat{a}_0 + e(k) \qquad (9.3.10)$$

Equation 9.3.10 is obtained from Equation 9.3.1 by substituting the estimated numerical values for the unknown parameters.

The problem is solved by expressing the predicted value of $y(k+1)$ in terms of the available information up to time kT, substituting it into Equation 9.3.9 and performing the minimization of $I_k[u]$ with respect to $u(k)$. The details are presented in Appendix D.2.1. The resulting controller is determined by

$$Ru(k) + \hat{B}_0'Q[\hat{y}(k+1|k) - y^d(k+1)] = 0 \qquad (9.3.11)$$

where $\hat{y}(k+1|k)$ denotes the predicted value (optimal in the sense of the least-squares error criterion) of $y(k + 1)$ based on the information available at time kT. Specifically,

$$\hat{y}(k + 1|k) = \hat{\alpha}'(k)\phi(k) \qquad (9.3.12)$$

The controller may also be expressed as follows:

$$u(k) = [R + \hat{B}_0' Q \hat{B}_0]^{-1} \hat{B}_0' Q[y^d(k + 1) - z\hat{A}_r(z^{-1})y(k) - \hat{a}_0 - z\hat{B}_r(z^{-1})u(k-1)] \qquad (9.3.13)$$

where $\hat{A}_r(z^{-1}) = \hat{A}_1 z^{-1} + \ldots + \hat{A}_n z^{-n}$, $\hat{B}_r(z^{-1}) = \hat{B}_1 z^{-1} + \ldots + \hat{B}_n z^{-n}$ and the caret refers to the numerical values calculated from Equations 9.3.7 and 9.3.8.

To make the variables of the manipulator joints track the desired trajectory, $\{y^d(k)\}$, a self-tuning control algorithm based on the time-series MIMO-model given in Equation 9.3.10 can now be summarized:

Realization of MIMO self-tuning LQG-controller

1. Determine initial approximations for $\hat{\alpha}(0)$, $P(0)$, $u(0)$, ..., $u(n - 1)$ and choose the forgetting factor γ and the weighting matrices R and Q. Set $k = n$.

2. Compute the parameter estimate $\hat{\alpha}(k)$ using Equations 9.3.7 and 9.3.8. Determine the control input $u(k)$ from Equation 9.3.13.

3. Apply input $u(k)$ to the system. Obtain the current measurements. Replace k by $k + 1$.

4. Repeat the calculations from step 2 on until the final time of the motion is reached.

The multivariable self-tuning controller presented can be used as a secondary controller, if sufficiently accurate information about the manipulator dynamics is available to form a primary controller. Alternatively, the self-tuning controller can be used by itself as a sole controller for the manipulator system. Interactions between the joint variables can be incorporated in the model by properly choosing the off-diagonal elements of matrices A_i and B_i in the polynomials of Equations 9.3.2 and 9.3.3.

The procedure described will first be applied to the design of a multivariable self-tuning controller in the joint space for the gross motion of a manipulator operating at moderately high speeds and then for controlling the variables of the end-effector motion of a manipulator in the Cartesian base coordinate system.

9.3.1 MULTIVARIABLE SELF-TUNING CONTROLLER FOR INTERACTING JOINT DYNAMICS

The ARX-models used for independent joint dynamics in the design of self-tuning controllers can be represented as multivariable systems. Indeed, the set of difference Equations 9.2.6, $i = 1, 2, 3$ in Example 9.2.1 describing the input–output relations of the three joints can be expressed as a multivariate ARX-model:

$$
\begin{bmatrix} \omega_1(k) \\ \omega_2(k) \\ \omega_3(k) \end{bmatrix} - \begin{bmatrix} \hat{a}_2^1 & 0 & 0 \\ 0 & \hat{a}_2^2 & 0 \\ 0 & 0 & \hat{a}_2^3 \end{bmatrix} \begin{bmatrix} \omega_1(k-2) \\ \omega_2(k-2) \\ \omega_3(k-2) \end{bmatrix} = \begin{bmatrix} (\hat{b}_0^1 + \hat{b}_1^1 z^{-1}) u_1(k-1) \\ (\hat{b}_0^2 + \hat{b}_1^2 z^{-1}) u_2(k-1) \\ (\hat{b}_0^3 + \hat{b}_1^3 z^{-1}) u_3(k-1) \end{bmatrix}
$$

$$
+ \begin{bmatrix} \hat{a}_0^1 \\ \hat{a}_0^2 \\ \hat{a}_0^3 \end{bmatrix} + \begin{bmatrix} e_1(k) \\ e_2(k) \\ e_3(k) \end{bmatrix} \quad (9.3.14)
$$

Equation 9.3.14 can be written in a concise form as Equation 9.3.10, where output $y(k) = [\omega_1(k) \ \omega_2(k) \ \omega_3(k)]'$ and input $u(k) = [u_1(k) \ u_2(k) \ u_3(k)]'$; moreover, the bias vector $\hat{a}_0 = [\hat{a}_0^1 \ \hat{a}_0^2 \ \hat{a}_0^3]'$ and the equation error $e(k) = [e_1(k) \ e_2(k) \ e_3(k)]'$. The (3×3) matrix polynomials $\hat{A}(z^{-1})$ and $\hat{B}(z^{-1})$ are

$$
\hat{A}(z^{-1}) = I - \hat{A}_2 z^{-2}
$$
$$
\hat{B}(z^{-1}) = \hat{B}_0 + \hat{B}_1 z^{-1}
$$
$$(9.3.15)$$

$$
\text{where} \quad \hat{A}_2 = \text{diag}[\hat{a}_2^1 \ \hat{a}_2^2 \ \hat{a}_2^3],
$$

$$
\hat{B}_0 = \text{diag}[\hat{b}_0^1 \ \hat{b}_0^2 \ \hat{b}_0^3] \quad \text{and} \quad \hat{B}_1 = \text{diag}[\hat{b}_1^1 \ \hat{b}_1^2 \ \hat{b}_1^3].
$$

The multivariate time-series model in Equation 9.3.14 describes the input–output relation of the independent joint dynamics, since matrix $\hat{A}_2$ and thus the $\hat{A}$-matrix are diagonal, that is, any row in Equation 9.3.14 contains variables of only one joint. To account for the interacting effects between the joint variables, off-diagonal elements of the A-matrix in Equation 9.3.1 are chosen to be nonzero; naturally, they also need to be estimated.

The general time-series model in the form of Equation 9.3.10 will next be used to design a multivariate self-tuning controller for the gross motion of a manipulator operating at moderately high speeds.

EXAMPLE 9.3.1

The task here is to determine an adaptive control for the joint variables of a Stanford/JPL manipulator so that the motion of the manipulator joints follow a desired path as closely as possible. The desired trajectory is described by the discrete points $\{\theta^d(k), \omega^d(k)\}$ in the joint space where $\theta^d(k)$ and $\omega^d(k)$ specify the desired position and velocity at time kT, respectively.

The problem is solved by using a multivariable autoregressive model in Equation 9.3.10 for the input–output relation of the joints. In order to restrict the complexity of the multivariable model, the variables of the manipulator joints may be decomposed into two groups on the basis of the mechanical structure of the manipulator. The first three joints are mainly associated with the positional movement and the last three joints with the orientation. The coupling terms between the variables of the two groups will be neglected. The terms describing the interactions between the variables within each group are included. Thus, matrix A_j, $j = 1, \ldots, n$ in Equation 9.3.10 is written as

$$
A_j = \left[\begin{array}{c|c} A_j^1 & 0 \\ \hline 0 & A_j^6 \end{array} \right]
\tag{9.3.16}
$$

where A_j^1 and A_j^6 are (3×3) matrices with unknown elements to be estimated. Matrices B_j in 9.3.10 may be chosen to be diagonal.

The autoregressive model for the velocity vector $\omega(k) = [\omega_1(k) \; \omega_2(k) \; \omega_3(k); \; \omega_4(k) \; \omega_5(k) \; \omega_6(k)]'$ is chosen on the basis of simulation studies:

$$
\omega(k) = A_2\omega(k - 2) + B_0 u(k - 1) + B_1 u(k - 2) + a_0 + e(k)
\tag{9.3.17}
$$

where vector $a_0 = [a_0^1 \; a_0^2, \ldots, a_0^6]'$ accounts for the gravitational forces, $u(k-1) = [u_1(k-1) \; u_2(k-1), \ldots, u_6(k-1)]'$ represents the voltages applied at time $(k-1)T$ to the motors of the joints. In the equation error vector, $e(k) = [e_1(k) \; e_2(k) \ldots e_6(k)]'$, each $e_i(k)$, $i = 1, \ldots, 6$ represents a white Gaussian zero-mean noise process, which is independent of $\omega(k-j), u(k-j)$ for $j \geq 1$ and $e_\ell(k)$ for $\ell \neq i$.

The comparison of Equations 9.3.17 and 9.3.1 reveals that A_1 is chosen as a null matrix to simplify calculations, and A_2 can be chosen as the expression in Equation 9.3.16 with $j = 2$ and

$$
A_2^1 = \begin{bmatrix} a_2^{11} & a_2^{12} & a_2^{13} \\ a_2^{21} & a_2^{22} & a_2^{23} \\ a_2^{31} & a_2^{32} & a_2^{33} \end{bmatrix}
\tag{9.3.18}
$$

$$
A_2^6 = \begin{bmatrix} a_2^{44} & a_2^{45} & a_2^{46} \\ a_2^{54} & a_2^{55} & a_2^{56} \\ a_2^{64} & a_2^{65} & a_2^{66} \end{bmatrix}
\tag{9.3.19}
$$

For computing the parameter estimates in Equation 9.3.17, expressions 9.3.4 and 9.3.5 are written:

$$
\alpha = \left[\begin{array}{cc|ccc|ccc|c} A_2^1 & 0_{3\times3} & b_0^{11}, & \ldots, & 0 & b_1^{11}, & \ldots, & 0 & a_0^1 \\ & & & \vdots & & & \vdots & & \vdots \\ 0_{3\times3} & A_2^6 & 0 & \ldots, & b_0^{66} & 0 & \ldots, & b_1^{66} & a_0^6 \end{array} \right]
\tag{9.3.20}
$$

$$
\phi(k-1) = [\omega_1(k-2)\ldots\omega_6(k-2); u_1(k-1)\ldots u_6(k-1),
$$

$$
u_1(k-2)\ldots u_6(k-2); 1]'
\tag{9.3.21}
$$

The parameters on each row of α in Equation 9.3.20 are then estimated at each sampling instance by applying Equations 9.3.7 and 9.3.8 successively for $i = 1, \ldots, 6$. Thus, six parameters are determined at a time. The calculated numerical values are substituted for the unknown coefficients in Equation 9.3.17.

The controller with self-tuning is determined by minimizing the position and velocity errors squared and the energy as specified by the performance criterion similar to Equation 9.2.2:

$$I_k[u] = E\{\|[\omega(k+1) - \omega^d(k+1)] + \rho[\theta(k+1) - \theta^d(k+1)]\|_Q^2 + \|u(k)\|_R^2 |\Sigma(k)\}$$
(9.3.22)

where position $\theta(k+1)$ is approximated using the velocity vector and $\theta(k)$ as shown in Equation 9.2.3. The weighting matrices $\rho = \rho' = \mathrm{diag}[\rho_1 \ldots \rho_6] \geq 0$, $R = R' \geq 0$, and $Q = Q' > 0$ are chosen in view of the specific operational objectives.

The minimization of expression 9.3.22 subject to the constraint Equation 9.3.17 with the estimates substituted for the unknown parameters determines the controller

$$u(k) = [R + \hat{B}_0'Q\hat{B}_0]^{-1}\hat{B}_0'Q\{\omega^d(k+1) - \hat{A}_2\omega(k-1) - \hat{B}_1u(k-1) - \hat{a}_0$$

$$+ \rho[\theta^d(k+1) - \theta(k+1)]\} \quad (9.3.23)$$

The controller in Equation 9.3.23 is of the same form as the controller in Equation 9.3.13. The algorithm described by steps 1–5 can now be applied to realize the self-tuning control scheme given by Equation 9.3.23.

An alternative but equivalent formulation of the control problem could be obtained by expressing first the constraint Equations 9.3.17 and 9.2.3 in a multivariable form similar to Equation 9.3.10:

$$\begin{bmatrix} (1 - z^{-1})I & -Tz^{-1} \\ 0 & \hat{A}(z^{-1}) \end{bmatrix}\begin{bmatrix} \theta(k) \\ \omega(k) \end{bmatrix} = \begin{bmatrix} 0 \\ \hat{B}(z^{-1}) \end{bmatrix}u(k-1) + \begin{bmatrix} 0 \\ \hat{a}_0 \end{bmatrix} + \begin{bmatrix} 0 \\ e(k) \end{bmatrix}$$
(9.3.24)

where $[\theta'(k)\omega'(k)]'$ represents $y(k)$ of Equation 9.3.10. The performance criterion in Equation 9.3.22 could then be written in the form of Equation 9.3.9 by specifying the symmetric weighting matrix $\overline{Q}$ for the tracking error vector as

$$\overline{Q} = \begin{bmatrix} \rho Q\rho & \rho Q \\ Q\rho & Q \end{bmatrix}$$
(9.3.25)

The design problem would then be in the same form as specified by Equations 9.3.9 and 9.3.10. Instead of Equations 9.3.24 and 9.3.25, the formulation of Equations 9.3.17–9.3.22 is used here to reduce the dimensionality of the equations and the number of mathematical operations.

Digital simulations of a Stanford/JPL arm were performed using the same numerical values for the manipulator as in Example 9.2.1. The desired trajectory is a straight line segment joining point $P_1 = (10, 0, 2)$

to point $P_2 = (10, 20, 2)$ expressed in inches in the Cartesian base coordinate system. The velocity profile is assumed to have the shape of an equilateral trapezoid with the maximum velocity of v_0^m and acceleration of a_0^m, where $v_0^m = 105.3$ in./s and $a_0^m = 1052.6$ in./s^2. The desired trajectory $\{\theta^d(k), \omega^d(k)\}$ is calculated by following the procedure discussed in Chapter 7.

The input–output relation of the manipulator is modeled by Equation 9.3.17. The parameter vector α_i with six unknowns appearing on each row of Equation 9.3.20 is estimated by using Equations 9.3.7 and 9.3.8. The initial approximations for the parameters are $\hat{A}_2(0) = I$, $\hat{a}_0(0) = 0$, $\hat{B}_0(0) = I$, $\hat{B}_1(0) = 0$, and $P(0) = 10^4 I$. To design the self-tuning controller, the following weighting matrices in Equation 9.3.22 are used: $R = \|\hat{B}_0\|^2$, $Q = I$, and $\rho = 0.5I/T$ where T is the sampling period. As an illustration of the typical graphs recorded in the simulations, the joint variables θ_3 and θ_5 versus time are displayed in Figures 9.12a and 9.12b. They demonstrate the common characteristics of the response of the system with a self-tuner: after a learning period, the tracking of the desired trajectory is acceptable.

When the operating speed of the gross motion is slow, no appreciable improvements in the tracking errors may be noticed using a multivariate

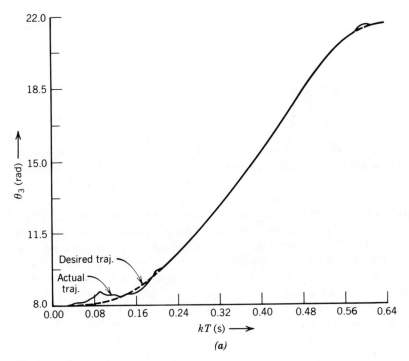

(a)

FIGURE 9.12a

Actual and desired trajectories of Stanford/JPL arm with multivariable self-tuning controller. Example 9.3.1.

a. Joint three positions θ_3 and θ_3^d versus time.

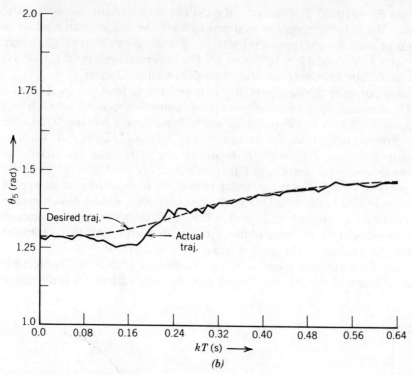

FIGURE 9.12*b*
Joint five positions θ_5 and θ_5^d versus time.

self-tuning controller as compared with those in using a SISO self-tuner [3]. However, when the operating speed is sufficiently increased, the self-tuning controller designed under the assumption of independent joint dynamics may not be able to track the desired trajectory, and tracking errors can become very large with time. It may be remedied by using a multivariate self-tuning controller as described here to obtain a good tracking of the desired trajectory.

The example demonstrates that interactions between the joint variables of a manipulator can be incorporated in a multivariate time-series model used in designing a controller of self-tuning type. It is usually difficult to determine which joint interactions should be included in the multivariate model, that is, which off-diagonal terms in the plant matrices should be chosen to be estimated and still keep the number of these parameters small. Some experimental simulations are often needed before the choice is made, and it may change when the desired trajectory is changed. Another alternative to account for the interactions between the joint variables is described in Problem 9.11. Improvements in the tracking errors of the manipulator motion at high operating speeds can indeed be achieved by means of an appropriately designed multivariate self-tuning controller as compared with the performance of a self-tuning controller designed on the basis of the independent joint dynamics.

9.3.2 DIRECT CONTROL OF GROSS MOTION IN CARTESIAN BASE COORDINATES WITH SELF-TUNING CONTROLLER

The time-series model used in the previous sections to design a self-tuning controller for the gross motion of a manipulator relates the inputs of the actuators and the joint variables. This system includes the actuator and manipulator dynamics. The variables of the system are in the joint space. As described in Chapters 2–4, the discrete values of the joint variables can be converted (mapped) into the Cartesian base coordinate system by means of the kinematic equations. It seems feasible that a multivariate time-series model could be constructed for the inputs of the actuators and the output variables of the end-effector in the Cartesian base coordinate system and used in designing a controller with self-tuning for the manipulator motion. The design of such a self-tuning controller is next discussed [6].

The kinematic equations relate the joint variables to the variables of Cartesian base coordinate system. If vector $p = [p_x \quad p_y \quad p_z]'$ and $\psi = [\psi_x \quad \psi_y \quad \psi_z]'$ denote the position and orientation of the end-effector, respectively, in the Cartesian base coordinate system, then the kinematic equations may be expressed as

$$W = \begin{bmatrix} p \\ \psi \end{bmatrix} = \Lambda(q) \quad \text{or} \quad q = \Lambda^{-1}(W) \tag{9.3.26}$$

where the vector-valued function Λ contains usually sin- and cos-terms and the structural kinematic parameters of the manipulator, and $\Lambda^{-1}(W)$ represents the inverse function of $\Lambda(q)$. The relationship for the velocities can also be established:

$$V = \begin{bmatrix} v \\ \omega \end{bmatrix} = J(q)\dot{q} \quad \text{or} \quad \dot{q} = J^{-1}[\Lambda^{-1}(W)]V \tag{9.3.27}$$

where J signifies the Jacobian matrix for the manipulator and the three dimensional vectors $v = \dot{p}$ and ω represent the actual translational and rotational velocities of the end-effector, respectively, in the Cartesian coordinate system. The inverse of $J(q)$ is assumed to exist in the operating range. The acceleration vectors are then determined as follows:

$$\dot{V} = \dot{J}(q)\dot{q} + J(q)\ddot{q} \quad \text{or} \quad \ddot{q} = J^{-1}[\Lambda^{-1}(W)]\{\dot{V} - \dot{J}[\Lambda^{-1}(W)]J^{-1}[\Lambda^{-1}(W)]V\} \tag{9.3.28}$$

Equations 9.3.26–9.3.28 establish the relationships between the joint variables and the variables of the Cartesian base coordinate system.

The equations of motion of a serial link N-joint manipulator can be expressed in Lagrange's formulation as given by Equation 8.2.14. If Equations 9.3.26–9.3.28 are substituted for q, $\dot{q}$, and $\ddot{q}$, the dynamical model in Equation 8.2.14 is obtained in terms of the variables of the Cartesian coordinate system:

$$\dot{V} - (\dot{J}J^{-1}V) + JD^{-1}(C + G) = JD^{-1}F \qquad (9.3.29)$$

where the arguments have been dropped for clarity. Equation 9.3.29 describes the equation of motion of a manipulator in terms of the generalized position W, velocity V, and acceleration $\dot{V}$ expressed in the Cartesian base coordinate system. The dynamical model in Equation 9.3.29 for the motion of a manipulator is, however, untractable and hardly usable for designing a controller for the system.

To avoid using a complex dynamical model such as the one in Equation 9.3.29, a discrete linear multivariate time-series model in the form of Equation 9.3.1 can be assumed to describe the relationship between the inputs of the actuators and the variables of the Cartesian base coordinate system [3, 6]. It will serve as the basis to design a multivariable self-tuning controller operating directly on the variables of the Cartesian base coordinate system.

The output vector of the multivariate model can be chosen to consist of the independent variables of the Cartesian (world) coordinate system, while the components of the input vector are the actuator voltages, or the generalized torques. For example, the designer may describe the end-effector motion using the components v_x, v_y, and v_z of the translational velocity vector v as the first three components and ω_x, ω_y, and ω_z of the rotational velocity vector ω as the last three components in the output vector expressed in the Cartesian coordinate system. Having selected the input and output variables, a multivariate linear time-series model can then be specified for the design of a self-tuning controller. Thus, a general discrete-time autoregressive MIMO-model is chosen in the form of Equation 9.3.1 where output $y(k) = V(k) = [v'(k) \quad \omega'(k)]'$ and vector $u(k)$ specifies the voltage inputs to the motors of the joints applied at time kT.

The parameters in Equation 9.3.1 are calculated by Equations 9.3.7 and 9.3.8 using the notations defined in Equations 9.3.4–9.3.6 where

$$\phi(k - 1) = [v'(k - 1)\omega'(k - 1), \ldots, v'(k - n)\omega'(k - n);$$
$$u'(k - 1)\ldots u'(k - n - 1); 1]' \quad (9.3.30)$$

In order to design a controller for the manipulator, the desired trajectory for the motion of the manipulator must be specified. It can be described, for example, by the discrete points $\{W^d(k), V^d(k)\}$ in the Cartesian coordinate system specifying the desired position and velocity values. The goal in the controller design is to minimize the (average) squared deviations from the desired values and the energy consumed. This can be accomplished by selecting a performance criterion similar to Equation 9.3.22 that is to be minimized:

$$I_k[u] = E\{\|V(k + 1) - V^d(k + 1) + \rho[W(k + 1) - W^d(k + 1)]\|_Q^2$$
$$+ \|u(k)\|_R^2|\Sigma(k)\} \quad (9.3.31)$$

where $V(k + 1)$ and $V^d(k + 1)$ represent the actual and desired velocity vectors at time $(k + 1)T$, respectively. Vectors $W(k + 1) = \text{col}[p_x(k + 1)\ p_y(k + 1)\ p_z(k + 1),\ \psi_x(k + 1)\ \psi_y(k + 1)\ \psi_z(k + 1)] = [p'(k + 1),\ \psi'(k + 1)]$ and $W^d(k + 1)$ describe the corresponding positions and orientations relative to the Cartesian coordinate system at time $(k + 1)T$. The symmetric matrix ρ expresses the importance of the position error relative to the velocity error. The weighting matrices Q and R are symmetric and positive definite. $E\{\cdot\,|\Sigma(k)\}$ is an expectation operation conditioned on the current and past measurements of $V(k), W(k)$ and the past controls. The control $u(k)$ will be admissible when it is a function of the current and past measurements and past controls.

The problem is to minimize $I_k[u]$ with respect to admissible control vectors while satisfying the difference Equation 9.3.1 with the parameter estimates substituted for the unknown coefficients.

In order to determine the minimizing controller, Equation 9.3.1 with k replaced by $(k + 1)$ is substituted for $V(k + 1)$ in Equation 9.3.31 and the position vector $W(k + 1)$ in Equation 9.3.31 is replaced by

$$W(k + 1) = W(k) + V(k)T \qquad (9.3.32)$$

where the sampling interval T should be sufficiently small for Equation 9.3.32 to be valid. Performing the expectation operation and the minimization of $I_k[u]$ with respect to admissible controls the following equation is obtained:

$$\text{R}u(k) + \hat{B}_0'Q\{V(k + 1|k) - V^d(k + 1) + \rho[W(k) + V(k)T - W^d(k + 1)]\} = 0 \qquad (9.3.33)$$

where $V(k + 1|k)$ represents the predicted value of $V(k)$ based on the measurements available at time kT, and $\hat{B}_0$ is the first term in polynomial $\hat{B}(z^{-1})$. The predicted value $V(k + 1|k)$ can be determined from Equation 9.3.1. Equation 9.3.33 specifies the (optimal) admissible control $u(k)$ as a feedback controller; it is expressed as a function of the current velocity $V(k)$ and position $W(k)$ that can be calculated from the measurements.

Two examples using digital simulations are next presented to illustrate the direct control of the gross motion in the Cartesian base coordinate system whereby the calculation of the inverse kinematic solution is avoided.

EXAMPLE 9.3.2

The dynamics of a Stanford/JPL arm are simulated as discussed in Example 9.2.1. The angular positions of the first three joints are used in the kinematic equations to calculate the position $p = [p_x\ p_y\ p_z]'$ of the wrist in the Cartesian base coordinate system. The corresponding velocity components in the same coordinate system are also determined. Thus, the output measurements $p(k)$ and $v(k)$ at time kT are available. They are contaminated by white Gaussian noise as in Example 9.2.1. The orientation of the gripper is not considered in this study. The ith component of the input is voltage u_i applied to the motor of joint i, $i = 1, 2, 3$.

The problem is to design a self-tuning controller for the first three joints in such a way that the gross motion of the wrist follows a desired discrete trajectory in the Cartesian coordinate system [6]. The desired trajectory $\{p^d(k)\} = \{p_x^d(k)\ p_y^d(k)\ p_z^d(k)\}$ is specified so that the wrist moves along a circular path located on the plane $x = -30$ in. with the center point at $(-30$ in., 8.5 in., 20 in.) and a diameter of 25 in. The motion starts at $(-30$ in., 8.5 in., 32.5 in.) with the zero speed. The maximum speed is 98 in./s for the gripper. The sampling period T used in the simulation is 0.028 s.

In order to design the controller, the following multivariate model is chosen for the Cartesian velocities:

$$v(k) = a_0 + A_1 v(k-1) + B_0 u(k-1) + e(k) \qquad (9.3.34)$$

where velocity $v(k) = [v_x(k)\ v_y(k)\ v_z(k)]'$. Vector a_0 is three-dimensional with unknown components, and A_1 and B_0 represent (3×3) matrices containing unknown elements. The equation error $e(k)$ is a random sample from a white Gaussian noise process describing modeling errors.

The following matrices of the multivariate model in Equation 9.3.34 are assumed:

$$A_1 = \begin{bmatrix} a_{11} & 0 & 0 \\ 0 & a_{22} & 0 \\ 0 & 0 & a_{33} \end{bmatrix} \quad B_0 = \begin{bmatrix} b_{11} & b_{12} & b_{13} \\ b_{21} & b_{22} & b_{23} \\ b_{31} & b_{32} & b_{33} \end{bmatrix} \quad a_0 = \begin{bmatrix} a_{01} \\ a_{02} \\ a_{03} \end{bmatrix} \quad (9.3.35)$$

Thus, the component $v_x(k)$ in Equation 9.3.34 is dependent only on the previous value $v_x(k-1)$ and the control inputs and not explicitly on the other velocity components; $v_y(k)$ and $v_z(k)$ are similarly specified. The control matrix is such that all components of the velocity vector are influenced by u_1, u_2, and u_3. The unknown elements of the parameters A_1, B_0, and a_0 are estimated using Equations 9.3.7 and 9.3.8.

The adaptive controller will be designed so that the following performance criterion is minimized:

$$I_k[u] = E\{\|v(k+1) - v^d(k+1) + \rho[p(k) + v(k)T - p^d(k+1)]\|_Q^2$$
$$+ \|u(k)\|_R^2 |\Sigma(k)\} \quad (9.3.36)$$

The weighting matrices are selected as follows:

$$\rho = 100\,I \qquad R = \|\hat{B}_0\|^2 = \hat{B}_0'\hat{B}_0 \qquad Q = \begin{bmatrix} 10 & 0 & 0 \\ 0 & 1 & 0 \\ 0 & 0 & 1 \end{bmatrix} \quad (9.3.37)$$

The numerical values in the ρ- and Q-matrices are such that the positional and velocity errors along the path of the first component are weighted more heavily than those of the others. In general, the selection of proper numerical values for the weighting matrices Q, ρ, and R requires some experimental simulation runs.

The adaptive controller is specified by

$$u(k) = (R + \hat{B}_0'Q\hat{B}_0)^{-1}\hat{B}_0'Q\{v^d(k+1) - \hat{A}_1 v(k) - \hat{a}_0 + \rho[p^d(k+1) - p(k) - v(k)T]\}$$
(9.3.38)

where the parameter values $\hat{B}_0$, $\hat{A}_1$, and $\hat{a}_0$ represent the estimates of B_0, A_1, and a_0, respectively.

Digital simulations of the system were performed on CDC 6600. Typical simulation results are displayed in Figures 9.13a–9.13c. The desired (dotted line) and the actual simulated trajectories of the wrist are shown on the (yz)-, (xz)- and (yx)-planes. The simulation results indicate that it takes a while before the estimator and the controller are able to determine proper parameter values so that the actual path starts tracking closely the desired path.

As mentioned previously, the initial approximations of the parameters have a strong influence on the initial behavior of the manipulator motion. If the motion is repeated, the last estimates of the parameters can be used as the initial approximations in the next run. Figures 9.14a–9.14c show the simulation results of the same manipulator when the motion of the wrist is rerun using the last estimates of the previous run. They illustrate considerable improvements in the system performance during tracking of the desired trajectory.

EXAMPLE 9.3.3

The gross motion of the Stanford/JPL manipulator discussed in Example 9.3.2 is to be directly controlled in the Cartesian base coordinate system using all six joints. The manipulator is to move so that the center point of the gripper is transferred from point $P_0 = (10, 0, 2)$ in. to point $P_f = (10, 10, 2)$ in. along a straight line segment in the Cartesian base coordinate system. The desired trajectory points are generated by assuming an equilateral trapezoidal profile for the velocity. The gripper starts at rest, reaching the maximum velocity v_0^m with a constant acceleration in j sampling periods ($T = 10$ ms). It maintains the constant speed v_0^m for $(n - 2j)$ sampling periods and then decelerates to the zero speed at P_f.

The problem is to design a self-tuning controller so that the gross motion of the gripper tracks the desired discrete trajectory given in the Cartesian coordinate system. The controller operates directly on the Cartesian coordinates of the gripper.

To design a self-tuning controller, a multivariate ARX-model of order one for the velocity vector is chosen. The unknown coefficients that are estimated according to Equations 9.3.7 and 9.3.8 are substituted into the model to obtain

$$V(k) = \hat{A}_1 V(k-1) + \hat{a}_0 + \hat{B}_0 u(k-1) + e(k)$$
(9.3.39)

where the generalized Cartesian velocity at time kT is $V(k) = [v_x(k)\, v_y(k)\, v_z(k)\, \omega_x(k)\, \omega_y(k)\, \omega_z(k)]'$, the bias vector $\hat{a}_0 = [\hat{a}_0^1 \ldots \hat{a}_0^6]'$, the input to the actuators at time $(k-1)T$ is $u(k-1) = [u_1(k-1) \ldots u_6(k-1)]'$, the error

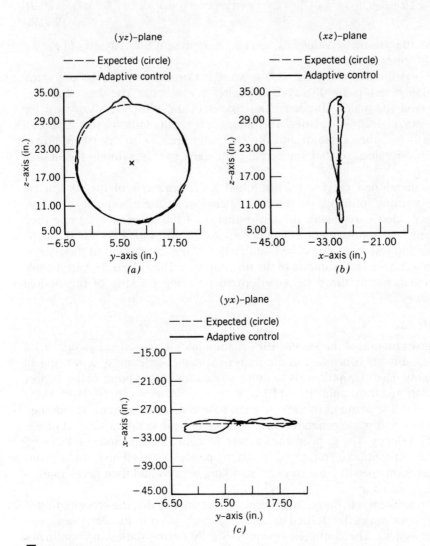

FIGURE 9.13 *a–c*
Direct control of Cartesian position of wrist with multivariable self-tuning controller. Initial parameter values arbitrary. Example 9.3.1.
a. Actual and desired trajectories of wrist on (*yz*)-plane.
b. Actual and desired trajectories of wrist on (*xz*)-plane.
c. Actual and desired trajectories of wrist on (*yx*)-plane.

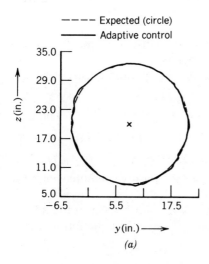

Expected (circle)
Adaptive control

z(in.) —→
y(in.) —→

(a)

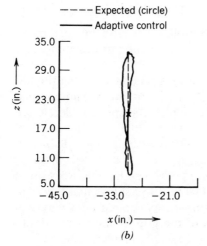

Expected (circle)
Adaptive control

z(in.) —→
x(in.) —→

(b)

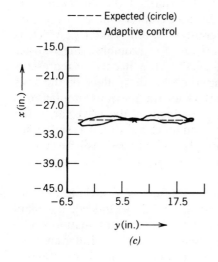

Expected (circle)
Adaptive control

x(in.) —→
y(in.) —→

(c)

FIGURE 9.14 *a–c*

Direct control of Cartesian position of wrist with multivariate self-tuning controller. Initial parameter estimates equal to last estimates of previous run. Example 9.3.1.

a. Actual and desired trajectories of wrist on (yx)-plane.

b. Actual and desired trajectories of wrist on (xz)-plane.

c. Actual and desired trajectories of wrist on (yx)-plane.

vector $e(k) = [e_1^v(k) \ldots e_6^v(k)]'$ and the (6×6) matrices $\hat{A}_1$ and $\hat{B}_0$ contain the numerical values of the estimated parameters.

To achieve the design objectives, the following performance criterion is minimized:

$$I_k[u] = E\{\|[V(k+1) - V^d(k+1)] + \rho[W(k) + V(k)T - W^d(k+1)]\|_Q^2 \\ + \|u(k)\|_R^2|\Sigma(k)\} \quad (9.3.40)$$

where the generalized position vector at time kT is $W(k) = [p'(k) \quad \psi'(k)]'$, its desired value is $W^d(k)$ and ρ, Q, and R are chosen positive definite weighting matrices.

The minimization of $I_k[u]$ in Equation 9.3.40 with respect to the admissible controls determines the controller for the system in Equation 9.3.39:

$$u(k) = [R + \hat{B}_0'Q\hat{B}_0]^{-1}\hat{B}_0'Q\{[V^d(k+1) - \hat{A}_1V(k) - \hat{a}_0] \\ + \rho[W^d(k+1) - W(k) - V(k)T]\} \quad (9.3.41)$$

Equation 9.3.41 specifies the self-tuning controller. The estimated parameters in Equations 9.3.39 and 9.3.41 are calculated according to Equations 9.3.7 and 9.3.8.

For digital simulations, the desired trajectory is generated so that the time for the acceleration is $7T = 70$ ms ($j = 7$), the constant maximum speed is maintained until time $100T = 1$ s, after which the speed is decelerated to zero in 70 ms. The maximum speed is $v_0^m = 20$ in./s. The weighting matrices in the performance index are $Q = 50I$, $\rho = 10I$, and $R = \hat{B}_0'\hat{B}_0$. The matrices in Equation 9.3.39 are $\hat{A}_1 = \text{diag}(\hat{a}_{11} \ldots \hat{a}_{66})$, and $\hat{B}_0$ is a general (6×6) matrix. The digital simulations were performed in FORTRAN on the CDC 6600 computer. A typical graph in Figure 9.15a shows that the actual position p_x (solid line) tracks the desired trajectory (dotted line) very well. The other Cartesian coordinate variables of the gripper follow the desired trajectory equally well.

Another simulation of the same task was performed after increasing the speed of the gripper [11]. In this case, the acceleration time is $10T = 100$ ms ($j = 10$), the operating time at the constant speed is 0.60 s, which is followed by the deceleration time of 100 ms. To accomplish the transfer from P_0 to P_1, the maximum velocity of 28.57 in./s is needed; it represents about 43% increase in speed. A self-tuning controller determined on the basis of the diagonal $\hat{A}_1$-matrix in Equation 9.3.39 for the previous simulation is first applied. The tracking error of each component of the W-vector in the simulations performed increases with time and quickly diverges as shown for the p_x-component in Figure 9.15b. Apparently, the interactions between the joints in the system are significant and should not be omitted at the operating speed used.

A possible approach to modify the system design is to include nonzero off-diagonal terms in the $\hat{A}_1$-matrix of Equation 9.3.39. Indeed, the plant matrix $\hat{A}_1$ can be chosen to have the block-diagonal form of Equation 9.3.16 with $j = 1$. The resulting control system is then simulated. Typical graphs are displayed in Figures 9.16a and 9.16b for the positional components p_x and p_y

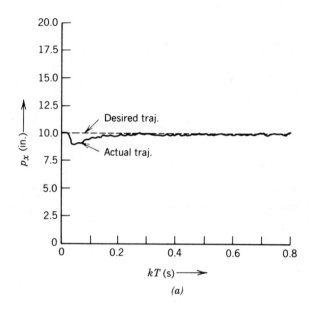

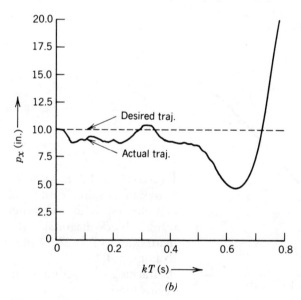

Figure 9.15 a and b

Tracking of end-effector position with self-tuning
control at different velocities. Example 9.3.3.
a. Tracking in p_x-component versus time when
$v_0^m = 20$ in./s.
b. Tracking in p_x-component versus time when
$v_0^m = 28.57$ in./s.

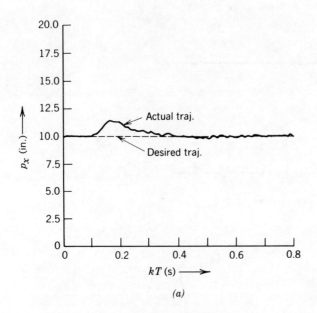

(a)

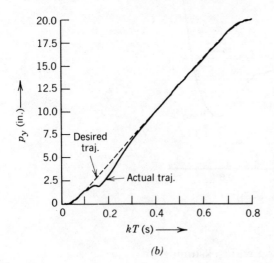

(b)

FIGURE 9.16 *a* AND *b*
Cartesian position control of end-effector with self-tuner using block-diagonal A_1-matrix, and $v_0^m = 28.57$ in./s. Example 9.4.2.
a. Tracking of p_x-component versus time.
b. Tracking of p_y-component versus time.

of the end-effector versus time. All coordinate variables of the end-effector exhibit good tracking of the desired trajectories. The improvement in the performance is obtained on the cost of additional mathematical operations and an increase in computation time. In this example, no a priori information is assumed to be available for the construction of a primary controller to cancel nonlinear terms in the dynamical model.

An alternative approach to achieve an improved tracking at high operating speeds in the problem discussed is to increase the sampling rate (or decrease the sampling period) or the order of the model [11].

As demonstrated here, a self-tuning controller for the gross motion of a manipulator can be designed to operate directly on the position and orientation of the end-effector expressed in the Cartesian base coordinate system. Its design is based on an appropriately chosen MIMO-model of the ARX-type. The controller is designed as a self-tuning LQG-controller. The proper selection of the weighting matrices for Equation 9.3.9 or 9.3.31 will usually require experimentations to achieve a proper operation and acceptable responses of the closed-loop system. The resulting manipulator system with the multivariable self-tuning controller will track the desired trajectory at a reasonable accuracy. We should note that the designed controller may not exhibit the self-tuning property (in the true sense) because the duration of the manipulator motion is finite.

In view of the implementation of the described control scheme, we should observe that the position and orientation of the end-effector cannot, at present, directly be measured accurately at a reasonable cost. On the other hand, the measurements needed in the control algorithm described can be computed from the encoder readings using the kinematic equations in the forward direction. This is usually simpler than the calculations of the kinematic solution in the backward direction, which are presently often performed.

9.4 COMMENTS ON ADAPTIVE CONTROLLER DESIGN FOR MANIPULATOR GROSS MOTION

In designing adaptive self-tuning controllers for the gross motion of robot manipulators, the designer needs to select a discrete-time model for the input–output relation. For convenience, we have used autoregressive exogenous time-series (ARX-) models, although autoregressive moving average (ARMA-) models could equally well be applied. In view of the computational time constraints imposed by the short sampling period, the order of the model should be chosen low and increased if difficulties are encountered in the operation of the self-tuner. If the calculation of the P-matrix at every sampling instant poses a computational burden, it may be updated only at each second or third sampling instant. Moreover, the gain multiplying the prediction error in the correction term in Equations 9.1.16 and 9.3.7 may be approximated by some simple expressions that behave like $1/k$ as integer k increases (similar to the gain in the stochastic approximation). Thus, time-consuming computations to determine the gain in question can be reduced if the capabilities of the available computer so require. To keep the number of the parameters that need to be estimated small, the designer should be careful in selecting the unknown elements of the matrices in the polynomials of the multivariate ARX-model. Simulation experiments are usually helpful to determine the order of the model for the input–output relation of the manipulator and to select the unknown parameters, particularly in the multivariate model. The order of a time-series model can also be determined by statistical hypothesis testing.

If some physical parameters in the dynamical model of the manipulator are known *a priori* at a sufficient degree of accuracy, the terms involving these parameters in the model should be isolated and used [13] in constructing a primary controller. This is a typical step in the estimation problems. It will reduce the burden on the computer of the manipulator.

If no accurate *a priori* information about the dynamics is available, an adaptive self-tuning controller can still be constructed for the gross motion. It can be designed on the basis of an ARX-model describing the relationship between the joint inputs and the position and/or velocity variables. If the self-tuning controller designed on the basis of a position ARX-model has difficulties to function properly, an advisable alternative is to design a self-tuning controller using a velocity ARX-model.

If independent joint dynamics can be assumed, the ARX-model is chosen as a SISO-model. This assumption holds at slow to moderate operating speeds. A self-tuning controller based on SISO-models of the joint variables has successfully been implemented in the laboratory on a Stanford/JPL arm [14] and on a PUMA arm.

When the coupling effects between the joints play an important role in the dynamical behavior of the manipulator, a MIMO-model needs to be chosen to describe the input–output relation that is used in designing a self-tuning controller. Thus, the plant matrices in the ARX-model will contain nonzero off-diagonal elements. It is also possible to introduce cross-product terms in a (nonlinear) difference equation model describing the input–output relation of a manipulator system (see Problem 9.11). If the stochastic noise terms are neglected, a controller can be designed by minimizing a deterministic one-stage performance criterion. Thus, the interacting effects between the joints can be incorporated in the design of a self-tuning controller for the manipulator motion.

The main advantage in using a self-tuning controller is that the dynamics of the manipulator need not be known very accurately. Moreover, it can adapt to varying loads of unknown masses being carried. Since the controller gains are adjusted at the sampling instances on the basis of the current measurements, sensory information available at a sampling time in a proper form can be taken into account in the control strategy. Thus, the self-tuning controller can adapt to the changing environmental conditions. Its performance can be improved on the basis of its past performance on a task; thus, the manipulator is "trainable." This may suggest the construction of a task-dependent knowledge-base system for the initial parameter estimates and the weighting factors (or the pole-zero assignments). Moreover, an expert system could determine the adjustments usually performed by the designer on the basis of the experimental runs. Because the duration of the motion of a manipulator is finite, the controller may not exhibit the self-tuning property in the usual sense. However, a good tracking of the desired trajectory can usually be achieved by using a controller of self-tuning type.

The stability of a manipulator system with self-tuning controller has not been addressed. For a general six-joint manipulator such as a Stanford/JPL or PUMA manipulator, the determination of the global stability (for example, in the sense of Lyapunov) is not within the realm of the designer, at least at present. When no sufficiently accurate information about the dynamics of the manipulator is available to design a primary controller, the stability of the manipulator system containing a self-tuning controller as a sole controller (that is, there is no explicit cancellation of nonlinear terms) can hardly be determined. If the nonlinear terms are known at a sufficient degree of accuracy so that a primary controller can be designed to cancel the nonlinear terms in the model, the stability of the (linearized) system with a self-tuning controller can be studied [15].

The gross motion of a robot manipulator poses an interesting framework for stability studies because the motion usually lasts only a finite time. In most stability studies, the primary interest is in the asymptotic behavior of the system, that is, in the response as the time approaches infinity (excluding finite escape times). To circumvent this conceptual paradox, the concept of finite-time stability could be adopted for the manipulator motion. Although it is conceptually attractive, it is unlikely that the stability analysis of a general manipulator system in this framework would become more tractable.

If the nonlinear dynamical model of a manipulator is linearized using feedforward–feedback loops, but the cancellation of the nonlinear terms is not exact, the stability studies are quite involved. If the linearized model is determined by the variational method, the resulting linear equation is, in general, time-varying because of nonconstant nominal trajectories, as discussed in Section 8.2.3. The determination of stability of the time-varying system is difficult. An alternative approach is to impose (assume) bounds on the error terms resulting from imperfect cancellations [13, 16]. Then, the designer can investigate the stability of the linear system by standard methods. It should be mentioned that most published results on the stability of the manipulator motion apply to infinite-time motions and are only local in nature.

An alternative to the design of a self-tuning controller for the gross motion of a manipulator is to construct a model reference adaptive controller (MRAC). It can be accomplished by first determining a linearized model using a variational method. Then, the inverse dynamics of the linearized (constant parameter) model is used in the feedforward loop, and an adaptive controller is designed as a feedback loop on the basis of Lyapunov's function [16, 17]. The design of the system with MRAC is usually performed using continuous-time models. An interesting exposition on the comparison of the model-reference adaptive controllers and self-tuning controllers is presented in [9].

The number of reports on various versions of adaptive control schemes and their applications to the control of the gross motion of robot manipulators is enormous; only a few are referenced here. In selecting an adaptive controller for a manipulator system designed for a specific task, the familiarity of the designer with a particular scheme is often the decisive factor. The literature offers very few guidelines to select for a manipulator system the adaptive control scheme that is most suitable to a specific task. Most reports on the use of adaptive controllers for robot manipulators describe simulation studies, and seldom actual laboratory experiments. The use of adaptive controllers in industrial robot applications is at present minimal. We can expect the number of these applications to increase with the experience and knowledge gathered in laboratory experiments [14, 17] and by the education of the students on various uses of adaptively controlled robot manipulators.

9.5 SUMMARY

The goal in the gross motion control of a manipulator is to make the end-effector track a specified trajectory. If the dynamical model of the manipulator including the numerical values of the parameters is accurately known, a primary controller can be constructed such that the end-effector tracks the desired trajectory under ideal circumstances. To compensate for the imperfect cancellation of nonlinear terms and for the effects of modeling errors and disturbances, a secondary controller can be designed as a self-tuning controller.

The complete dynamical equations for the motions of many manipulator systems are usually not known accurately. If some terms in the dynamical model are exactly known and can be computed, they should be used in the construction of a primary controller. As a consequence, the known nonlinear terms can at least approximately be canceled in the system. A self-tuning controller is then designed on the basis of the model representing the input–output relation of the system involving the remaining terms. If no accurate information is available about the manipulator dynamics, a controller with self-tuning can still be designed so that the manipulator will track the specified desired trajectory.

We have presented the design of a self-tuning controller by the explicit (indirect) method: the parameters in a chosen time-series ARX-model are first estimated by the least-squared error method, and then the controller gains are determined. The equations for the recursive least-squares algorithm are presented. Although they are here used for the design of a self-tuning controller, they can also be applied to the determination of some specific physical parameters such as masses, second moments of inertias, and friction using an appropriate model and measurements.

The controller with self-tuning is constructed by solving a LQG-problem or alternatively a pole-zero placement problem. The minimization of a well-defined quadratic criterion subject to the linear model constraint with Gaussian additive disturbance determines the structure of the LQG-controller. When the design of the controller is performed by the pole-zero placement method, the structure of the controller is assumed by the designer and the pulse transfer function of the controller will shift the poles and zeros of the closed-loop system to the desired locations. A self-tuning controller for a manipulator can be designed by either method.

The controllers for the gross motions of manipulators are usually designed to function in the joint space. The variables used in time-series ARX-models are the joint positions or the joint velocities. The proper function of a manipulator system with the self-tuner can often be achieved effectively by using a velocity model instead of a position model.

A multivariate controller of self-tuning type for the gross motion of a manipulator can also be constructed to operate directly on the position (velocity) of the end-effector expressed in terms of the variables of the Cartesian base coordinate system. Thus, the calculations involving the inverse kinematic equations can be avoided, and the efficiency of the calculations is improved.

REFERENCES

[1] G. N. Saridis, "Toward the Realization of Intelligent Controls," *Proc. of the IEEE,* Vol. 67, No. 8, August 1979, pp. 1115–1133.

[2] K. J. Aström, "Design Principles for Self-Tuning Regulators," *Proc. of International Symposium on Methods and Applications in Adaptive Control,* editors A. V. Balakrishnan and M. Thoma, Bochan, W. Germany, 1980.

[3] A. J. Koivo and T. H. Guo, "Adaptive Linear Controller for Robotic Manipulators," *Proc. of the 20th IEEE Conference on Decision and Control,* San Diego, CA, December 1981; also *IEEE Transactions on Automatic Control,* Vol. AC-28, No. 2, February 1983, pp. 162–172.

[4] A. J. Koivo and R. Paul, "Manipulator with Self-Tuning Controller," *Proc. of the IEEE 1980 International Conference on Systems, Man and Cybernetics Society,* Cambridge, MA, October 1980.

[5] R. G. Walters and M. M. Bayoumi, "Application of a Self-Tuning Pole-Placement Regulator to an Industrial Manipulator," *Proc. of the 21st IEEE Conference on Decision and Control,* Orlando, FL, December 1982.

[6]   A. J. Koivo, "Self-Tuning Manipulator Control in Cartesian
      Base Coordinate System," *Journal of Dynamical Systems, Mea-
      surements and Control (Trans. of ASME)*, Vol. 107, December
      1985, pp. 316–323.

[7]   P. E. Wellstead, J. M. Edmunds, D. Prager, and P. Zanker,
      "Self-Tuning Pole/Zero Assignment Regulators," *Interna-
      tional Journal of Control*, Vol. 30, No. 1, 1979, pp. 1–26.

[8]   K. J. Aström and B. Wittermark, "Self-Tuning Controllers
      Based on Pole-Zero Placement," *IEE Proceedings*, Vol. 27,
      Pt. D, No. 3, 1980, pp. 120–130.

[9]   I. D. Landau, "Model Reference Adaptive Controllers and
      Stochastic Self-Tuning Regulators—A Unified Approach,"
      Journal of Dynamical Systems, Measurements and Control, Vol.
      103, 1981, pp. 404–416.

[10] A. K. Bejczy, "Robot Arm Dynamics and Control," *JPL
      Technical Memo*, No. 33–669, February 1974.

[11] R. Souissi and A. J. Koivo, "Design of Adaptive Self-Tuning
      Controller for High Speed Manipulators," *Proc. of IEEE
      1987 International Conference on Systems, Man and Cybernetics*,
      Alexandria, VA, October 1987, pp. 897–902.

[12] D. L. Prager and P. E. Wellstead, "Multivariable Pole-
      Assignment Self-Tuning Regulators," *IEE Proceedings*, Vol.
      128, Pt. D, No. 1, January 1980, pp. 9–18.

[13] J. J. Craig, P. Hsu and S. S. Sastry, "Adaptive Control of
      Mechanical Manipulators," *Proc. of the 1986 IEEE Interna-
      tional Conference on Robotics and Automation*, San Francisco,
      April 1986.

[14] T. H. Guo and A. J. Koivo, "On a Linearized Model
      and Adaptive Controller Implementation for Manipulator
      Motion," *Journal of Robotic Systems*, Vol. 1, No. 2, 1984, pp.
      141–156.

[15] P. J. Gawthrop, "On the Stability and Convergence of a Self-
      Tuning Controller," *International Journal of Control*, Vol. 31,
      No. 5, 1980, pp. 973–998.

[16] M. Tomizuka, R. Horowitz, and G. Anwar, "Adaptive Tech-
      niques for Motion Controls of Robotic Manipulators," *Proc.
      of Japan-USA Symposium on Flexible Automation*, Osaka,
      Japan, 1986, pp. 217–224.

[17] H. Seraji, "Adaptive Independent Joint Control of Manipu-
      lators: Theory and Experiment," *Proc. of 1988 IEEE Interna-
      tional Conference on Robotics and Automation*, Vol. 2, Philadel-
      phia, PA, April 1988, pp. 854–861.

PROBLEMS

9.1   In the planar manipulator described in Example 9.1.1,
      assume that $\theta_3 = -\pi/2$ rad for the duration of motion.

   a.   Write the explicit expression for the primary controller
        given in Equation 9.1.9 for this particular case, that is,
        assume that the Coriolis and gravity terms are accu-
        rately known.

   b.   Determine the model for which a secondary controller
        needs to be constructed, that is, express Equation
        9.1.12 in an explicit form.

   c.   If the steady-state values of the joint variables are $\bar{\theta}_1 =$
        $\pi/4$ rad, $\bar{\theta}_2 = -\pi/2$ rad, $\bar{\theta}_3 = \pi/2$ rad, determine the
        (steady-state) values of the torques (system inputs).

   d.   Repeat parts *(a)* and *(b)* under the assumption that
        only the gravity terms are used in forming the primary
        controller.

9.2   The controller given in Equation 9.1.29 is used in the
      manipulator system described by Equation 9.1.28.

   a.   Determine the steady-state value of the output $y_{ss} =$
        $\lim y(k)$ as $k \to \infty$.

   b.   Suppose $\epsilon = 1000$, and $\hat{b}_0$ is in the range $[0.5,5]$.
        Approximate the percentage of the (largest) steady-
        state error, that is, $(y_{ss} - y^d)/y^d$ when y^d equals a
        constant value.

   c.   Compare the controller in Equation 9.1.29 with the
        minimum variance controller ($\epsilon_i = 0$).

9.3   To determine the condition on the (asymptotic) stability of
      the manipulator model consisting of Equations 9.2.3 and
      9.2.5, the equations are to be written as a vector difference
      equation ($e_i \equiv 0$):

$$\begin{bmatrix} \theta_i(k+1) \\ \omega_i(k+1) \end{bmatrix} = \bar{A} \begin{bmatrix} \theta_i(k) \\ \omega_i(k) \end{bmatrix} + \begin{bmatrix} 0 \\ u_i^1 \end{bmatrix}$$

   a.   Specify the $\bar{A}$-matrix and u_i^1. Determine the eigen-
        values of the plant matrix $\bar{A}$. Give condition for the
        stability.

   b.   Solve the equation for $[\theta_i(N), \omega_i(N)]'$ in terms of the
        initial conditions and the desired trajectory values.

    c.   Discuss the asymptotic behavior of the solution
       obtained in *(b)*.

9.4   A manipulator is to turn a screwdriver as shown in Figure
9.17. The following constraints are imposed on the motion
relative to the coordinate system shown in Figure 9.17:

$$v_x = 0, \ v_y = 0, \ v_z = c, \ p_x = 0, \ p_y = 0, \ p_z = ct$$

where c is a constant and vectors $v = [v_x \ \ v_y \ \ v_z]'$ and
$p = [p_x \ \ p_y \ \ p_z]'$. Specify the velocity and position of the
end-effector tool, respectively.

    Suppose that the input–output relation of the end-
effector tool of the manipulator system is modeled by Equation 9.3.1.

    a.   Formulate an optimization problem for designing a
       self-tuning LQG-controller so as to satisfy the forego-
       ing constraints. Specifically, determine an appropriate
       performance criterion to be minimized.

    b.   Design the controller that solves the optimization
       problem formulated in *(a)*.

9.5   The design specifications for a manipulator response are
given in terms of the second-order transfer function $G^d(s)$
of a model-reference system: the desired damping ratio 0.7
and the undamped natural frequency 35 ms.

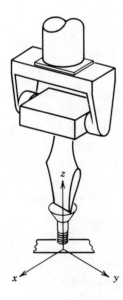

FIGURE 9.17
Manipulator turns a screwdriver; Problem 9.4.

a. Determine the pulse transfer function $G^d(z)$ of the model-reference system corresponding to $G^d(s)$ when the sampling period is 30 ms. Determine the steady-state gain. Introduce an open-loop gain so that the steady-state (DC) gain of this pulse transfer function is one.

b. Repeat part (a) for the critical damping when the input is a step-function, while the other design parameters are the same as in part (a).

c. Explain what is meant by a nonminimum phase pulse transfer function. Is either of the transfer functions determined in parts (a) and (b) of nonminimum phase type?

9.6 The following pulse transfer function is given:

$$G^d(z^{-1}) = \frac{0.25z^{-1}}{1 - z^{-1} + 0.25z^{-2}}$$

and the sampling period is 20 ms.

a. Can the transfer function $G^d(s)$ of the model reference system be determined? Discuss.

b. Determine the values of the desired damping ξ^d and the desired (undamped) natural frequency ω_n^d in an underdamped reference model that could give rise to $G^d(z^{-1})$.

c. Determine the poles of the given pulse transfer function.

d. If $G^d(z^{-1}) = Y(z^{-1})/U(z^{-1})$, determine the difference equation relating the input $u(k)$ and the output $y(k)$.

9.7 The input–output relation of a manipulator system is modeled as

$$(1 - \hat{a}_1 z^{-1})y(k) = z^{-1}\hat{b}_0 u(k) + e(k)$$

where $y(k)$ is the output and $u(k)$ the input of the system. The values $y(k)$ and $u(k)$ are measured.

a. Determine the equations that can be used to calculate the estimated parameters $\hat{a}_1$ and $\hat{b}_0$ on-line. Draw a flow-chart for an algorithm.

b. In the model given above, suppose that $e(k) \equiv 0$, and the errors in the parameter estimates can be ignored. A controller is to be designed for the manipulator system on the basis of the given model. Determine the controller by minimizing

$$I_k[u] = [y(k + 1) - y^d(k + 1)]^2 + \epsilon u^2(k)$$

c. Express the controller in the feedback–feedforward form:

$$u(k) = K_1 y(k) + K_2$$

Specify K_1 and K_2. How is this controller different from a PID-controller?

9.8 A discrete-time model for the input–output relation of a manipulator joint is given:

$$(1 - 0.81z^{-2})y(k) = z^{-1}(1 + 0.5z^{-1})u(k)$$

a. Determine a feedback controller so that the closed-loop system has a pulse transfer function given by

$$G^d(z^{-1}) = K/z^k$$

for some positive integer k.

b. Discuss the step response of the joint variable when its motion is described by $G^d(z^{-1})$. Graph the response. Is it a desirable response for a joint variable? Is the controller practical in view of its implementation?

9.9 A self-tuning controller for the input–output relation of a manipulator joint is to be designed by the pole-zero placement method on the basis of the following ARX-model:

$$(1 - \hat{a}_2 z^{-2})y(k) = z^{-1}(\hat{b}_0 + \hat{b}_1 z^{-1})u(k) + \hat{a}_0 + e(k)$$

The desired pulse transfer function $G^d(z^{-1})$ is given as

$$G^d(z^{-1}) = \frac{1 - 0.4z^{-1}}{1 + 0.42z^{-1} - 0.46z^{-2}}$$

a. Determine the general form of the controller when the joint variable is to track the values $\{y^d(k)\} = \{1 - 0.2k\}$ over the duration of 5 s and the sampling period $T = 50$ ms.

b. Specify the controller explicitly so that the pulse transfer function of the input–output relation of the overall system is $G^d(z^{-1})$.

c. Write down the equations for the parameter estimation based on the recursive least-squares method.

d. List the steps to implement the designed controller.

9.10 To account for the interacting effects between the joint variables in a two-joint serial link planar (vertical) manipulator, a MIMO-model of the ARX-type is proposed for the design of a self-tuning controller. The joint velocities $\omega_1(k)$ and $\omega_2(k)$ at time kT are modeled as follows:

$$\hat{A}(z^{-1})\omega(k) = z^{-1}\hat{B}(z^{-1})u(k) + \hat{a}_0 + e(k)$$

where $\omega(k) = [\omega_1(k) \quad \omega_2(k)]'$, input $u(k) = [u_1(k) \quad u_2(k)]'$, $\hat{a}_0 = [\hat{a}_0^1 \quad \hat{a}_0^2]'$, $e(k) = [e_1(k) \quad e_2(k)]'$ is the equation error, $\hat{A}(z^{-1}) = I - \hat{A}_1 z^{-1}$ and $\hat{B}(z^{-1}) = \hat{B}_0$, where the caret refers to the estimated numerical values.

a.   Suggest a form for $\hat{A}(z^{-1})$ and $\hat{B}(z^{-1})$ such that interactions between the two joint variables can be accounted for. Keep the number of the parameters to be calculated by the least-squares estimation algorithm minimal.

b.   What is the role of $\hat{a}_0$ in the model? Is it needed in this case?

c.   Formulate a minimization problem using the LQG-formulation so that the joint variables θ_1 and θ_2 will track a sequence $\{\theta_1^d(k), \theta_2^d(k)\}$ of specified points in the joint space.

d.   Determine the solution to the optimization problem posed in part (c) by specifying all necessary equations and initial conditions needed in the computations.

e.   Modify the problem posed in (c) so that the minimum variance controller will be obtained. Specify the minimum variance controller.

f.   Substitute your minimum variance controller into the ARX-model given. Determine the steady-state values of the variables. Discuss. What is the main drawback of this controller?

9.11   A mathematical model is given in Table 5.4 for a two-joint serial link planar manipulator. To account for the joint interactions, an engineer proposes an ARX-model for the self-tuning LQG-controller design in the joint space:

$$\omega_1(k) = \hat{a}_0^1 + \hat{a}_2^1\omega_1(k-2) + \hat{a}_{12}^1\omega_1(k-1)\omega_2(k-1) + \hat{b}_0^1 u_1(k-1) + e_1(k)$$

$$\omega_2(k) = \hat{a}_0^2 + \hat{a}_2^2\omega_2(k-2) + \hat{a}_{12}^2\omega_1(k-1)\omega_2(k-1) + \hat{b}_0^2 u_2(k-1) + e_2(k)$$

a.   Is the difference equation model proposed linear or nonlinear? Give reasons. Which terms in the model represent possible interacting effects between the joints?

b.   Suppose that the measurements of $\{\omega_1(k), u_1(k)\}$ and $\{\omega_2(k), u_2(k)\}$ for $k = 0, 1, \ldots, 100$ are available.

Determine, if possible, the equations for estimating the numerical values of the parameters in the model indicated by the carets by an off-line procedure.

c.   Assume that the equation errors $e_1(k) \equiv 0$ and $e_2(k) \equiv 0$. Using the parameter values obtained in (b), design a controller so that the position and velocity variables $\{\theta_1(k), \omega_1(k); \theta_2(k), \omega_2(k)\}$ of the joints track the desired values $\{\theta_1^d(k), \omega_1^d(k); \theta_2^d(k), \omega_2^d(k)\}$. Use the LQ-formulation.

d.   Suppose $e_1(k) \neq 0$ and $e_2(k) \neq 0$. Can you solve the LQG-problem that is formulated in part (b)? Discuss and explain.

e.   The numerical values of the parameters in the proposed model can be determined by an off-line procedure. Can the equations used be converted to a recursive form? Discuss.

COMPUTER PROBLEMS

9.C.1   The input voltage and velocity measurements of a joint in a Stanford manipulator are given in Table D.1 of Appendix D. Assume that the measurements become available sequentially in time.

a.   Assume the time-series ARX-model given in Equation D.1.16. Determine the recursive equations for estimating the parameters of the model on-line.

b.   Implement the parameter estimation algorithm on a digital computer.

c.   Graph $\hat{a}_i(k)$ versus k, $i = 0, 1$, and 2, and $\hat{b}_0$ versus k. Graph a component of the P-matrix versus time.

d.   Repeat (c) using a different initial approximation.

9.C.2   For a cylindrical robot manipulator, the mathematical model and the numerical values of the parameters are given in Computer Problem 8.C.1. The desired trajectory of the end-effector in the Cartesian base coordinates for a task is specified in Table 7.1.

a.   Determine the desired trajectory of the joint position variables.

b.   Determine the desired velocity trajectory using the data of Table 7.1.

c. Construct a self-tuning controller using the LQG-formulation for each joint under the assumption of independent joint dynamics.

d. Simulate the system with the self-tuning controller. Graph the actual and desired trajectories for each joint variable as a function of time. Also, graph the tracking errors in these variables versus time.

e. Present the simulation results mentioned in *(d)* for other initial approximations of the parameters being estimated.

f. Suggest means for improving the initial approximations of the parameters.

9.C.3 a. Construct a self-tuning controller by the pole-zero placement method for the manipulator system in Computer Problem 8.C.1. Assuming the independent joint dynamics, the desired pulse transfer function corresponds to the transfer function $G^d(s)$ of a model reference system with the desired damping factor $\xi_i^d = 0.8$ and $\omega_{ni}^d = 0.50T$ where the sampling period $T = 30$ ms.

b. Perform digital simulations of the system with the designed controller. Repeat the simulation experiment in parts *(d)–(f)* of Computer Problem 9.C.2 by the pole-zero placement method.

9.C.4 The equations of motion for a planar manipulator described in Computer Problem 5.C.1 is given in Table 5.4. For a specific task, the desired trajectory $\{\theta_i^d(t), \dot{\theta}_i^d(t)\}$, $i = 1, 2$ is generated in Computer Problem 5.C.2 corresponding to the motion of the end-effector along a straight line.

a. Design a LQG-controller with self-tuning that will operate directly on the position or velocity of the end-effector expressed in the Cartesian base coordinate system.

b. Simulate the manipulator motion with the self-tuning controller designed in *(a)*.

c. Graph the trajectory obtained by simulations and the desired trajectory for (i) p_x versus k, p_x^d versus k in the same set of coordinate axes, (ii) p_y^d, p_y^d versus k on another graph, and (iii) the actual and desired trajectory on the $p_x p_y$-plane (phase plane). Graph also the tracking errors for both coordinates versus time.

CHAPTER TEN

CONTROL OF GENERALIZED CONTACT FORCES EXERTED BY ROBOT MANIPULATOR

We have previously discussed the gross motion control of robotic manipulators by assuming implicitly that the manipulator can move freely in the workspace. In many applications such as in sliding on a table top, pulling or pushing a drawer, deburring, and assembling parts, constraints are imposed on the motion of a robot by the environment. The constraints may be expressed in terms of the position, orientation, and/or velocity of the end-effector depending on the particular task of the manipulator. When the motion of a manipulator is restrained by the task constraints, the manipulator motion must comply with the restrictions imposed by the environment. Thus, the manipulator exhibits *compliance* in its motion [1].

The compliance in a manipulator is *passive* when it is inherent in the structure of the manipulator. For example, a passive compliance is represented by a spring or a damper built in the hand, or by the flexibility of joints (gears) or a link. It has a restrictive use in applications, although the compliance of special structures in a system has potential to be used, for example, to compensate for jamming or for inaccuracies of processed pieces in assembly. The compliance in a robotic manipulator is *active*, when it can be generated by means of software, for example, using force control loops. The active compliant motion has the advantage of reprogrammability. It can also be made adaptable to various applications by appropriate program changes. It is cost-effective and economical, particularly in manufacturing of diverse and often changeable products.

An essential variable in the compliant motion is the generalized force (force and moment) exerted by the end-effector. The actual external generalized force acting between the end-effector and the environment can be measured continuously, for example, by a wrist-force sensor, as is discussed in Chapter 1. The generalized force measurements represent a form of sensory information that is essential in order to enhance the autonomy and intelligence of a robot system.

The external forces and moments at the contact point of the end-effector (or its tool) and the environment impose constraints on the system motion due to the aforementioned restrictions. In order to study the control of the generalized forces exerted by the end-effector, the interfacing contact of the end-effector with the environment can be considered as either soft or hard. The *soft contact* between the end-effector and the environment can be modeled by means of a spring, a damper, a mass, and/or an inertial moment. This implies that the environmental soft (surface) constraint is not a strict inequality; instead, the restriction can be violated by a small amount without stopping the motion. The forces and moments can be large even for small movements. The *hard contact* between the end-effector and the environment is such that it is modeled as a strict mathematical relation, that is, the constraint is described as an inequality or as an equality when contact occurs between the end-effector and the environment. The variables of the motion must satisfy the constraints at any instant of time.

We will discuss in this chapter the modeling and control of the generalized contact forces exerted by the end-effector of a robot manipulator. We present first the control of generalized external forces for the soft contact case. We will study the use of multivariable PID-controllers and self-tuning controllers to make the end-effector track the desired position (velocity) while exerting some specified generalized force on the environment. We will then present the control problem for the hard contact case, and discuss algorithms applicable to the control of the generalized contact force and position (velocity) of the end-effector.

10.1 MODELING OF SOFT CONTACT AND CONTROL OF EXTERNAL GENERALIZED FORCE EXERTED BY END-EFFECTOR

When the end-effector of a manipulator is to perform a specific task, it usually moves first in a free (unconstrained) space under the influence of a gross motion control and then makes a contact with the constraint surface. During this compliant motion, the contact between the end-effector and the environment is assumed. Force control can take place on the hyperplane normal to the constraint surface, and position (velocity) control occurs on the tangential hyperplane of the constraint surface.

The nature of contact between the end-effector of a manipulator and environment depends on the particular application. For example, when a tool such as a (metal) brush is attached to the gripper for deburring, the contact between this tool and the workpiece (environment) represents a spring-like phenomenon. The contact force can appropriately be described as a spring force that is dependent on the displacement of the spring from its resting position. When the end-effector moves in a predominantly viscous environment, the interaction of the robot gripper with the environment can be modeled as a dashpot (damper) representing damping. The external force effect is then modeled as being dependent on the velocity changes in the motion. The characteristic contact forces in this case are similar to those appearing when the end-effector presses a flat surface supported by pistons mounted in cylinders (dampers). In many of these applications, the movement associated with changes in the contact forces also exhibits sluggishness. This indicates that inertial masses should be included in the model of contact interactions. The preceding cases involve translational motion.

Although the foregoing discussion is addressed to the translational motion, similar models can be applied to the rotational motion. In the sequel, the generalized force at the contact refers to the force and moment, and the generalized displacements from the desired values of the variables describing the position and orientation of the end-effector are composed of the translational and rotational errors.

In view of the foregoing considerations, a model for a contact between the end-effector of a manipulator and the environment that involves translational motion can consist of a (fictitious) concentrated mass m_{ei}^t that is connected through a damper (dashpot) with a coefficient B_{ei}^t and a spring with a stiffness coefficient K_{ei}^t to the environment, as shown in Figure 10.1. Then, a model

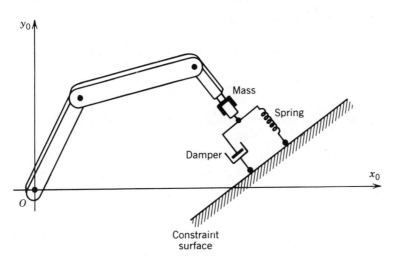

FIGURE 10.1
Model for a soft contact between the end-effector and environment.

for the contact in the case that an incremental translational motion is possible
may be written as

$$m^t_{ei}\Delta\ddot{p}_{ei} + B^t_{ei}\Delta\dot{p}_{ei} + K^t_{ei}\Delta p_{ei} = \Delta F^c_{ei} \tag{10.1.1}$$

where integer i refers to the ith constraint equation, $i = 1, \ldots, n$, $\Delta F^c_{ei} = F^c_{ei} - F^{cd}_{ei}$ describes the difference in the ith component of the force from its nominal
value at the interface and $\Delta p_{e_i} = p_{ni} - p^d_{ni}$ represents the ith component of
the displacement from its nominal value p^d_{ni} on the normal hyperplane of
the contact surface. Polynomial $(m^t_{ei}s + B^t_{ei} + K^t_{ei}/s)$ with $s = d(\cdot)/dt$ may be
called the impedance (admittance) at the interface. The contact forces can
be controlled either indirectly by controlling displacement Δp_{ei} or directly by
controlling force ΔF^c_{ei}.

If the task of a manipulator involves rotational motion during the contact between the end-effector (tool) and the environmental object, as in the
insertion of a screw, the contact interface may be modeled by a rotational
dynamic equation similar to Equation 10.1.1. Such a soft contact model contains the damping and (torsional) spring coefficients B^r_{ei} and K^r_{ei}, respectively,
the fictitious second-order moment I_{ei}, the change $\Delta\psi_{ei}$ in the rotation angle
and the change ΔM^c_{ei} in the moment (torque). This model of a soft contact for
rotational movement and that for translational motion (Equation 10.1.1) can
be combined to obtain a general second-order model for a soft contact interface. Indeed, the designer may choose a linear model for the soft contact in
the Cartesian base coordinate system so that the changes $\Delta W_e = [\Delta p_e{}'\Delta\psi_e{}']'$
in the motion at the interface are related to the external force and moment
by means of matrices K_c, B_c and m_c. These matrices contain the coefficients
of the spring, damper and inertias associated with the constrained motion:

$$[m_c s^2 + B_c s + K_c]\Delta W_e = \Delta F_e \tag{10.1.2}$$

where the generalized external force $\Delta F_e = [(\Delta F^c_e)'(\Delta M^c_e)']'$ has components
$\Delta F^c_e = [\Delta F^c_{e1} \cdots F^c_{ek_1}]'$ and $\Delta M^c_e = [\Delta M^c_{e1} \cdots \Delta M^c_{ek_2}]'$ with $k_1, k_2 < 3$. Moreover, $\Delta W_e = [\Delta p_e{}'\Delta\psi_e{}']'$ has components $\Delta p_e{}' = [\Delta p_{e1} \cdots \Delta p_{en_1}]$
and $\Delta\psi_e{}' = [\Delta\psi_{e1} \cdots \Delta\psi_{en_2}]$ with $n_1, n_2 \le 3$. The impedance matrix $[m_c s^2 + B_c s + K_c]$ of the contact interface is specified by

$$m_c = \begin{bmatrix} m^t_e & 0 \\ 0 & I_e \end{bmatrix} \qquad B_c = \begin{bmatrix} B^t_e & 0 \\ 0 & B^r_e \end{bmatrix} \qquad \begin{bmatrix} K^t_e & 0 \\ 0 & K^r_e \end{bmatrix}$$

where the superscripts in the coefficient matrices refer to the translational
and rotational motions. These submatrices can usually be chosen as diagonal
matrices, for example, $B^t_e = \text{diag}[B^t_{e1} \cdots B^t_{en_1}]$. The contact model in Equation
10.1.2 contains some important special cases to be discussed next.

When m^t_e and B^t_e are zero in the contact force model involving translational
motion ($\Delta\psi_e \equiv 0$) then the generalized force exerted by the manipulator
can be controlled indirectly by regulating Δp_e through the following vector
relationship:

$$K^t_e\Delta p_e = \Delta F^c_e \tag{10.1.3}$$

where the $(k_1 \times n_1)$-matrix K_e^t is usually diagonal and $k_1 \leq 3$ specifies the number of the constraint equations (i.e. the number of variables to be force controlled). By properly selecting matrix K_e^t the components of the force error ΔF_e^c can be controlled loosely corresponding to the small values of the elements in matrix K_e^t or tightly represented by large values of the elements. An algorithm for the control of the spring force error ΔF_e^c contained in Equation 10.1.3 is sometimes referred to as the stiffness control [2].

Similarly, the translational motion model of soft contact represents a damper or a dashpot, if m_e^t and K_e^t in Equation 10.1.2 are zero matrices. In this case with $\Delta \psi_e \equiv 0$

$$B_e^t \Delta \dot{p}_e = \Delta F_e^c \qquad (10.1.4)$$

By properly selecting matrix B_e^t the components of the contact force error ΔF_e^c can also be controlled loosely or tightly. A strategy for controlling the viscous-like contact force ΔF_e^c contained in Equation 10.1.4 is termed damping or damper control [3].

The general model for the contact interfacing involving translational motion is given by Equation 10.1.1. The control of contact force errors in this case is called the impedance (admittance) control [4]. We should notice that in the aforementioned cases the external contact forces or displacements are to be controlled, and not the stiffness or damping coefficient nor the impedance.

If the external generalized force exerted by the end-effector of a manipulator through a mass, dashpot, and/or spring on the environmental surface is not perpendicular to the constraint surface, it is convenient for the analysis to decompose the force into the components that lie on the tangential and normal hyperplanes of the constraint surface. The components of the contact force (F_e^c) that are on the hyperplane perpendicular to the constraint surface describe the motion for the exerted force and those on the tangential hyperplane the motion for the position (velocity). The motion is thus restricted on the normal hyperplane, but it is unconstrained on the tangential hyperplane. The position (velocity) can be controlled on the tangential hyperplane and the force on the normal hyperplane of the constraint surface. By the assumption of the soft contact, the force changes on the normal hyperplane of the constraint surface are directly related to the position, velocity, and/or acceleration changes on this plane. A desired value assigned to the external force specifies the corresponding values for the desired position, velocity, and/or acceleration components of the model, or vice versa.

The free motion of a serial link N-joint manipulator before a contact occurs is assumed to be modeled in Lagrange's formulation by Equation 8.2.14, where q is an N-dimensional vector describing the joint positions at time t, $D(q)$ is the inertia matrix, $C(q, \dot{q})$ describes the Coriolis and centripetal effects, $G(q)$ accounts for gravitational terms and F is an N-dimensional joint input vector. After the end-effector touches the environmental surface, the equations of motion during the time of contact include the generalized torque expressed in the joint space that corresponds to the generalized contact force F_e. Thus, the manipulator dynamics are governed by

$$D(q)\ddot{q} + C(q, \dot{q}) + G(q) = F - J'(q)F_e \qquad (10.1.5)$$

where matrix $J'(q)$ signifies the transpose of the Jacobian matrix evaluated at $q = q(t)$. The components of the generalized force vector F_e in Equation 10.1.5 are expressed along the axes of the base coordinate system. When the interface between the end-effector and the environment can be described as a soft constraint, the model for designing a force controller must include the soft contact model such as Equation 10.1.2.

The problem to be studied in detail is the control of certain components of the position and the generalized force in Equation 10.1.5 while satisfying the constraint of the soft contact such as the differential Equation 10.1.2. We will present first an algorithm that will control in the joint space certain components of the force vector and some components of the position vector. The control algorithm is basically a multivariable PID-controller operating on the generalized force and/or position (velocity) errors in the joint space. Then we will discuss control algorithms that operate directly on force and position (orientation) errors in the Cartesian base coordinate system.

10.1.1   INDIRECT POSITION FORCE (PF) CONTROL IN JOINT SPACE WITH MULTIVARIABLE PID-CONTROLLER

The problem to be addressed is to make the end-effector track a desired trajectory in the direction of some coordinate axes and exert a desired force in the directions of the other coordinate axes. If the task geometry had the property that force and position constraints were separately aligned with the manipulator joint axis, then the forces and positions could be servoed independently, assuming that the system is completely controllable. If the condition is only approximately met, acceptable results may still be obtained. If the force and position servoing are performed independently in the Cartesian base coordinate system, the force servoing actuator that produces the largest changes in the force exerted could be isolated. The output of this actuator would then cause the desired value of the force without regard for the contributions of the other manipulator joints. Furthermore, the components of the position could be servoed without regard for the tangential contributions in the motion by the other joints. However, this approach [5] due to the aforementioned approximation can cause considerable errors, restricting its usefulness.

The external generalized force control in the case of soft contact can be achieved indirectly by controlling variable ΔW_e in Equation 10.1.2. The generalized contact force error corresponds to the generalized joint torque error that is obtained from Equation 10.1.2:

$$J'(m_c s^2 + B_c s + K_c)J\Delta q = J'\Delta F_e \qquad (10.1.6)$$

where s is a differential operator, the small change ΔW_e of the end-effector position and orientation in the Cartesian space is converted into the change $J\Delta q$ in the joint space, and $J'\Delta F_e$ is equal to the change in the joint torque. The problem of controlling the contact force F_e can be achieved by regulating $\Delta q(t)$, where $\Delta q(t) = q(t) - q^d(t)$ specifies the difference between the actual (q) and desired (q^d) joint variable values.

The control system for the manipulator can be constructed by designing first a primary controller specified by $u^d(t)$ and then a secondary controller determined by $\hat{u}(t)$. The former control (u^d) will make the output of the manipulator track the desired position (velocity) and force trajectories under ideal circumstances, and the latter $(\hat{u})$ will compensate for the effects caused by inaccurate modeling and disturbances. Thus, the composite controller generates the input vector $F(t)$ for the system in Equation 10.1.5 by the following relationship:

$$F(t) = u^d(t) + D(q)\hat{u}(t) \qquad (10.1.7)$$

where $D(q)$, the inertia matrix in Equation 10.1.5, is introduced as a gain matrix.

The primary controller is determined so that the centripetal, Coriolis, and gravitational effects are canceled, and the desired value of the external generalized force F_e^d is achieved. Thus

$$u^d(t) = D(q^d)\ddot{q}^d + C(q^d, \dot{q}^d) + G(q^d) + J'(q^d)F_e^d \qquad (10.1.8)$$

where the superscript d refers to the values of the desired trajectory. Under the assumption of the perfect cancellation, that is

$$D(q) = D(q^d) \qquad C(q^d, \dot{q}^d) = C(q, q) \qquad G(q) = G(q^d) \qquad J(q) = J(q^d) \qquad (10.1.9)$$

the substitution of Equations 10.1.7–10.1.9 into the dynamical model of the manipulator in Equation 10.1.5 gives

$$\Delta\ddot{q} = \hat{u}(t) - D^{-1}(q)J'(q)(F_e - F_e^d) \qquad (10.1.10)$$

where the external generalized force error $F_e - F_e^d = \Delta F_e$ is determined by Equation 10.1.6. Equation 10.1.10 will be next used to construct a secondary controller for tracking the desired position and force.

The secondary control $\hat{u}(t)$ is chosen as a multivariable PID-controller operating on the position error and a P-controller operating on the joint force error

$$\hat{u}(t) = -K_p\Delta q - K_I\int \Delta q(\sigma)\,d\sigma - K_D\Delta\dot{q} - K_fD^{-1}(q)J'(q)\Delta F_e \qquad (10.1.11)$$

where K_p, K_I, K_D, and K_f signify the controller gains to be chosen, for example, on the basis of the desired step responses of the closed-loop system.

The substitution of the secondary control $\hat{u}(t)$ from Equation 10.1.11 into Equation 10.1.10 gives the error dynamics

$$\Delta\ddot{q} + K_D\Delta\dot{q} + K_p\Delta q + K_I\int \Delta q(\sigma)\,d\sigma + (K_f + I)D^{-1}(q)J'(q)\Delta F_e = 0 \qquad (10.1.12)$$

The position error Δq and the force error $\Delta F_e(t)$ between the generalized contact force $F_e(t)$ and the desired value $F_e^d(t)$ are governed by Equation 10.1.12. The elements in the tuning matrices K_D, K_p, K_I and K_f should be selected so that the transients of the system exhibit acceptable behavior. The coefficient of the force error that depends on $q(t)$ in Equation 10.1.12 can be regarded as constant over the sampling period for the first-order approximation. Then the eigenvalues of the system in Equation 10.1.12 may be used to describe the dynamic behavior over a sampling period. Thus, the controller gains may be determined, for example, by the eigenvalue assignment (pole placement) method, or by tuning them experimentally. The controller for the system in Equation 10.1.5 is determined by Equations 10.1.7, 10.1.8, and 10.1.11. The control strategy described for the regulation of the position and force errors can be realized as shown in Figure 10.2, where the relations in Equation 10.1.9 have been utilized in the implementation.

In order to gain insight in the behavior of the force control scheme, two special cases will be discussed. Specifically, we will study first the case that $m_c = 0$, $B_c = 0$, and $K_c \neq 0$, that is, the contact interface is modeled as a spring. We will then present the force control scheme for the case that $m_c = 0$, $K_c = 0$, but $B_c \neq 0$, that is, the contact interface is regarded as a damper.

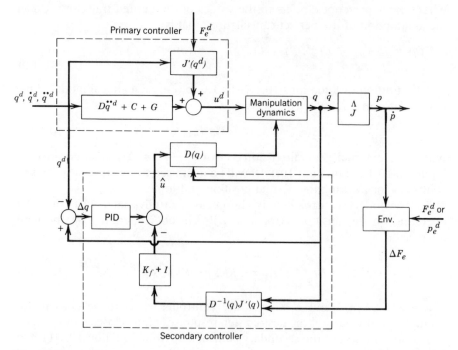

FIGURE 10.2

Realization of primary controller and secondary multivariable PID-controller for position (velocity) and force control.

Environmental Contact Modeled as a Spring

When the interface between the end-effector and the environmental surface is assumed to be modeled as a linear spring [2] that is perpendicular to the constraint surface, the contact force is directly proportional to the spring displacement. The force error is $\Delta F_e = K_c[(W_e - W_{eo}) - (W_e^d - W_{eo})] = K_c \Delta W_e$, where W_{eo} is the unstretched position of the spring, and ΔW_e is a change in the spring position vector from the desired value along the normal to the environmental surface of the contact. Hence, $m_c = 0$ and $B_c = 0$ in Equation 10.1.6 and $\Delta \psi_e = 0$ for the case under consideration.

It is possible to select the coordinate system in view of the environmental constraints in such a way that matrix K_c is diagonal. Each diagonal element is associated with a definite coordinate direction. If a particular diagonal entry of the stiffness matrix K_c is small, then the contact force in this direction is small. On the other hand, when a specific diagonal entry is chosen as a large number, indicating a stiff spring, even a small displacement in this direction results in a considerable force. A large stiffness coefficient requires an accurate position control. The desired environmental generalized force can be achieved indirectly by controlling the corresponding positions accurately.

When the control input specified by Equations 10.1.7–10.1.9 and 10.1.11 is applied to the manipulator system, the dynamical model for the joint position error $\Delta q(t) = q(t) - q^d(t)$ obeys Equation 10.1.12. Since $\Delta F_e = K_c J(q) \Delta q$ in this particular case, Equation 10.1.12 assumes the following form:

$$\Delta \ddot{q} + K_D \Delta \dot{q} + [K_p + \overline{K_f} D^{-1}(q) J'(q) K_c J(q)] \Delta q + K_I \int \Delta q(\sigma) d\sigma = 0 \quad (10.1.13)$$

where $\overline{K_f} = K_f + I$, K_p, K_I, and K_D are the matrices that are used in tuning for the desired system response. The gain K_f of the force error appears in the coefficient of Δq only. The entries in matrix K_c are selected in view of the specific task to be performed by the manipulator.

Over a sufficiently short sampling period, the coefficients in Equation 10.1.13 can be considered as constants. Then, the characteristic equation for the system in Equation 10.1.13 can be written as

$$| s^2 I + K_D s + [K_p + \overline{K_f} D^{-1} J' K_c J] + K_I/s | = 0 \quad (10.1.14)$$

where the determinant of matrix A is denoted by $|A|$. By properly selecting the elements of the gain matrices, the dynamical behavior of $\Delta q(t)$ in Equation 10.1.13, and thus the contact force error ΔF_c, can be made acceptable, for example, relative to the overshoot and settling time of the step response. The gain matrices are thus determined by Equation 10.1.14. They should be recalculated for each sampling interval.

Large values of the product of K_f and K_c can cause vibrations and even local instability in the force control scheme when the contact is modeled as a stiff spring. This has been shown [6] for some simple examples using the root locus method. Equations 10.1.13 and 10.1.14 indicate that matrices K_f and K_c appear in the same product multiplying the position error. This product plays an important role in determining the transient behavior of Equation 10.1.13 over a sampling period. If the integrating component is absent ($K_I = 0$), then the natural frequency of the system is determined by the term containing the product of the spring constant K_c and the proportional gain K_f of the force feedback.

Environmental Contact Modeled as a Damper

When the interface between the end-effector and the environment is modeled as a linear damper [3, 7], the force normal to the contact surface is directly proportional to the derivative of the displacement, that is, $\Delta F_e = B_c \Delta \dot{W}_e$, where $\Delta \dot{W}_e$ is the velocity change associated with a small displacement to the direction perpendicular to the environmental contact surface and $\Delta \psi_e$ is assumed to be zero. In this particular case, matrices m_c and K_c in Equation 10.1.6 are zero.

The contact force will be controlled indirectly by regulating the velocity changes in the direction that is normal to the environmental surface. The entries in the damping matrix B_c are selected using arguments similar to the ones used in selecting the matrix K_c of the stiffness coefficients. That is, the coordinates are first selected so that matrix B_c can be made diagonal. Large numbers chosen for the elements of B_c dictate that the corresponding velocity changes in these directions will be held under tight control, that is, as close as possible to the desired values. Small values in the diagonal of matrix B_c imply that the corresponding velocity errors do not cause strong corrective actions in the actuator inputs.

The generalized joint torque $F(t)$ given by Equations 10.1.7–10.1.9 and 10.1.11 is applied to the manipulator joints. Since $\Delta \dot{W}_e = J(q)\Delta \dot{q} + \dot{J}(q)\Delta q$, the displacement dynamics in Equation 10.1.12 give

$$\Delta \ddot{q} + [K_D + \overline{K}_F D^{-1}(q)J'(q)B_c J(q)]\Delta \dot{q} + [K_p + \overline{K}_f D^{-1}(q)J'(q)B_c \dot{J}(q)]\Delta q$$

$$+ K_I \int \Delta q dt = 0 \quad (10.1.15)$$

where matrices K_p, K_I, K_D and $\overline{K}_f$ serve as the tuning parameters. The gain $\overline{K}_f$ of the force error appears together with matrix B_c in the coefficients of Δq and $\Delta \dot{q}$ in Equation 10.1.15 of the position error.

The characteristic equation associated with system 10.1.15 is obtained by assuming that matrices D, J, and $\dot{J}$ are approximately constant over a sampling period:

$$|s^3 I + [K_D + \overline{K}_f D^{-1}J'B_c J]s^2 + [K_p + \overline{K}_f D^{-1}J'B_c \dot{J}]s + K_I| = 0 \quad (10.1.16)$$

The roots of Equation 10.1.16 determine the dynamic behavior of $\Delta q(t)$ governed by Equation 10.1.15 over a sampling period. The desired locations for these roots can be assigned by means of the parameters in the aforementioned tuning matrices. Thus, the gain matrices can be determined by the eigenvalue assignment method. They should be recalculated for each sampling interval.

A comparison of the characteristic Equations 10.1.14 and 10.1.16 reveals that the basic form of these two equations is similar. However, Equation

10.1.14 does not depend on $\dot{J}(q)$, the derivative of the Jacobian matrix, and Equation 10.1.16 does. The gain K_f of the force error feedback and the damping coefficient B_c of the contact model appear in the same product in Equation 10.1.15.

The generalized force error ΔF_e is modeled as a soft contact by Equation 10.1.2. It permits the designer to eliminate the force error ΔF_e in Equation 10.1.12 and derive Equations 10.1.13 and 10.1.15 for the position error. The control of the force error is thus performed indirectly by regulating the position error.

An example is next presented to illustrate the approach described.

EXAMPLE 10.1.1

The motion of the end-effector of a three-joint planar manipulator on a vertical plane is constrained by $p_x \leq 0.4$, and the position of the end-effector is (p_x, p_y) in the base coordinate system. The system is shown in Figure 10.3a.

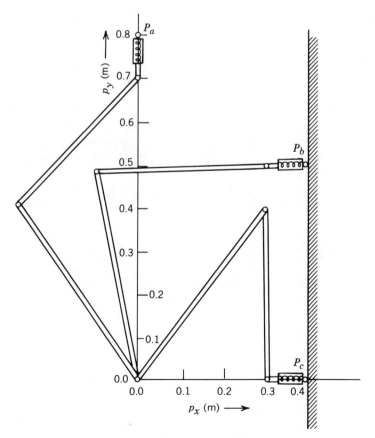

FIGURE 10.3a
Free and constraint motion of a planar manipulator. Example 10.1.1.

The end-effector is to move along a straight line from point P_a (0, 0.8m) to point P_b (0.4m, 0.5m) and then on the constraint surface to point P_c (0.4m, 0) while exerting a force $F_e^d = [F_{ex}^d \ 0]'$ in the normal direction of the constraint surface. The velocity profile in the Cartesian base coordinate system is composed of two equilateral trapezoids shown in Figure 10.3b. The free motion from P_a to P_b lasts $t_f = 1.2$ s with the acceleration time of $t_1 = 0.18$ s and the constraint motion from P_b to P_c lasts also 1.2 s with acceleration time of $t_2 = 0.18$ s. The orientation of the third link specified by $\theta_1 + \theta_2 + \theta_3$ is initially $\pi/2$ rad and is changed to and kept at 0 rad from point P_b on to point P_c. The contact force between the end-effector and the constraint surface is modeled as a spring with coefficient K_c^t. It is appropriate, for example, in deburring applications. The external moment M_e^c is zero.

The problem is to design a controller for the manipulator so that its end-effector moves along the specified trajectory and maintains the given value for the external force F_e exerted on the constraint surface.

The dynamical model for the motion of the manipulator may be written (see Equation 9.1.7)

$$D(\theta)\begin{bmatrix} \ddot{\theta}_1 \\ \ddot{\theta}_1 + \ddot{\theta}_2 \\ \ddot{\theta}_1 + \ddot{\theta}_2 + \ddot{\theta}_3 \end{bmatrix} + C(\theta, \dot{\theta}) + B(\dot{\theta}) + G(\theta) =$$

$$\begin{bmatrix} \tau_1 \\ \tau_2 \\ \tau_3 \end{bmatrix} - \begin{bmatrix} \ell_1 s_1 + \ell_2 s_{12} + \ell_3 s_{123} & -\ell_1 c_1 - \ell_2 c_{12} 1\ell_3 c_{123} \\ \ell_2 s_{12} + \ell_3 s_{123} & -\ell_2 c_{12} - 1\ell_3 c_{123} \\ \ell_3 s_{123} & -1\ell_3 c_{123} \end{bmatrix} \begin{bmatrix} F_{2,3x} \\ F_{2,3y} \end{bmatrix} \quad (10.1.17)$$

where $\theta(t) = [\theta_1(t) \ \theta_1(t) + \theta_2(t) \ \theta_1(t) + \theta_2(t) + \theta_3(t)]'$, $B(\dot{\theta}) = [B_1\dot{\theta}_1 \ B_2\dot{\theta}_2 \ B_3\dot{\theta}_3]'$ and $D(\theta)$, $C(\theta, \dot{\theta})$ and $G(\theta)$ are defined by a direct comparison of Equation 10.1.17 with Equation 9.1.7. The external force $[F_{2,3x}F_{2,3y}]'$ is multiplied by the transposed Jacobian matrix to reflect it to the joint space. This force is zero during the unconstrained motion and $F_{2,3x} = F_{ex}$ and $F_{2,3y} = 0$

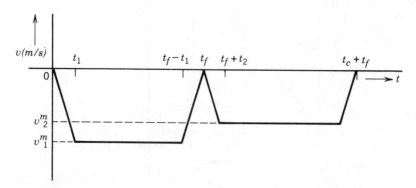

FIGURE 10.3b
Velocity profile for planar manipulator motion. Example 10.1.1.

during the constraint motion. The contact force F_{ex} is determined by the aforementioned spring.

To make the manipulator track the desired values, the inputs to the joints are generated by means of the primary (u^d) and secondary ($\hat{u}$) controllers according to

$$\begin{bmatrix} \tau_1(t) \\ \tau_2(t) \\ \tau_3(t) \end{bmatrix} = u^d(t) + D(\theta)\hat{u}(t) \tag{10.1.18}$$

where $D(\theta)$ represents the inertia matrix of Equation 10.1.17.

The primary controller $u^d(t)$ is next chosen as

$$u^d(t) = D(\theta^d)\begin{bmatrix} \ddot{\theta}_1^d \\ \ddot{\theta}_1^d + \ddot{\theta}_2^d \\ \ddot{\theta}_1^d + \ddot{\theta}_2^d + \ddot{\theta}_3^d \end{bmatrix} + C(\theta^d, \dot{\theta}^d) + B(\dot{\theta}^d) + G(\theta^d)$$

$$+ \begin{bmatrix} \ell_1 s_1^d + \ell_2 s_{12}^d + \ell_3 s_{123}^d \\ \ell_2 s_{12}^d + \ell_3 s_{123}^d \\ \ell_3 s_{123}^d \end{bmatrix} F_{ex}^d \tag{10.1.19}$$

where the inertia matrix (D) and the Coriolis (C) and gravity (G) vectors defined in Equation 10.1.17 are evaluated along the desired position trajectory θ^d and $u^d(t) = [\tau_1^d(t)\ \tau_2^d(t)\ \tau_3^d(t)]$. Equations 10.1.18 and 10.1.19 are substituted into Equation 10.1.17. Under the assumption of Equation 10.1.9, the nonlinear terms cancel in the dynamical model. The tracking error $\Delta\theta(t) = [\theta(t) - \theta^d(t)]$ will then be governed by Equation 10.1.10, where the Jacobian matrix transposed is specified in Equation 10.1.17 and $\Delta F_e = F_e - F_e^d = [F_{ex} - F_{ex}^d, 0]'$, since $\Delta M_e^c = 0$.

The secondary controller is selected as a PD-controller operating on the position error $\Delta\theta$ and a P-controller operating on the force error ΔF_e. Thus,

$$\hat{u}(t) = -K_p\Delta\theta - K_D\Delta\dot{\theta} - K_f D^{-1}(\theta)J'(\theta)\Delta F_e \tag{10.1.20}$$

The substitution of Equation 10.1.20 for the secondary controller $\hat{u}(t)$ into Equation 10.1.10 gives with $\Delta\theta = \Delta q$

$$\Delta\ddot{\theta} + K_D\Delta\dot{\theta} + K_p\Delta\theta + (K_f + I)D^{-1}(\theta)J'(\theta)\Delta F_e = 0 \tag{10.1.21}$$

where the force error ΔF_e is determined by the soft contact model as

$$\Delta F_e = K_c^t J(\theta)\Delta\theta \tag{10.1.22}$$

Equations 10.1.21 and 10.1.22 are combined to obtain

$$\Delta\ddot{\theta} + K_D\Delta\dot{\theta} + [K_p + \overline{K}_f D^{-1}(\theta)J'(\theta)K_c^t J(\theta)]\Delta\theta = 0 \tag{10.1.23}$$

where $\overline{K}_f = K_f + I$. The task is now to tune the controller gains K_D and K_p so that the response $\Delta\theta(t)$ governed by Equation 10.1.23 exhibits desired characteristics.

The gain matrices K_D and K_p are determined so that each component in Equation 10.1.23 exhibits the characteristics of a linear underdamped (ξ_i^d, ω_{ni}^d) second order reference model. For a sampling period, one may require that

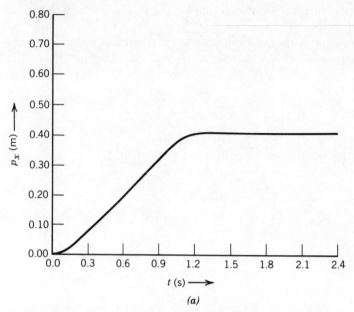

FIGURE 10.4a

The X-coordinate of the end-effector position p_x versus time.
Example 10.1.1.

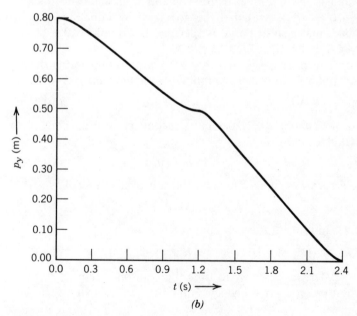

FIGURE 10.4b

The y-coordinate of end-effector position p_y versus time.
Example 10.1.1.

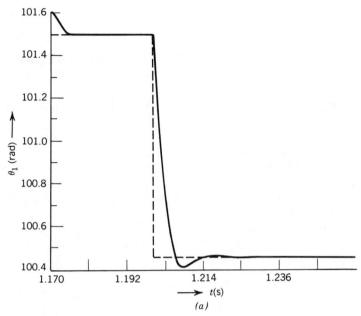

FIGURE 10.5a
Enlarged response of joint variable θ_1 versus time about 1.2 s.
Example 10.1.1.

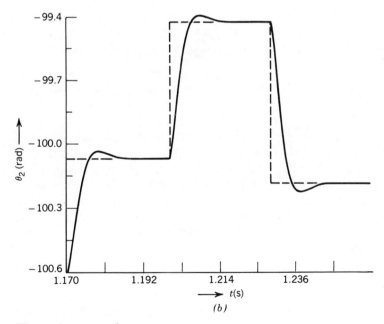

FIGURE 10.5b
Enlarge response of joint variable θ_2 versus time about 1.2 s.
Example 10.1.1.

$$K_D = \text{diag}[2\xi_1^d \omega_{n1}^d \quad 2\xi_2^d \omega_{n2}^d \quad 2\xi_3^d \omega_{n3}^d] \qquad (10.1.24)$$

$$\overline{K}_f = [\Omega - K_p][J'(q)K_c^t J(q)]^{-1}D(q) \qquad (10.1.25)$$

where $\Omega = \text{diag}[(\omega_{n1}^d)^2 \ (\omega_{n2}^d)^2 \ (\omega_{n3}^d)^2]$ and $K_f = \overline{K}_f - I$. The control gain K_p and the spring coefficient matrix K_c^t are chosen in regard to the task to be performed.

Digital simulations were performed for the system described using the following numerical values: $m_1 = 5.0$ kg, $m_2 = 4.0$ kg, $m_3 = 1.0$ kg, $\ell_1 = 0.5$ m, $\ell_2 = 0.4$ m, $\ell_3 = 0.1$ m, the diameter of the circular cross-section of the links is 0.02 m and the frictional coefficients $B_1 = B_2 = B_3 = 8.0$ Nms. The desired damping and (undamped) natural frequency of the reference model are given by $\xi_i^d = 0.707$, $(\omega_{ni}^d)^2 = 471.5$, that correspond to about 5% overshoot and $0.4T$ settling time in the step response (sampling period $T = 30$ ms). Moreover, $F_{ex}^d = 10N$, the spring coefficient $K_c^t = \text{diag}[10^3 \ 10^6]$, and the gain for the position error $K_p = \text{diag}[0 \ 10^3 \ 0]$. The gains K_D and K_f for the velocity and force errors, respectively, can now be computed by Equations 10.1.24 and 10.1.25 at every sampling instant. They need to be recalculated for each sampling period.

Digital simulation results for p_x and p_y versus time are shown in Figures 10.4a and 10.4b. The actual and desired responses overlap in the figures. The behavior of the joint variables θ_1 and θ_2 versus time over a short time period about 1.2 s is enlarged and displayed in Figures 10.5a and 10.5b. The transients exhibit a slight overshoot and a short settling time, as required by the design specifications.

10.1.2   SEPARATE POSITION FORCE (PF) CONTROL USING MULTIVARIABLE PID-CONTROLLER (HYBRID CONTROLLER)

In the algorithms for the position, velocity, and force control described in the previous section, the force control is achieved indirectly by servoing on a variable that is linearly related to the external generalized force. Thus, the force and position servoing are not explicitly separated. In the hybrid control scheme [8] to be discussed, the error is formed in the force acting on the hyperplane that is normal to the constraint surface in the Cartesian base coordinate system, while the error in the position (velocity) is determined on the tangential hyperplane. By introducing a selection matrix consisting of binary numbers, the components of the force and position to be controlled are explicitly specified. The error vectors are then mapped into the joint space where appropriate corrective actions are generated, for example, by means of a PID-controller.

Suppose that in a Cartesian base coordinate system the end-effector can move freely on the tangential hyperplane of the constraint surface, but it is restricted in its motion on the hyperplane normal to the constraint surface. The desired values for the components of the force and the position on their respective hyperplanes are specified. The position and force errors between the actual and desired values can then be defined. To represent these errors

by means of N-dimensional position and force vectors for an N-joint manipulator, the components of the positional error vector Δp_e on the tangential
hyperplane and those of the force vector ΔF_e^c on the normal hyperplane can
be specified by using selection matrix S. The entries of matrix S consist of
binary numbers. The diagonal entries of the selection matrix S associated with
the components of the force to be controlled are chosen equal to one, and
the remaining elements are zero. Thus, the selection matrix S specifies the
directions in the Cartesian base coordinate system along which the forces are
controlled. If I signifies the identity matrix, then matrix $(I - S)$ has nonzero
elements on the diagonal that specify the directions along which the position
is controlled. The remaining components of $(I - S)$ are zero.

The generalized force exerted by the manipulator at the contact point can
still be assumed to be described by the soft contact model given in Equation
10.1.2. It can be incorporated in the expression of the force error.

An example is next presented to illustrate the choice of the selection
matrix.

EXAMPLE 10.1.2

To demonstrate the use of the selection matrix S, suppose that the generalized force error vector ΔF_e and the generalized position error vector
ΔW are six-dimensional, that is, $\Delta F_e = [\Delta F_{e1}\ \Delta F_{e2}\ \Delta F_{e3}\ \Delta M_{e1}\ \Delta M_{e2}\ \Delta M_{e3}]'$,
where the first three components correspond to the external force and
the last three components the external moment (torque), and $\Delta W_e = $
$[\Delta p_{e1}\ \Delta p_{e2}\ \Delta p_{e3}\ \Delta \psi_{e1}\ \Delta \psi_{e2}\ \Delta \psi_{e3}]'$, where the first three components are associated with the position and the last three with the orientation. The force com-
ponents ΔF_{e3} and ΔM_{e2} of the error vector ΔF_e and the position components
Δp_{e1}, Δp_{e2}, $\Delta \psi_{e1}$, and $\Delta \psi_{e3}$ of the error vector ΔW_e are to be controlled.

The problem is to describe the specified error components in terms of six-
dimensional force error ΔF_e and position error ΔW_e by means of the selection
matrix S.

The components of the generalized force vector ΔF_e that are force controlled can be isolated from ΔF_e by means of the selection matrix S. It follows
that

$$S\Delta F_e = \begin{bmatrix} 0 & 0 & 0 & 0 & 0 & 0 \\ 0 & 0 & 0 & 0 & 0 & 0 \\ 0 & 0 & 1 & 0 & 0 & 0 \\ 0 & 0 & 0 & 0 & 0 & 0 \\ 0 & 0 & 0 & 0 & 1 & 0 \\ 0 & 0 & 0 & 0 & 0 & 0 \end{bmatrix} \begin{bmatrix} \Delta F_{e1} \\ \Delta F_{e2} \\ \Delta F_{e3} \\ \Delta M_{e1} \\ \Delta M_{e2} \\ \Delta M_{e3} \end{bmatrix} \tag{10.1.26}$$

$$(I - S)\Delta W_e = \begin{bmatrix} 1 & 0 & 0 & 0 & 0 & 0 \\ 0 & 1 & 0 & 0 & 0 & 0 \\ 0 & 0 & 0 & 0 & 0 & 0 \\ 0 & 0 & 0 & 1 & 0 & 0 \\ 0 & 0 & 0 & 0 & 0 & 0 \\ 0 & 0 & 0 & 0 & 0 & 1 \end{bmatrix} \begin{bmatrix} \Delta p_{e1} \\ \Delta p_{e2} \\ \Delta p_{e3} \\ \Delta \Psi_{e1} \\ \Delta \Psi_{e2} \\ \Delta \Psi_{e3} \end{bmatrix}$$

Thus, the selection matrix S allows the designer to use six-dimensional force and position error vectors while specifying the components that will be force controlled and those that will be position controlled.

In the previous section, the force error ΔF_e and the position error ΔW_e are related by the model of soft contact between the end-effector and the environment. The controller then regulates the joint error corresponding to ΔW_e. The selection matrix introduced can keep the components of the force and position errors separated in the Cartesian space. When they are converted into the joint space, the joint actuators together contribute to compensate for the portions of the force and position errors that are determined by the selection matrix.

The system in Equation 10.1.5 is next designed for force and position servoing using the generalized force and position error vectors. The controller design will be described after converting the error vectors from the Cartesian base coordinate system into the joint space. Then, a multivariate PID-controller will be determined in terms of the joint variables and the selection matrix.

The generalized force error $S\Delta F_e$ expressed in the Cartesian base coordinate system can be compensated by a corrective joint torque change $\Delta\Gamma_e$ expressed in the joint space, that is,

$$\Delta\Gamma_e = J'(q)S\Delta F_e \tag{10.1.27}$$

where $J'(q)$ is the usual transpose of the Jacobian matrix. The position error $(I - S)\Delta W_e$ can also be converted into the joint space

$$\Delta q_p = J^{-1}(q)(I - S)\Delta W_e$$

$$= J^{-1}(q)(I - S)J\Delta q \tag{10.1.28}$$

where the inverse of the Jacobian matrix J is assumed to exist, and Δq_p signifies the positional error in the joint space corresponding to ΔW_e. The incremental changes ΔW_e and Δq are assumed to be sufficiently small at all times so that Equation 10.1.28 is valid.

A primary controller $u^d(t)$ is first constructed for ideal operating conditions so that the end-effector tracks the desired trajectory $(q^d, \dot{q}^d)$ while exerting a specified external generalized force F_e^d in the given directions on the environment. By replacing the desired force F_e^d by SF_e^d, Equation 10.1.8 gives the primary controller

$$u^d(t) = D(q^d)\ddot{q}^d + C(q^d, \dot{q}^d) + G(q^d) + J'(q^d)SF_e^d \tag{10.1.29}$$

When control $F = u^d + \hat{u}$ is substituted into Equation 10.1.5, the nonlinear terms are exactly canceled under the assumption that the relations in Equation 10.1.9 hold. The dynamical model for error $\Delta q(t)$ is then given by Equation 10.1.10, where the selection matrix S is added to multiply the force error, that is,

$$\Delta\ddot{q} = \hat{u}(t) - D^{-1}(q)J'(q)S\Delta F_e(t) \tag{10.1.30}$$

where $\Delta F_e(t) = F_e(t) - F_e^d(t)$ represents the force error in the Cartesian base coordinate system.

The secondary controller $\hat{u}(t)$ that compensates for modeling errors and disturbances is chosen as the combination of a PID-controller operating on the position error $\Delta q_p(t)$ given in Equation 10.1.28 and a PI-controller operating on the force error $S\Delta F_e(t)$ expressed in the joint space. The positional control may be expressed as

$$\hat{u}_p(t) = -K_3 \Delta q_p - K_2 \int \Delta q_p(\sigma) \, d\sigma - K_1 \Delta \dot{q}_p \qquad (10.1.31)$$

The secondary controller is then determined as

$$\hat{u}(t) = \hat{u}_p(t) - [K_4 J' S \Delta F_e + K_5 \int J' S \Delta F_e(\sigma) \, d\sigma] \qquad (10.1.32)$$

The gain matrices $K_i, i = 1, \ldots, 5$ are tuned, for example, so that acceptable step responses of the joint variables are obtained.

The composite control $u(t)$ is given by Equation 10.1.7. The primary control $u^d(t)$ is specified in Equation 10.1.29 and the secondary control $\hat{u}(t)$ in Equation 10.1.32. The implementation of the control strategy is shown in Figure 10.6.

When the composite controller is connected into the manipulator system, the model for the position and force errors is obtained from Equations 10.1.30–10.1.32.

$$\Delta \ddot{q} + [K_1 \Delta \dot{q}_p + K_3 \Delta q_p + K_2 \int \Delta q_p(\sigma) \, d\sigma] + [K_4 + D^{-1}(q)] J'(q) S \Delta F_e$$

$$+ K_5 \int J' S \Delta F_e(\sigma) \, d\sigma] = 0 \qquad (10.1.33)$$

where Δq_p is given by Equation 10.1.28. The selection matrix S in Equation 10.1.33 determines which components of the position and force errors are present in a particular row of Equation 10.1.33. The force error can still be modeled as soft contact given by Equation 10.1.1.

Since the composite control depends on the errors in the position of end-effector and the external force it exerts, the multivariable PID-controller is often called a hybrid controller [8]. The control strategy for the hybrid controller is described and implemented in the joint space, but the control scheme is designed using the variables expressed in the Cartesian coordinate system.

The components of ΔF_e and ΔW_e are separated in the Cartesian base coordinate system in the sense that no two components are controlled independently on the same coordinate axis. Their loading effects on the joints, however, are distributed among the joint actuators. Thus, the joint actuators contribute in such portions that the separation of the components of ΔF_e and ΔW_e results in the Cartesian base coordinate system.

An example is presented next to illustrate the approach.

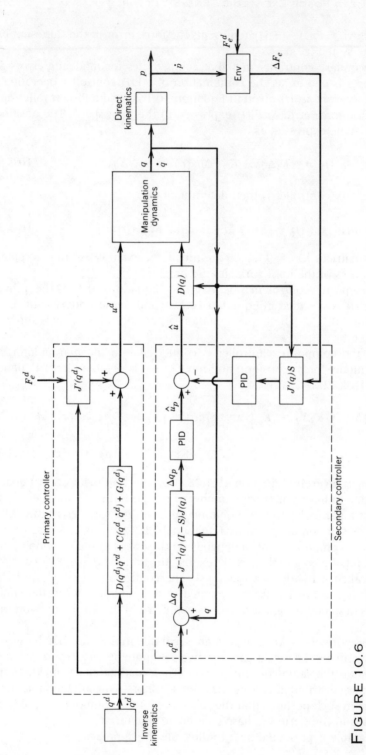

FIGURE 10.6

Realization of primary and secondary multivariable PID-controller (hybrid controller) for position and force control.

EXAMPLE 10.1.3

Joints one and three in a Stanford/JPL manipulator are moved while the
remaining joints are "frozen" at the specific positions by means of their
breaks. The motion of the end-effector takes place on a plane. It is constrained by a horizontal surface. The configuration of the system and the
Cartesian base coordinate system are shown in Figure 10.7. When a rotating
brush attached to the gripper touches the horizontal surface shown in Figure
10.7, a reaction force F_{ey} will act on the gripper. The contact force is modeled
as a spring, that is

$$F_{ey}(t) = K_c^t[p_y(t) - p_{y0}(t)] \tag{10.1.34}$$

where K_c^t is a spring constant, p_y is the y-component of the end-effector
position, and p_{y0} the unstretched length of the spring. After the spring touches
the surface, it will maintain the contact over the duration of the motion. The
positions of the revolute and prismatic joints are q_1 and q_2, respectively.

The dynamic behavior of the joints in the manipulator [8] may be approximated by

$$\begin{bmatrix} I(q_2) & 0 \\ 0 & m \end{bmatrix} \begin{bmatrix} \ddot{q}_1 \\ \ddot{q}_2 \end{bmatrix} + \begin{bmatrix} B_1(\dot{q}_1) \\ B_2(\dot{q}_2) \end{bmatrix} = \begin{bmatrix} F_1(t) \\ F_2(t) \end{bmatrix} - J'(q) \begin{bmatrix} 0 \\ F_{ey} \end{bmatrix}$$

$$\tag{10.1.35}$$

$$J(q) = \begin{bmatrix} q_2 \cos q_1 + \ell \sin q_1 & \sin q_1 \\ \ell \cos q_1 - q_2 \sin q_1 & \cos q_1 \end{bmatrix} \tag{10.1.36}$$

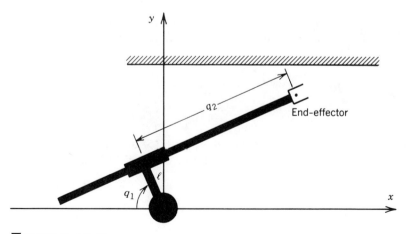

FIGURE 10.7

Two-joint planar manipulator and a horizontal constraint surface.
Example 10.1.3.

$$\begin{bmatrix} p_x \\ p_y \end{bmatrix} = \begin{bmatrix} q_2 \sin q_1 - \ell \cos q_1 \\ q_2 \cos q_1 + \ell \sin q_1 \end{bmatrix} \tag{10.1.37}$$

where $[p_x\ p_y]'$ is the position of the end-effector in the base coordinate system, $I(q_2)$ describes the moment of inertia of the rotating parts about the origin, m is the mass of the prismatic link with payload, and the link offset is ℓ. The input is torque F_i acting on joint $i, i = 1,2$. The friction on joint i is modeled as $B_i(\dot{q}_i)$, $i = 1,2$. If τ_{si} denotes static friction (constant) and τ_{ci} Coulomb friction (constant), then $B_i(\dot{q}_i) = -sqn(\dot{q}_i)[\min(\tau_{si}, |\tau_i|)]$ for $|\dot{q}_i| \leq$ 0.1 rad/s and $B_i(\dot{q}_i) = -sqn(\dot{q}_i)[\tau_{ci}]$ otherwise. Equations 10.1.35–10.1.37 describe the model for the motion of the two-joint manipulator in the joint coordinate space.

The problem is to design a multivariate (hybrid) PID-controller for the position and force of the foregoing manipulator. The end-effector touching a horizontal surface shown in Figure 10.6 should exert a constant force of 10 N in the y-direction while the x-position is kept stationary.

To design a multivariate PID-controller for the system, the force and joint position errors are defined as $\Delta F_e = F_e - F_e^d$ and $\Delta q = q - q^d$ where $F_e = [F_{ex} F_{ey}]'$, $q = [q_1\ q_2]'$ and the superscript refers to the desired value of the variable. The (compliance) selection matrix S is chosen so that (compare Equations 10.1.26 and 10.1.28)

$$S\Delta F_e = \begin{bmatrix} 0 & 0 \\ 0 & 1 \end{bmatrix} \Delta F_e = \begin{bmatrix} 0 \\ \Delta F_{ey} \end{bmatrix} \tag{10.1.38}$$

$$\Delta q_p = J^{-1}(q)\begin{bmatrix} 1 & 0 \\ 0 & 0 \end{bmatrix} J(q) \Delta q \tag{10.1.39}$$

The controller will be composed of the primary (u^d) and secondary ($\hat{u}$) controllers. The primary controller $u^d(t) = F^d(t)$ is determined by Equation 10.1.29 to obtain

$$u^d(t) = \begin{bmatrix} I(q_2^d) & 0 \\ 0 & m \end{bmatrix} \begin{bmatrix} \ddot{q}_1^d \\ \ddot{q}_2^d \end{bmatrix} + \begin{bmatrix} B_1(\dot{q}_1^d) \\ B_2(\dot{q}_2^d) \end{bmatrix} + J'(q^d)\begin{bmatrix} 0 \\ F_{ey}^d \end{bmatrix} \tag{10.1.40}$$

By substituting $u = u^d + \hat{u}$ into Equation 10.1.35, and assuming that the nonlinear expressions in Equations 10.1.35 and 10.1.40 cancel, Equation 10.1.30 assumes the following form:

$$\Delta\ddot{q} = \hat{u}(t) - \begin{bmatrix} 1/I(q_2) & 0 \\ 0 & 1/m \end{bmatrix} J'(q)\begin{bmatrix} 0 \\ \Delta F_{ey} \end{bmatrix} \tag{10.1.41}$$

The secondary controller $\hat{u}(t)$ is next designed on the basis of the force and position errors.

The secondary controller is determined by Equation 10.1.32 by choosing
a PID-controller for the position error and a PI-controller for the force error

$$\hat{u}(t) = -[K_3\Delta q_p + K_2\int \Delta q_p(\sigma)\,d\sigma + K_1\Delta\dot{q}_p] - \{K_4 J'(q)\begin{bmatrix} 0 \\ \Delta F_{ey} \end{bmatrix}$$

$$+ K_5\int J'(q)\begin{bmatrix} 0 \\ \Delta F_{ey} \end{bmatrix}d\sigma\} \quad (10.1.42)$$

where the gain matrices $K_i, i = 1, \ldots, 5$ are determined, for example, experimentally so that an acceptable step response behavior is obtained. When
Equation 10.1.42 is substituted into Equation 10.1.41, the model in Equation
10.1.33 for the position and force errors is obtained.

The composite control $u(t) = u^d(t) + \hat{u}(t)$ for the manipulation is determined by Equations 10.1.40 and 10.1.42.

The manipulator system in Equations 10.1.35–10.1.37 was simulated with
the following numerical values: $\ell = 15.3$ cm, $m = 7.26$ kg, $I(q_2) = (3.98 -$
$5.25q_2 + 6.47q_2^2)$ kgm², $\Delta F_{ey} = K_c^t(p_y - p_y^d)$ with $K_c^t = 100$ N/m, $[\tau_{c_1}\,\tau_{c_2}] = [5.0$
Nm 3.5 N], $[\tau_{s_1}\,\tau_{s_2}] = [5.75$ Nm 4.6 N], and the sampling period 16.7 ms.
In the implementation of the controller, the controller gains were chosen
empirically to obtain stable and accurate system responses [8]. Figures 10.8a
and 10.8b show step responses of position p_x and force F_{ey} versus time. The
maximum force error is less than 1.0 N and the corresponding position error
less than 0.5 cm.

The simulations demonstrate a good performance, although the tuning
may require considerable experimentation so that the transients will be properly damped.

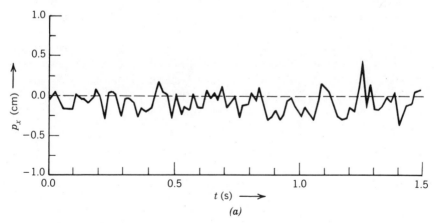

(a)

FIGURE 10.8a

The x-coordinate of end-effector position p_x versus time. Example 10.1.3.

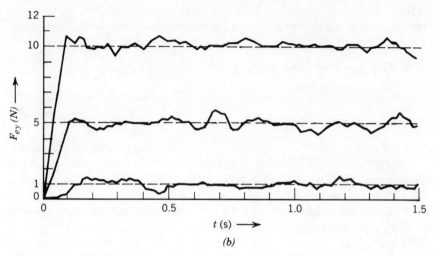

(b)

FIGURE 10.8*b*

Force responses versus time when step inputs of height 1, 5, and 10 N are applied at $t = 0$. Example 10.1.3.

10.1.3   DECOUPLING POSITION AND FORCE CONTROL MODEL IN CARTESIAN BASE COORDINATE SYSTEM

The use of the selection matrix S discussed in the previous section allows the separation of the components of the force and position error vectors in the Cartesian base coordinate system. When the hybrid controller is implemented in the joint space, this decomposition is not apparent. An alternative approach is to design and implement the entire control algorithm in the Cartesian base coordinate system [9].

In order to design a controller that functions in the Cartesian base coordinate system, the dynamical model given by Equation 10.1.5 is first expressed in terms of the variables of this coordinate system. Equation 9.3.26 defines vector W for position p and orientation ψ, that is, $W' = [p' \; \psi']$. Then, the velocity and acceleration vectors are determined by

$$\dot{W} = J(q)\dot{q} \tag{10.1.43}$$

$$\ddot{W} = J(q)\ddot{q} + \dot{J}(q)\dot{q} \tag{10.1.44}$$

where q is the joint variable vector and $J(q)$ is the Jacobian matrix. The dynamical model in Equation 10.1.5 can be rewritten

$$\ddot{W} = J(q)D^{-1}(q)[F - J'(q)F_e - C(q,\dot{q}) - G(q)] + \dot{J}(q)J^{-1}(q)\dot{W} \tag{10.1.45}$$

where the generalized contact force F_e of dimension n is assumed to be perpendicular to the contact surface and properly aligned with the base coordinate system. The position vector in Equation 10.1.45 can be decomposed so that the components in the directions that are force controlled are combined into an n-dimensional vector W_1 and the components that are position controlled into an $(N - n)$-dimensional vector W_2. The force vector is correspondingly expressed as $F'_e = [(F_e^1)' \; 0]$ since the components of the W_2-vector are position controlled. Thus, Equation 10.1.45 gives

$$
\begin{bmatrix} \ddot{W}_1 \\ \ddot{W}_2 \end{bmatrix} = \begin{bmatrix} J_1(q) \\ J_2(q) \end{bmatrix} D^{-1}(q) \left\{ F - C(q, \dot{q}) - G(q) - [J_1'(q) J_2'(q)] \begin{bmatrix} F_e^1 \\ 0 \end{bmatrix} \right\}
$$

$$
+ \begin{bmatrix} \dot{J}_1(q) \\ \dot{J}_2(q) \end{bmatrix} J^{-1}(q) \dot{W} \quad (10.1.46)
$$

where the Jacobian matrix $J(q)$ has been partitioned according to the vector decomposition to submatrices $J_1(q)$ of dimension $(n \times N)$ and $J_2(q)$ of dimension $[(N - n) \times N]$.

Equation 10.1.46 represents the dynamical model that has been decomposed into two subsystems, when the end-effector is in contact with the environmental object. The equation governing the motion of W_1 contains the external force F_e^1. On the other hand, the motion of W_2 taking place on the hyperplane that is tangent to the contact surface is independent of the external force, since the external force is zero on this hyperplane. Thus, the position vector W has been decomposed into two components W_1 and W_2 of which the former is force-servoed on the hyperplane perpendicular to the contact surface and the latter is position-servoed on the tangential hyperplane of the contact points.

A controller is next designed so that (1) the external force $F_e^1(t)$ assumes the desired values $F_e^{1d}(t)$ specified in the directions of W_1 and (2) the position of the end-effector in the direction of W_2 tracks the given trajectory $W_2^d(t)$ as closely as possible. Moreover, the controller should decouple the two subsystems W_1 and W_2 so that the equation governing the W_1-vector depends only on the force error $\Delta F_e^1 = F_e^1 - F_e^{1d}$ but not on the W_2-vector and the equation describing the W_2-vector is a function of only the position error $\Delta W_2 = W_2 - W_2^d$ but not of the external force. Thus, vector W_1 will describe the motion that involves the force control and vector W_2 the position control on their respective hyperplanes.

Before specifying the controller, it is convenient to introduce a decomposition for the inverse J^{-1} of the Jacobian matrix as

$$
J^{-1}(q) = [\chi_1(q) \; \chi_2(q)] \quad (10.1.47)
$$

where submatrix χ_1 is of dimension $(N \times n)$ and χ_2 of dimension $N \times (N-n)$. By the definition of the inverse of J, it follows that $J_1 \chi_1 = I$, an $(n \times n)$-matrix, $J_1 \chi_2 = 0$ that is of dimension $n \times (N - n)$, $J_2 \chi_1 = 0$ of dimension $(N - n) \times n$ and $J_2 \chi_2 = I$, a $(N - n) \times (N - n)$-matrix. These relations will be helpful in the subsequent development.

The controller will be composed of the primary (u^d) and secondary ($\hat{u}$) controllers. The primary controller $u^d(t) = F^d(t)$ is determined so that the nonlinear terms in the dynamical model given in Equation 10.1.46 will cancel under ideal operating conditions. It may be chosen as

$$u^d(t) = \hat{C}(q,\dot{q}) + \hat{G}(q) + \hat{J}_1'(q)F_e^1 - \hat{D}(q)\hat{J}^{-1}(q)\dot{\hat{J}}(q)[\hat{J}^{-1}(q)\dot{W}] + \hat{D}(q)\hat{J}^{-1}(q)W^d$$
(10.1.48)

where the caret refers to the values of the given expressions calculated on the desired trajectory.

When the control input $F(t) = u^d(t) + \hat{u}(t)$ is applied to system 10.1.46, exact cancellation of the nonlinear terms is achieved if

$$D(q) = \hat{D}(q) \qquad C(q,\dot{q}) = \hat{C}(q,\dot{q}) \qquad G(q) = \hat{G}(q)$$

$$J(q) = \hat{J}(q) \qquad \dot{J}(q) = \dot{\hat{J}}(q)$$
(10.1.49)

As a consequence, Equation 10.1.46 becomes

$$\begin{bmatrix} \Delta\ddot{W}_1 \\ \Delta\ddot{W}_2 \end{bmatrix} = \begin{bmatrix} J_1(q) \\ J_2(q) \end{bmatrix} D^{-1}(q)\hat{u}(t)$$
(10.1.50)

where $\Delta\ddot{W}_i = \ddot{W}_i - \ddot{W}_i^d, i = 1, 2$ describes the acceleration error.

The secondary controller $\hat{u}(t)$ will be chosen so that variable W_1 will be regulated by the force error and variable W_2 by the position error. The designer may select

$$\hat{u}(t) = D(q)\chi_1(q)[K_1\Delta F_e^1 + K_2\int \Delta F_e^1(\sigma)d\sigma] - D(q)\chi_2(q)[K_4\Delta W_2 + K_3\Delta\dot{W}_2]$$
(10.1.51)

In Equation 10.1.51, the force regulation is performed by a PI-controller and the position regulation by a PD-controller.

The substitution of Equation 10.1.51 into Equation 10.1.50 leads to

$$\Delta\ddot{W}_1 = K_1\Delta F_e^1 + K_2\int \Delta F_e^1(\sigma)d\sigma$$
(10.1.52)

$$\Delta\ddot{W}_2 + K_3\Delta\dot{W}_2 + K_4\Delta W_2 = 0$$
(10.1.53)

The gain matrices $K_i, i = 1, 2, 3$ and 4 can now be tuned for the desired output response. Equations 10.1.52 and 10.1.53 reveal that the force and position regulation are now performed independently in the Cartesian base coordinate system.

The composite control $u(t) = u^d(t) + \hat{u}(t)$ is determined by Equations 10.1.48 and 10.1.51. It controls directly the position and force in the Cartesian base coordinate system. The subsystems expressed in terms of the position and force errors are decoupled. Thus, the position and force servoing can be performed independently.

10.2 ADAPTIVE POSITION VELOCITY FORCE (PVF) CONTROL WITH SELF-TUNING IN BASE COORDINATE SYSTEM

The task to be performed by a manipulator determines usually the directions for the desired force, position, and/or velocity in the Cartesian base coordinate system. When the interfacing between the environment and the end-effector can be modeled as a soft contact, the desired force, position, and/or its derivatives in the directions of the properly chosen coordinates are related by the model in Equation 10.1.2.

The multivariate PID-controller presented in Sections 10.1.1 and 10.1.2 for the force control is designed by converting the external force into the joint space. The controller is then implemented using the variables in the joint space. An alternative approach discussed in section 10.1.3 is to construct a PID-controller in the Cartesian base coordinate system after decomposing the model into two uncoupled subsystems that can be regulated independently of each other. Thus, the position and force servoing are accomplished separately.

In both approaches, a primary controller is designed for the manipulator system to function under ideal operating conditions. It is complimented by a secondary controller that compensates for modeling errors and disturbances. Thus, it is implicitly assumed that the manipulator model is accurately known. In many applications, however, the accurate information about the equations of motion is not available.

When the dynamical model of a manipulator is not accurately known, an adaptive self-tuning controller may be designed for position, velocity, and force servoing by the methods discussed in Section 9.4 for the gross motion control. To construct a self-tuning controller for the gross motion of a manipulator, an autoregressive exogenous (ARX) time-series model is chosen to describe the input–output relation in the system. The output variables in Section 9.4 are the positions and/or velocities of the joint variables or the position, orientation, and/or velocity of the end-effector expressed in the Cartesian base coordinate system. Since the desired values of the external force exerted by the end-effector on the environment is usually specified in the Cartesian base coordinate system, it is natural to express the entire system of the input–output relations for the force, position, orientation, and/or velocity in the Cartesian base coordinate system.

When the external generalized force and the end-effector position are to be controlled, the approach to design a self-tuning controller for the gross motion of the manipulator may be adopted to the compliant motion. A performance criterion can be defined to account for the design specifications. The ARX-model can be modified, although it may not be necessary. The variables in the directions along which the desired forces are specified can be replaced in the ARX-model by the corresponding force variables in the case of a soft contact. Thus, the dependent variables in the ARX-model written in the Cartesian base coordinate system are the variables of those base coordinates for which the desired values are specified. For example, if the generalized external force components F_{ey}^d and M_{ex}^d in the y- and x-directions and the position and orientation components p_x^d, p_z^d, ψ_y^d, ψ_z^d in the directions indicated by the subscripts are given, then the output variables in the ARX-model can be chosen as $[p_x F_{ey} p_z, M_{ex} \psi_y \psi_z]'$. Since these variables are assumed to be available from the measurements, the method of determining the parameters in the assumed ARX-model and the controller gains for the self-tuning controller is directly applicable. Thus, equations 9.3.1–9.3.13 can be used to design a self-tuning LQG-controller. Alternatively, a self-tuning controller can be constructed by the pole-zero placement method, as discussed in Section 9.1.3.

An example will be presented to illustrate the design of a self-tuning position velocity force (PVF) controller for manipulators.

EXAMPLE 10.2.1

The end-effector of the manipulator shown in Figure 10.7 holds a tool as the extension of the last link. At the end of the tool is a rotating metal brush for deburring. The position of the brush is initially at point P_a (−8.8 cm, 30.4 cm). It is moved in 0.5 s along a straight line to point P_b (−2.61 cm, 50.0 cm), where the spring-like brush touches the horizontal constraint surface with zero contact force. While moving from point P_b to point P_c (56.25 cm, 60.0 cm), the end-effector is to exert a force 10 N in the y-direction on the constraint surface via the brush. The contact between the metal brush and the constraint surface can be modeled as a linear spring with coefficient $K_c^t = 100$ N/m. The velocity is assumed to be constant during the motion. The duration of the motion is 5 s.

The problem is to construct a self-tuning LQG-controller for the free and constraint motion so that the end-effector (and the brush) track the specified trajectory connecting points P_a–P_b–P_c.

The dynamical model for the manipulator motion is given by Equations 10.1.35 and 10.1.36. The position of the brush is calculated by Equation 10.1.37, where q_2 is measured from the brush to the axis of the offset link. The input to the system is $u(t) = [F_1(t) \ F_2(t)]'$ and the output is the position vector $[p_x p_y]'$ during the unconstrained motion and the position force vector $[p_x F_{ey}]'$ during the constrained motion.

To design a self-tuning controller, a velocity model of the ARX-type is chosen to describe the input–output relation of the system

$$v(k) = a_0 + A_1 v(k-1) + B_0 u(k-1) + e(k) \qquad (10.2.1)$$

where kT, $k = 1, 2, 3, \ldots$ is the sampling instant, the output at time kT is the translational velocity $v(k) = [v_x(k) \ v_y(k)]'$, the input at time kT is $u(k)$, and the modeling error $e(k)$ represents a white Gaussian noise process. The parameters are chosen as follows:

$$a_0 = \begin{bmatrix} a_0^1 \\ a_0^2 \end{bmatrix} \quad A_1 = \begin{bmatrix} a_{11} & 0 \\ 0 & a_{22} \end{bmatrix} \quad B_0 = \begin{bmatrix} b_{11} & b_{12} \\ b_{21} & b_{22} \end{bmatrix} \quad (10.2.2)$$

The parameters in Equation 10.2.1 are calculated recursively by equations 9.1.16 and 9.1.17 on the basis of the available observations.

A self-tuning controller is then designed on the basis of two criteria: $I_k^1[u]$ for the unconstrained motion and $I_k^2[u]$ for the constrained motion. The former criterion is defined during the unconstrained motion as

$$I_k^1[u] = E\{\|v(k+1) - v^d(k+1) + \rho \Delta p(k+1)\|_Q^2 + \|u(k)\|_R^2 \mid \Sigma(k)\} \tag{10.2.3}$$

where $\Delta p(k+1) = p(k) + Tv(k) - p^d(k+1)$ and $v(k+1) - v^d(k+1)$ are the position and velocity errors, respectively, and the desired position $p^d(k+1) = [p_x^d(k+1)p_y^d(k+1)]'$ and the desired velocity $v^d(k+1) = [v_x^d(k+1)v_y^d(k+1)]'$. Criterion $I_k^2[u]$ is defined during the constrained motion as

$$I_k^2[u] = E\{\|v(k+1) - v^d(k+1) + \rho \begin{bmatrix} \Delta p_x(k+1) \\ \Delta F_{ey}(k+1) \end{bmatrix}\|_Q^2 + \|u(k)\|_R^2 \mid \Sigma(k)\} \tag{10.2.4}$$

where the force error $\Delta F_{ey}(k+1) = F_{ey}(k+1) - F_{ey}^d(k+1) = K_c^t[p_y(k+1) - p_y^d(k+1)]$. The symmetric weighting matrices ρ and Q in Equations 10.2.3 and 10.2.4 determine the relative importance of the position, velocity, and force errors in the task being performed. R, ρ and Q are chosen as positive semidefinite matrices. $E\{\cdot \mid \Sigma(k)\}$ refers to the expectation operation conditioned on the information Σ available at time kT.

The problem is to minimize the criterion in Equation 10.2.3 during the unconstrained motion and the criterion in Equation 10.2.4 during the constrained motion while satisfying the equation that results after the estimated values are substituted for the unknown parameters.

The minimization of the defined performance functional in Equations 10.2.3 and 10.2.4 determines the self-tuning controller

$$u(k) = -[R + \hat{B}_0'Q\hat{B}_0]^{-1}\hat{B}_0'Q[\hat{a}_0 + \hat{A}_1v(k) - v^d(k+1) + \rho \Delta y(k+1)] \quad (10.2.5)$$

where $\Delta y(k+1) = [\Delta p_x(k+1)\Delta p_y(k+1)]'$ during the unconstrained motion and $\Delta y(k+1) = [\Delta p_x(k+1)\Delta F_{ey}(k+1)]'$ during the constrained motion. The caret refers to the estimated parameter values. Equation 10.2.5 specifies the controller in a feedback form.

The system was simulated using the model given by Equations 10.1.34–10.1.37 with the numerical values specified in Example 10.1.3. The weighting matrices in Equations 10.2.3 and 10.2.4 are chosen so that during the unconstrained motion

$$R = \begin{bmatrix} 0.6 & 0 \\ 0 & 0.6 \end{bmatrix} \quad Q = \begin{bmatrix} 3.10^3 & 0 \\ 0 & 3.10^3 \end{bmatrix} \quad \rho = \begin{bmatrix} 100 & 0 \\ 0 & 100 \end{bmatrix} \quad (10.2.6)$$

and during the constrained motion

$$R = \begin{bmatrix} 10^{-3} & 0 \\ 0 & 10^{-3} \end{bmatrix} \quad Q = \begin{bmatrix} 6 & 0 \\ 0 & 10^5 \end{bmatrix} \quad \rho = \begin{bmatrix} 100 & 0 \\ 0 & 100 \end{bmatrix} \quad (10.2.7)$$

The initial estimates of the parameters in the estimation Equations 9.1.16 and 9.1.17 are $\hat{\alpha}(0) = 0$, $P(0) = 10I$ and $\gamma = 0.95$.

Digital simulations were performed on the manipulator system with the self-tuning controller. Figures 10.9a–10.9c show the graphs for p_x, p_y and F_{ey} versus time. The graph for force F_{ey} indicates that after the contact of the brush with surface, the external force quickly increases to 10 N, exhibits overshoot, and then settles down quickly. The tracking of the desired values in Figures 10.9a–10.9c is very good. The graphs for the errors of the foregoing variables from their desired values are displayed in Figures 10.10a–10.10c. The trajectories recorded are typical for a system driven by a self-tuning controller: after an initial learning period, a good tracking of the desired values is achieved. It should be mentioned that the approach described is applicable also to the case in which the end-effector is stationary in the x-direction while exerting a force F_{ey}^d in the y-direction.

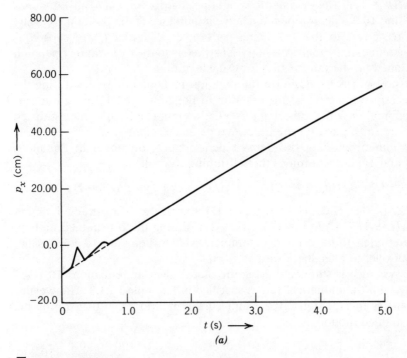

FIGURE 10.9a
The x-coordinate of end-effector position p_x versus time. Example 10.2.1.

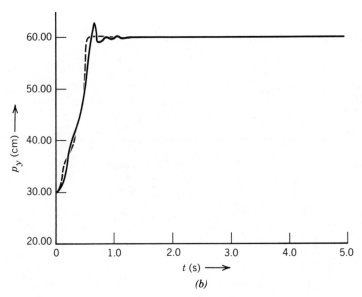

FIGURE 10.9*b*
The *y*-coordinate of end-effector position p_y versus time.
Example 10.2.1.

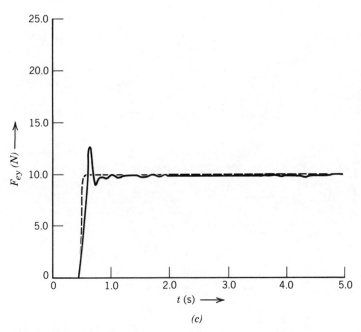

FIGURE 10.9*c*
Force F_{ey} exerted by end-effector versus time. Example 10.2.1.

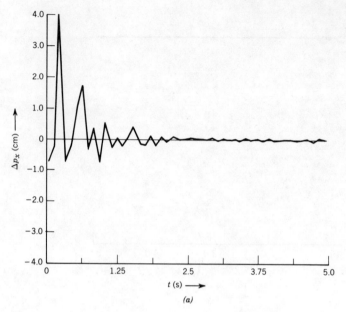

FIGURE 10.10a
Position error Δp_x in p_x versus time. Example 10.2.1.

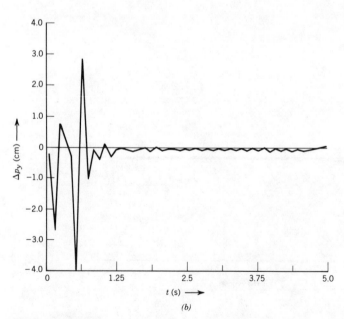

FIGURE 10.10b
Position error Δp_y in p_y versus time. Example 10.2.1.

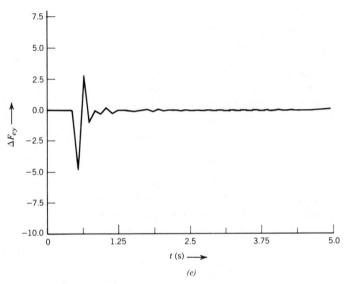

FIGURE 10.10c
Force error ΔF_{ey} in F_{ey} versus time. Example 10.2.1.

10.3 HARD ENVIRONMENTAL CONTACT FORCES EXERTED BY A MANIPULATOR

In the previous sections, the contact between an end-effector and the environment was assumed to be soft. Such an interface can be modeled by means of various combinations of a spring, a damper, and a mass (inertia). However, an environmental surface is often hard, for example, a metal surface. At the contact point, the end-effector and the environmental surface will maintain their shapes regardless of the value of the force exerted. The contact surface represents a hard constraint. It can be described by a strict mathematical relation (equality during the contact), which must be satisfied. This relation limits the changes in some variables of the system as long as the end-effector maintains the contact.

Suppose that the generalized coordinates and torques of a nonconservative system are described in the joint space by the N-vectors $q = [q_1 \cdots q_N]'$ and $F = [\tau_1 \cdots \tau_N]'$, respectively. Using the expressions for the kinetic and potential energies, the system equations of an unconstrained motion can be written in the following scalar-valued form (compare with Equation 10.1.5):

$$\delta q'[D(q)\ddot{q} + C(q,\dot{q}) + G(q) - F] = 0 \qquad (10.3.1)$$

where δq signifies a weak variation of q, and $C(q,\dot{q})$ and $G(q)$ represent the Coriolis, centripetal torque and the gravitational torque, respectively.

The generalized coordinates are assumed to be constrained through the set of the following k relations expressed in the joint space:

$$H(q)\dot{q} = 0 \qquad (10.3.2)$$

where the $(k \times N)$-matrix H is assumed to have full rank k $(k < N)$. If Equation 10.3.2 can be integrated so that a constraint equation for the generalized coordinates is obtained, then the system constraint 10.3.2 is called *holonomic*; otherwise, the constraint is nonholonomic. Equation 10.3.2 can be used to obtain k equations for the virtual displacement δq [10]:

$$H\delta q = 0 \qquad (10.3.3)$$

The number of the degrees of freedom (DOF) in the system described by Equations 10.3.1 and 10.3.2 is $m = N - k$. In forming Lagrange's equations on the basis of the energy functions, the generalized coordinates must be independent. In case the motion of a system is restrained by k holonomic constraints, then k dependent variables should be eliminated in Lagrange's energy function before determining the equations of motion.

Instead of solving the constraint equations for the dependent generalized variables and eliminating them explicitly from the energy expressions, an alternative approach is to use Lagrange's multipliers with the constraint equations [9]. By introducing the k-dimensional Lagrange multiplier $\lambda = [\lambda_1 \ldots \lambda_k]'$, the equality constraint 10.3.3 can be combined with expression 10.3.1 to obtain

$$\delta q'[D(q)\ddot{q} + C(q,\dot{q}) + G(q) - F - H'\lambda] = 0 \qquad (10.3.4)$$

Since k components of the N-dimensional vector q are dependent, it follows that the corresponding components of the variational vector δq are also dependent. In order to be more specific, let the first m components $q_1, \ldots, q_m$ of the N-dimensional vector be independent, and the last components $q_{m+1}, \ldots, q_{m+k}$ dependent in Equation 10.3.4 (note: $m + k = N$). Thus, Equation 10.3.4 contains m independent variables $q_1, \ldots, q_m$, k unknown variables $\lambda_1, \ldots, \lambda_k$ to be determined, and k dependent variables $q_{m+1}, \ldots, q_N$ specified by Equation 10.3.3.

Equation 10.3.4 gives the dynamical model for the manipulator motion during hard contact

$$D(q)\ddot{q} + C(q,\dot{q}) + G(q) = F + H'\lambda \qquad (10.3.5)$$

Equation 10.3.5 contains the m independent joint variables, the k Lagrange multipliers, and the dependent joint variables determined by the constraint Equation 10.3.3. The term $H'\lambda$ in Equation 10.3.5 can be considered as the generalized torque of the constrained system expressed in the joint space.

In order to solve explicitly for λ, it is convenient to differentiate Equation 10.3.2 with respect to time:

$$\dot{H}(q)\dot{q} + H(q)\ddot{q} = 0 \qquad (10.3.6)$$

Equation 10.3.5 is next combined with Equation 10.3.6. The joint acceleration $\ddot{q}$ is solved from Equation 10.3.5 and substituted into Equation 10.3.6.

The resulting equation can be solved for the k-dimensional Lagrange's multiplier λ:

$$\lambda = [H(q)D^{-1}(q)H'(q)]^{-1}\{H(q)D^{-1}(q)[C(q,\dot{q}) + G(q) - F] - \dot{H}(q)\dot{q}\} \quad (10.3.7)$$

Equation 10.3.7 indicates that the constraint torque $H'(q)\lambda$ depends on the joint position q, velocity $\dot{q}$, and the input torque F of the manipulator. When Lagrange's multiplier λ from Equation 10.3.7 is substituted into Equation 10.3.5, the resulting expression describes the equations of motion when the end-effector is in contact with a hard surface.

The constraint Equation 10.3.2 is given in the joint space. Suppose now that the environmental constraint on the motion of an end-effector (or a tool) is specified in the Cartesian base coordinate system as follows:

$$\overline{C}\begin{bmatrix} \dot{p}_e \\ \omega_e \end{bmatrix} = 0 \quad (10.3.8)$$

where $\dot{p}_e$ and ω_e represent the translational and rotational speeds of the end-effector, respectively, expressed in the based coordinate system. The constraint matrix $\overline{C}$ is of dimension $(k \times 6)$ and of full rank $k(k < 6)$. Its entries can depend on the position and orientation of the end-effector.

The velocities in the Cartesian base coordinate system and in the joint space are related through the Jacobian matrix $J(q)$, that is

$$\begin{bmatrix} \dot{p}_e \\ \omega_e \end{bmatrix} = J(q)\dot{q} \quad (10.3.9)$$

where the Jacobian matrix is of dimension $(6 \times N)$. The constraint Equation 10.3.8 can be expressed in the joint space as

$$\overline{C}J(q)\dot{q} = 0 \quad (10.3.10)$$

Equation 10.3.10 is in the same form as the environmental constraint Equation 10.3.2; thus

$$H(q) = \overline{C}J(q) \quad (10.3.11)$$

The rank of matrix H is assumed to be k, that is, the same as that of $\overline{C}$.

To determine the generalized constraint force in the Cartesian base coordinate system, $H'\lambda$ is first determined. The solution of λ is given by Equation 10.3.7, for which $\dot{H}\dot{q}$ is calculated from the following expression:

$$\dot{H}(q)\dot{q} = [\dot{\overline{C}}J(q) + \overline{C}\dot{J}(q)]\dot{q} \quad (10.3.12)$$

Thus, Equation 10.3.7 determines Lagrange's multiplier vector for this case. The generalized torque of the constraint during the contact is given by $H'\lambda$ in the joint space, that is

$$H'\lambda = J'(\overline{C}'\lambda) \quad (10.3.13)$$

By Equation 10.3.13, the generalized constraint force in the Cartesian base coordinate system is $\overline{C}'\lambda$, a 6-dimensional vector, when the constraints on the motion are in the form of Equation 10.3.8.

In some applications, the hard constraints are specified in the Cartesian base coordinate system as the translational position (p_e) and rotational velocity (ω_e) constraints. That is

$$C_p(p_e) = 0 \qquad \bar{C}_r \omega_e = 0 \tag{10.3.14}$$

where C_p is a differentiable vector-valued function of dimension k_p ($1 \leq k_p \leq 3$) and $\bar{C}_r$ a constant matrix of dimension $(k_r \times 3)$ ($1 \leq k_r \leq 3$). The constraints given by Equation 10.3.14 in the Cartesian base coordinate system can be cast into the form of Equation 10.3.8 by differentiating expression $C_p(p_e)$ in Equation 10.3.14 with respect to time. It follows that matrix $\bar{C}$ is now specified as

$$\bar{C} = \begin{bmatrix} \dfrac{dC_p}{dp_e} & 0 \\ 0 & \bar{C}_r \end{bmatrix} \tag{10.3.15}$$

and matrix dC_p/dp_e in Equation 10.3.15 is

$$\frac{dC_p}{dp_e} = \begin{bmatrix} \dfrac{\partial C_{p1}}{\partial p_x} & \dfrac{\partial C_{p1}}{\partial p_y} & \dfrac{\partial C_{p1}}{\partial p_z} \\ \cdot & \cdot & \cdot \\ \dfrac{\partial C_{pk_p}}{\partial p_x} & \dfrac{\partial C_{pk_p}}{\partial p_y} & \dfrac{\partial C_{pk_p}}{\partial p_z} \end{bmatrix} \tag{10.3.16}$$

Each row i in Equation 10.3.16 represents the gradient of the function $C_{pi}, i = 1, \ldots, k_p$ with respect to p_e. In this particular case, term $\dot{\bar{C}}J\dot{q} = \dot{\bar{C}}\dot{p}_e$ for Equation 10.3.12 is evaluated from Equation 10.3.15 to obtain

$$\dot{\bar{C}}J\dot{q} = \left[\dot{p}_e' \frac{d^2 C_{p1}}{dp_e^2} \dot{p}_e \cdot \dot{p}_e' \frac{d^2 C_{pk_p}}{dp_e^2} \dot{p}_e, \; \omega_e' \bar{C}_r' \right]' \tag{10.3.17}$$

where the second-order derivatives signify the Hessian matrices of the particular function.

The generalized constraint torque is determined again by Equation 10.3.13. It can be written in a convenient form if Lagrange's multiplier is decomposed as $\lambda = [\lambda_p' \lambda_r']'$. The generalized constraint force in the Cartesian base coordinate system is given by Equation 10.3.13

$$\bar{C}'\lambda = \begin{bmatrix} F_e^c \\ M_e^c \end{bmatrix} = \begin{bmatrix} \left(\dfrac{dC_p}{dp_e} \right)' \lambda_p \\ \bar{C}_r' \lambda_r \end{bmatrix} \tag{10.3.18}$$

Equation 10.3.18 specifies the generalized contact force exerted by the end-effector on the hard environmental surface, when the constraints are in the form of Equation 10.3.14.

When the constraint is a hard surface in the Cartesian base coordinate system, then $k_p = 1$ and the constraint force in Equation 10.3.18 is normal to the surface, assuming that there is no friction at the point of contact. For the case that the constraint is a curve on the surface, then $k_p = 2$ and the constraint force acts on the hyperplane normal to the surface; there is no constraint force component along the tangent of the curve at the point of contact. For a point constraint, $k_p = 3$ and the constraint force has three components. Similar considerations apply to the generalized constraint torque M_e^c in Equation 10.3.18.

EXAMPLE 10.3.1

When a manipulator is turning a crank shown in Figure 10.11, the rotational axis may be assumed to be parallel to the z-axis of the Cartesian base coordinate system. The end-effector moves along a circular path that is parallel to the xy-plane. The radius of this circle is R and the center at point (y_0, y_0, z_0). The problem is (1) to determine the constraint matrix $\bar{C}$ for this particular task and (2) to obtain $\dot{H}(q)\dot{q}$ in Equation 10.3.12. They are needed to compute Lagrange's multiplier vector given by Equation 10.3.7 and the generalized contact force specified by Equation 10.3.18.

(1) The vector-valued function $C_p(p_e)$ in Equation 10.3.14 is first determined. The constraint motion of the end-effector is described by

$$C_p = \begin{bmatrix} C_{p1} \\ C_{p2} \end{bmatrix} = \begin{bmatrix} p_z - z_0 \\ (p_x - x_0)^2 + (p_y - y_0)^2 - R^2 \end{bmatrix} = 0 \qquad (10.3.19)$$

where the end-effector position is $p_e = [p_x \; p_y \; p_z]'$ in the base coordinate system.

Turning crank

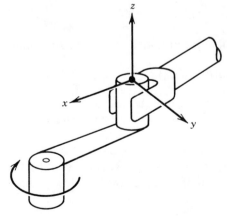

FIGURE 10.11
Crank turned by manipulator. Example 10.3.1.

The rotational constraint is given in the form of $\bar{C}_r \omega_e = 0$. Since the rotation takes place about the z-axis, it follows that $\omega_{ex} = 0$ and $\omega_{ey} = 0$.

The constraint equation can now be expressed in the form of Equation 10.3.8:

$$\bar{C} \begin{bmatrix} \dot{p}_e \\ \omega_e \end{bmatrix} = \begin{bmatrix} \dfrac{dC_p}{dp_e} & 0 \\ 0 & \bar{C}_r \end{bmatrix} \begin{bmatrix} \dot{p}_e \\ \omega_e \end{bmatrix} = 0 \qquad (10.3.20)$$

The matrices are then determined:

$$\frac{dC_p}{dp_e} = \begin{bmatrix} \dfrac{dC_{p1}}{dp_e} \\ \dfrac{dC_{p2}}{dp_e} \end{bmatrix} = \begin{bmatrix} \dfrac{\partial C_{p1}}{\partial p_x} & \dfrac{\partial C_{p1}}{\partial p_y} & \dfrac{\partial C_{p1}}{\partial p_z} \\ \dfrac{\partial C_{p2}}{\partial p_x} & \dfrac{\partial C_{p2}}{\partial p_y} & \dfrac{\partial C_{p2}}{\partial p_z} \end{bmatrix} \qquad (10.3.21)$$

$$\bar{C} = \begin{bmatrix} \dfrac{dC_p}{dp_e} & 0 \\ 0 & \bar{C}_r \end{bmatrix} = \begin{bmatrix} 0 & 0 & 1 & 0 & 0 & 0 \\ 2(p_x - x_0) & 2(p_y - y_0) & 0 & 0 & 0 & 0 \\ 0 & 0 & 0 & 1 & 0 & 0 \\ 0 & 0 & 0 & 0 & 1 & 0 \end{bmatrix} \qquad (10.3.22)$$

Equation 10.3.22 gives the constraint matrix $\bar{C}$ for Equation 10.3.8 that specifies the constraint for the described task. Moreover, the constraint matrix in Equation 10.3.2 is $H(q) = \bar{C}J$ by Equation 10.3.11, where J is the Jacobian matrix of the manipulator.

(2) The expression for $\bar{C}J\dot{q}$ is specified by Equation 10.3.17. The second derivatives in this particular case are directly evaluated to give: $\partial^2 C_{p2}/\partial p_x^2 = 2$, $\partial^2 C_{p2}/\partial p_y^2 = 2$ and the other entries are zero. Hence,

$$\dot{\bar{C}}J(q)\dot{q} = \begin{bmatrix} \dot{p}_e'[0]\dot{p}_e & \dot{p}_e' \begin{bmatrix} 2 & 0 & 0 \\ 0 & 2 & 0 \\ 0 & 0 & 0 \end{bmatrix} \dot{p}_e & 0 & 0 \end{bmatrix}'$$

$$= \begin{bmatrix} 0 & 2\dot{p}_x^2 + 2\dot{p}_y^2 & 0 & 0 \end{bmatrix}' \qquad (10.3.23)$$

Expression 10.3.23 is used to evaluate $\dot{H}\dot{q}$ in Equation 10.3.12 to obtain

$$\dot{H}(q)\dot{q} = \dot{\bar{C}}J(q)\dot{q} + \bar{C}\dot{J}(q)\dot{q} \qquad (10.3.24)$$

Lagrange's multiplier in Equation 10.3.7 can now be computed. Thus, the generalized contact force $\bar{C}'\lambda$ in the Cartesian base coordinate system can be determined.

The crank is considered here as restraining the motion of the manipulator. If the dynamics of the crank were included, the system would form a closed chain and should be studied in the framework of closed chain dynamics.

10.4 INDIRECT PVF CONTROL USING PSEUDOVELOCITIES AND A REDUCED-ORDER MODEL

The motion of a serial link manipulator with N joints has N degrees of freedom (DOF). When the system motion is restricted by k constraints $(0 \le k < N)$

$$H(q)\dot{q} = 0 \qquad (10.4.1)$$

where q and $\dot{q}$ are the joint position and velocity, respectively, N DOF will be reduced to $m = N - k$. Suppose that the components of the joint position vector q are decomposed and rearranged into two groups: m independent components $q_1, \ldots, q_m$ and k dependent components $q_{m+1}, \ldots, q_N$. In principle, one can obtain explicit expressions for k dependent variables, and then eliminate them in the dynamical model given by Equation 10.1.5. Thus, a dynamical model with m independent variables results. However, the determination of the explicit expressions for the dependent variables is often tedious, if not impossible.

An attractive alternative is to introduce new variables, so-called pseudovelocities $v_1(k), \ldots, v_m(t)$ defined by

$$v = P(q)\dot{q} \qquad (10.4.2)$$

where $v = [v_1, \ldots, v_m]'$ and the dimensionality of matrix P is $(m \times N)$. Matrix P is so selected that matrix $[H'P']'$ has full rank N. The pseudovelocity components may not represent any physical variables in the system. Although matrix $P(q)$ is somewhat arbitrary, it should be so selected that the generalized coordinate q can be determined from Equation 10.4.2 by integration. A possible choice is to choose v as a subset of the components of the $\dot{q}$-vector. In this case P is a constant.

The differentiation of Equation 10.4.2 with respect to time gives for pseudoacceleration

$$\dot{v} = P(q)\ddot{q} + \dot{P}(q)\dot{q} \qquad (10.4.3)$$

The m-dimensional pseudoacceleration $\dot{v}$ in Equation 10.4.3 is in terms of the joint variables and their derivatives. The joint variables are constrained by Equation 10.4.1 and Equation 10.3.6, which is obtained from Equation 10.4.1 by differentiation.

Equations 10.4.2 and 10.4.3 relate the pseudovelocity v and pseudoacceleration $\dot{v}$ to the joint position, velocity, and acceleration. These equations can be solved for the joint velocity $\dot{q}$ and acceleration $\ddot{q}$. Indeed, Equations 10.4.1–10.4.3 can be combined to obtain

$$\begin{bmatrix} H \\ P \end{bmatrix} \dot{q} = \begin{bmatrix} 0 \\ v \end{bmatrix} \qquad (10.4.4)$$

$$\begin{bmatrix} H \\ P \end{bmatrix} \ddot{q} + \begin{bmatrix} \dot{H} \\ \dot{P} \end{bmatrix} \dot{q} = \begin{bmatrix} 0 \\ \dot{v} \end{bmatrix} \qquad (10.4.5)$$

where the arguments have been dropped in the notations, H is a $(k \times N)$ matrix and P a $(m \times N)$ matrix. To solve for $\dot{q}$ and $\ddot{q}$, it is convenient to define

$$\begin{bmatrix} H \\ P \end{bmatrix}^{-1} = [\pi_1 \ \pi_2] \qquad (10.4.6)$$

where π_1 is a $(N \times k)$ matrix and π_2 a $(N \times m)$ matrix. Equation 10.4.6 implies that $H\pi_1 = I, H\pi_2 = 0, P\pi_2 = I$ and $P\pi_1 = 0$.

Expressions for $\dot{q}$ and $\ddot{q}$ can be solved from Equations 10.4.4 and 10.4.5

$$\dot{q} = \pi_2 \nu \qquad (10.4.7)$$

$$\ddot{q} = \pi_2 \dot{\nu} - (\pi_1 \dot{H} + \pi_2 \dot{P}) \pi_2 \nu \qquad (10.4.8)$$

The joint velocity $\dot{q}$ and acceleration $\ddot{q}$ are related to the pseudovelocity ν and pseudoacceleration $\dot{\nu}$ by Equations 10.4.7 and 10.4.8.

A reduced-order model in terms of the pseudovelocity and pseudoacceleration is next derived. The dynamical model for the end-effector of a manipulator touching an environmental hard surface is given in Equation 10.3.5. If both sides of Equation 10.3.5 are premultiplied $[\pi_1 \pi_2]'$, the resulting equation can be separated into two equations by utilizing the properties of Equation 10.4.6:

$$\lambda = \pi_1'(q) [D(q)\ddot{q} + C(q,\dot{q}) + G(q) - F] \qquad 10.4.9)$$

$$\pi_2'(q)D(q)\ddot{q} = \pi_2'(q)[F - C(q,\dot{q}) - G(q)] \qquad (10.4.10)$$

Equation 10.4.10 describes m dynamical equations. Equation 10.4.9 determines the k-dimensional λ-vector. Alternatively, the λ-vector can be calculated by Equation 10.3.7.

A reduced-order model is obtained by substituting expressions 10.4.7 and 10.4.8 for $\dot{q}$ and $\ddot{q}$ in Equation 10.3.5

$$D[\pi_2\dot{\nu} - (\pi_1\dot{H} + \pi_2\dot{P})\pi_2\nu] + C + G = F + H'\lambda \qquad (10.4.11)$$

where the arguments have been dropped in the notations.

Both sides of Equation 10.4.11 are next premultiplied by π_2'. It leads to the system of m equations $(m \leq N)$:

$$\pi_2'D\pi_2\dot{\nu} = \pi_2'[F - C - G + D(\pi_1\dot{H} + \pi_2\dot{P})\pi_2\nu] \qquad (10.4.12)$$

In Equation 10.4.12, the rank of the $(m \times N)$-matrix π_2' in m, and the inertia matrix D is positive definite; therefore, the square matrix $\pi_2'D\pi_2$ has rank m and is invertible. Equation 10.4.12 can be rewritten as follows:

$$\dot{\nu} = (\pi_2'D\pi_2)^{-1}\pi_2'[F - C - G + D(\pi_1\dot{H} + \pi_2\dot{P})\pi_2\nu] \qquad (10.4.13)$$

The pseudoacceleration vector $\dot{\nu}$ is governed by Equation 10.4.13. In order to use Equation 10.4.13, submatrices π_1 and π_2 specified in Equation 10.4.6 must be known. Equation 10.4.13 can be used to solve for pseudoacceleration $\dot{v}$ and pseudovelocity v. After they have been determined for the constrained motion, the joint acceleration $\ddot{q}$ and velocity $\dot{q}$ can be computed from Equations 10.4.7 and 10.4.8, respectively. By integrating Equation 10.4.7, the joint position $q(t)$ can also be determined.

Equation 10.4.13 represents an m-dimensional model for an N-joint manipulator whose motion is limited by k constraints $(k + m = N)$. This reduced-order model is described in terms of the pseudovelocity and pseudoacceleration vectors. They are related to the joint variables by Equations 10.4.7 and 10.4.8. The generalized constraint torque $H'(q)\lambda$ is determined by Equations 10.4.1 and 10.4.9.

When the desired trajectory is specified in terms of the joint variables and their derivatives, the corresponding desired trajectory for the variables in the reduced-order model is calculated by means of Equations 10.4.2 and 10.4.3. A controller can then be designed on the basis of Equations 10.4.13 for tracking the pseudovelocity and Equations 10.4.9 and 10.4.1 for controlling the constraint torque $H'\lambda$. The controller $(F = u)$ can be composed of a primary controller (u^d) that is designed to function under ideal operating conditions and a secondary controller $(\hat{u})$ that compensates for modeling errors and disturbance effects.

The structure for the primary controller $u^d(t) = F^d(t)$ can be chosen on the basis of Equation 10.4.13 as

$$u^d(t) = \hat{C} + \hat{G} - \hat{D}(\hat{\pi}_1\dot{H} + \hat{\pi}_2\dot{P})\hat{\pi}_2 v \qquad (10.4.14)$$

where the caret refers to the value of the function evaluated along the desired trajectory.

The composite control $u(t) = F(t)$ is determined by the outputs of the primary and secondary controllers: $F(t) = u^d(t) + \hat{u}(t)$, where the secondary controller $\hat{u}(t)$ is still to be determined. The composite control $F(t)$ is substituted into Equation 10.4.13. The equalities of the nonlinear terms in the resulting equation are assumed, that is, in addition to the relations given in Equation 10.1.49 for the pseudo-inertia and Jacobian matrices, the gravity and Coriolis terms, the following relations are assumed to hold for $i = 1, 2$:

$$\pi_i = \hat{\pi}_i \quad \dot{H} = \dot{\hat{H}} \quad \dot{P} = \dot{\hat{P}} \qquad (10.4.15)$$

Then the reduced-order model in Equation 10.4.13 can be expressed as follows:

$$\dot{v} = (\pi_2'D\pi_2)^{-1}\pi_2'\hat{u}(t) \qquad (10.4.16)$$

The secondary control $\hat{u}(t)$ in Equation 10.4.16 is next chosen so that it is composed of two components: one of which will control the pseudovelocity $v(t)$ and the other will control the constraint force by controlling the Lagrange multiplier. Thus, it may be chosen as

$$\hat{u}(t) = H'\hat{u}_1(t) + D\pi_2\hat{u}_2(t) \qquad (10.4.17)$$

where the k-dimensional vector $\hat{u}_1(t)$ and the m-dimensional vector $\hat{u}_2(t)$ are control inputs to be determined. The gain matrices in Equation 10.4.17 are such that the pseudovelocity $\nu(t)$ and the generalized constraint torque $H'\lambda$ can be controlled independently by means of $\hat{u}_1(t)$ and $\hat{u}_2(t)$, respectively. This will be shown next by determining the dynamical model for $\nu(t)$ and $\lambda(t)$ when the composite control $u(t) = F(t)$ is applied to the manipulator system.

The substitution of Equation 10.4.17 into Equation 10.4.16 leads to

$$\dot{\nu} = \hat{u}_2(t) \tag{10.4.18}$$

where the properties of Equation 10.4.6 have been utilized. Equation 10.4.18 describes the dynamics of the pseudoacceleration as a function of control $\hat{u}_2(t)$. This control is so chosen that the pseudovelocity tracks the specified trajectory.

The control of the generalized constraint torque $H'\lambda$ can be determined by first combining Equations 10.4.9 and 10.4.8 to obtain

$$\lambda = \pi_1' D \pi_2 \dot{\nu} - \pi_1' D (\pi_1 \dot{H} + \pi_2 \dot{P}) \pi_2 \nu + \pi_1'(C + G - F) \tag{10.4.19}$$

The composite control $F = u^d + \hat{u}$ determined by Equations 10.4.14 and 10.4.17 is next substituted into Equation 10.4.19. The assumption on the exact cancellation of the nonlinear terms by Equations 10.1.49 and 10.4.15 leads to

$$\lambda = \pi_1' D \pi_2 \dot{\nu} - \pi_1' \hat{u}$$
$$= -\hat{u}_1 \tag{10.4.20}$$

Equation 10.4.20 shows that Lagrange's multiplier λ and thus the constraint generalized torque $H'\lambda$ can be controlled independently by means of $\hat{u}_1(t)$.

Equations 10.4.18 and 10.4.20 indicate that there is no interaction between the m-dimensional pseudoacceleration $\dot{\nu}$ and the k-dimensional constraint torque $H'\lambda$. Thus, the control law given by Equations 10.4.14 and 10.4.17 decouples the pseudoacceleration and the constraint generalized torque. The control variables $\hat{u}_1$ and $\hat{u}_2$ can be determined for the desired transient behavior of the system.

EXAMPLE 10.4.1

The motion of a tool at the end-effector of a six-joint manipulator is constrained by the following velocity relation:

$$S_1 \begin{bmatrix} \dot{p}_t \\ \omega_t \end{bmatrix} = S_1 \dot{W}_t = 0 \tag{10.4.21}$$

where $\dot{W}_t = [(\dot{p}_t)'(\omega_t)']'$ and the velocity components that are zero determine the elements of the $(k \times 6)$ selection matrix S_1, which is thus composed of the zeros and ones. The three-dimensional vectors $\dot{p}_t$ and ω_t represent the translational and rotational velocity vectors of the tool, respectively, in the tool coordinate frame.

The problem is (1) to determine a reduced-order model for the system in terms of pseudoacceleration and pseudovelocity, (2) to obtain an expression for the constraint force, and (3) to determine a control strategy for the system.

(1) The constraint Equation 10.4.21 must first be expressed in the Cartesian base coordinate system in the form of Equation 10.3.8. This is accomplished by means of the rotation submatrix $(A_0^6)_R$ of the homogeneous transformation matrix A_0^6 that transforms a vector in the tool coordinate frame to a vector in the base coordinate system. Thus, the application of the rotation matrix $(A_0^6)_R$ to vector $\dot{p}_t$ rotates it to vector $\dot{p}_e$ which has the components on the base coordinate system. Similarly, W_t is related to W_e. Thus, Equation 10.4.21 becomes

$$S_1 \begin{bmatrix} (A_0^6)_R & O_{3\times3} \\ O_{3\times3} & (A_0^6)_R \end{bmatrix}^{-1} \dot{W}_e = S_1 (A_0^t)_R^{-1} \dot{W}_e = 0 \qquad (10.4.22)$$

Comparing Equations 10.4.22 and 10.3.8 gives $\bar{C} = S_1(A_t^0)_R$. Equation 10.4.22 can be expressed in the joint space using the Jacobian matrix $J = J(q)$

$$S_1(A_t^0)_R J \dot{q} = 0 \qquad (10.4.23)$$

Equation 10.4.23 containing k constraints is now in the form of Equation 10.3.10. It is convenient to write $J_c = (A_t^0)_R J = (A_0^t)_R' J$. Then, $H = S_1 J_c = S_1(A_0^t)_R' J$ is the constraint matrix in Equation 10.3.2. It is composed of the rows of matrix J_c that correspond to the constraint directions of the motion.

To define the pseudovelocity vector v in Equation 10.4.2, the P-matrix is to be specified. This can be accomplished by introducing a $(6-k)\times6$ selection matrix denoted as S_2. It has the rows of the identity matrix that are not included in the S_1-matrix in Equation 10.4.21. Since Equation 10.4.23 specifies the H-matrix constraint of Equation 10.3.2, the m-dimensional pseudovelocity vector in Equation 10.4.2 is chosen as follows:

$$v = S_2 J_c \dot{q} \qquad (10.4.24)$$

The comparison of Equations 10.4.24 and 10.4.2 gives

$$P = S_2 J_c \qquad (10.4.25)$$

The value of the P-matrix in Equation 10.4.25 varies with the joint position vector. Equation 10.4.24 defines the pseudovelocity vector $v = [v_1, \ldots, v_{6-k}]'$, where $6 - k = m$. The components of the pseudovelocity v represent the components of $\dot{W}_t$ that are nonzero.

Before Equation 10.4.12 written for the case under consideration, matrices π_1 and π_2 are identified in Equation 10.4.6. Indeed,

$$\begin{bmatrix} H \\ P \end{bmatrix}^{-1} = \begin{bmatrix} S_1 J_c \\ S_2 J_c \end{bmatrix}^{-1} = [\pi_1 \ \pi_2] \qquad (10.4.26)$$

By the definition of the inverse, Equation 10.4.26 implies that

$$S_1 J_c \pi_1 = I \qquad (10.4.27)$$

$$S_2 J_c \pi_2 = I \qquad (10.4.28)$$

Equations 10.4.27 and 10.4.28 can be solved for matrices π_1 and π_2 to obtain $\pi_1 = J_R^{-1} S_1'$ and $\pi_2 = J_R^{-1} S_2'$, since $S_1 S_1' = I$ and $S_2 S_2' = I$ in this particular example.

The reduced-order model for the pseudoacceleration can now be written from Equation 10.4.12. By assuming that the inverse of J_c exists and by denoting $(J_c^{-1})'DJ_c^{-1} = D_c$, Equation 10.4.12 can be expressed as

$$(S_2 D_c S_2') \dot{\nu} = S_2 (J_c^{-1})'[F - C - G + D(J_c^{-1}\dot{J}_c)J_c^{-1}S_2'\nu]$$

$$= S_2 D_c \dot{J}_c J_c^{-1} S_2'\nu + S_2 (J_c^{-1})'(F - C - G) \tag{10.4.29}$$

Equation 10.4.29 represents the reduced-order model governing the dynamics of the pseudoacceleration and -velocity. The comparison of Equation 10.4.29 with the manipulator model in Equation 10.1.45 reveals that the solution matrix S_2 isolates from Equation 10.1.45 the equations that correspond to the unconstrained directions. Thus, the structure of the model for the pseudoaccelerations become specified.

(2) The generalized constraint force in the Cartesian coordinate system is equal to $\bar{C}'\lambda$ determined by Equation 10.3.18. Since $\bar{C}' = (A_t^0)'_R S_1'$ by Equation 10.4.22, the constraint force $[F_t' M_t']'$ in the tool frame can be expressed as

$$\begin{bmatrix} F_t \\ M_t \end{bmatrix} = (A_t^0)_R \bar{C}'\lambda$$

$$= S_1'\lambda \tag{10.4.30}$$

Thus, the components of the generalized constraint force in the directions of the constrained coordinates are determined by the Lagrange multipliers.

(3) The architecture for the primary (u^d) and secondary $(\hat{u})$ controllers is determined by Equations 10.4.14 and 10.4.17. In this particular case, $\pi_1 \dot{H} + \pi_2 \dot{P} = J_c^{-1}\dot{J}_c$. Thus, the expressions for the foregoing controllers are

$$u^d(t) = C + G - DJ_c^{-1}\dot{J}_c J_c^{-1} S_2'\nu \tag{10.4.31}$$

$$\hat{u}(t) = J_c' S_1'\hat{u}_1 + DJ_c^{-1} S_2'\hat{u}_2 \tag{10.4.32}$$

where the equalities given in Equations 10.1.49 and 10.1.15 for the nonlinear terms have been assumed to hold.

The composite control input to the manipulator is $F(t) = u^d(t) + \hat{u}(t)$. The control variables $\hat{u}_1(t)$ and $\hat{u}_2(t)$ can be determined so that the desired system response is obtained.

An alternative approach to design a controller for the system in Equation 10.4.13 is to determine the primary controller so that the dynamics for the error $\Delta\nu(t)$ in the pseudovelocity and the control input $\hat{u}(t)$ are obtained. The secondary controller given by Equation 10.4.17 can still be used. The control components $\hat{u}_1$ and $\hat{u}_2$ can be chosen, for example, as PID-controllers (see Problem 10.12).

10.5 DISCUSSION

The movement of a manipulator is usually partly in a free space and partly in a limited space. When a manipulator comes into contact with the environment the impact is assumed in this presentation to be negligible, for example, due to a spring-like contact or reduced speed. During the contact between the end-effector and environment, the motion occurs in the restricted space. The motion of the manipulator has to comply with the constraint imposed by the surface. It is usually assumed that the end-effector (or a tool) maintains contact with the environment during the constrained motion. Moreover, the end-effector is assumed to exert an external generalized force on the environment during the motion in the directions that are constrained.

The contact between the end-effector and the environment may be considered as soft or hard contact. A soft contact can be modeled by means of a spring, damper, moment of inertia and/or mass. The contact force model is augmented to the dynamical model of the manipulator. A movement in the constrained direction determines the generalized force exerted by the end-effector (or a tool) on the environmental object. The control of the generalized force in the case of soft contact thus becomes a problem of servoing accurately on the generalized position variables or their derivatives in the directions that are constrained by the surface of contact. Thus, the entire system can be servoed on the position. An alternative approach is to design a controller so that the manipulator system that includes the generalized contact force will make its linearized model behave like the contact force model [12].

A hard contact, on the other hand, is described by (holonomic) constraint equations during the time when the end-effector is in contact with the environmental object. The system must comply with these hard constraints during the contact motion. Because of the hard constraint equations, the DOF and thus the number of independent variables in the system are reduced. A reduced-order model in terms of pseudovelocities can be determined to describe the constrained motion of the system. The selection of the new variables poses several challenging problems. A control architecture can be designed so that the control of the pseudovelocities is decoupled from the control of the generalized constraint force.

The stability of compliant motion has been treated in the literature for some simple cases [3, 6, 12]. Either the equations of the compliant motion are linearized about a fixed point that results in a time invariant linear model, or a primary controller is assumed to cancel the nonlinear terms in the equations of motion to obtain a tractable model for stability studies. The determination of global stability for the compliant motion of a manipulator is difficult in general without simplifying assumptions, even in the case of planar manipulators. Researchers are actively studying these problems, and new results on the subject can be expected.

10.6 SUMMARY

We have discussed the control of the position and its derivatives on the tangential hyperplane and the force on the normal hyperplane of the constraint surface at the point of contact. Using a selection matrix, the coordinates that are position controlled can be separated from those that are force controlled.

The variables to be controlled can be mapped onto the joint space. We have presented the design of a primary controller for compliant motion to make the manipulator track the desired position and force values under ideal operating conditions. A secondary controller is designed as a multivariable PID-controller to compensate for the inaccurate modeling and disturbance effects and to complement the primary controller. The composite control operates on the joint space variables so as to generate appropriate corrections for the variables in the Cartesian base coordinate system. Thus, the desired tracking of the end-effector is achieved.

An alternative approach is to separate the directions of the variables along which the position and force are controlled. Then a primary and a secondary controller (a hybrid position/force controller) can be designed in the Cartesian base coordinate system to make the designated variables of the manipulator track the desired positions and the remaining variables the generalized force as closely as possible. The secondary controller can again be designed as a multivariate PID-controller. Thus, the control of position and/or its derivatives and force can be performed separately in the Cartesian base coordinate system.

If sufficiently accurate information is not available to design a primary controller to operate on position and contact force, a multivariable self-tuning controller can be constructed on the basis of the input–output measurements. The variables for which the desired values are given by the design specifications are included in performance criterion and/or used to modify an ARX-model. The controller with self-tuning is then designed as an LQG-controller by minimizing a quadratic criterion or as a feedback–feedforward controller based on the zero-pole placement method.

When the motion of the manipulator is subject to hard constraints, the number of DOF in the manipulator during the contact motion is usually reduced. We have discussed the determination of the dynamical model for the reduced-order system by introducing pseudovelocities and pseudoaccelerations. Moreover, we have obtained an expression for the generalized constraint force acting at the contact point in terms of Lagrange's multiplier vector. We have discussed the design of a primary controller for the end-effector during the contact motion and presented the construction of a secondary controller as a combination of two control inputs. The secondary controller is determined so that the pseudoacceleration is controlled by one of the input variables and the generalized constraint force by the other. Moreover, these two control subsystems can be decoupled, which facilitates the design of the control inputs that are generated by the secondary controller.

REFERENCES

[1]   M. T. Mason, "Compliance and Force Control for Computer Controlled Manipulators," *IEEE Transactions on Sys-*

tems Man and Cybernetics, Vol. SMC-11, No. 6, June 1981, pp. 418–432.

[2] J. K. Salisbury, "Active Stiffness Control of a Manipulator in Cartesian Coordinates," *Proc. of the 19th IEEE Conference on Decision and Control*, Albuquerque, NM, November 1980, pp. 95–100.

[3] D. E. Whitney, "Force Feedback Control of Manipulator Fine Motions," *Journal of Dynamic Systems, Measurement and Control, ASME Transactions*, June 1977, pp. 91–97.

[4] N. Hogan, "Impedance Control: An Approach to Manipulation: Part I, Part II, Part III," *Journal of Dynamical Systems, Measurement, and Control*, Vol. 107, March 1985, pp. 1–24.

[5] R. Paul and B. Shimano, "Compliance and Control," *Proc. of 1976 Joint Automatic Control Conference*, San Francisco, CA, July 1976, pp. 694–699.

[6] C. H. An and J. N. Hollerbach, "Dynamic Stability Issues in Force Control of Manipulators," *Proc. of 1987 IEEE International Conference on Robotics and Automation*, Raleigh, NC, March 1987, pp. 890–896.

[7] D. E. Whitney, "Historical Perspective and State of the Art in Robot Force Control," *Proc. of 1985 IEEE International Conference on Robotics and Automation*, St. Louis, MO, April 1985, pp. 262–268.

[8] M. H. Raibert and J. J. Craig, "Hybrid Position/Force Control of Manipulators," *Journal of Dynamic Systems, Measurement and Control*, Vol. 102, June 1981, pp. 126–133.

[9] R. Kankaanranta and H. N. Koivo, "Dynamics and Simulation of Compliant Motion of a Manipulator," *IEEE Journal on Robotics and Automation*, Vol. 4, No. 2, April 1988, pp. 163–173.

[10] A. J. Koivo, "Force-Position-Velocity Control with Self-Tuning for Robotic Manipulators," *Proc. of 1986 International Conference on Robotics and Automation*, San Francisco, California, April 1986, pp. 1563–1568.

[11] H. Goldstein, *Classical Mechanics*, Addison-Wesley, Reading, MA, 1980.

[12] H. Kazerooni, T. B. Sheridan and P. K. Houpt, "Robust Compliant Motion for Manipulators, Part I: Fundamental Concepts of Compliant Motion," *IEEE Journal of Robotics and Automation*, Vol. RA-1, No. 2, June 1986, pp. 83–92.

PROBLEMS

10.1   The generalized velocity vector $\dot{W} = [\dot{p}'\,\omega']'$ expressed in the Cartesian base coordinate system is related to the joint velocity vector $\dot{q}$ by means of the Jacobian matrix, that is, $\dot{W} = J(q)\dot{q}$.

  a.   Starting from the foregoing velocity relation, show that $\Delta W = J(q)\Delta q$ where Δq and ΔW are small changes in the joint position q and Cartesian position W, respectively.

  b.   Show that $\Delta \dot{W} = J(q)\Delta \dot{q} + \dot{J}(q)\Delta q$ by starting from the velocity relation.

10.2   Assume a soft contact between the end-effector of a manipulator and the environment. The model of the manipulator is given by Equation 10.1.5. Instead of using the control $F(t)$ as given by Equation 10.1.7, the designer selects $F(t) = u^d(t) + \hat{u}(t)$.

  a.   Determine the primary controller for the manipulator system so that the tracking of the desired joint trajectories is achieved.

  b.   Determine the equations of motion of the joint error vector Δq in terms of the secondary control $\hat{u}$ and the force error expressed in the Cartesian base coordinate system.

  c.   Suggest a secondary controller for the position force (PF) control.

  d.   Determine the characteristic equation associated with the dynamics of the joint position error when the soft contact is modeled as a spring with coefficient K_c.

10.3   When a manipulator is in contact with an environmental surface, the constraint motion is described by Equation 10.1.5. For a specified trajectory of the end-effector given in the Cartesian base coordinate system and for the given generalized force to be exerted, a controller is proposed in the following form:

$$F(t) = u^d(t) + \hat{u}(t)$$
$$= C + G + DJ^{-1}\dot{W}^d - DJ^{-1}\dot{J}J^{-1}\dot{W} + J'F_e^d + \hat{u}(t)$$

where $u^d(t)$ and $\hat{u}(t)$ specify the primary and secondary controllers.

a. Show the dynamical model for the position error ΔW
    expressed in the Cartesian base coordinate system can
    be written as

$$\Delta \ddot{W} + K_1 \Delta F_e = K_2 \hat{u}(t)$$

where matrices K_1 and K_2 are to be determined, and
ΔF_e represents the force error.

b. Suggest a secondary controller for the system by
    assuming that the soft contact force is modeled as a
    spring with coefficient K_c.

10.4 The secondary controller ($\hat{u}$) appearing in Equation
      10.1.11 can be chosen as the combination of two PID-
      controllers of which the one is operating on the joint posi-
      tion error and the other on the force error expressed in
      the Cartesian coordinate system.

a. Give an explicit expression for the secondary con-
    troller in terms of the foregoing errors.

b. Determine the differential equation that results when
    the secondary controller specified in (a) is used in
    Equation 10.1.10.

c. Assume the generalized contact force between the
    end-effector and the environment is modeled as a
    damper. Determine the dynamical model for the joint
    position error and the characteristic equation of this
    system. Which variables can be used in tuning the
    secondary controller?

10.5 The three-joint planar manipulator described in Example
      10.1.1 moves in the subspace of the plane defined by
      $p_y \leq -0.4p_x + 0.6$ where (p_x, p_y) specifies the position of
      the end-effector in the Cartesian base coordinate system.
      The gripper starts from point P_a (0.8 m, 0) and is moved
      to point P_b (0.5 m, 0.4 m) along the straight line. The
      gripper is then moved from point P_b to point P_c (0, 0.6
      m) while maintaining contact with the constraint surface
      and exerting the specified force F_e^d to the direction that is
      normal to the constraint surface. The contact is modeled
      as a spring with coefficient K_c^t.

a. Determine the dynamical model for the manipulator
    system (i) during the free motion and (ii) during the
    constraint motion.

b. Design a primary controller (u^d) and a secondary
    controller ($\hat{u}$) so as to make the gripper of the manip-

ulator follow the specified trajectory P_a–P_b–P_c. The orientation of the end-effector during the constrained motion is to be controlled so that the force exerted by the manipulator stays perpendicular to the contact surface.

10.6   Assume that the gripper of the manipulator in Problem 10.5 can move in the restricted area of the plane defined by $(p_x - x_0)^2 + (p_y - y_0)^2 \geq R^2$ where $R = 0.3$ m, $(x_0, y_0) = (0.299$ m, 0.623 m$)$. Repeat parts (a) and (b) of Problem 10.5 in this case.

10.7   The Jacobian matrix of a planar manipulator is given

$$J(\theta) = \begin{bmatrix} J_1(\theta) \\ J_2(\theta) \end{bmatrix} = \begin{bmatrix} \ell_1 s_1 + \ell_2 s_{12} & -\ell_1 c_1 - \ell_2 c_{12} \\ \ell_2 s_{12} & -\ell_2 c_{12} \end{bmatrix}$$

The inverse of the Jacobian matrix is $J^{-1}(\theta) = [\chi_1(\theta) \ \chi_2(\theta)]$ as in Equation 10.1.47.

   a.   Determine $\chi_1(\theta)$ and $\chi_2(\theta)$.

   b.   Show that $J_1(\theta)\chi_1(\theta) = 1$, $J_1(\theta)\chi_2(\theta) = 0$, $J_2(\theta)\chi_1(\theta) = 0$ and $J_2(\theta)\chi_2(\theta) = 1$.

10.8   The end-effector of the planar manipulator shown in Figure 10.7 (Example 10.2.1) moves on the surface described by $p_y = -0.4p_x + 0.6$ while exerting a force F_e on the surface. The contact is considered as a hard contact. The orientation of the gripper need not be considered.

   a.   Determine the constraint equation for the motion of the manipulator.

   b.   Determine the generalized contact force F_e^c exerted by the manipulator in the direction normal to the surface (i) in the joint space and (ii) in the Cartesian base coordinate system in terms of the joint variables.

10.9   The planar manipulator in Example 10.2.1 moves a tool attached to its end-effector on the surface described $(p_x - x_0)^2 + (p_y - y_0)^2 = R^2$ in the clockwise direction. Assume that the tool is a rigid extension of the prismatic link. Repeat parts (a) and (b) of Problem 10.8 in this case.

10.10   When the contact between the end-effector and an environmental object can be considered as a soft contact, the changes in the external force acting on the hyperplane normal to the contact point can be modeled by Equation 10.1.1. A similar model may be assumed when the interaction between the end-effector and the environment is represented by an external moment (torque).

a.   Write the model for a soft contact torque ΔM_e when
the movement $\Delta\psi$ represents a change in the rotation
angle. Define and explain the meaning of the coefficients in such a model. Specify the hyperplane in
which the contact moment ΔM_e acts.

b.   Draw a schematic representing the model for the soft
contact torque.

10.11   The insertion of a peg into a hole [9] is assumed to take
place under ideal conditions: the peg maintains a continuous contact with the environmental surface, the fit is
perfect, and the friction is ignored (Figure 10.12). The z-
axis is parallel with the axes of the hole and the peg, and
the hole is located on the xy-plane at (x_0, y_0, z_0). The posi-

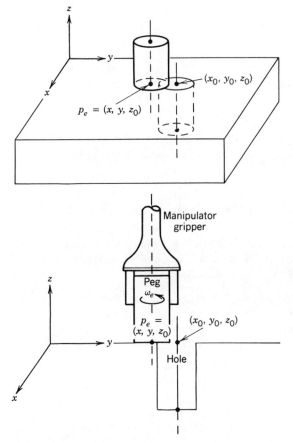

FIGURE 10.12
Inserting peg into a hole.

tion of a point on the peg is p_e, and the angular velocity of the peg ω_e. The constraints of the motion of the six-joint manipulator used for the insertion are: $C_{p1} = x - x_0 = 0$, $C_{p2} = y - y_0 = 0$, $\omega_{ex} = 0$, $\omega_{ey} = 0$.

a. Show that the constraint matrix $\bar{C}$ in Equation 10.3.15 is

$$\bar{C} = \begin{bmatrix} 1 & 0 & 0 & 0 & 0 & 0 \\ 0 & 1 & 0 & 0 & 0 & 0 \\ 0 & 0 & 0 & 1 & 0 & 0 \\ 0 & 0 & 0 & 0 & 1 & 0 \end{bmatrix}$$

b. Use the selection matrix S_1 which has the rows of the (6×6) identity matrix that are not used in forming the $\bar{C}$-matrix to specify H in Equation 10.3.2 (10.3.11).

c. Determine the P-matrix in Equation 10.4.2 to define the pseudovelocity vector in terms of the selection matrix S_1, the Jacobian matrix and the joint velocities.

d. Determine the components of the pseudovelocity vector in terms of the translational and rotational velocities of the peg in the Cartesian base coordinate system.

e. Determine the reduced-order model for the manipulator in the general form in terms of the pseudo-variables.

10.12 The model in Equation 10.4.13 describes the evolution of the pseudovelocity $\nu(t)$ in time. A primary and secondary controller can be designed so that the pseudovelocity tracks the specified trajectory $\nu^d(t)$ while a desired known force F_e^d given in the Cartesian base coordinate system is exerted on a hard surface.

a. A controller in the form $K_1 u^d(t) + K_2 \hat{u}(t)$ for the reduced-order model given in Equation 10.4.13 can be determined so that the error in the pseudovelocity is governed by an equation of the following form:

$$\Delta \dot{\nu} = A \Delta \nu + B \hat{u}(t)$$

Specify K_1 and K_2, A and B.

b. If the secondary controller ($\hat{u}$) is now chosen as specified by Equation 10.4.17, determine the equation for the error $\Delta \nu$ in the pseudovelocity and the equation for the Lagrange multiplier. How are these equations different from Equations 10.4.18 and 10.4.20?

    c.   Apply the solution determined in parts (a) and (b) to the problem described in Example 10.4.1.

COMPUTER PROBLEMS

10.C.1   The motion of the planar manipulator shown in Figure 10.7 is to be simulated using Equations 10.1.35–10.1.37 with the numerical values of Example 10.1.3. The desired trajectory of the end-effector is from point P_a (0, 0.6 m) to point P_b (0.8 m, 0.28 m) along the constraint surface $p_y = -0.4p_x + 0.6$. The model for the contact force is a spring with constant $K_c^t = 100$ N/m. The desired value of the contact force is 8 N.

    a.   Determine the desired trajectory in the joint space.

    b.   Decompose the contact force into two components along the *x*- and *y*-axes. Design a primary and a secondary controller for the manipulator system to meet the design specifications.

    c.   Simulate first the system described by Equations 10.1.35–10.1.37 when $F_{ey}^d = 0$ and PID-controller operating the position error is used. Graph (i) p_x versus time, (ii) p_y versus time and (iii) p_y versus p_x.

    d.   Simulate the system when $F_{ey}^d = 8$ N and the primary and secondary controllers given in Equations 10.1.40 and 10.1.42 are used. Tune the controller gains so that the position and force errors are less than 10% of the desired values during the motion.

    e.   Graph (i) p_x versus time, (ii) p_y versus time, (iii) p_y versus p_x on the phase plane and (iv) F_{ey} versus time. Comment on the performance of the controller.

10.C.2   The planar manipulator and its task (design) specifications are the same as in Computer Problem 10.C.1.

    a.   Design a self-tuning controller by using the velocity variables expressed in the Cartesian base coordinate system in the ARX-model. Specifically, (i) specify the ARX-model, and write the parameter estimation equations for your model; (ii) specify the perfor-

mance criterion for the design of a controller; (iii)
determine the self-tuning LQG-controller for your
system.

b. Simulate the system designed to demonstrate that
your controller satisfies the design specifications.
(You may assume first that $e(k) = 0$. Then super-
impose samples from Gaussian noise process $[(\bar{e} =$
$0, \sigma = 0.01)]$ on the measurements.)

c. Graph (i) p_x versus time, (ii) F_{ey} versus time, (iii) p_y
versus time, (iv) p_y versus p_x on the phase plane.
Graph also the desired trajectory for each case.
(v) one of the parameter estimates versus time,
(vi) the control input versus time. Comment on the
performance of the controller.

APPENDIX A

HAMILTON'S PRINCIPLE AND EULER–LAGRANGE EQUATIONS FOR DYNAMICAL SYSTEMS

We will first illustrate Hamilton's principle by starting from Newton's law of dynamics to obtain a mathematical expression for Hamilton's principle. Using variational calculus, the integral representing Hamilton's principle is then converted into a set of differential equations, which are recognized as the equations of Euler–Lagrange.

A.1 HAMILTON'S PRINCIPLE

The motion of a point mass m moving under the influence of a force (field) $\mathbf{F}$ and located at the position $\mathbf{r} = \mathbf{r}(t)$ relative to a fixed coordinate system is governed by Newton's law:

$$m\frac{d^2\mathbf{r}(t)}{dt^2} - \mathbf{F} = 0 \qquad (\text{A.1})$$

Suppose that there is slight variation (perturbation) about the nominal path $\mathbf{r}(t)$ resulting in a new path $\mathbf{r}(t) + \delta\mathbf{r}(t)$, where both $\mathbf{r}(t)$, $\delta\mathbf{r}(t)$ and their time derivatives are assumed to be continuous. The motion of the particle starts at time t_0 and ends at time t_1. If the initial point $\mathbf{r}(t_0)$ and the terminal point $\mathbf{r}(t_1)$ are fixed, then $\delta\mathbf{r}(t_0) = \delta\mathbf{r}(t_1) = 0$. In the sequel, the vectors are considered as column vectors and denoted in lightface rather than in boldface.

The inner product of the terms appearing in Equation A.1 is next formed with δr, and integrated with respect to time to obtain

$$\int_{t_0}^{t_1}\left[m\frac{d^2r'}{dt^2}\delta r - F'\delta r\right]dt = 0 \tag{A.2}$$

where $F'\delta r$ describes the work done by force F over a small displacement δr. The first term in Equation A.2 can be integrated by parts to obtain

$$\int_{t_0}^{t_1}m\frac{d^2r'}{dt^2}\delta r\, dt = \left[m\frac{dr'}{dt}\delta r\right]_{t_0}^{t_1} - \int_{t_0}^{t_1}m\frac{dr'}{dt}\frac{d(\delta r)}{dt}dt \tag{A.3}$$

The term outside the integral on the right of Equation A.3 vanishes because the endpoints of the path are fixed. The second term can be recognized as a first-order variation δK in the kinetic energy $K = \frac{1}{2}m\dot{r}'\dot{r}$; namely

$$\delta K = m\dot{r}'\delta\dot{r}$$

$$= m\frac{dr'}{dt}\frac{d(\delta r)}{dt} \tag{A.4}$$

Equation A.2 can now be rewritten as follows:

$$-\int_{t_0}^{t_1}[\delta K + F'\delta r]dt = 0 \tag{A.5}$$

Equation A.5 states *Hamilton's principle* for a single mass.

When the total energy of the system does not change along the motion, that is, for conservative systems, Equation A.5 can be developed to an alternative form, which provides additional insight in Hamilton's principle.

In a conservative system, the potential energy of the system does not depend on time explicitly. Then, the second term $F'\delta r$ can be expressed as a differential of the force potential. The negative of the force potential is called the potential energy. Thus

$$F'\delta r = -\delta P \tag{A.6}$$

where P represents the potential energy. Equation A.6 shows that the force F is the negative gradient of the potential energy.

Hamilton's principle expressed in Equation A.5 may now be rewritten in terms of the kinetic and potential energies for a *conservative* system by interchanging the operators for the integration and variation:

$$\delta\int_{t_0}^{t_1}(K - P)dt = 0 \tag{A.7}$$

where the symbol δ in front of the integral signifies a first-order variation.

Hamilton's principle expressed in Equation A.7 states that the integral of the difference between the kinetic and potential energies is stationary (i.e., minimum or maximum) along the nominal path. A necessary condition for a stationary point is given in Equation A.7. When the motion of the system satisfies Equation A.7, the *principle of stationary action* [1,2] holds for the system. The stationary value of the integral $\int (K - P)\,dt$ is minimum at least for sufficiently small time intervals for most applications. When this is the case and Equation A.7 is satisfied by the motion of the given system moving under the influence of forces acting on it, then the *principle of least action* applies to the system.

The integrand in Equation A.7 is called Lagrange's (energy) function $\mathscr{L}$, that is,

$$\mathscr{L} = K - P \tag{A.8}$$

The integral of Lagrange's function is sometimes referred to as the action functional. In terms of the Lagrangian function, Hamilton's principle states that

$$\delta \int_{t_0}^{t_1} \mathscr{L}\,dt = 0 \tag{A.9}$$

Equation A.9 represents a necessary condition for a stationary point of the integral.

The presentation above is applied to a single mass point m. When there are several point masses in the system, the approach described is still applicable. The terms in Hamilton's principle represent then the total kinetic energy and the total potential energy of the system. Moreover, the general principle applies not only to point masses but also to rigid bodies subject to interconnections and constraints.

A.2   EULER–LAGRANGE EQUATIONS

When a dynamical system such as a serial link manipulator has N degrees of freedom (DOF), N independent variables, that is, N generalized coordinates, can be chosen. They can serve as the basis vectors in the N-dimensional space. These variables denoted by $q_1, \ldots, q_N$ uniquely specify the position of all points in the system. Moreover, they can be used to express the kinetic (K) and potential (P) energies for the given system. Thus, Lagrange's energy function $\mathscr{L}$ can be determined in terms of $\{q_i\}, i = 1, \ldots, N$.

To determine a differential equation model for the motion of a dynamical system, it is advantageous to develop Hamilton's principle expressed in Equation A.9 to an alternative form known as Euler–Lagrange's differential equations. These equations are obtained by considering the first-order variation of the Lagrangian function $\mathscr{L}$ about a nominal path specified by the generalized coordinates $q_1 = q_1(t), \ldots, q_N = q_N(t)$. Euler–Lagrange's equations for conservative systems will next be presented by starting from Equation A.9 [1,3].

Suppose that a nominal path for a system is given in terms of the generalized coordinates $q_1(t), \ldots, q_N(t)$, where each $q_i(t)$, $i = 1, \ldots, N$ has continuous derivatives. The endpoints of the path are assumed to be specified. The value of Lagrange's function along the nominal path is

$$\mathcal{L}_1 = \mathcal{L}[q_1(t), \ldots, q_N(t); \dot{q}_1(t), \ldots, \dot{q}_N(t)] \qquad \text{(A.10)}$$

The generalized coordinates are next subjected to (sufficiently) small perturbations resulting in a new path $q_1(t) + \delta q_1(t), \ldots, q_N(t) + \delta q_N(t)$, where each $\delta q_i(t)$ represents a weak variation. This means that each $\delta q_i(t)$ possesses a bounded continuous first-order derivative; in other words, $\delta q_i(t)$ has a continuously turning tangent. The corresponding value of Lagrange's function becomes

$$\mathcal{L}_2 = \mathcal{L}[q_1(t) + \delta q_1(t), \ldots, q_N(t) + \delta q_N(t); \dot{q}_1(t) + \delta \dot{q}_1(t), \ldots, \dot{q}_N(t) + \delta \dot{q}_N(t)] \qquad \text{(A.11)}$$

The total change in Lagrange's energy function due to the weak variations is expressed by $\mathcal{L}_2 - \mathcal{L}_1$. By determining Taylor's series expansion of $(\mathcal{L}_2 - \mathcal{L}_1)$ about the nominal trajectory, and maintaining only the first-order variational terms, the principal linear variational part of $\int (\mathcal{L}_2 - \mathcal{L}_1)\,dt$ is obtained, that is, the first-order variation $\int \delta \mathcal{L}\,dt$ of Lagrange's function. The higher order terms are assumed to be negligible. This presumes that the norm of each $\delta q_i(t)$ is sufficiently small for all $t \epsilon [t_0, t_1]$ and the derivatives of $\mathcal{L}$ are bounded. Thus,

$$\int_{t_0}^{t_1} \delta \mathcal{L}[q_1(t), \ldots, q_N(t); \ \dot{q}_1(t), \ldots, \dot{q}_N(t)]\,dt$$

$$= \int_{t_0}^{t_1} \left[\frac{\partial \mathcal{L}}{\partial q_1} \delta q_1 + \cdots + \frac{\partial \mathcal{L}}{\partial q_N} \delta q_N + \frac{\partial \mathcal{L}}{\partial \dot{q}_1} \delta \dot{q}_1 + \cdots + \frac{\partial \mathcal{L}}{\partial \dot{q}_N} \delta \dot{q}_N \right] dt \qquad \text{(A.12)}$$

where each partial derivative is evaluated along the nominal path.

Hamilton's principle expressed in Equation A.9 must be satisfied along the motion. That is, the expression in Equation A.12 must vanish for any arbitrary variation of $q_i(t)$. Since δq_i and $\delta \dot{q}_i$ are not independent, expression A.12 is rewritten after integrating by parts the terms with the generalized velocities:

$$\left[\frac{\partial \mathcal{L}}{\partial \dot{q}_1} \delta q_1(t) \right]_{t_0}^{t_1} + \cdots + \left[\frac{\partial \mathcal{L}}{\partial \dot{q}_N} \delta q_N(t) \right]_{t_0}^{t_1} + \int_{t_0}^{t_1} \left\{ \left[\frac{\partial \mathcal{L}}{\partial q_1} - \frac{d}{dt}\left(\frac{\partial \mathcal{L}}{\partial \dot{q}_1} \right) \right] \delta q_1(t) \right.$$

$$\left. + \cdots + \left[\frac{\partial \mathcal{L}}{\partial q_N} - \frac{d}{dt}\left(\frac{\partial \mathcal{L}}{\partial \dot{q}_N} \right) \right] \delta q_N(t) \right\} dt = 0 \qquad \text{(A.13)}$$

Since the endpoints of the nominal path are fixed, it follows that $\delta q_i(t_0) = 0$, and $\delta q_i(t_1) = 0$ and the terms outside the integral sign vanish. Necessary conditions for the vanishing of the integral require that the terms in the brackets of the integrand must be zero, since the variations $\delta q_i(t)$ are arbitrary. Hence,

$$\frac{d}{dt}\left(\frac{\partial \mathcal{L}}{\partial \dot{q}_i}\right) - \frac{\partial \mathcal{L}}{\partial q_i} = 0 \qquad (A.14)$$

where $i = 1, \ldots, N$, and $\mathcal{L} = \mathcal{L}(q_1, \ldots, q_N; \dot{q}_1, \ldots, \dot{q}_N)$. Equation A.14 is *Euler–Lagrange's equation* for a conservative system.

If the system is nonconservative, Equation A.5 can be applied to obtain Euler–Lagrange's equation, since it is feasible that the force cannot directly be derived from the potential energy.

When external forces are acting on the system, they are included on the right side of Equation A.14. Thus, Euler–Lagrange's equation assumes the following form:

$$\frac{d}{dt}\left(\frac{\partial \mathcal{L}}{\partial \dot{q}_i}\right) - \frac{\partial \mathcal{L}}{\partial q_i} = F_i \qquad (A.15)$$

where $i = 1, \ldots, N$, and F_i represents the generalized forces acting in the direction of the q_i-coordinate. Frictional forces that oppose the motion in the direction of the coordinate q_i are included (with negative signs) in the right side of Equation A.15. In using Equation A.15, the generalized coordinates $q_i, i = 1, \ldots, N$ must represent *independent* variables. If the initially chosen variables are not independent, a set of independent variables should first be determined before Lagrange's Equation A.15 is applied.

Equation A.15 is Euler–Lagrange's equation that determines the dynamical model of a manipulator in Lagrange's formulation.

REFERENCES

[1] I. M. Gelfand and S. V. Fomin, *Calculus of Variations*, Prentice-Hall, Englewood Cliffs, NJ, 1963.

[2] F. Gantmacher, *Lectures in Analytical Mechanics*, Mir Publishers, Moscow, 1975.

[3] H. Goldstein, *Classical Mechanics*, Addison-Wesley, Reading, MA, 1980.

APPENDIX B

EQUATIONS OF MOTION FOR A SERIAL LINK MANIPULATOR IN LAGRANGE FORMULATION

In a serial link manipulator, the homogeneous transformation A_{i-1}^i relates the ith and $(i-1)$st coordinates, where $i = 1, \ldots, N$, and N specifies the number of the joints in the manipulator. The position of an incremental mass dm_h on the ith link is assumed to be specified by vector p_{ih} in the ith coordinate frame. The same point in the base coordinate system is

$$p_{0h} = A_0^1 A_1^2 \ldots A_{i-1}^i p_{ih}$$

$$= A_0^i p_{ih} \qquad (B.1)$$

where the homogeneous transformation matrix $A_0^i = A_0^1 \ldots A_{i-1}^i$. The velocity vector associated with the position vector p_{ih} is

$$\dot{p}_{0h} = \frac{d}{dt}(A_0^i p_{ih})$$

$$= \sum_{j=1}^{i} \frac{\partial A_0^i}{\partial q_j} \dot{q}_j P_{ih} \qquad (B.2)$$

where q_j is the generalized coordinate for the jth joint.

We will next determine the kinetic and potential energies of the aforementioned incremental mass on the ith link. By an appropriate integration operation, Lagrange's energy function for the entire link of the manipulator can be determined.

The expression for the kinetic energy of the incremental mass dm_h on link i is

$$dK_h = \frac{1}{2}\dot{p}_{0h}{}'\,\dot{p}_{0h}\,dm_h$$

$$= \frac{1}{2}\left(\sum_{k=1}^{i} \frac{\partial A_0^i}{\partial q_k}\dot{q}_k p_{ih}\right)'\left(\sum_{j=1}^{i} \frac{\partial A_0^i}{\partial q_j}\dot{q}_j\,p_{ih}\right)dm_h \qquad (B.3)$$

$$= \frac{1}{2}\mathrm{tr}\left\{\left(\sum_{k=1}^{i} \frac{\partial A_0^i}{\partial q_k}\dot{q}_k\,p_{ih}\right)\left(\sum_{j=1}^{i} \frac{\partial A_0^i}{\partial q_j}\dot{q}_j\,p_{ih}\right)'\right\}dm_h$$

where $\mathrm{tr}(\,\cdot\,)$ signifies the trace of the matrix; it equals the sum of the diagonal elements. When the transposition operation in Equation B.3 is changed from the first parenthesis to the second, the trace operation must be introduced. Indeed, direct calculations show that $q'q = \mathrm{tr}(qq')$ where q is a multidimensional vector. The change in the transposition operations allows us to obtain an expression in which the vectors p_{ih} and $p_{ih}{}'$ are next to each other; thus the integral of $p_{ih}p_{ih}{}'dm_h$ can be related to the incremental inertia matrix, which could not be accomplished without the trace operation (why not?).

The kinetic energy K_i associated with link i is obtained by integrating the expression in Equation B.3 with respect to dm_h, and observing that $\dot{q}_j$ and $\dot{q}_k$ are scalars:

$$K_i = \int_{\text{link }i} dK_h = \frac{1}{2}\mathrm{tr}\left[\sum_{j=1}^{i}\sum_{k=1}^{i} \frac{\partial A_0^i}{\partial q_k}\left(\int_{\text{link }i} p_{ih}p_{ih}'dm_h\right)\left(\frac{\partial A_0^i}{\partial q_j}\right)'\dot{q}_j\dot{q}_k\right]$$

$$= \frac{1}{2}\mathrm{tr}\left[\sum_{j=1}^{i}\sum_{k=1}^{i} \frac{\partial A_0^i}{\partial q_j}I_i\left(\frac{\partial A_0^i}{\partial q_k}\right)'\dot{q}_j\dot{q}_k\right] \qquad (B.4)$$

where the pseudo-inertial matrix for link i has been defined as

$$I_i = \int_{\text{link }i} p_{ih}p_{ih}'dm_h \qquad (B.5)$$

Vector $p_{ih} = [p_{x_i}, p_{y_i}, p_{z_i}, 1]'$ has the components expressed in the ith coordinate frame $(x_i\,y_i\,z_i)$. Thus, I_i describes the mass moments of link i about the origin of the ith coordinate frame. It may be shown (see Problem 5.5) that the matrix of the pseudo-inertia of the ith link may be expressed as

$$I_i = \begin{bmatrix} \frac{1}{2}(-I_{x_i} + I_{y_i} + I_{z_i}) & I_{x_iy_i} & I_{x_iz_i} & \bar{p}_{x_i}m_i \\[2mm] I_{x_iy_i} & \frac{1}{2}(I_{x_i} - I_{y_i} + I_{z_i}) & I_{y_iz_i} & \bar{p}_{y_i}m_i \\[2mm] I_{x_iz_i} & I_{y_iz_i} & \frac{1}{2}(I_{x_i} + I_{y_i} - I_{z_i}) & \bar{p}_{z_i}m_i \\[2mm] m_i\bar{p}_{x_i} & m_i\bar{p}_{y_i} & m_i\bar{p}_{z_i} & m_i \end{bmatrix}$$

$$(B.6)$$

where the link mass m_i is concentrated at the point $\bar{p}_i = [\bar{p}_{x_i}, \bar{p}_{y_i}, \bar{p}_{z_i}, 1]'$. The second order mass moments in Equation B.6 are determined in accordance with the subscripts, that is,

$$I_{x_i} = \int (p_{y_i}^2 + p_{z_i}^2)\, dm_i \qquad I_{y_i} = \int (p_{x_i}^2 + p_{z_i}^2)\, dm_i \qquad I_{z_i} = \int (p_{x_i}^2 + p_{y_i}^2)\, dm_i$$

$$I_{wv} = \int p_w p_v\, dm_i, \qquad w \neq v; \qquad w, v = x_i, y_i, z_i \tag{B.7}$$

where the integration is over the ith link. The terms I_{x_i}, I_{y_i}, and I_{z_i} are the centroidal (second) mass moments of inertia and I_{wv} is the centroidal mass product of inertia of the link. The fourth row and columns in the matrix of Equation B.6 contain the first-order mass moments of the link.

Equation B.4 gives the kinetic energy of the ith link. If the manipulator has N links, then the total kinetic energy K is

$$K = \sum_{i=1}^{N} K_i$$

$$= \frac{1}{2} \sum_{i=1}^{N} \mathrm{tr}\left[\sum_{k=1}^{i} \sum_{j=1}^{i} \frac{\partial A_0^i}{\partial q_k} I_i \left(\frac{\partial A_0^i}{\partial q_j} \right)' \dot{q}_j \dot{q}_k \right] \tag{B.8}$$

The kinetic energy K given by Equation B.8 is needed in the Lagrangian function.

The potential energy associated with the mass m_i of link i is

$$P_i = -m_i g' A_0^i \bar{p}_i \tag{B.9}$$

where the vector $g = [g_{0x}, g_{0y}, g_{0z}, 0]'$ describes the gravitational acceleration with components in the directions of the coordinates x_0, y_0, z_0 of the base coordinate system. The total potential energy associated with N links of the manipulator is

$$P = \sum_{i=1}^{N} P_i$$

$$= -\sum_{i=1}^{N} m_i g' A_0^i \bar{p}_i \tag{B.10}$$

Lagrange's energy function $\mathscr{L} = K - P$ needed for Equation 5.1.2 is now determined by Equations B.8 and B.10, which are expressed in terms of N *independent* variables $q_1, \ldots, q_N$ for the manipulator.

The equations of motion are obtained by forming the appropriate derivatives appearing in Euler–Lagrange's Equation 5.1.2. Particularly, the dynamical model for the nth link is

$$\frac{d}{dt}\left(\frac{\partial \mathscr{L}}{\partial \dot{q}_n}\right) - \frac{\partial \mathscr{L}}{\partial q_n} = F_n \qquad (B.11)$$

where F_n is the *generalized* force acting in the direction of the q_n-coordinate. For each joint $n, n = 1, \ldots, N$, an equation in the form of Equation B.11 is obtained.

The derivatives in Equation B.11 are next determined:

$$\frac{\partial \mathscr{L}}{\partial q_n} = \sum_{i=1}^{N} \sum_{k=1}^{i} \sum_{j=1}^{i} \text{tr}\left[\frac{\partial A_0^i}{\partial q_k} I_i \left(\frac{\partial^2 A_0^i}{\partial q_j \partial q_n}\right)'\right] \dot{q}_j \dot{q}_k + \sum_{i=1}^{N} m_i g' \frac{\partial A_0^i}{\partial q_n} \bar{p}_i \qquad (B.12)$$

$$\frac{\partial \mathscr{L}}{\partial \dot{q}_n} = \sum_{i=1}^{N} \sum_{k=1}^{i} \text{tr}\left[\frac{\partial A_0^i}{\partial q_k} I_i \left(\frac{\partial A_0^i}{\partial q_n}\right)'\right] \dot{q}_k \qquad (B.13)$$

Since $A_0^i = A_0^1 A_1^2 \ldots A_{i-1}^i$ depends on only $q_1, \ldots, q_i$, it follows that $\partial A_0^i / \partial q_n$ and $\partial A_0^i / \partial \dot{q}_n$ for $n > i$ equal to zero. Hence, the lower limit one in the first summations of Equations B.12 and B.13 can be replaced by n.

The substitution of the expressions of Equations B.12 and B.13 into Equation B.11 leads to

$$\sum_{i=n}^{N} \left\{ \sum_{k=1}^{i} \text{tr}\left[\frac{\partial A_0^i}{\partial q_n} I_i \left(\frac{\partial A_0^i}{\partial q_k}\right)'\right] \ddot{q}_k + \sum_{k=1}^{i} \sum_{j=1}^{i} \text{tr}\left[\frac{\partial A_0^i}{\partial q_n} I_i \left(\frac{\partial^2 A_0^i}{\partial q_k \partial q_j}\right)'\right] \dot{q}_k \dot{q}_j - \right.$$

$$\left. m_i g' \frac{\partial A_0^i}{\partial q_n} \bar{p}_i \right\} = F_n \qquad (B.14)$$

By introducing new notations for the variables in the summations, Equation B.14 can be rewritten

$$\sum_{k=1}^{N} D_{nk} \ddot{q}_k + \sum_{k=1}^{N} \sum_{j=1}^{N} D_{nkj} \dot{q}_k \dot{q}_j + G_n = F_n \qquad (B.15)$$

where the following abbreviations have been introduced:

$$D_{nk} = \sum_{i=\max(n,k)}^{N} \text{tr}\left[\frac{\partial A_0^i}{\partial q_n} I_i \left(\frac{\partial A_0^i}{\partial q_k}\right)'\right] \qquad (B.16)$$

$$D_{nkj} = \sum_{i=\max(n,k,j)}^{N} \text{tr}\left[\frac{\partial A_0^i}{\partial q_n} I_i \left(\frac{\partial^2 A_0^i}{\partial q_k \partial q_j}\right)'\right] \qquad (B.17)$$

$$G_n = -\sum_{i=n}^{N} m_i g' \frac{\partial A_0^i}{\partial q_n} \bar{p}_i \qquad (B.18)$$

Equation B.15 represents the dynamical model for the nth link of a manipulator in Lagrange's formulation. Thus, N nonlinear second order differential equations describe the dynamics of the manipulator with N joints.

REFERENCES

[1] A. K. Bejczy, "Robot Arm Dynamics and Control," Technical Memo 33-669, Jet Propulsion Lab., February 1974.

[2] R. P. Paul, *Robot Manipulators, Mathematics, Programming and Control*, The MIT Press, Cambridge, MA, 1982.

APPENDIX C

EQUATIONS OF NEWTON AND EULER FOR DYNAMICAL MODELING

For a translational (linear) motion, Newton's law states that the resultant $\mathbf{F}$ of the external forces acting on a rigid body equals the acceleration of the body times its mass, that is,

$$m\ddot{\mathbf{r}} = \mathbf{F} \tag{C.1}$$

where the mass m is concentrated at the center of gravity (centroid), and the acceleration vector $\ddot{\mathbf{r}}$ emanates from the origin of the base coordinate system to the centroid of the body. Equation C.1 can also be interpreted as the rate of change in the linear momentum $\mathbf{L} = m\ddot{\mathbf{r}}$ being equal to the sum of the external forces.

For a rotational (angular) motion, Euler's equations describe the dynamics. To obtain the equation of motion, the basic principles of rotational dynamics are applied first to a set of particles (point masses), and then to a rigid body, which may be considered as a limit when the number of particles increases. This approach provides us with the basic equations to study the rotation of rigid bodies, such as that of the links in a manipulator [1].

The angular momentum $\mathbf{H}$ of a set of points $\Delta m_i, i = 1, \ldots, n$ is defined as follows:

$$\mathbf{H} = \sum_{i=1}^{n} (\mathbf{p}_i^* \times \dot{\mathbf{p}}_i^* \Delta m_i) \tag{C.2}$$

where the position vector $\mathbf{p}_i^*$, and the velocity vector $\dot{\mathbf{p}}_i^*$ of the incremental point mass Δm_i are expressed relative to the fixed (xyz)-coordinate frame with the origin at the centroid. If the angular velocity of the point mass Δm_i is $\boldsymbol{\omega}_{oi}$, then $\dot{\mathbf{p}}_i^* = \boldsymbol{\omega}_{oi} \times \mathbf{p}_i^*$ can be substituted into Equation C.2.

The angular momentum of a homogeneous rigid body (such as a link) rotating at the angular velocity $\boldsymbol{\omega}_0 = \omega_{0x}\,\mathbf{i} + \omega_{0y}\,\mathbf{j} + \omega_{0z}\,\mathbf{k}$ about the centroid G at the instant considered can be determined by a limiting process ($\Delta m_i \to 0, n \to \infty$). When the cross products $\mathbf{p} \times (\boldsymbol{\omega}_0 \times \mathbf{p}^*)$ in Equation C.2 are evaluated and the limiting processes are performed, the following expressions for the components of the angular momentum $\mathbf{H}_G$ in the x-, y- and z-directions are obtained:

$$H_{G_x} = I_x\,\omega_{0x} \quad - \; I_{xy}\,\omega_{0y} - I_{zx}\,\omega_{0z} \tag{C.3}$$

$$H_{G_y} = -I_{xy}\,\omega_{0x} + I_y\,\omega_y \quad - \; I_{yz}\,\omega_{0z} \tag{C.4}$$

$$H_{G_z} = -I_{zy}\,\omega_{0x} - I_{yz}\,\omega_{0y} + I_z\,\omega_{0z} \tag{C.5}$$

where the following abbreviations have been adopted for the centroidal mass moments of inertia: $I_x = \int (p_y^2 + p_z^2)\,dm$, $I_y = \int (p_z^2 + p_x^2)\,dm$, $I_z = \int (p_x^2 + p_y^2)\,dm$; and for the mass products of inertia: $I_{xy} = \int p_x p_y\,dm$, $I_{zx} = \int p_z p_x\,dm$, $I_{zy} = \int p_z p_y\,dm$.

Equations C.3–C.5 may be written more concisely by writing $\omega = [\omega_{0x}, \omega_{0y}, \omega_{0z}]'$:

$$\begin{bmatrix} H_{G_x} \\ H_{G_y} \\ H_{G_z} \end{bmatrix} = \begin{bmatrix} I_x & -I_{xy} & -I_{zx} \\ -I_{xy} & I_y & -I_{yz} \\ -I_{zx} & -I_{yz} & I_z \end{bmatrix} \begin{bmatrix} \omega_{0x} \\ \omega_{0y} \\ \omega_{0z} \end{bmatrix} \tag{C.6}$$

or simply

$$H_G = I\boldsymbol{\omega}_0 \tag{C.7}$$

Equations C.6 and C.7 describe the angular momentum of a homogeneous link about its centroid.

If Equation C.7 is differentiated with respect to time, the rate of change in the angular momentum is obtained. It is equal to the resultant M of the external torques acting on the rigid body:

$$\frac{dH_G}{dt} = \frac{d}{dt}(I\boldsymbol{\omega}_0) = M \tag{C.8}$$

The component M_i of torque M in some cases can be computed by evaluating the cross-product of the position vector $\mathbf{r}_G^*$ and the force vector $\mathbf{F}$, where $\mathbf{r}_G^*$ is a vector from the centroid to the point where the force $\mathbf{F}$ is acting, that is, $\mathbf{M} = \mathbf{r}_G^* \times \mathbf{F}$.

In order to use Equation C.8, the rate of change $\dot{H}_G$ in the angular momentum must be determined. The angular momentum H_G of the rigid body describes the motion relative to the centroidal coordinate frame (xyz), which maintains a fixed orientation. If the components of H_G and ω_0 were described along the x-, y-, and z-axes, the centroidal moments and cross-products of inertia in the inertia matrix would change continuously with time, since the body rotates. On the other hand, when the mass moments and mass products of inertia are described relative to the *moving coordinate frame* that rotates with the rigid body, they will have the same values during the motion.

Vector $\mathbf{H}_G$ represents the angular momentum about the centroid of the body relative to the fixed (xyz)-frame. The rate of change $\dot{\mathbf{H}}_G$ of the angular momentum vector with respect to the fixed coordinate frame (xyz) is equal to the rate of change in the angular momentum relative to the moving coordinate frame added to the cross-product $\omega_0 \times \mathbf{H}_G$ (compare with Equation 4.1.8):

$$\frac{d\mathbf{H}_G}{dt} = \frac{d^*\mathbf{H}_G}{dt} + \omega_0 \times \mathbf{H}_G \tag{C.9}$$

where the starred derivative signifies the rate of change of $\mathbf{H}_G$ in the starred coordinate frame, and vector ω_0 is the angular velocity of the rotating coordinate frame expressed in the fixed coordinate system.

Since the rate of change in the angular momentum equals the resultant $\mathbf{M}_0$ of the external moments (torques), Equation C.9 can be rewritten to obtain:

$$\frac{d^*\mathbf{H}_G}{dt} + \omega \times \mathbf{H}_G = \mathbf{M}_0 \tag{C.10}$$

If the moving coordinate frame is chosen so that the axes coincide with the principal axes of inertia, then the mass products of inertia in Equation C.6 are zero, and the inertia matrix becomes diagonal. In this particular case, Equation C.10 can be decomposed into the following equations describing the motions about the coordinate axes x, y, and z:

$$I_x \dot{\omega}_{0x} - (I_y - I_z) \omega_{0y} \omega_{0z} = M_{0x} \tag{C.11}$$

$$I_y \dot{\omega}_{0y} - (I_z - I_x) \omega_{0z} \omega_{0x} = M_{0y} \tag{C.12}$$

$$I_z \dot{\omega}_{0z} - (I_x - I_y) \omega_{0x} \omega_{0y} = M_{0z} \tag{C.13}$$

Equations C.11–C.13 are called Euler's equations of motion. They are useful in the analysis of the rotational motion of rigid bodies.

After the substitution of Equation C.8 into Equation C.10, Euler's equation can be rewritten:

$$\mathbf{M}_c + \omega_0 \times (\mathbf{H}_G) = \mathbf{M}_0 \tag{C.14}$$

where $M_c = I\dot{\omega}_0$.

Equations C.1 and C.14 can be applied to the solid body in a free-body configuration represented by the ith link of a serial link manipulator. It leads to a dynamical model for the motion of the ith link. The model equations are thus described in recursive forms relative to the serial links. This is a typical feature in Newton–Euler's formulation.

REFERENCES

[1] H. Goldstein, *Classical Mechanisms*, Addison-Wesley Publishers, Reading, MA, 1980.

APPENDIX D

REVIEW OF SELF-TUNING CONTROLLER DESIGN BY THE EXPLICIT METHOD

A model for the input–output relation of a system expressed as an equation can be used to analyze the behavior of the system under various conditions and to design a control strategy so as to make the system behave in a desirable manner. A system model can be determined on the basis of physical laws; however, it may in some cases be too complicated to be used for designing controllers. An attractive and practical alternative is then to assume a sufficiently simple model in which the parameters are determined so as to obtain a good fit between the measured input–output data and the model-generated values. Such a model can be chosen as a difference equation. It can serve as the basis for designing a controller. We will review the estimation of the parameters in such a discrete-time model and then using this discrete-time model construct an adaptive self-tuning controller.

D.1 DETERMINATION OF DIFFERENCE EQUATION MODEL

Although several powerful methods to estimate model parameters have been described in the literature [1], only the least-squares error method is presented here. It is applied first to a single-input single-output (SISO) system and then to a multiple-input multiple-output (MIMO) system. The parameter estimation algorithm is described as difference equations for on-line computations.

D.1.1 DISCRETE TIME-SERIES MODEL

The system is assumed to be described in terms of the minimal number of parameters (the canonical representation) as a SISO time series model:

$$y_m(k) = a_0 + a_1 y_m(k-1) + \cdots + a_n y_m(k-n) + b_0 u(k-d)$$

$$+ \cdots + b_n u(k-n-d) \quad (D.1.1)$$

where the sampling period T is omitted in the arguments, $u(k)$ is the input, $y_m(k)$ the output at time kT, $k = n, n+1, \ldots$, integer $n > 0$ specifies the order of the model and $a_i, b_i, i = 0, 1, \ldots, n$ are constant parameters. The positive integer d represents the time delay dT between the input and the output of the system. By introducing a delay operator z^{-1}, that is, $z^{-1} y_m(k) = y_m(k-1)$, Equation D.1.1 may be written as

$$A(z^{-1}) y_m(k) = B(z^{-1}) u(k-d) + a_0 \quad (D.1.2)$$

where $A(z^{-1}) = 1 - (a_1 z^{-1} + \cdots + a_n z^{-n})$ and $B(z^{-1}) = b_0 + b_1 z^{-1} + \cdots + b_n z^{-n}$. The poles of the model are assumed to be inside the unit circle on the z-domain.

Since the measured values $\{y(k)\}$ usually do not satisfy Equation D.1.2 exactly, an equation error $e(k)$ is introduced:

$$e(k) = A(z^{-1})[y(k) - y_m(k)] \quad (D.1.3)$$

where $y(k)$ is the measured noise-corrupted value of the output. Equation D.1.3 may be rewritten using Equation D.1.2:

$$e(k) = A(z^{-1}) y(k) - B(z^{-1}) u(k-d) - a_0 \quad (D.1.4)$$

Equation D.1.4 can also be expressed explicitly in the following form:

$$y(k) = a_0 + a_1 y(k-1) + \cdots + a_n y(k-n) + b_0 u(k-d) +$$

$$\cdots + b_n u(k-n-d) + e(k) \quad (D.1.5)$$

In Equation D.1.5, residual $e(k)$ signifies modeling errors. Equation D.1.5 represents a discrete time series model of autoregressive type with external excitation. For given values of $\{y(k)\}$ and $\{u(k)\}$, Equation D.1.5 is linear in the parameters.

The system model in Equation D.1.5 is assumed to (a) be (asymptotically) stable, (b) have all poles excited by the input, and (c) have an output that contains all (dominant) natural frequencies of the system. Moreover, residuals $\{e(k)\}, k = n, n+1, \ldots$ represent independent, zero-mean samples that are not influenced by the input. The numerical values of the unknown parameters b_i and $a_i, i = 0, \ldots, n$ are determined by minimizing a well-defined criterion. If such a parameter estimation problem has a unique solution, the models considered are called identifiable.

D.1.2 PARAMETER ESTIMATION BY THE LEAST SUM OF SQUARED ERRORS METHOD

A suitable criterion that measures the accuracy of the fitting of the model-generated values to the measured data can be chosen as a quadratic func-

tional of the equation errors. An attractive choice is the (sample) mean value
of the squared errors:

$$E(\alpha) = \frac{1}{N+1} \sum_{k=n}^{n+N} e^2(k) \tag{D.1.6}$$

where the parameter vector $\alpha = [a_1 \cdots a_n; b_0 \cdots b_n; a_0]'$ and $(n+N)T$
indicates the time of the last measurement. Equation D.1.6 is a quadratic
function of α, as is seen by substituting $e(k)$ from Equation D.1.5 into
Equation D.1.6. The error criterion D.1.6 possesses a minimum with respect
to α.

The problem is now to minimize $E(\alpha)$ in Equation D.1.6 with respect
to α while satisfying the model Equation D.1.5. The minimizing value of
parameter α provides the best fit to the measured data with respect to the
criterion $E(\alpha)$.

The choice of $E(\alpha)$ as a quadratic criterion makes it possible to solve the
minimization problem explicitly. It is convenient first to define

$$\phi(k-1)$$
$$= \mathrm{col}[y(k-1)y(k-2) \cdots y(k-n); u(k-d) \cdots u(k-n-d); 1] \tag{D.1.7}$$

The constraint Equation D.1.5 can be rewritten:

$$y(k) = \phi'(k-1)\alpha + e(k) \tag{D.1.8}$$

To minimize $E(\alpha)$, Equation D.1.8 is substituted for $e(k)$ in Equation D.1.6
to obtain

$$E(\alpha) = \frac{1}{N+1} \sum_{k=n}^{n+N} [y(k) - \phi'(k-1)\alpha]^2$$
$$= \frac{1}{N+1}\left\{ [y(n) - \phi'(n-1)\alpha]^2 + \cdots \right.$$
$$\left. + [y(n+N) - \phi'(n+N-1)\alpha]^2 \right\} \tag{D.1.9}$$

Equation D.1.9 may be written as the square of the components in the
Euclidean norm

$$E(\alpha) = \frac{1}{N+1}\left\| \begin{matrix} y(n) - \phi'(n-1)\alpha \\ \vdots \\ y(n+N) - \phi'(n+N-1)\alpha \end{matrix} \right\|^2 \tag{D.1.10}$$

Equation D.1.10 can be rewritten as

$$E(\alpha) = \frac{1}{N+1}\left\| \begin{bmatrix} y(n) \\ \vdots \\ y(n+N) \end{bmatrix} - \begin{bmatrix} \phi'(n-1) \\ \vdots \\ \phi'(n+N-1) \end{bmatrix}\alpha \right\|^2$$
$$= \frac{1}{N+1}\| Y(n+N) - S(n+N-1)\alpha \|^2 \tag{D.1.11}$$

where

$$Y(n + N) = \text{col}[y(n)y(n + 1) \cdots y(n + N)] \qquad \text{(D.1.12)}$$

$$S(n + N - 1) = \begin{bmatrix} \phi'(n - 1) \\ \phi'(n) \\ \vdots \\ \phi'(n + N - 1) \end{bmatrix}$$

$$= \begin{bmatrix} y(n - 1) & \cdots & y(0) & u(n - d) & \cdots & u(-d) & 1 \\ y(n) & \cdots & y(1) & u(n + 1 - d) & \cdots & u(1 - d) & 1 \\ \vdots & & & & & \vdots & \vdots \\ y(n + N - 1) & \cdots & y(N) & u(n + N - d) & \cdots & u(N - d) & 1 \end{bmatrix}$$

$$\text{(D.1.13)}$$

The minimum of $E(\alpha)$ in Equation D.1.11 relative to α is determined by setting the derivative of $E(\alpha)$ with respect to α equal to zero; it gives

$$\hat{\alpha} = [S'(n + N - 1)S(n + N - 1)]^{-1}S'(n + N - 1)Y(n + N) \qquad \text{(D.1.14)}$$

where $\hat{\alpha}$ represents the value of α that minimizes the error criterion.

The corresponding minimum value of the error criterion $E(\alpha)$ is

$$E(\hat{\alpha}) = \frac{1}{N + 1} \sum_{k=n}^{n+N} [y(k) - \phi'(k - 1)\hat{\alpha}]^2 \qquad \text{(D.1.15)}$$

Equation D.1.14 specifies the parameter vector $\hat{\alpha}$ that provides the best fit of the model values to the measured data points. Since the last measurements are collected at $k = n + N$, the minimizing parameter value may be denoted as $\hat{\alpha}_{N+n}$. Under certain conditions, it can be shown that $\hat{\alpha}_{n+N}$ given by Equation D.1.14 converges in the mean squared error sense to the true parameter value as $N \rightarrow \infty$ [2]. The best estimate $\hat{\alpha}$ of the parameter vector can be computed by Equation D.1.14 after matrix $S(n + N - 1)$ and $Y(n)$ have been formed on the basis of the measurements. Thus, the calculations must be performed off-line, after the measurements have been collected.

An example is now presented to demonstrate the least-squares method.

EXAMPLE D.1.1

In a Stanford/JPL manipulator, independent joint dynamics is assumed to apply. The voltages into a joint motor and the joint velocities have been measured. They are given in Table D.1.

The following model for the joint velocity $\{v(k)\}$ and the input voltage $\{u(k)\}$ is assumed

$$v(k) = a_0 + a_1v(k - 1) + a_2v(k - 2) + b_0u(k - 1) + e(k) \qquad \text{(D.1.16)}$$

The unknown parameters a_0, a_1, a_2, and b_0 must be determined so that the best fit of the model-generated values to the measurements is obtained in the sense of the least-squared errors. The equation error $e(k)$ is assumed to have zero mean.

The unknown parameter vector α is first defined: $\alpha = \text{col}[a_1 a_2; b_0; a_0]$. The problem formulation is as follows: Determine the optimal parameter vector $\hat{\alpha}_{101}$ that minimizes the error criterion

$$E(\alpha) = \frac{1}{100} \sum_{k=2}^{101} e^2(k) \qquad (D.1.17)$$

where $e(k)$ is governed by Equation D.1.16. Thus, $d = 1, b_1 = b_2 = 0, n = 2$, and $N = 99$. The model Equation D.1.16 is rewritten in the form of Equation D.1.8:

$$v(k) = [v(k-1)v(k-2); u(k-1); 1]\alpha + e(k) \qquad (D.1.18)$$

where $[v(k-1)v(k-2); u(k-1); 1] = \phi'(k-1)$. The expressions in Equations D.1.12 and D.1.13 assume the following forms:

$$Y(101) = \text{col}[v(2)v(3) \cdots v(101)] \qquad (D.1.19)$$

$$S(100) = \begin{bmatrix} v(1) & v(0) & u(1) & 1 \\ v(2) & v(1) & u(2) & 1 \\ \cdot & \cdot & \cdot & \cdot \\ \cdot & \cdot & \cdot & \cdot \\ v(100) & v(99) & u(100) & 1 \end{bmatrix} \qquad (D.1.20)$$

The best estimate of the parameter vector α is computed by Equation D.1.14 to obtain

$$\hat{\alpha}_{101} = \begin{vmatrix} 0.598\ 0.289; -2.554; 0.070 \end{vmatrix}' \qquad (D.1.21)$$

The corresponding minimum value of the error criterion $E(\alpha)$ is

$$E(\hat{\alpha}) = \frac{1}{100} \sum_{k=2}^{101} e^2(k) = 1.514 \qquad (D.1.22)$$

When the value of $\hat{\alpha}$ is substituted into Equation D.1.16, the resulting model can be used to compute the one-step ahead predicted values generated by the model. This response is graphed with the data points in Figure D.1. Equation D.1.18 gives the sample sequence of the error $e(k)$ for $2 \le k \le 101$:

$$e(k) = v(k) - \phi'(k-1)\hat{\alpha} \qquad (D.1.23)$$

An interesting interpretation for the equation error $e(k)$ can be given on the basis of Equation D.1.23. The term $\phi'(k-1)\hat{\alpha}$ represents the one-step ahead predicted value $y(k|k-1)$ of the output when the past values of the output and input are known. Thus, $e(k)$ describes the difference between the measured value $y(k)$ and the predicted value. The sample sequence of

TABLE D.1
Measured Data for Example D.1.1

k	0	1	2	3	4	5	6
u(k)	-1.00000	-1.00000	-1.00000	-1.00000	-1.00000	-1.00000	-1.00000
v(k)	.00000	1.42857	3.42857	3.71429	5.85714	6.57143	8.28571

k	7	8	9	10	11	12	13
u(k)	-1.00000	-1.00000	-1.00000	.00000	.00000	.00000	.00000
v(k)	7.42857	11.00000	11.57143	10.85714	10.42857	10.14286	8.00000

k	14	15	16	17	18	19	20
u(k)	.00000	.00000	.00000	.00000	.00000	.00000	.10000
v(k)	6.42857	5.42857	3.71429	3.42857	2.14286	.85714	.00000

k	21	22	23	24	25	26	27
u(k)	.20000	.30000	.40000	.50000	.60000	.70000	.80000
v(k)	-.14286	.00000	-.14286	-.57143	-1.00000	-1.85714	-2.42857

k	28	29	30	31	32	33	34
u(k)	.90000	1.00000	1.00000	1.00000	1.00000	1.00000	1.00000
v(k)	-3.28571	-4.57143	-5.57143	-5.57143	-8.42857	-8.28571	-11.14286

k	35	36	37	38	39	40	41
u(k)	1.00000	1.00000	1.00000	1.00000	1.00000	.90000	.80000
v(k)	-10.00000	-10.42857	-14.00000	-12.85714	-15.42857	-17.00000	-17.14286

k	42	43	44	45	46	47	48
u(k)	.70000	.60000	.50000	.30000	.20000	.10000	
v(k)	-16.14286	-18.42857	-19.00000	-18.42857	-15.71429	-17.57143	-16.57143

k	49	50	51	52	53	54	55
u(k)	.00000	-.10000	-.20000	-.30000	-.40000	-.50000	-.60000
v(k)	-13.71429	-14.00000	-12.57143	-9.85714	-6.57143	-6.14286	-2.42857

k	56	57	58	59	60	61	62
u(k)	-.70000	-.80000	-.90000	-1.00000	-1.00000	-1.00000	-1.00000
v(k)	-.71429	1.00000	2.14286	3.42857	4.42857	5.58143	7.14286

k	63	64	65	66	67	68	69
u(k)	-1.00000	-1.00000	-1.00000	-1.00000	-1.00000	-1.00000	-1.00000
v(k)	8.71429	8.14286	11.14286	13.00000	11.42857	15.00000	15.71429

k	70	71	72	73	74	75	76
u(k)	-.90000	-.80000	-.70000	-.60000	-.50000	-.40000	-.30000
y(k)	15.57143	18.28572	18.71428	17.14286	19.57143	20.00000	19.00000

k	77	78	79	80	81	82	83
u(k)	-.20000	-.10000	.00000	.00000	.00000	.00000	.00000
y(k)	16.57143	18.28572	16.85714	15.57143	12.28571	13.28571	11.71429

k	84	85	86	87	88	89	90
u(k)	.00000	.00000	.00000	.00000	.00000	.00000	1.00000
v(k)	.00000	-2.14286	-3.00000	-5.00000	-5.14286	-6.85714	-8.71429

k	91	92	93	94	95	96	97
u(k)	1.00000	1.00000	1.00000	1.00000	1.00000	1.00000	1.00000
v(k)	.00000	-2.14286	-3.00000	-5.00000	-5.14286	-6.85714	-8.71429

k	98	99	100	101	102	103	104
u(k)	1.00000	1.00000	.00000	.00000	.00000	.00000	.00000
v(k)	-8.57143	-9.14286	-12.57143	-8.28571	-8.42857	-7.14286	-4.14286

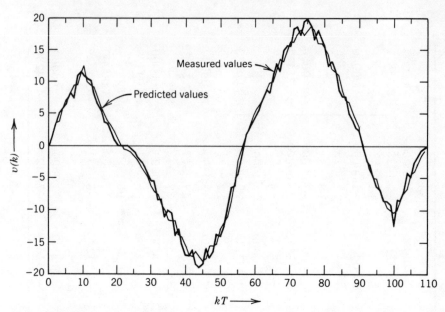

FIGURE D.1
One-step ahead predicted and measured values versus time. Example D.1.1.

the prediction errors can be used to test the accuracy and the validity of the model determined on the basis of the measurements [1,2].

The least-squared error method may give biased estimates. This is inconsequential when the model is used with a self-tuning controller that inherently compensates for the bias. However, if some physical parameters such as masses, inertias, or friction coefficients in a manipulator are estimated, then the bias must be corrected to avoid erroneous estimates.

D.1.3 ON-LINE PARAMETER ESTIMATION BY RECURSIVE EQUATIONS

In section D.1.2, the least-squares method is applied to determine the optimal parameters in a given linear difference equation model. In using Equation D.1.14, all measurements must be available before computing $\hat{\alpha}_{n+N}$. Thus, the computations are performed off-line. Suppose now that a new measurement $[y(n + N + 1), u(n + N + 1 - d)]$ is obtained, and a new (improved) parameter value must be computed on the basis of all known measurements. If Equation D.1.14 is used, the entire sequence of mathematical operations should be repeated. It is very inefficient. It is possible, however, to manipulate Equation D.1.14 into a different form that requires fewer mathematical operations. This form is recursive, and it will only update the last estimate. The equations for recursive calculations will next be derived.

(i) *SISO Model*

Suppose that $2n$ measurements $(N = n)$ have been obtained for a single-input single-output (SISO) model. These measurements satisfy the following equation:

$$Y(2n) = S(2n - 1)\alpha + e(n, 2n) \qquad \text{(D.1.24)}$$

where $e(n, 2n) = \text{col}[e(n) \ldots e(2n)]$. The best estimate $\hat{\alpha}_{2n}$ of the parameter vector on the basis of $2n$ measurements is specified by Equation D.1.14:

$$\hat{\alpha}_{2n} = P(2n) S'(2n - 1) Y(2n) \qquad \text{(D.1.25)}$$

where

$$P(2n) = \left[S'(2n - 1) S(2n - 1) \right]^{-1} \qquad \text{(D.1.26)}$$

When a new measurement $[y(2n + 1), u(2n + 1 - d)]$ is received, it is governed by the following equation:

$$y(2n + 1) = \phi'(2n)\alpha + e(2n + 1) \qquad \text{(D.1.27)}$$

The model Equations D.1.24 and D.1.27 can be rewritten in a combined form:

$$\begin{bmatrix} Y(2n) \\ y(2n + 1) \end{bmatrix} = \begin{bmatrix} S(2n - 1) \\ \phi'(2n) \end{bmatrix} \alpha + \begin{bmatrix} e(n, 2n) \\ e(2n + 1) \end{bmatrix} \qquad \text{(D.1.28)}$$

The best estimate $\hat{\alpha}_{2n+1}$ of α on the basis of $(2n + 1)$ measurements is given by Equation D.1.14 with appropriate arguments:

$$\hat{\alpha}_{2n+1} = P(2n + 1) S'(2n) Y(2n + 1)$$

$$= P(2n + 1)\left[S'(2n - 1)\, \phi(2n) \right] \begin{bmatrix} Y(2n) \\ y(2n + 1) \end{bmatrix} \qquad \text{(D.1.29)}$$

Moreover, Equation D.1.26 becomes

$$P(2n + 1) = \left\{ \left[S'(2n - 1)\phi(2n) \right] \begin{bmatrix} S(2n - 1) \\ \phi'(2n) \end{bmatrix} \right\}^{-1}$$

$$= \left[S'(2n - 1) S(2n - 1) + \phi(2n)\phi'(2n) \right]^{-1}$$

$$= \left[P^{-1}(2n) + \phi(2n)\phi'(2n) \right]^{-1} \qquad \text{(D.1.30)}$$

The well-known matrix inversion lemma is next applied to Equation D.1.30. It can be proven by the definition of the inverse matrix. Equation D.1.30 becomes

$$P(2n + 1) = \left[P^{-1}(2n) + \phi(2n)\phi'(2n) \right]^{-1}$$

$$= P(2n) - P(2n)\phi(2n)\left[\phi'(2n) P(2n)\phi(2n) + 1 \right]^{-1}\phi'(2n) P(2n) \qquad \text{(D.1.31)}$$

If Equation D.1.31 is substituted into Equation D.1.29, the following manipulations may be performed:

$$\hat{\alpha}_{2n+1} = P(2n + 1)\left[S'(2n - 1)Y(2n) + \phi(2n)y(2n + 1)\right]$$

$$= \hat{\alpha}_{2n} - P(2n)\phi(2n)\left[\phi'(2n)P(2n)\phi(2n) + 1\right]^{-1}\phi'(2n)\hat{\alpha}_{2n}$$

$$+ P(2n)\phi(2n)y(2n + 1) - P(2n)\phi(2n)\left[\phi'(2n)P(2n)\phi(2n) + 1\right]^{-1}$$

$$\phi'(2n)P(2n)\phi(2n)y(2n + 1) \quad (D.1.32)$$

The last two terms may be combined to obtain III + IV terms =

$$P(2n)\phi(2n)\{I - \left[\phi'(2n)P(2n)\phi(2n) + 1\right]^{-1}\phi'(2n)P(2n)\phi(2n)\}y(2n + 1)$$

$$= P(2n)\phi(2n)\left[\phi'(2n)P(2n)\phi(2n) + 1\right]^{-1}\left[\phi'(2n)P(2n)\phi(2n) + 1\right.$$

$$\left. - \phi'(2n)P(2n)\phi(2n)\right]y(2n + 1)$$

$$= P(2n)\phi(2n)\left[\phi'(2n)P(2n)\phi(2n) + 1\right]^{-1}y(2n + 1) \quad (D.1.33)$$

Substituting D.1.33 into Equation D.1.32 gives

$$\hat{\alpha}_{2n+1} = \hat{\alpha}_{2n} + K(2n)\left[y(2n + 1) - \phi'(2n)\hat{\alpha}_{2n}\right] \quad (D.1.34)$$

where

$$K(2n) = P(2n)\phi(2n)\left[\phi'(2n)P(2n)\phi(2n) + 1\right]^{-1}$$

Equation D.1.34 describes the recursive equation that specifies how the estimated value $\hat{\alpha}_{2n}$ must be updated to obtain $\hat{\alpha}_{2n+1}$ when an additional measurement $[y(2n + 1), u(2n + 1)]$ has become available.

The last term on the right of Equation D.1.34 specifies the correction term needed to change $\hat{\alpha}_{2n}$ to $\hat{\alpha}_{2n+1}$. The term $[y(2n + 1) - \phi'(2n)\hat{\alpha}_{2n}]$ describes the difference between the predicted value $\phi'(2n)\hat{\alpha}_{2n}$ and the measured value $y(2n + 1)$. This difference is multiplied by a time-varying gain. The algorithm for the on-line computation consists of Equations D.1.31 and D.1.34. These equations may be rewritten by replacing $2n$ by k, and denoting $\hat{\alpha}_k$ as $\hat{\alpha}(k)$:

$$\hat{\alpha}(k + 1) = \hat{\alpha}(k) + K(k)\left[y(k + 1) - \phi'(k)\hat{\alpha}(k)\right] \quad (D.1.35)$$

$$K(k) = P(k)\phi(k)\left[\phi'(k)P(k)\phi(k) + 1\right]^{-1} = P(k + 1)\phi(k) \quad (D.1.36)$$

$$P(k) = P(k - 1) - P(k - 1)\phi(k)\left[\phi'(k)P(k - 1)\phi(k) + 1\right]^{-1}\phi'(k)P(k - 1)$$

$$= P(k - 1) - K(k - 1)\left[\phi'(k)P(k)\phi(k) + 1\right]K'(k - 1) \quad (D.1.37)$$

where $k = 2n, 2n + 1, \ldots$ Equation D.1.36 can be proven by writing Equation D.1.37 for $P(k + 1)$, substituting it on the right side of Equation D.1.36 and simplifying. An example is next presented to illustrate the application of the on-line parameter estimation algorithm.

EXAMPLE D.1.2

The model of a system describing the relation between input $\{u(k)\}$ and output $\{y(k)\}$ is assumed:

$$y(k) = a_1 y(k-1) + b_1 u(k-2) + a_0 + e(k) \tag{D.1.38}$$

where a_1, b_1, and a_0 are unknown constant parameters, and $e(k)$ signifies modeling errors. The problem is to calculate the constants a_1, b_1, and a_0 on-line on the basis of the measured values of $\{y(k)\}$ and $\{u(k)\}$.

For this example, the data were generated for $a_1 = -0.8, b_1 = 4.0$, and $a_0 = 6.0$ when the inputs $\{u(k)\}$ and $\{e(k)\}$ for $k = 0, 1, \ldots, 200$ were applied to the system. The inputs were chosen as follows: $u(k) = 0.1k$ for $0 \le k < 100$ and $u(k) = -0.1k + 20$ for $100 \le k \le 200$. Residual $e(k)$ for each k was sampled from a process described by a Gaussian distribution with zero mean and unit variance. The resulting data points $\{y(k)\}$ and $\{u(k)\}$ are then used as the given measurement of a plant to recover the original parameters.

To determine the unknown constants a_1, b_1, and a_0, Equation D.1.8 for $n = 1$ is first written:

$$y(k) = [y(k-1) u(k-2) 1] \begin{bmatrix} a_1 \\ b_1 \\ a_0 \end{bmatrix} + e(k) \tag{D.1.39}$$

Thus, $\phi'(k) = [y(k-1); u(k-2); 1]$, and $\alpha = \text{col}[a_1; b_1; a_0]$.

Equations D.1.35, D.1.36, and D.1.27 are used in the on-line computational scheme. Matrix P in Equation D.1.37 is a symmetric (3×3) matrix:

$$P(k) = \begin{bmatrix} p_{11}(k) & p_{12}(k) & p_{13}(k) \\ p_{12}(k) & p_{22}(k) & p_{23}(k) \\ p_{13}(k) & p_{23}(k) & p_{33}(k) \end{bmatrix} \tag{D.1.40}$$

and $\hat{\alpha}(k) = [\hat{a}_1(k); \hat{b}_1(k); \hat{a}_0(k)]'$. In Equation D.1.36, the inversion of the matrix reduces to the division by a scalar. The initial values of the parameters $\hat{\alpha}(0)$ and $P(0)$ are assumed: $\hat{\alpha}(0) = [1 1 1]$ and $P(0) = 30I$, where I is the identity matrix.

The results of the on-line identification are graphed in Figures D.2a–D.2c. The simulation results demonstrate the convergence of the algorithm.

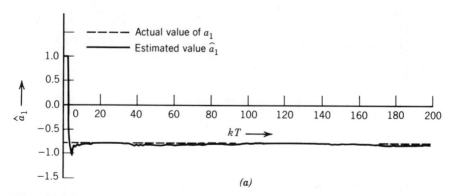

FIGURE D.2a–c
On-line estimation of parameters. Example D.1.2.
a. Parameter $\hat{a}_1$ versus time.

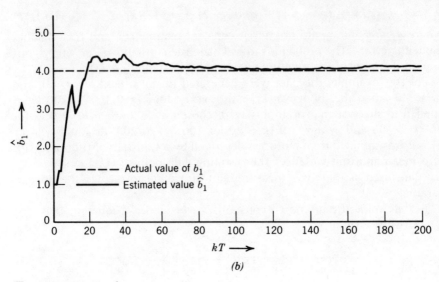

FIGURE D.2*b*
Parameter $\hat{b}_1$ versus time.

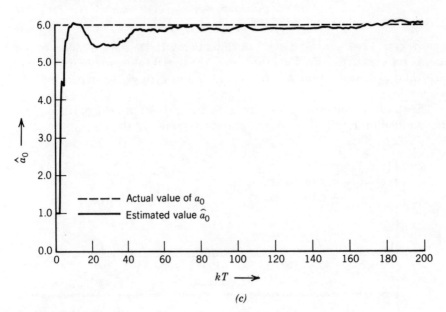

FIGURE D.2*c*
Parameter $\hat{a}_0$ versus time.

(ii) *MIMO Model*

A multiple-input multiple-output (MIMO) difference equation can be expressed as

$$A(z^{-1})y(k) = B(z^{-1})u(k - d) + a_0 + e(k) \tag{D.1.41}$$

where the m-dimensional output vector $y(k)$ has as the ith component $y_i(k)$, and the m-dimensional input vector $u(k)$ has as the jth component $u_j(k)$, $i, j = 1, \ldots, 2N$. The residual $e(k)$ represents a white Gaussian zero-mean noise with the finite covariance. The sampling period T has been dropped in the arguments. The delay dT is such that d is a positive integer. The constant vector a_0 is m-dimensional. Matrices $A(z^{-1})$ and $B(z^{-1})$ are matrix polynomials defined as

$$A(z^{-1}) = I - (A_1 z^{-1} + \cdots + A_n z^{-n}) \tag{D.1.42}$$

$$B(z^{-1}) = B_0 + B_1 z^{-1} + \cdots + B_n z^{-n} \tag{D.1.43}$$

where the positive integer n determines the order of the model, and A_i, B_0, and B_i, $i = 1, \ldots, n$ are matrices with unknown constant elements. Matrix B_0 is nonsingular, and det $B(z^{-1})$ has all zeros strictly inside the unit circle.

The parameters in the MIMO-model are estimated by considering the equation of each row at a time and applying the algorithm described in (i). To estimate the parameters in Equation D.1.41, matrix α and vector $\phi(k-1)$ are defined:

$$\alpha = \left[A_1 \cdots A_n; B_0 \cdots B_n; a_0\right]' = \left[\alpha_1 \cdots \alpha_{m'}\right] \tag{D.1.44}$$

where

$$\alpha_i' = \left[a_1^{i1} \cdots a_1^{im} a_2^{i1} \cdots a_2^{im} \cdots; b_0^{i1} \cdots b_0^{im} \cdots b_n^{i1} \cdots b_n^{im}; a_0^i\right] \tag{D.1.45}$$

$$\phi(k-1) = \left[y'(k - 1) \cdots y'(k - n); u'(k - d) \cdots u'(k - d - n); 1\right]' \tag{D.1.46}$$

Equation D.1.41 can be rewritten as follows:

$$y(k) = \alpha'\phi(k - 1) + e(k) \tag{D.1.47}$$

The algorithm for the parameter estimation is constructed so that one vector α_i at a time in matrix α of Equation D.1.44 is estimated. The error criterion is chosen for each vector α_i as follows:

$$E(\alpha_i) = \frac{1}{N + 1} \sum_{k=n}^{N+n} \gamma_i^{N+n-k} e_i^2(k) \tag{D.1.48}$$

where $(N + n)$ represents the number of the measurements, the weighting term γ_i is called a forgetting factor, and $e_i(k)$ is the ith component of the e-vector. The problem is to minimize $E(\alpha_i)$ relative to the parameter vector α_i.

The solution to the least-squares problem is furnished by equations similar to Equations D.1.35–D.1.37. These equations are presented here as the recursive equations. The caret refers to estimated values, and the estimate of α_i at time k is written as $\hat{\alpha}_i(k)$. Thus,

$$\hat{\alpha}_i(k) = \hat{\alpha}_i(k-1) + P(k)\phi(k-1)\big[y_i(k) - \phi'(k-1)\hat{\alpha}_i(k-1)\big] \quad \text{(D.1.49)}$$

$$P(k) = \frac{1}{\gamma}\bigg[P(k-1) - P(k-1)\phi(k-1)\big[\gamma + \phi'(k-1)P(k-1)\phi(k-1)\big]^{-1}$$
$$\phi'(k-1)P(k-1)\bigg] \quad \text{(D.1.50)}$$

where matrix $P(k)$ is symmetric. The parameter estimates of the multivariable time-series model can be calculated on-line using Equations D.1.49 and D.1.50.

The accuracy of the estimated time-series model can be considered from several viewpoints. The residuals $\{e(k)\}$ in Equations D.1.8 and D.1.47 can be used for computing figures of merits for the specific class of models. In fact, the sample sequence $\{e(k)\}$ can be generated from Equations D.1.8 and D.1.47 after the optimal parameters have been determined. Then, the sample sequence can be checked by different methods, for example by a zero-mean test, a serial independence test, and/or a test for the absence of sinusoidal terms to validate the model. Extensive expositions on the tests for the validity of the models can be found in the literature [1,2].

D.2 ADAPTIVE SELF-TUNING CONTROLLER DESIGN

The discrete time series models presented in Equations D.1.5 and D.1.41 with known parameters can serve as the basis for designing controllers for a system such as a multijoint manipulator. A controller may be constructed by minimizing a well-defined performance criterion or by solving a pole-zero placement (eigenvalue assignment) problem. If the true values of the model parameters are not known, the design of a controller can be based on an approximate discrete time series model. It is obtained from Equation D.1.5 or D.1.41 by substituting the estimated values for the parameters of the model. The controller constructed on the basis of this approximate model will have approximate values for the gains. If the parameter estimates in the time series model converge to the true (unknown) parameter values, then under certain conditions the controller gains converge to the optimal gain values that could have been determined if the true parameter values had been known [3–5]. This property in a controller is called the self-tuning property.

The design of a self-tuning controller will next be discussed by minimizing a well-defined performance criterion.

D.2.1 SELF-TUNING CONTROLLER BASED ON MINIMUM OF PERFORMANCE CRITERION

An adaptive self-tuning controller will be designed so that the system with this controller will track the desired trajectory specified by the discrete points $\{y^d(k)\}$, while the energy associated with the input is kept at a minimum. These goals are achieved by choosing the following one-stage performance criterion to be minimized:

$$I_k[u] = E\{\| y(k + d) - y^d(k + d) \|_Q^2 + \| u(k) \|_R^2 |\Sigma(k)\} \qquad (D.2.1)$$

where $\| \cdot \|_R$ indicates the generalized norm with weight R, that is, $\| u \|_R^2 = u' Ru$, and matrices R and Q are symmetric and positive semidefinite. The expectation operation $E\{\cdot |\Sigma(k)\}$ is conditioned on the available measurements up to and including time kT. The inputs considered in the minimization   of $I_k[u]$ are in the set Ω of admissible functions that can depend on available measurements and the past controls.

The problem is to minimize the performance criterion D.2.1 relative to the admissible controls while satisfying the constraint equation

$$y(k) = \hat{\alpha}'\phi(k - 1) + e(k) \qquad (D.2.2)$$

Since the plant Equation D.2.2 is *linear*, and the criterion D.2.1 is *quadratic* subject to *Gaussian* disturbances, the problem is often called the LQG-problem.

The problem is solved by expressing the predicted value of $y(k + d)$ in terms of the available information up to time kT, substituting it into Equation D.2.1 and performing the minimization of $I_k[u]$ with respect to $u(k)$. The control that minimizes D.2.1 is

$$Ru(k) + B_0'Q\big[\hat{y}(k + d|k) - y^d(k + d)\big| = 0 \qquad (D.2.3)$$

where $\hat{y}(k + d|k)$ denotes the optimal predicted value (in the sense of least-squares error) of $y(k + d)$ given the past measurements and controls up to and including time kT. For $r = 1, \ldots, d$ the r-step-ahead predicted value may be calculated recursively using the prediction equations obtained from Equation D.2.2:

$$\hat{y}(k + r|k) = \hat{\alpha}'(k - 1)\phi(k - 1 + r|k) \qquad (D.2.4)$$

$$\phi(k - 1 + r|k) = \big[\hat{y}'(k - 1 + r|k) \cdots \hat{y}'(k + 1|k), y'(k), \cdots, y'(k + r - n);$$

$$u'(k - d + r) \cdots u'(k - d - n + r); 1\big|' \qquad (D.2.5)$$

The algorithms for the adaptive self-tuning control scheme consists of Equations D.1.35–D.1.37 or D.1.49–D.1.50 for the parameter estimation and D.2.2–D.2.5 for computing the controller gains. Thus, the parameters are first estimated and then the controller gains are computed. This approach to designing a self-tuning controller is called the indirect, or explicit method. The two operations can also be combined to obtain directly the values of the controller; this approach is termed the direct (implicit) method [3,4].

EXAMPLE D.2.1

The single-input single-output relation of a system is modeled as

$$y(k) = \hat{a}_1 y(k - 1) + \hat{a}_2 y(k - 2) + \hat{b}_1 u(k - 2) \qquad (D.2.6)$$

where $k = 2, 3, \ldots$ and $y(k) = 0$ for $k \leq 0$. The unknown constant parameters a_1, a_2, and b_1 have been replaced by their estimates $\hat{a}_1$, $\hat{a}_2$, and $\hat{b}_1$, respectively. They are determined by the least-squares error method as described by Equations D.1.35–D.1.37.

The design specification is that output $\{y(k)\}$ should track the desired values $\{y^d(k)\}$ as closely as possible. It is accomplished by selecting the following single-stage performance criterion:

$$I_k(u) = \left[y(k+2) - y^d(k+2)\right]^2 + \epsilon u^2(k) \qquad (D.2.7)$$

where a cost has been assigned to deviations of $y(k+2)$ from the desired value $y^d(k+2)$ and the use of the control energy (u^2). The weighting factor ϵ is a nonnegative number. It can be used to weigh the relative importance of the squared error and the control energy consumption; moreover, it can also be used to convert the control energy and the squared error into appropriate cost values. The expectation operation is not needed in $I_k[u]$, since the model Equation D.2.6 is deterministic, that is, $e(k) = 0$.

The problem is to design a controller so as to minimize $I_k[u]$ for each k with respect to the admissible controls $u(k) \in \Omega$. The control input $u(k)$ is to be expressed in the feedback form that can be implemented, that is, $u(k)$ can be a function of $y(i)$, $i \le k$ and the past inputs.

Since the (predicted) value $y(k+2)$ and control $u(k)$ in $I_k[u]$ are related, the approach is to compute first $y(k+2)$ in terms of $y(i)$, $i \le k$ and the past inputs. After expression $y(k+2)$ is available, it is substituted into Equation D.2.7, which is then minimized relative to $u(k)$.

The predicted value $y(k+2)$ of the output is expressed in terms of the information available at time k using Equation D.2.6. At time $(k+2)$, Equation D.2.6 becomes

$$y(k+2) = \hat{a}_1 y(k+1) + \hat{a}_2 y(k) + \hat{b}_1 u(k) \qquad (D.2.8)$$

If expression D.2.8 is substituted into Equation D.2.7, the minimization results in an unrealizable expression for the control $u(k)$ (why?). By substituting for $y(k+1)$ in Equation D.2.8 the expression from Equation D.2.6 one obtains the following equation:

$$y(k+2) = (\hat{a}_1^2 + \hat{a}_2)y(k) + \hat{a}_1\hat{a}_2 y(k-1) + \hat{b}_1 u(k) + \hat{a}_1\hat{b}_1 u(k-1) \quad (D.2.9)$$

The right side of Equation D.2.9 is the predicted value of the output at time $(k+2)$ based on the available information at time k.

Equation D.2.9 is substituted into expression D.2.7 to obtain

$$I_k[u] = \left[(\hat{a}_1^2 + \hat{a}_2)y(k) + \hat{a}_1\hat{a}_2 y(k-1) + \hat{b}_1 u(k)\right.$$
$$\left. + \hat{a}_1\hat{b}_1 u(k-1) - y^d(k+2)\right]^2 + \epsilon u^2(k) \quad (D.2.10)$$

Expression $I_k(u)$ possesses a minimum relative to $u(k)$. The minimizing (optimal) value of the input is

$$u(k) = \frac{-\hat{b}_1}{\hat{b}_1^2 + \epsilon}\left[(\hat{a}_1^2 + \hat{a}_2)y(k) + \hat{a}_1\hat{a}_2 y(k-1)\right.$$
$$\left. + \hat{a}_1\hat{b}_1 u(k-1) - y^d(k+2)\right| \quad (D.2.11)$$

The control law in Equation D.2.11 is in the feedback form. It is realizable, since all parameters, the values of $y(k)$, $y(k-1)$, and $u(k-1)$ appearing in Equation D.2.11 are known. The parameters are first estimated at the sampling instant k according to Equations D.1.35–D.1.37. Then, the control input is computed by Equation D.2.11. These calculations are repeated at each sampling instant. Thus, the controller gains vary with the parameter estimates. The minimum value of the performance criterion is calculated by substituting Equation D.2.11 into Equation D.2.7.

EXAMPLE D.2.2

An adaptive self-tuning controller is next designed for the stochastic case, that is, $e(k) \neq 0$ in Equation D.2.2. The model of a system is assumed to be

$$y(k) = \hat{a}_1 y(k-1) + \hat{a}_2 y(k-2) + \hat{b}_1 u(k-2) + e(k) \qquad (D.2.12)$$

where $e(k)$ signifies a white Gaussian zero-mean random variable that is assumed to be uncorrelated with inputs $u(k), k \geq 0$ and output $y(i), i < k$. The estimates have been substituted for the parameters to obtain an approximate model in Equation D.2.12, in which estimation errors have been ignored.

The performance criterion to be minimized is chosen as

$$I_k[u] = E\left\{ \left[y(k+2) - y^d(k+2) \right]^2 + \epsilon u^2(k) \middle| \Sigma(k) \right\} \qquad (D.2.13)$$

where $E\{\cdot | \Sigma(k)\}$ represents the conditional expectation operation when the measurements $y(i), i \leq k$ and the past inputs $u(i), i < k$ indicated by $\Sigma(k)$ are given. The minimizing value of $I_k(u)$ is such that the predicted value of $y(k+2)$ should be as close to the desired value as possible (in the mean-square sense), and that the average energy used is minimal. The task is to minimize the cost criterion D.2.13 with respect to $u(k)$ while satisfying the difference Equation D.2.12 constraint, that is, to solve this LQG-problem.

To determine the minimizing value of input $u(k)$, the expression of $y(k+2)$ needs to be computed from Equation D.2.12 in terms of $y(i)$ and $u(k), i \leq k$. The predicted value $y(k+2|k)$ of the output $y(k+2)$ based on the information up to and including time kT is evaluated in the average sense; it is the conditional mean value of $y(k+2)$.

Equation D.2.12 can be used recursively to express $y(k+2)$ in terms of $y(i)$ and $u(i)$, for $i \leq k$. The result can be written as

$$y(k+2) = (\hat{a}_1^2 + \hat{a}_2)y(k) + \hat{a}_1\hat{a}_2 y(k-1) + \hat{b}_1 u(k) + \hat{a}_1\hat{b}_1 u(k-1)$$
$$+ \hat{a}_1 e(k+1) + e(k+2) \qquad (D.2.14)$$

By performing the conditional expectation operation on both sides of Equation D.2.14, one obtains

$$y(k+2|k) = (\hat{a}_1^2 + \hat{a}_2)y(k) + \hat{a}_1\hat{a}_2 y(k-1) + \hat{b}_1 u(k) + \hat{a}_1\hat{b}_1 u(k-1) \qquad (D.2.15)$$

where $y(k+2|k) = E[y(k+2)|y(i), u(i), i \leq k]$ is the predicted value in the sense of the least-squared prediction error. Equation D.2.14 may be rewritten:

$$y(k+2) = y(k+2|k) + \hat{a}_1 e(k+1) + e(k+2) \qquad (D.2.16)$$

The substitution of Equation D.2.16 into Equation D.2.13 and performing the conditional expectation result in

$$I_k[u] = \left[y(k+2|k) - y^d(k+2)\right]^2 + \epsilon u^2(k) + (\hat{a}_1^2 + 1)\sigma^2 \qquad \text{(D.2.17)}$$

where the constant σ^2 signifies the variance of the error term.

When Equation D.2.15 is substituted into Equation D.2.17, the performance criterion can be minimized with respect to $u(k)$. The minimizing value of input $u(k)$ will be in the same form as the one given by Equation D.2.11. The minimum value of $I_k[u]$ can be obtained by substituting $u(k)$ from Equation D.2.11 into Equation D.2.17. The comparison of the resulting expression with the minimum performance criterion in Equation D.2.7 shows that the expression for the cost function in the stochastic case is larger due to the variance term.

REFERENCES

[1]   R. L. Kashyap and A. R. Rao, *Dynamic Stochastic Models from Empirical Data,* Academic Press, New York, pp. 180–218, 1976.

[2]   G. M. Jenkins and D. G. Watts, *Spectral Analysis and Its Applications,* Holden Day, Inc., San Francisco, 1968.

[3]   K. J. Aström and B. Wittermark, "Self-Tuning Controllers Based on Pole-Zero Placement," *IEE Proceedings,* Vol. 127, Pt. D, No. 3, pp. 120–130, 1980.

[4]   D. W. Clarke and P. J. Gawthrop, "Self-Tuning Control," *IEE Proceedings,* Vol. 126, No. 6, pp. 633–640, June 1979.

[5]   P. E. Wellstead, J. M. Edmunds, D. Prager, and P. Zanker, "Self-Tuning Pole/Zero Assignment Regulators," *International Journal of Control,* Vol. 30, No. 1, pp. 1–26, 1979.

INDEX